IAASTD

International Assessment of Agricultural Knowledge, Science and Technology for Development

Latin America and the Caribbean (LAC) Report

IAASTD

International Assessment of Agricultural Knowledge, Science and Technology for Development

UNDP

UNEP

UNESCO

THE WORLD BANK

WHO

GLOBAL ENVIRONMENT FACILITY

IAASTD

International Assessment of Agricultural Knowledge, Science and Technology for Development

Latin America and the Caribbean (LAC) Report

Library of Congress Cataloging-in-Publication data.

International assessment of agricultural knowledge, science and technology for development (IAASTD) : Latin America and the Caribbean (LAC) report / edited by Beverly D. McIntyre . . . [et al.].
p. cm.
Includes bibliographical references and index.
ISBN 978-1-59726-546-1 (cloth : alk. paper) —
ISBN 978-1-59726-547-8 (pbk. : alk. paper)
1. Agriculture—Latin America—International cooperation.
2. Agriculture—Caribbean Area—International cooperation.
3. Sustainable development—Latin America. 4. Sustainable development—Caribbean Area. I. McIntyre, Beverly D. II. Title: Latin America and the Caribbean (LAC) report.
HD1428.I545 2008
338.98′07—dc22 2008046047

British Cataloguing-in-Publication data available.

Printed on recycled, acid-free paper

Interior and cover designs by Linda McKnight, McKnight Design, LLC.

Manufactured in the United States of America

10 9 8 7 6 5 4 3 2 1

Contents

Statement by Governments

In the view of all the countries, the Report makes a valuable and important contribution to our understanding of knowledge, science, and technology for development, based on recognition of the need to deepen our understanding of the challenges that lie ahead. This assessment is a constructive exercise and makes an important contribution that all countries need to develop further in order to ensure that agricultural knowledge, science, and technology achieve their potential, with a view to attaining the goals of sustainable development and poverty and hunger reduction, thereby improving the quality of rural life and human health and facilitating equitable development in a way that is socially, economically, and environmentally sustainable.

Based on this declaration, the following governments accept the Latin America and the Caribbean Report:

Belize, Brazil, Costa Rica, Cuba, Dominican Republic, El Salvador, Honduras, Panama, Paraguay, Uruguay
(10 countries).

Foreword

The objective of the International Assessment of Agricultural Knowledge, Science and Technology for Development (IAASTD) was to assess the impacts of past, present and future agricultural knowledge, science and technology on the:

- reduction of hunger and poverty,
- improvement of rural livelihoods and human health, and
- equitable, socially, environmentally and economically sustainable development.

The IAASTD was initiated in 2002 by the World Bank and the Food and Agriculture Organization of the United Nations (FAO) as a global consultative process to determine whether an international assessment of agricultural knowledge, science and technology was needed. Mr. Klaus Töepfer, Executive Director of the United Nations Environment Programme (UNEP) opened the first Intergovernmental Plenary (30 August-3 September 2004) in Nairobi, Kenya, during which participants initiated a detailed scoping, preparation, drafting and peer review process.

The outputs from this assessment are a Global and five Sub-Global reports; a Global and five Sub-Global Summaries for Decision Makers; and a cross-cutting Synthesis Report with an Executive Summary. The Summaries for Decision Makers and the Synthesis Report specifically provide options for action to governments, international agencies, academia, research organizations and other decision makers around the world.

The reports draw on the work of hundreds of experts from all regions of the world who have participated in the preparation and peer review process. As has been customary in many such global assessments, success depended first and foremost on the dedication, enthusiasm and cooperation of these experts in many different but related disciplines. It is the synergy of these interrelated disciplines that permitted IAASTD to create a unique, interdisciplinary regional and global process.

We take this opportunity to express our deep gratitude to the authors and reviewers of all of the reports—their dedication and tireless efforts made the process a success. We thank the Steering Committee for distilling the outputs of the consultative process into recommendations to the Plenary, the IAASTD Bureau for their advisory role during the assessment and the work of those in the extended Secretariat. We would specifically like to thank the cosponsoring organizations of the Global Environment Facility (GEF) and the World Bank for their financial contributions as well as the FAO, UNEP, and the United Nations Educational, Scientific and Cultural Organization (UNESCO) for their continued support of this process through allocation of staff resources.

We acknowledge with gratitude the governments and organizations that contributed to the Multidonor Trust Fund (Australia, Canada, the European Commission, France, Ireland, Sweden, Switzerland, and the United Kingdom) and the United States Trust Fund. We also thank the governments who provided support to Bureau members, authors and reviewers in other ways. In addition, Finland provided direct support to the Secretariat. The IAASTD was especially successful in engaging a large number of experts from developing countries and countries with economies in transition in its work; the Trust Funds enabled financial assistance for their travel to the IAASTD meetings.

We would also like to make special mention of the organizations who hosted the regional coordinators and staff and provided assistance in management and time to ensure success of this enterprise: the African Center for Technology Studies (ACTS) in Kenya, the Inter-American Institute for Cooperation on Agriculture (IICA) in Costa Rica, the International Center for Agricultural Research in the Dry Areas (ICARDA) in Syria, and the WorldFish Center in Malaysia.

The final Intergovernmental Plenary in Johannesburg, South Africa was opened on 7 April 2008 by Achim Steiner, Executive Director of UNEP. This Plenary saw the acceptance of the Reports and the approval of the Summaries for Decision Makers and the Executive Summary of the Synthesis Report by an overwhelming majority of governments.

Signed:

Co-chairs
Hans H. Herren,
Judi Wakhungu

Director
Robert T. Watson

Preface

In August 2002, the World Bank and the Food and Agriculture Organization (FAO) of the United Nations initiated a global consultative process to determine whether an international assessment of agricultural knowledge, science and technology (AKST) was needed. This was stimulated by discussions at the World Bank with the private sector and nongovernmental organizations (NGOs) on the state of scientific understanding of biotechnology and more specifically transgenics. During 2003, eleven consultations were held, overseen by an international multistakeholder steering committee and involving over 800 participants from all relevant stakeholder groups, e.g., governments, the private sector and civil society. Based on these consultations the steering committee recommended to an Intergovernmental Plenary meeting in Nairobi in September 2004 that an international assessment of the role of AKST in reducing hunger and poverty, improving rural livelihoods and facilitating environmentally, socially and economically sustainable development was needed. The concept of an International Assessment of Agricultural Knowledge, Science and Technology for Development (IAASTD) was endorsed as a multi-thematic, multi-spatial, multi-temporal intergovernmental process with a multistakeholder Bureau cosponsored by the FAO, the Global Environment Facility (GEF), United Nations Development Programme (UNDP), United Nations Environment Programme (UNEP), United Nations Educational, Scientific and Cultural Organization (UNESCO), the World Bank and World Health Organization (WHO).

The IAASTD's governance structure is a unique hybrid of the Intergovernmental Panel on Climate Change (IPCC) and the nongovernmental Millennium Ecosystem Assessment (MA). The stakeholder composition of the Bureau was agreed at the Intergovernmental Plenary meeting in Nairobi; it is geographically balanced and multistakeholder with 30 government and 30 civil society representatives (NGOs, producer and consumer groups, private sector entities and international organizations) in order to ensure ownership of the process and findings by a range of stakeholders.

About 400 of the world's experts were selected by the Bureau, following nominations by stakeholder groups, to prepare the IAASTD Report (comprised of a Global and five Sub-Global assessments). These experts worked in their own capacity and did not represent any particular stakeholder group. Additional individuals, organizations and governments were involved in the peer review process.

The IAASTD development and sustainability goals were endorsed at the first Intergovernmental Plenary and are consistent with a subset of the UN Millennium Development Goals (MDGs): the reduction of hunger and poverty, the improvement of rural livelihoods and human health, and facilitating equitable, socially, environmentally and economically sustainable development. Realizing these goals requires acknowledging the multifunctionality of agriculture: the challenge is to simultaneously meet development and sustainability goals while increasing agricultural production.

Meeting these goals has to be placed in the context of a rapidly changing world of urbanization, growing inequities, human migration, globalization, changing dietary preferences, climate change, environmental degradation, a trend toward biofuels and an increasing population. These conditions are affecting local and global food security and putting pressure on productive capacity and ecosystems. Hence there are unprecedented challenges ahead in providing food within a global trading system where there are other competing uses for agricultural and other natural resources. AKST alone cannot solve these problems, which are caused by complex political and social dynamics, but it can make a major contribution to meeting development and sustainability goals. Never before has it been more important for the world to generate and use AKST.

Given the focus on hunger, poverty and livelihoods, the IAASTD pays special attention to the current situation, issues and potential opportunities to redirect the current AKST system to improve the situation for poor rural people, especially small-scale farmers, rural laborers and others with limited resources. It addresses issues critical to formulating policy and provides information for decision makers confronting conflicting views on contentious issues such as the environmental consequences of productivity increases, environmental and human health impacts of transgenic crops, the consequences of bioenergy development on the environment and on the long-term availability and price of food, and the implications of climate change on agricultural production. The Bureau agreed that the scope of the assessment needed to go beyond the narrow confines of S&T and should encompass other types of relevant knowledge (e.g., knowledge held by agricultural producers, consumers and end users) and that it should also assess the role of institutions, organizations, governance, markets and trade.

The IAASTD is a multidisciplinary and multistakeholder enterprise requiring the use and integration of information, tools and models from different knowledge paradigms including local and traditional knowledge. The IAASTD does not advocate specific policies or practices; it assesses the major issues facing AKST and points towards a range of AKST options for action that meet development and sustainability goals. It is policy relevant, but not policy prescriptive. It integrates scientific information on a range of topics that are critically interlinked, but often addressed independently, i.e., agriculture, poverty, hunger, human health, natural resources, environment, development and innovation. It will enable decision makers to bring a richer base of knowledge to bear on policy and management decisions on issues previously viewed in isolation. Knowledge gained from historical analysis (typically the past 50 years) and an analysis of some future development alternatives to 2050 form the basis for assessing options for action on science and technology, capacity development, institutions and policies, and investments.

The IAASTD is conducted according to an open, transparent, representative and legitimate process; is evidence-based; presents options rather than recommendations; assesses different local, regional and global perspectives; presents different views, acknowledging that there can be more than one interpretation of the same evidence based on different world views; and identifies the key scientific uncertainties and areas on which research could be focused to advance development and sustainability goals.

The IAASTD is composed of a Global assessment and five Sub-Global assessments: Central and West Asia and North Africa – CWANA; East and South Asia and the Pacific – ESAP; Latin America and the Caribbean – LAC; North America and Europe – NAE; and Sub-Saharan Africa – SSA. It (1) assesses the generation, access, dissemination and use of public and private sector AKST in relation to the goals, using local, traditional and formal knowledge; (2) analyzes existing and emerging technologies, practices, policies and institutions and their impact on the goals; (3) provides information for decision makers in different civil society, private and public organizations on options for improving policies, practices, institutional and organizational arrangements to enable AKST to meet the goals; (4) brings together a range of stakeholders (consumers, governments, international agencies and research organizations, NGOs, private sector, producers, the scientific community) involved in the agricultural sector and rural development to share their experiences, views, understanding and vision for the future; and (5) identifies options for future public and private investments in AKST. In addition, the IAASTD will enhance local and regional capacity to design, implement and utilize similar assessments.

In this assessment agriculture is used to include production of food, feed, fuel, fiber and other products and to include all sectors from production of inputs (e.g., seeds and fertilizer) to consumption of products. However, as in all assessments, some topics were covered less extensively than others (e.g., livestock, forestry, fisheries and the agricultural sector of small island countries, and agricultural engineering), largely due to the expertise of the selected authors.

The IAASTD draft Report was subjected to two rounds of peer review by governments, organizations and individuals. These drafts were placed on an open access web site and open to comments by anyone. The authors revised the drafts based on numerous peer review comments, with the assistance of review editors who were responsible for ensuring the comments were appropriately taken into account. One of the most difficult issues authors had to address was criticisms that the report was too negative. In a scientific review based on empirical evidence, this is always a difficult comment to handle, as criteria are needed in order to say whether something is negative or positive. Another difficulty was responding to the conflicting views expressed by reviewers. The difference in views was not surprising given the range of stakeholder interests and perspectives. Thus one of the key findings of the IAASTD is that there are diverse and conflicting interpretations of past and current events, which need to be acknowledged and respected.

The Global and Sub-Global Summaries for Decision Makers and the Executive Summary of the Synthesis Report were approved at an Intergovernmental Plenary in April 2008. The Synthesis Report integrates the key findings from the Global and Sub-Global assessments, and focuses on eight Bureau-approved topics: bioenergy; biotechnology; climate change; human health; natural resource management; traditional knowledge and community based innovation; trade and markets; and women in agriculture.

The IAASTD builds on and adds value to a number of recent assessments and reports that have provided valuable information relevant to the agricultural sector, but have not specifically focused on the future role of AKST, the institutional dimensions and the multifunctionality of agriculture. These include: FAO State of Food Insecurity in the World (yearly); InterAcademy Council Report: Realizing the Promise and Potential of African Agriculture (2004); UN Millennium Project Task Force on Hunger (2005); Millennium Ecosystem Assessment (2005); CGIAR Science Council Strategy and Priority Setting Exercise (2006); Comprehensive Assessment of Water Management in Agriculture: Guiding Policy Investments in Water, Food, Livelihoods and Environment (2007); Intergovernmental Panel on Climate Change Reports (2001 and 2007); UNEP Fourth Global Environmental Outlook (2007); World Bank World Development Report: Agriculture for Development (2008); IFPRI Global Hunger Indices (yearly); and World Bank Internal Report of Investments in SSA (2007).

Financial support was provided to the IAASTD by the cosponsoring agencies, the governments of Australia, Canada, Finland, France, Ireland, Sweden, Switzerland, US and UK, and the European Commission. In addition, many organizations have provided in-kind support. The authors and review editors have given freely of their time, largely without compensation.

The Global and Sub-Global Summaries for Decision Makers and the Synthesis Report are written for a range of stakeholders, i.e., government policy makers, private sector, NGOs, producer and consumer groups, international organizations and the scientific community. There are no recommendations, only options for action. The options for action are not prioritized because different options are actionable by different stakeholders, each of whom has a different set of priorities and responsibilities and operates in different socioeconomic and political circumstances.

1

Agriculture in Latin America and the Caribbean: Context, Evolution and Current Situation

Coordinating Lead Authors:
Elsa Nivia (Colombia), Ivette Perfecto (Puerto Rico)

Lead Authors:
Mario Ahumada (Chile), Karen Luz (USA), Rufino Pérez (Dominican Republic), Julio Santamaría (Panama)

Contributing Authors:
Jahi Michael Chappell (USA), Michelle Chauvet (Mexico), Luis Fernando Chávez (Venezuela), Clara Cruzalegui (Peru), Dalva Maria da Mota (Brazil), Edson Gandarillas (Bolivia), Rosa Luz González (Mexico), Tirso Gonzales (Peru), Eric Holt Jiménez (USA), Carlos J. Pérez (Nicaragua), Ericka Prentice-Pierre (Trinidad and Tobago)

Review Editor:
Amanda Gálvez (Mexico)

Key Messages

1. Latin American agriculture is characterized by its heterogeneity and diversity of cultures and actors. Its heterogeneity is expressed by reference to agroecological conditions, resource endowment and means of production and access to information and other services. The diversity of cultures and actors implies differences in the systems for producing, generating and using knowledge, resource management and stewardship, worldviews, survival strategies and forms of social organization.

2. For purposes of this evaluation, three agricultural systems are considered: the traditional-indigenous system, the conventional system and the agroecological system. The traditional/indigenous system is based on local/ancestral knowledge and is very much tied to the territory and includes peasant systems. The conventional system has a market-based approach, is focused on intensive production practices and tends towards monoculture and the use of external inputs. The agroecological/organic system is based on the combination of agroecology and traditional knowledge and favors the use of organic inputs and the integration of natural processes.

3. The environmental and social vulnerability of Latin American agriculture is one of the results of implementing the development models prevalent in the last 50 years. The development models of the last 50 years have accorded priority to capital- and technology-intensive production systems that consume large quantities of fuels from non-renewable sources, are oriented to the external market, with limited social benefits. In the traditional/indigenous production systems the effects of those models are expressed mainly in their displacement towards the agricultural frontier causing deforestation, erosion of resources and loss of biodiversity. The agroecological/organic systems, in the context of the predominant models, are geared to market segments with high purchasing power, which excludes large social sectors from their benefits.

4. Agricultural productivity has increased in the last 50 years; nonetheless, this has not resulted in a reduction of poverty or hunger. There are 54 million people suffering malnutrition in the region, while the amount of food produced is three times the amount consumed. Although agricultural knowledge, science and technology (AKST) systems have been aimed at the goal of increasing agricultural production, factors such as the lack of access to and distribution of foods and the low purchasing power of a large sector of the population have stood in the way of this translating into less hunger. Hunger and malnutrition in LAC are not the result of the inability to produce enough food; therefore, increasing production will not solve the problem of hunger and malnutrition in the region. To the contrary, one of the main problems in the rural sector has been food importation from other countries where production is subsidized. This supply of food products drives down the price of local products and has a direct negative impact on the standard of living and the ability to make a living of the rural population.

5. LAC has abundant natural resources but they are not used efficiently and are highly degraded. Latin America and the Caribbean represent the most extensive reserve of arable land in proportion to the population. The region has 576 million ha, which is equivalent to 30% of the world's arable land and 28.5% of the region's land. In addition, the region contains five of the 10 richest countries in terms of biodiversity, with 40% of the world's genetic reserves (plant and animal). Nonetheless, natural resource use and management has been characterized by the underutilization of the arable lands, with a high proportion of latifundia with absentee owners, resulting in the use of only 25% of available lands. Moreover, there is a steady loss of soil and diversity due to problems of erosion, urbanization, pollution and expansion of agriculture.

6. Most of the region's rural population has lost or experienced a diminution of their access to and control over the use and conservation of the natural resources (land, water, genetic resources) in the last 50 years. This situation is an effect of the implementation of the agricultural policies of exploitation, privatization and patenting of natural resources stemming from the use of the neoliberal agroexport model that has been adopted by most countries in Latin America and the Caribbean. There has been a great concentration of wealth, natural resources and entrepreneurial resources, among others, with growing marginalization, exclusion, poverty and migration from rural to urban areas and to other countries. Special mention should be made of the mounting conflicts in the region brought about by the concentration of land tenure and the loss of the right to land of thousands of peasant and indigenous families.

7. While the policies favoring the opening up of trade have created market opportunities for the countries of the region, they have increased the vulnerability of small- and medium-scale producers in the region, benefiting almost exclusively the large-scale producers. The free trade agreements and structural adjustment programs fostered by the international financial institutions and adopted by the national governments have created an unlevel playing field in which local producers have to compete with imported products subsidized in their countries of origin. This has resulted in the displacement of many small-scale producers, creating a rural exodus in many countries. In some cases, the producers have reacted by forming cooperatives and developing alternative markets, in particular the fair trade market and the market for organic produce. Many large producers have successfully inserted themselves in the international market.

8. In LAC, approximately 25% of the inhabitants live on less than US$2 a day. These levels of poverty have persisted despite economic growth in the region. Per capita GDP in Latin America and the Caribbean declined 0.7% in the 1980s and increased 1.5% in the 1990s, without poverty levels changing significantly.

9. Malnutrition and hunger have a detrimental impact on the potential for development of the countries of the region and increase susceptibility to disease. In percentage terms, the undernourished population in Latin America and the Caribbean fell from 13 to 10% from 1992 to 2003. Nonetheless, the region continues to have a population of 54 million people who are undernourished, with stark regional differences. For example, in Mesoamerica undernourishment increased from 22 to 25% during that same period. This number of undernourished inhabitants means vulnerability to disease, the impossibility of having a normal educational performance and therefore the inability to participate efficiently and productively in development processes.

10. In LAC, food dependency has been exacerbated as a result of neoliberal globalization. The importation of subsidized food products has dismantled local production systems, creating dependence on food produced in other countries. The situation is aggravated as the poorest, especially rural, inhabitants whose main source of income is agriculture, have to face the progressive difficulty of the decreasing purchasing power for acquiring food, whether locally produced or imported. This has resulted in the loss of food sovereignty, especially in the most vulnerable sectors of the region.

11. The performance of agricultural systems is mixed in terms of production and sustainability, as well as environmental impacts. The traditional/indigenous system is characterized by diversity with variable levels of production (from high to very low). The conventional system has high levels of production and competitiveness in external markets, yet under current conditions is not sustainable or efficient in terms of energy use. The agroecological system has high productivity and sustainability and a market niche for certified organic products, yet has been limited by the lack of governmental-institutional support and there is a debate as to whether it can satisfy the world demand for food.

12. The development of agriculture over the last 50 years in LAC has caused critical environmental impacts. Among the impacts, mention should be made first of the deforestation of vast areas high in biodiversity, especially in the tropical forests of Central America and the Amazon. In addition, the use of agrochemicals and soil erosion caused by farming have had a major negative impact on terrestrial, aquatic and marine biodiversity. More diversified agricultural systems can mitigate these impacts up to a point, providing habitats and also connectivity between fragments of natural habitats.

13. In LAC, emigration is on the increase as is the vulnerability of the rural population. This is due to the substitution of a large part of the agricultural labor force by machinery and technologies, provoking a reduction in the number of farms due to the concentration of landholdings; the loss of land tenure by peasants and indigenous communities; rural violence; and population increase.

14. In LAC, cultural diversity, local/traditional knowledge and agrobiodiversity are being lost. Specifically, local or traditional customs and knowledge are hardly taken into account in the vertical model of technological development prevailing in the region. The predominant technologies, which are displacing local or traditional knowledge and wisdom, are generally selected with scant participation of the peasant and indigenous communities. This process of cultural and technological erosion has been casting aside an ancestral rural cultural heritage, with local content, adapted to its surroundings, yielding to external, more uniform knowledge and cultures.

15. The health of rural communities in LAC has been detrimentally affected by problems of acute and chronic intoxications in the countryside due to the indiscriminate use of agrochemicals. For example, in Central America, the Plagsalud program of PAHO/WHO estimated 400,000 acute intoxications per year; underregistration is estimated at 98%. The problems of intoxication are worse in rural areas because no occupational health programs have been put in place for farmers, nor are there health services specifically geared to treating intoxications due to exposure to pesticides, causing several chronic diseases that reduce the capacity to generate income. Children, the elderly, the infirm and the malnourished are the most vulnerable, compromising the right to life and human dignity.

16. The population of women who are poor, wage earners and heads of household is growing as a proportion of the total population living in poverty in rural areas. Although there are particularities in different subregions of Latin America and the Caribbean, in general, as the participation of men in agriculture diminishes, the role of women increases. Male migration is one of the main reasons for the increase of the female population in the rural economy. The expansion of non-traditional export crops, wars, violence and forced displacement are other causes of the so-called "feminization of agriculture."

17. Transgenic crops have been progressively adopted in LAC, with impacts perceived by some as negative and by others as positive, in relation to the goals of sustainability, poverty reduction and equity. Transgenic crops are used in commercial production, especially of cotton, soybean, maize and canola. The social and environmental repercussions are differentiated for each of these crops and by countries of the region. The technology has been adopted quickly by the producers of the conventional/productivist system, increasing profitability, but in some regions it has also accentuated the above-mentioned social and environmental deterioration. Biosafety policies are recommended that impede the consumption and cultivation of transgenic organisms in countries that are the centers of origin of those crops, so as to avoid contamination and preserve genetic diversity. In regions that are not centers of origin, regulatory arrangements should be guided by the precautionary principle. The possibility of genetic contamination in some species has been demonstrated and it should be an essential part of biosafety policies, which should also take into account transgenic edible crops used for the production of non-edible nutraceuticals, biopharmaceuticals, or industrial products.

18. Policies for alternative energy supply based on renewable resources motivated by the worldwide energy crisis present opportunities and threats to the agricultural sector, thus their externalities should be carefully analyzed. Agricultural production for use in alternatives to fossil fuels has increased quickly in recent years in LAC, benefiting some economic sectors and providing alternative markets to the agroindustrial sector. Although the development of these crops offers an opportunity for rural revitalization, there are risks of negative environmental and social impacts. The expansion of crops for biofuels, such as sugar cane, oil palm, soybean and timber, is diminishing food production with a negative impact on food security in some regions and with a detrimental impact mainly on small-scale producers, indigenous populations and other traditional communities. The use of by-products or animal and plant waste is another source of biofuels whose use attenuates environmental problems.

19. The structures of agricultural regulation in LAC are not institutionally adequate, resulting in regional weaknesses such as low competitiveness and the vulnerability of the endemic natural patrimonies. There are some international agreements on biosafety, animal and plant quarantine, food safety, intellectual property and access to and management of genetic resources that have been important in other regions of the world as part of a sustainable agriculture development agenda. The understanding of these agreements by countries has not always meant that they adhere to them, but it has encouraged them to develop particular and appropriate regulatory strategies, for example, on the protection, access to and use and management of autochthonous natural patrimonies, independent of whether they adopt international regulatory frameworks.

1.1 Objectives and Conceptual Framework

Latin America and the Caribbean (LAC) has a population of 569 million people, 209 million of whom are poor and 81 million of whom suffer extreme poverty, most of whom live in rural areas (UNDP, 2005b; CEPAL, 2006b; FAO, 2006b). The region has great biodiversity and an abundance of natural resources, which contributes to the production of 36% of the cultivated foods and industrial species worldwide. Nonetheless, these resources are rapidly degrading (UNEP, 2006). The situation is all the more complicated since the region is one of those most affected by economic inequality in the world (CEPAL, 2004a; Ferranti et al., 2004). The region is facing the important task of improving rural livelihoods and ensuring nutritional security while reducing environmental degradation, addressing social and gender inequality and guaranteeing health and human welfare. Evaluating how AKST can contribute to these goals is a multisectoral task that requires paying attention to a wide variety of economic, environmental, ethical, social and cultural factors.

The authors of The Millennium Development Goals: A Latin American and Caribbean Perspective (UNDP, 2005a) conclude that the region produces sufficient food to meet the nutritional needs of all its inhabitants. Though this is not uniform across the region, all the countries, including those with a high rate of malnutrition, have a food energy supply of more than 2,000 kilocalories per person per day, which exceeds the minimum recommended for an adult (1,815 kilocalories) (Figure 1-1). In all, the region produces three times the quantity of food it consumes (UNDP, 2005a). These data suggest that hunger and malnutrition in the region today are not due exclusively to the failure to produce sufficient food and that the problem is more complex, hence the solution must go beyond technical aspects related to production. The divergence of opinions with respect to the causes and possible solutions underscores the need to undertake a critical international evaluation that makes it possible to analyze, using a comprehensive and multidisciplinary approach, aspects crucial for policy making.

It was with this purpose in mind that the International Assessment of Agricultural Knowledge, Science and Technology for Development (IAASTD) was undertaken. This evaluation is an initiative sponsored by different United Nations agencies, the World Bank and multilateral funds,[1] which seeks to analyze the complexities of the systems of knowledge, science and technology (KST) in Latin America and the Caribbean to understand how these systems can contribute to improving the living conditions of the poor in the region. The objectives of this chapter are: (1) to develop the conceptual framework for the evaluation, (2) to present the context (social, political, economic, environmental, cultural) that impacts on or is affected by agriculture in the region and (3) to undertake a critical assessment of the recent evolution and current situation of production systems, in particular an evaluation of the performance and impacts of the three main systems of production in the region: the indigenous/traditional, the conventional/productivist and the emerging agroecological system. The conceptual framework, context and current situation (Chapter 1), as well as the historical analysis of the role of knowledge, science and technology in agriculture (Chapter 2), will provide the elements needed for analyzing future scenarios (Chapter 3) and options for the future (Chapters 4 and 5). In particular, an effort is to be made to evaluate how agricultural knowledge, science and technology systems can contribute to the goals of sustainable development and in particular to reducing hunger and poverty, improving nutrition and human health, strengthening ways of life and equity and achieving environmental sustainability.

Reducing hunger and poverty, improving human nutrition, strengthening ways of life and achieving environmentally and socially sustainable economic development remain on the social and economic agenda of all local, national, regional and global strategies and interventions. Similarly, generating, accessing and using knowledge, science and technology are considered driving factors of and therefore fundamental components in such strategies and interventions, especially those geared to rural development and poverty reduction.

[1] World Bank (WB), Food and Agriculture Organization of the United Nations (FAO), World Health Organization (WHO), United Nations Environment Program (UNEP), United Nations Development Program (UNDP), United Nations Educational, Scientific and Cultural Organization (UNESCO), and the Global Environment Facility (GEF).

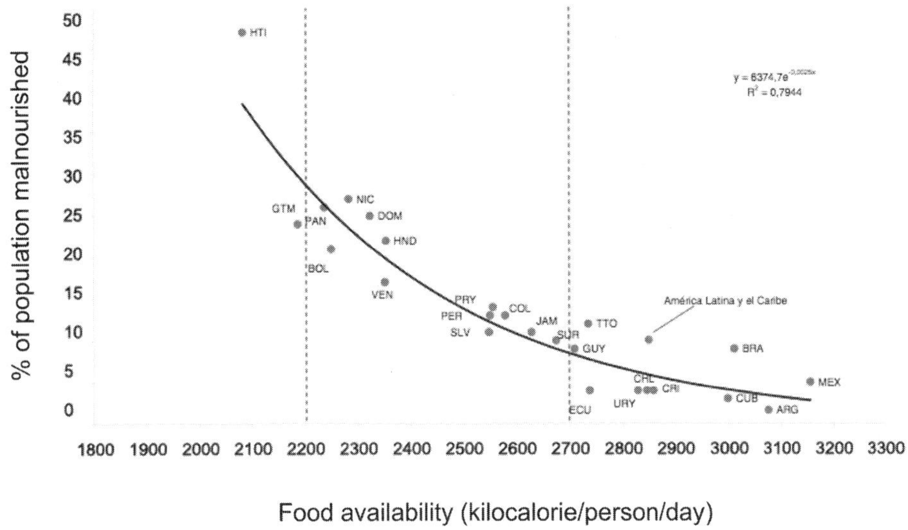

Figure 1-1. *Supply of food and percentage of population malnourished in LAC countries 2000-2002.*
Source: FAO, 2004.

The conceptual framework (Figure 1-2) taken as a reference for developing the content of this report seeks to understand and analyze the interrelations of the agricultural knowledge, science and technology systems, the agricultural production systems and the contextual factors and variables as a basis for retrospective and prospective analysis of their contribution to the attainment of the objectives of development and sustainability. The AKST systems can be understood as the set of actors (individuals and organizations), networks, configurations and interfaces among them that interact in generating, reconfiguring and disseminating information and technologies for innovation (institutional and technological) of agricultural production systems through processes of social learning regulated and guided by negotiated standards and rules for the purpose of improving the relationships among knowledge, technology, the environment and human development. The AKST systems aim to improve the performance indicators of agricultural production systems through processes of technological innovation.

In the conventional approaches to systems, the vulnerability of agricultural production systems is conceived of based on the worldview of the outside expert who acts under his or her universal conception of reality on the local views and interests and reproduces a division of labor in the process of generating, accessing and using knowledge that transforms producers to mere receptacles of values, concepts and paradigms generated far from their context and without any commitment to their needs, demands, or aspirations. This linear mode of intervention, in which just a few generate, others transfer and the thousands of producers adopt the technological innovations, has prevailed in the last 50 years. To the contrary, in the Agricultural Knowledge and Information Systems (AKIS) approach, the systems are considered to be a social construct in which the actors who constitute it perceive their interdependence, come to agreement on the present and future systematic vision, negotiate principles, premises, objectives, strategies and courses of action and systematize their experiences and lessons through semi-structured processes of interpretation and intervention negotiated through the integrated management of knowledge and innovation.

The integrated management of knowledge and innovation suggests identifying the worldview—conception of reality—that conditions the ways of thinking and acting of those who interact to transform their reality and therefore is centered on the changing web of relationships and meanings that influence perceptions, decisions and actions in human initiatives. Accordingly, this mode of intervention considers the actors of the social context in which the new technologies are generated and applied as being co-responsible at every stage of the process of generating, validating and using the relevant information and technologies for innovation in agriculture.

Agricultural production systems include all the activities for producing food, fibers, energy, biomass and environmental services such as landscape management and carbon sequestration. These productive and service activities entail the social and economic organization of the labor force, rural resources and information (direct drivers) with different performances in light of indicators such as efficiency, productivity, competitiveness, equity, quality and environmental sustainability.

In processes of innovation, science and technology are important but not sufficient components for attaining the objectives of development and sustainability, as they are conditioned by variables and factors from the regional and global context in their different dimensions (indirect drivers), including social, economic, institutional, cultural, political and environmental. The critical external factors are capable of bringing to bear strong influences on agricultural production systems, determining internal obsolescences, shortcomings of capacities and resources and flaws in their relationship with the external environment.

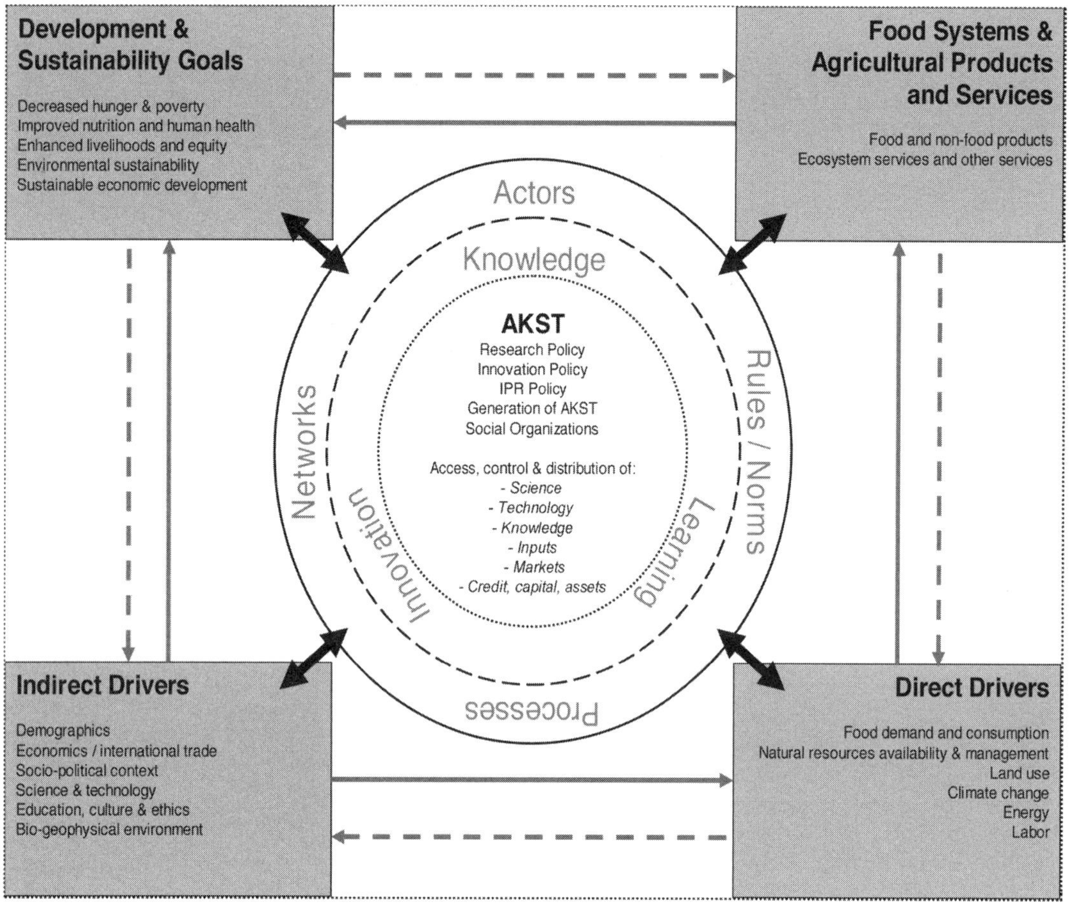

Figure 1-2. *IAASTD Conceptual Framework.*

In Latin America and the Caribbean, little progress has been made toward the millennium development goals (UNDP, 2005a). Based on the index of purchasing power parity of individuals and progress in fighting malnutrition and hunger, the region tends towards impoverishment, and the number of malnourished people in the region has diminished very slowly. In particular, in LAC in the last 10 years the number of poor and the rate of inequality has increased (Cardoso and Helwege, 1992; Rosenthal, 1996; Berry, 1998; O'Donnell and Tockman 1998; Hoffman and Centeno, 2003; Portes and Hoffman, 2003; CEPAL, 2004a; Ferranti et al., 2004).

Notwithstanding the great biodiversity and availability of natural resources, the rate of environmental degradation is the highest in the world, largely because of the type of agricultural development (industrial productivist model) pursued over the last 50 years. From 1970 to 2000, on average six hectares were deforested daily, only 60% of which was used for agricultural production; the remaining 40% were abandoned due to problems of degradation and land speculation (UNEP, 2002a). Increases in production and more intense use of the land, particularly in tropical areas, have led to problems of compaction, salinization, desertification, soil erosion, water pollution and negative effects on biodiversity and human health. The environmental, economic and social vulnerability of the planet, lifestyles, productive systems and ecosystems is associated with industrial development that has accorded priority to the mechanical and instrumental dimension over human, social and ethical considerations in human relations with other forms of life and with nature.

If this vulnerability reflects problems brought about by human action, sustainability can only emerge from social learning (Bhouraskar, 2005) and through human interaction (Röling, 2003) to create consensus-based actions that transcend particular private interests. Nonetheless, the proposals and solutions of the majority of development "experts" reveal that they themselves are held hostage to the mode of innovation (mode of interpretation + mode of intervention) that has prevailed in creating the problem that we need to grasp if we are to be able to overcome it. Following Albert Einstein, who said that it was not possible to overcome a complex problem using the same method that gave rise to it, this evaluation is done based on the premise that it is not possible to overcome complex situations using the same mode of interpretation and the same mode of intervention that gave rise to them. Therefore, it is urgent to undertake a critical analysis of the factors that gave rise to the present-day situation of poverty, hunger, inequality and environmental degradation so as to avoid falling once again into the same trap and to be able to propose options with real possibilities of change.

The schema for generating knowledge, the process of social learning and the innovation in agriculture which, it is hoped, will produce the conditions for and viability of human development is characterized and influenced by a dynamic context in which development processes are the result of policies formulated and applied based on the objectives and promises of the socioeconomic development models. In order for the AKST system to have a positive impact on the changes, leading to improvements in the standards of living and quality of life, the system has to be sensitive to stimuli and indicators that point to the degrees and nature of the changes demanded for attaining the development and sustainability objectives, taking into account alternative future scenarios.

Constructing scenarios is a methodology used to support the understanding of the future and decision-making on current policies and strategies. The scenarios offer a likely vision, distant in time, of the nature of complex phenomena and a model of how different sorts of phenomena will evolve (social, economic, environmental, technological) and interact. The use of scenarios makes it possible to manage the uncertainty that necessarily characterizes the future, depending on premises about the decisions of the social actors in relation to various macro variables.

Accordingly, applying the conceptual framework proposed entails, first, characterizing the global and regional context in which both the AKST systems and the agricultural production systems are found and analyzing the recent history and current situation of Latin American agriculture with special emphasis on the performance of production systems. This assessment, along with an assessment of the AKST systems (Chapter 2) and an elaboration of plausible future scenarios (Chapter 3) will be an input for proposing a series of realistic options that may contribute to attaining the goals of reducing poverty, hunger and inequity, as well as attaining environmentally sustainable development (Chapters 4 and 5).

1.2 Latin American and Caribbean Agricultural Production Systems

Recognizing the structural heterogeneity and diversity of actors, cultures and knowledge of Latin American agriculture both regionally and subregionally, it was decided to consider three agricultural systems for the purposes of this evaluation:

1. Traditional/indigenous (includes peasant);
2. Conventional/productivist;
3. Agroecological.

The importance of each of these systems varies not only among subregions, but also within each subregion and even within each country. The performance and impacts of three principal agricultural systems are presented in 1.7 (Table 1-1).

The *traditional/indigenous* system is a family agricultural system, primarily involving family consumption, under which one can distinguish the ethnic systems constituted by indigenous and Afro-descendant communities linked to the territory and the peasant systems. It is based on local/ancestral knowledge and is not very well articulated to the market for inputs and products, though today many peasants market part of their production. In general, this system is high in agrobiodiversity, outside inputs are used to a limited extent, if at all and labor is drawn from the family (Altieri, 1999; Toledo, 2005). The cosmovision of indigenous communities assumes a relationship with natural resources that goes beyond an economic-extractive activity: it implies an ecological-cultural-spiritual vision linked to the territory. (For the example of the Andean world view, see Figure 1-3.) This system stands out for sustainability with respect to the environment and energetic balance, with variable levels of production (Barrera-Bassols and Toledo, 2005). In several regions traditional/indigenous agriculture is displaced to marginal lands and much of the knowledge that undergirds it is being lost (David et al., 2001; Deere, 2005). In these conditions one finds low yields. In most countries of the region, governmental/institutional support has not fostered the strengthening of this system.

At the other end of the spectrum one finds the *conventional/productivist system,* also called the "industrial system." This system is characterized by a high degree of mechanization, monocultures and the use of external inputs, such as synthetic fertilizers and pesticides, as well as contract labor. It is based on technological knowledge and is highly articulated to the market and integrated to productive chains. This system has been supported by development models and it has benefited from support systems such as credit and technological capital (Chapter 2). Its prominence in the national and international markets makes the conventional/productivist system stand out for high levels of productivity and competitiveness. Nonetheless, it gives rise to significant negative externalities in terms of environmental, social and cultural costs (see 1.7).

As the environmental and human costs of conventional production have increased, the *agroecological* system is becoming more important. It is based on the knowledge of agroecology stemming from the interaction between scientific and traditional knowledge and aimed at reducing the negative impacts of the conventional systems through productive diversification and the use of ecologically-friendly technologies. This system is characterized by the search for sustainability in social, economic, cultural and environmental terms; scant articulation in productive chains; and a strong link to the market for differentiated products, especially organic products. The systems described are expressed in the subregions with differentiated nuances and through mixed forms or particular combinations.

1.3 Regionalization

Latin America and the Caribbean (LAC) is a very extensive and varied geographic region. It extends from Baja California (32 1/2° N) to Tierra del Fuego (55° S) and has a total of 2.050 billion ha (including internal bodies of water) in 45 countries with 569 million inhabitants. Given its great range of longitudes and altitudes, as well as its great biodiversity, LAC has a wide diversity of ecosystems including moist tropical jungles, dry forests, conifer forests, temperate forests, tropical savannahs, temperate savannahs, *páramos* and desert environments. To facilitate the analysis and characterization of the region in this evaluation we will refer to large geographic zones as follows: Southern Cone Andean Region, Central America, Mexico and the Caribbean (Table

Table 1-1. Main characteristics of agricultural systems considered in the assessment. Source: Authors' elaboration

	Indigenous/ traditional system	Conventional/ productivist system	Agroecological system
Main actors	Indigenous communities, Afro descendants and peasants.	Agribusiness, small, medium and large producers	Small, medium and large-scale producers, professionals
Inputs (type and origin)	Low external input, local technology	Chemical inputs, technological machinery and tools, externally bought fossil fuel	Low dependency on external inputs. Biological inputs produced from within the system. High technology integrated to endogenous, natural, physical and energetic processes
Knowledge and skills	Local/ancestral knowledge. Strongly rooted to the territory	Academic/ technological knowledge	Academic/ technological knowledge and know-how with emphasis on local/ancestral knowledge. Scientific knowledge strongly based on ecological science.
Diversification of production	Multi-crops; high biological diversity	Great scale monocultures with spatial and temporal rotations	Multi-crops, with spatial and temporal integration
Links to the market	Little or no linking with input/output markets. Production largely oriented to family consumption	Strong articulation with production chains and links to national and international markets.	Little articulation with production chains, but strong linking with markets of differentiated products.
Labor	Family and communal labor using different forms of labor exchanges.	Dominated by hired labor	Family and hired labor

Source: Authors' elaboration.

1-2). Nonetheless, on occasion it will be necessary to refer to the regions based on the natural ecosystems, such as tropical jungles, *pampas* and *cerrados*, mangroves, etc.

Due to the great diversity of ecosystems and climates in the region, LAC is characterized by a great diversity and complexity of agroecological zones, as well as types of production associated with these zones. Table 1-3 shows the agroecological zones of the region as well as the principal types of agriculture in these zones.

1.4 Global Context: Main Trends

To perform a critical evaluation of AKST systems and of agriculture in Latin America and the Caribbean, one must know the context in which these systems operate.

Since the 1950s, the combined effects of three revolutions—technological, economic and cultural—have been giving rise to new realities (Castells, 1996), shaped by old and new contradictions, which transform (in a differentiated manner) the many "worlds" that coexist in our region (Capra, 1982; Restivo, 1988; Dicken, 1992; Sachs, 1992; Barbour, 1993; Najmanovich, 1995; Castells, 1996, 1997, 1998; Chisholm, 1996; Escobar, 1998a; Wallerstein, 1999; Busch, 2000, 2001; Rifkin, 2000; Mooney, 2002; Santamaría-Guerra, 2003). The main global trends can be grouped as: (1) technological changes, (2) macroeconomic changes, especially globalization, (3) the emerging resistance movements with new outlooks and (4) environmental/natural changes.

Among the main technological changes we see the emergence of an immaterial economy dependent mainly on an intangible factor—*information*—and on the communications infrastructure. From this technology is emerging a digital hemisphere whose dynamic is dependent on virtual networks of power through which capital, decisions and information flow. The rise of the network concept, supported by new possibilities of digital technology and communications infrastructure, has implications for the management of interdisciplinary, inter-institutional and international projects. Also worthy of special note are the emerging scientific and technological possibilities (*robotics, new materials, nanotechnology, cellular and molecular genetics, information technology, etc.*) that point simultaneously to new advances important for humankind and to new inequalities within and among social groups and nations.

Globalization has accelerated the construction of a world economic and political order whose corporate and transnational nature is becoming consolidated under the dominant influence of actors with global interests and expansionist ambitions. This model has led to the decline of the

Figure 1-3. *Andean cosmovision.* Source: Gonzales, 1999; Gonzales et al., 1999.

sovereignty and autonomy of the nation-state, so as to give rise to the prevalence of transnational rules over national ones, giving rise to a crisis of representative democracy, with the emergence of a supranational state-network. Under this new model one notes, among other things, the end of the social contract between capital and labor under the notion of "labor flexibility," and the construction of transnational productive chains outside the control of nation-states and local actors through technological convergence and productive decentralization, as well as a process of homogenization that has led to the very fast erosion of cultural diversity.

The process of globalization has not been accepted passively by the governments and peoples of the region. The last decade has seen the formation of regional and subregional economic blocs for internal integration (economic, technological and political) and to counter external competition, as well as a struggle to establish a *global civil society* dependent on *participatory democracy* networks and emergence and proliferation of social movements to vindicate and uphold the importance of the interdependence among human, social and ecological considerations. These trends towards participatory democracy through social movements include the struggle for sustainable development mediated by the creation of a global civil society to monitor the excesses of transnational corporate capitalism; the rise of initiatives and dynamics that accord priority to local development as the starting point for transformations committed to human, social and ecological needs; the struggle for indigenous rights; and the struggle to control (and, in general, contest) the products of science and even the process of doing science (anti-GMO groups, anti-human cloning groups and groups to stop animal suffering, among others).

Finally, the environmental changes, particularly the loss of biodiversity and global warming, have assumed a

Table 1-2. **Geographic regions and countries in Latin America and the Caribbean. Source: CIA, 2008**

Region	Countries	Area (km^2)
Southern Cone	Argentina	2,766,890
	Brazil	8,514,876
	Chile	756,102
	*French Guiana	90,000
	*Guyana	214,969
	Malvinas Is.	1, 217
	Paraguay	406,752
	*Surinam	163,820
	Uruguay	176,215
	Subtotal	13,089,624
Andean region	Bolivia	1,098,581
	Colombia	1,138,914
	Ecuador	283,561
	Peru	1,285,216
	Venezuela, Rep. Bolivarian	912,050
	Subtotal	4,718,322
Central America and Mexico	Belize	22,966
	Costa Rica	51,000
	El Salvador	21,041
	Guatemala	108,889
	Honduras	112,088
	Mexico	1,964,375
	Nicaragua	120,340
	Panama	75,517
	Subtotal	2,476,216
The Caribbean	Anguilla	91
	Antigua and Barbuda	442
	Aruba	180
	Bahamas	13,878
	Barbados	430
	British Virgin Islands	153
	Cayman Islands	259
	Cuba	109,886
	Dominica	751
	Dominican Republic	48,671
	Grenada	344
	Guadeloupe	1,705
	Haiti	27,750
	Jamaica	10,991
	Martinique	1,102
	Montserrat	102
	Netherlands Antilles Is.	800
	Puerto Rico	8,870
	Saint Kitts and Nevis	261
	Saint Lucia	539
	Saint Martin	53
	Saint Vincent/Grenadines	389
	Trinidad and Tobago	5,130
	Turks and Caicos Is.	948
	US Virgin Islands	352
	Subtotal	234,341
	Total	***20,518,503***

* These countries, although located in South America, are frequently considered as part of the Caribbean due to their cultural affiliation with the rest of the Caribbean region.

central role in the different forms of international discourse. Climate change, for example, has been included as an item for discussion at the United Nations Security Council, even though not all the members of the Security Council approve of its inclusion. There are also multiple international agreements related to biodiversity and agriculture, which are crucial in an agricultural development agenda for the region, mainly when knowledge, science and technology are thought of as instruments for propelling such development. The most important initiatives for harmonizing regulatory frameworks in agriculture include (1) the Cartagena Protocol on Biosafety, which seeks to protect biodiversity in light of the risks associated with genetically modified organisms (transgenics); (2) the International Plant Protection Convention (IPPC), which seeks to prevent the dissemination and introduction of pests that affect plants and plant products and to promote appropriate measures for combating pests; (3) Codex Alimentarius, created in 1963 by the FAO and WHO to develop food standards, regulations and other related texts, such as codes of practices under the Joint FAO/WHO Food Standards Program; (4) the World Intellectual Property Organization (WIPO) established to foster the protection and effective use of intellectual property worldwide through cooperation with member states and other interested parties; (5) the International Union for the Protection of New Varieties of Plants (UPOV), an intergovernmental organization; and (6) the International Treaty on Phytogenetic Resources for Food and Agriculture.

There are other agreements related to controls on international trade and the use of potentially toxic substances, which largely have to do with agriculture because they include chemical pesticides that pose a high risk to the environment and human and animal health, such as: (1) the Basel Convention on the Transboundary Transport of Toxic Substances; (2) the FAO Code of Conduct on the Distribution and Use of Pesticides; (3) the Montreal Protocol for Substances that Deplete the Ozone Layer; (4) the Rotterdam Convention, which established the prior informed consent (PIC) procedure for trade in prohibited or severely restricted substances; and (5) the Stockholm Convention on Persistent Organic Pollutants (POPs), which includes more than a dozen organochlorinated pesticides, including DDT (UNEP, 2001; Bejarano, 2004).

As a result of these global changes, the swift restructuring of agriculture and the global food system is striking. Reflecting the nature, direction, priorities and contradictions of current global changes, both agriculture and the food system are being transformed by several changes. For example, agriculture and the food system are and will be profoundly restructured with the application of techniques associated with the revolutions in modern biotechnology (genetic engineering), nanotechnology, robotics and information technology and by the construction of transnational productive chains transforming the nature of productive and power relations, in which emerging global actors decide on the nature, direction and priorities of the new transnational agriculture. With the emergence of new scientific and technological revolutions, agribusiness, currently aimed at food production, is working on non-food products, such as energy products (biofuels, such as biodiesel and ethanol) and new fibers resulting from biotechnology and

Table 1-3. **Agroecological areas / types of production in Latin America and the Caribbean. Source: Dixon et al., 2001**

Agroecological areas/ Types of production	Countries or regions with these types of production or ecosystems	Total area (million ha)	Cropped area (% of region)	Population (millions)	Agricultural population (% of region)	Main subsistence forms	Prevalence of poverty
1. Irrigated	North of Mexico, coast and internal valleys of Peru and Chile, Argentina	200	3.7	11	9	Horticulture, fruit, cattle	Low-moderate
2. Forest based	Amazon River basin (Brazil, Bolivia, Peru, Ecuador, Colombia, Venezuela, Surinam and Guyana) and forested zones of Mexico and Central America	600	1.0	11	9	Subsistence, cattle ranching	Low-moderate
3. Coastal plantation and mixed	Central America, Mexico, the Caribbean and northeast coast and occidental area of South America	186	10.7	20	17	Export crops/tree crops, fishing, tubers, tourism	Highly variable
4. Intensive mixed	Central region of Brazil	81	16.0	10	8	Coffee, horticulture, fruit, off-farm work	Low
5. Mixed cereals and livestock	South of Brazil, north of Uruguay	100	18.0	7	6	Rice and livestock	Low-moderate
6. Moist temperate and mixed forest	Coastal area of the center of Chile	13	12.3	<1	1	Dairy, livestock, cereals, forestry and tourism	Low
7. Maize-beans	Mexico and Central America	65	9.2	<11	10	Maize, beans, coffee, horticulture and employment outside the farm	Extensive and severe
8. Intensive highlands and mixed (Northern Andes)	Andean region of Colombia, Ecuador and Venezuela	43	10.2	4	3	Vegetables, maize, coffee, cattle/pigs, cereals, potatoes, off-farm work	Low-generalized (smallholders)
9. Extensive mixed (Cerrados and Llanos)	Southeast of Amazonia in Brazil and Bolivia, north of Amazonia in Venezuela and Guyana	230	13.5	10	9	Livestock, oilseeds, grains, coffee	Low-moderate (small producers and landless)
10. Temperate mixed (Pampas)	East -central region of Argentina and part of Uruguay	100	20.0	7	6	Livestock, wheat, soybean	Low
11. Dryland mixed	North oriental coast of Brazil and the Yucatan Peninsula of Mexico	130	13.8	10	9	Livestock, maize, cassava, wage labor, seasonal migration	Extensive, especially drought-induced
12. Extensive dryland and mixed (Gran Chaco)	Central region of Argentina, north of Paraguay	70	11	<2	<2	Livestock, cotton, subsistence crops	Moderate
13. High altitude mixed (Central Andes)	East of Bolivia	120	1.1	>7	>7	Tubers, sheep, grains, llamas, horticulture, off-farm work	Extensive and severe
14. Pastoral	Andean region of Peru and Bolivia	67	—	<1	<1	Cattle and sheep	Low-moderate
15. Sparse (forest)	Chile and Argentina	60	<0.5	<4	3	Sheep, cattle, silviculture and tourism	low

drugs such as vaccines resulting from the combined activity of biotechnology and nanotechnology (Friedland et al., 1991; Goodman and Redclift, 1991; Friedmann, 1993; Bonnano et al., 1994; McMichael, 1994; Goodman and Watts, 1998; Busch, 2001; Mooney, 2002).

Countering these trends one finds the rise of very strong rural social movements and indigenous movements that propose alternatives for autonomy, food sovereignty, agroecology and peasant networks (Vía Campesina, MST and the World Social Forum, among others), as well as the growing number of consumers who demand local, organic, socially fair, diverse, nutritional and safe foods (Slow Food Movement and consumer associations).

Because of these and other changes, agriculture as we know it is facing a profound transformation, with implications for its protagonists whose impacts are not yet clear, much less understood. To understand the current situation of agriculture in LAC, one must review the history to understand the models, visions and development paradigms that shaped the strategies of intervention that gave rise to the consequences we are trying to overcome.

1.5 Regional Context

1.5.1 Evolution of development models

Development strategies in LAC were not designed in a political vacuum, but rather were decisively influenced by political events inside and outside the region that promoted and continue to promote development models that directly affect the agrarian policies of the region and AKST systems.

With the economic expansion of the United States after the Second World War came the need to expand external markets for its products, find new investment opportunities, access cheap raw materials to support growing industry and establish a global network of military power to ensure access for consumers, markets and raw materials. Consequently, the region's development was subordinated to U.S. interests and growth needs. To foster development and maintain economic stability internationally, the industrialized countries, led by the United States, assigned a new role to the World Bank and the International Monetary Fund, institutions originally created to rebuild Europe (Stiglitz, 2003). Yet the type of development promoted through the new international institutions is highly conditioned on the economic, political and military needs of the industrialized countries, especially the United States.

In the 1950s, President Harry Truman of the United States held great influence over the path of development in LAC. In his Fair Deal, Truman proposed the "technification" (intensification) of agriculture as one of the instruments for emerging from underdevelopment (a term he introduced in the international discourse). During his administration, a period marked by the proliferation of development projects began. In the 1960s, the program that most influenced the type of development in the region was the Alliance for Progress, a hemispheric initiative led by President John F. Kennedy to counter the potential influence of communist Cuba in the rest of LAC and to promote the U.S. economy (Smith, 1999); its development strategy entailed articulating the peasant sector to the market (Escobar, 1995). World Bank documents make clear that under this development strategy, the peasants of LAC had two options: (1) to become small entrepreneurs, or, (2) to disappear from the market (or from the agricultural sector). This strategy was focused on modernizing and monetizing the rural sector and making the transition from isolation to integration with the national economy. The technological vehicle for this strategy was the Green Revolution, yet its results in terms of improving the living conditions of the rural population have been much debated (Glaeser, 1987; Rosset et al., 2000; Evenson and Gollin, 2003). With the Green Revolution food production in LAC increased 8%, yet during the same period hunger in the region increased 19% (and this was not due to population increase, as the total amount of food per person also increased).

During the 1960s and 1970s, this conception of development held sway. To a certain point one can say that these development policies were successful since during these two decades Latin America and the Caribbean experienced unprecedented economic growth. Most of the countries attained per capita growth of 2.4% annually during the 1960s and some countries were able to maintain this rate in the 1970s (IDB, 1989). This growth was based largely on the import substitution model developed and promulgated by the United Nations Economic Commission for Latin America (ECLAC) (Bulmer-Thomas, 1987; Glaeser, 1987). This was a period of fast-paced industrialization and economic integration at the regional level. Yet once again the benefits of this growth were not distributed equitably and in many cases they did not even reach the most impoverished sectors of the region (ICCARD, 1989; Conroy et al., 1996). This period also saw the resurgence of military dictatorships in LAC. The increase in oil prices and the energy crisis of 1973 led to high levels of borrowing, which in turn resulted in an economic crisis in the 1980s. The collapse of the Latin American and Caribbean economies in the 1980s led the Inter-American Development Bank to name this period "The Lost Decade in Latin America" (IDB, 1989).

Given the threats of default by Mexico, Brazil and Peru, the international financial institutions, chiefly the World Bank and the International Monetary Fund, mobilized to impose structural adjustment programs on the economies of LAC. They also pressured the governments to impose austerity programs. The response to the crisis of the 1980s was the return to the liberal policies of the early years of the century, but now stronger than before and reinforced by a neoliberal program globally (IDB, 1989).

Guided by the international financial institutions' structural adjustment programs, the wave of liberalization and deregulation implemented in LAC in the 1990s extended to the rural sector. In addition to policies such as freeing up the economy and open markets geared to exports, the adjustment programs fostered a reduction in national industrial protection, lowering tariffs and cutting back on social spending and social development, including investment in agriculture.

In the last 15 years government economic policies have been geared to applying the rules of the so-called "Washington Consensus" (Stiglitz, 2003), in particular, policies to (1) ensure fiscal discipline (putting finances in order, fiscal responsibility, cutting public spending and voluntary retirement plans, among others); (2) implement tax reform (providing for universal incentives, tax reform); (3) free up imports (unilateral lowering of tariffs, free trade agreements); (4) privatize state enterprises and services (electricity, communications and ports); (5) deregulate the domestic market (freeing up the price system and eliminating subsidies); and (6) reform the state and introduce labor flexibility (reforms to the labor code and creating special regimes for foreign investment).

From an economic and commercial perspective, the United States, Canada and some Latin American governments gave impetus to the creation of the Free Trade Area of the Americas (FTAA) and subregional or bilateral variations of it. The FTAA is the regional expression of neoliberal globalization that is trying to become established through a process of asymmetric integration and under the leadership of the transnational companies. This asymmetric integration seeks to reorganize the economic factors and natural resources of Latin America and the Caribbean in keeping with the interests of U.S. corporate capital. The promoters of these free trade agreements argue that foreign investment will lead to economic development benefiting all, but these treaties, thus far, have had mixed effects (Gratius and Stiftung, 2002; Lederman et al., 2003; Gallagher, 2004). NAFTA, the free trade agreement among the United States, Canada and Mexico, exemplifies the mixed effects of these treaties. For example, a study by the World Bank concluded that due to NAFTA Mexico has come closer to the levels of development of the United States and Canada (Lederman et al., 2003). The study estimates that without NAFTA, the levels of exports and foreign investment would have been 25% and 40% (respectively) less than what was obtained with NAFTA. On the other hand, another study concludes that the environmental cost of economic growth in Mexico in the years in which NAFTA has been in force have been 10% of annual GDP, or US$50 billion annually in damages (Gallagher, 2004). In addition, it is argued that under NAFTA the government of Mexico has lost the capacity to protect the environment and human rights and that its citizens are losing the right to participate democratically in determining the course and priorities of their development (Gratius and Stiftung, 2002).

Following the neoliberal guidelines, IICA and other multilateral regional organizations in the Latin American countries are implementing the *New Rurality* approach, with three main components: competitiveness of agriculture and rural production, equity in the rural sector and the creation of a new institutional framework (IICA, 2000). The objectives of the *new rurality* are geared toward (1) improving and deepening the country involvment in international markets; (2) improving technically and professionalizing crop, livestock and forestry production and agribusiness development; (3) improving the capacity of the public sector to support sectoral development; (4) inducing gradually and with supervision the transfer of public services to the private sector.

The approach appears to take up anew some of the same guidelines of the previous models, with similar results. The recent data on economic growth and inequality in LAC in the first years of the millennium confirm this. Indeed, real per capita growth rates in the first four years of the millennium (2000-2004) were 2.1%, -1.1%, -2.1% and 0.5%, far below the averages attained in the 1960s and 1970s (CEPAL, 2004b) and economic inequality in the region continues to be the highest in the world (Ferranti et al., 2004).

In summary, the development models that have guided the economic policies and, therefore, agrarian policies, in LAC after the Second World War have answered mainly to the needs of the principal world power, the United States. With respect to agriculture and the development models, the role of the state is changing from producer and supervisor to organizer and facilitator of the development processes in the agricultural sector. Second, the multinational companies are already leading the process of technological development, especially in the area of biotechnology and consulting firms and NGOs are quickly filling the spaces being abandoned by the state in different technical, environmental and social areas. Finally, the privatization of utilities and resources associated with ecological services (such as water) distributes conservation costs locally among many, while the benefits are reaped by just a few, who generally are not part of the rural communities.

1.5.2 Social context

1.5.2.1 General situation of poverty in Latin America and the Caribbean

For the purposes of this evaluation, poverty is defined as a permanent condition of economic, social, political, health and environmental vulnerability stemming from asymmetrical property, trade and power relations, with reference to specific historical contexts and conditions that are ultimately determined by the economic relations of production and the development of the productive forces. Poverty is expressed in the lack or scarcity of goods and services (such as food, housing, education, health care, drinking water), resources (productive resources, employment, income) and sociopolitical conditions (human rights, economic, social and cultural rights, political rights) essential for meeting the basic needs that contribute to the loss or deterioration of the standard of living and quality of life resulting from the difficulty accessing, controlling and managing productive and natural resources.

There are two types of poverty in the region, structural and transitory. Structural poverty (or "hard poverty") affects mainly indigenous communities, rural women and ethnic minorities. The people affected by this type of poverty generally have little if any education, scant productive resources if any, limited productive knowledge and few technical skills and lack access to basic services. Transitory poverty affects peasant families and rural households that have limited or no access to land and which are especially vulnerable to the changes ushered in by the structural reforms, fluctuations in the economic cycle and social and political instability. Crises or sudden changes in economic policies have a detrimental impact on both agricultural and non-agricultural incomes, causing periodic declines in such incomes and deterioration in living conditions.

In 2005, Latin America and the Caribbean had a total population of 569 million people, 77.6% of whom are urban and 22.4% rural (CEPAL, 2006ab). At the same time, the region has a population of 209 million poor persons, 81 million of whom are living in extreme poverty (CEPAL,

2006). Of the poor, children and youth are hardest hit, as they accounted for approximately 60% of the poor as of 2002 (CEPAL, 2003; Dirven, 2004).

At the Millennium Summit, organized by the United Nations in 2000, the governments undertook to cut poverty in half in the following 15 years; even so, poverty reached the levels mentioned above. According to CEPAL (2006ab), the number of poor diminished in relative terms only 8.5% from 1990 to 2005, from 43.3% to 39.8% of the total population, whereas the number of people living in extreme poverty diminished, in the same period, from 22.5% to 15.4%. In the rural areas the downward trend is similar, yet poverty only declined in real terms from 65.4% to 58.8% of the rural population.

According to almost all indicators, LAC is the most unequal region in the world (Cardoso and Helwege, 1992; Rosenthal, 1996; Berry, 1998; O'Donnell and Tockman, 1998; Hoffman and Centeno, 2003; Portes and Hoffman, 2003; CEPAL, 2004; Ferranti et al., 2004). The Gini coefficient[2] for the region is 0.52, whereas for the industrialized countries of the OECD it is 0.332; in the Asian countries it is 0.40; and the Gini coefficient for Africa is 0.48. Note that the index of inequality is different from the poverty level: Africa is poorer than Latin America, but less unequal. The worst cases are Bolivia, Brazil, Honduras, Colombia, Nicaragua, Dominican Republic, Chile, Guatemala, Paraguay, Mexico and Argentina (Table 1-4).

In the late 1990s, six of every 10 poor lived in urban zones, making Latin America and the Caribbean the developing region that best exemplifies the worldwide process of the "urbanization of poverty" (in contrast with Asia and Africa, where most of the poor population is in the rural areas). Nonetheless, the impact of poverty in LAC continues to be greater among rural residents, especially among women. Economic globalization and neoliberal policies have affected the characteristics of the contemporary rural labor market, reducing to a minimum or eliminating government protection for workers, increasing unemployment and underemployment and displacing small-scale producers (Valdés, 2005). Nonetheless, there have been areas in which non-traditional export crops have expanded opportunities for rural employment, especially among women, though these jobs are often seasonal, poorly paid and under precarious conditions involving mistreatment and discrimination (Deere, 2005).

Most of the poor in the countries of the region were in the rural areas until the early 1980s. As a result of the negative social impact of the "crisis of the lost decade" and of the advance of the process of urbanization, poverty came to be located mostly in urban areas by the mid-1980s. During the subsequent period of economic and social improvement, the urbanization of poverty continued, until it stabilized at about 62% from 1994 to 1997 (as a result of a new increase in the number of rural poor) (Table 1-5).

The statement that poverty in LAC is mainly an urban phenomenon (Dirven, 2004) does not reflect the complexities of the situation. First, it should be noted that four large and relatively urbanized countries—Brazil, Mexico, Colombia and Argentina—dominate regional statistics. Second, surprisingly little is known of the degree of rural poverty in the region, since the estimates of poverty are incomplete, or little attention is paid in the analyses of poverty to rural poverty, especially as it affects the indigenous peoples of the region; they have higher poverty levels and have never been very well-represented in the statistics. Urban poverty in LAC has been better studied and documented through surveys. Nonetheless, there is information in the region that clearly illustrates the rural situation. For example, in three countries, the rural population is over half the national population (Guatemala, Haiti and Honduras). Since a much higher proportion of the rural population is poor, in at least 12 countries most of the poor live in rural areas. In at least five countries (Colombia, Brazil, Venezuela, Mexico and Panama) poverty is disproportionately distributed in rural areas. Finally, in all the countries of Latin America, the lowest income deciles, i.e., the extremely poor, are mostly made up of rural population. If one compares the average standard of living of the urban poor with that of the rural poor, it is clear that poverty is much more severe in rural areas.

According to CEPAL (2006ab), in absolute terms, the number of poor in urban areas has also increased, since in 1980 there were 73 million urban poor. The number of peasants in extreme poverty has climbed, over the last two decades, from 39.9 million to 46.4 million. The gains of the 1990s in terms of poverty alleviation have not offset the increase in poverty during the previous decade.

It is estimated that eight to ten million rural households are headed by women; some two or three million women perform seasonal work in agriculture or agroindustry; and 30 to 40 million women with spouses or partners are partly or entirely responsible for agricultural production and small-scale rural industry. Rural women have become part of the poorest population groups as a result of internal conflicts, the increase in the migration of men within and outside the country, natural disasters and the consequences of structural adjustment (see 1.6.2.6).

In terms of education, the illiterate population 15 years and over accounts for 9.5% of the total in this age group in LAC (CEPAL, 2004ab). Illiteracy is 10.3% among women and 8.8% among men. The drop-out rate is 37% for Latin American adolescents. Almost half drop out early, without finishing primary education, but in several countries most of those who drop out do so in the first year of secondary education; and most are in the lowest-income level, reinforcing the chain of inequality from childhood. Economic difficulties, work, or looking for employment are the main reasons young people give for dropping out of school. Among women, other reasons are household tasks, pregnancy and maternity.

[2] The *Gini coefficient* is a measure of inequality developed by Italian statistician Corrado Gini. Normally it is used to measure income inequality, but it can also be used to measure any form of unequal distribution. The Gini coefficient is a number between 0 and 1, where 0 corresponds to perfect equality (everyone has the same income) and 1 corresponds to perfect inequality (one person has all the income and everyone else has none). The *Gini index* is the Gini coefficient expressed as a percentage and is equal to the Gini coefficient multiplied by 100.

Table 1-4. **Gini coefficient of the income distribution around the years 1999, 2002 and 2005. Source: CEPAL 2006 based on special tabulation of the household surveys in each country.**

Inequality level	Around 1999	Around 2002	Around 2005
Very High 0.580–1	Brazil 0.640	Brazil 0.639	Brazil 0.613
	Bolivia 0.586	Bolivia 0.614	Honduras 0.587
	Nicaragua 0.584	Honduras 0.588	Colombia 0.584
High 0.520–0.579	Colombia 0.572	Nicaragua 0.579	Nicaragua (2001) 0.579
	Paraguay 0.565	Argentina[b] 0.578	Dominican Rep. 0.569
	Honduras 0.564	Paraguay 0.570	Chile 0.550
	Chile 0.560	Colombia 0.569	Guatemala (2002) 0.542
	Guatemala 0.560	Chile 0.559	Paraguay 0.536
	Dominican Rep. 0.554	Dominican Rep. 0.544	México 0.528
	Peru 0.545	Guatemala 0.542	Argentina[b] 0.526
	Argentina[b] 0.539	El Salvador 0.525	
	México 0.539	Peru 0.525	
	Ecuador[b] 0.521		
Medium 0.470–0.519	El Salvador 0.518	Panama[b] 0.515	Ecuador[b] 0.513
	Panama[b] 0.513	México 0.514	Peru 0.505
	Venez. (Rep. Bol.) 0.498	Ecuador[b] 0.513	Panama[b] 0.500
	Costa Rica 0.473	Venez. (Rep. Bol.) 0.500	El Salvador 0.493
		Costa Rica 0.488	Venez. (Rep. Bol.) 0.490
			Costa Rica 0.470
Low 0–0.469	Uruguay [b] 0.440	Uruguay [b] 0.455	Uruguay[b] 0.451

[a] The limit values of each category of the Gini coefficient are the same employed in chapter I of CEPAL, 2004.

[b] Urban areas.

In rural areas in particular, a very small percentage of the poor complete their secondary studies (UNDP, 2005a). In addition to the supply factors (availability of schools and quality of teaching), this may also reflect demand factors: with adolescents who work on the farm, or as wage-earning employees, the opportunity cost of sending them to school—without considering the costs of schooling and of room and board for those who must live in the town—is considerably greater than in urban areas.

On average, illiteracy in rural areas is two to six times greater than in urban areas and on average rural dwellers have three fewer years of schooling than urban dwellers. If one divides schooling into primary and secondary, it is clear that the difference is not so great at the primary level; nonetheless, the situation is completely different for the secondary level and the percentages are even lower in poor rural areas (World Bank, 1992; Psacharopoulos, 1993).

Table 1-5. **Evolution of urban and rural poverty in Latin America and the Caribbean (Absolute and relative numbers).**

Poor Population	Years					
	1970	1980	1986	1990	1994	1997
Total	119,800	135,900	170,200	200,200	201,500	204,000
Urban	44,200	62,900	94,400	121,700	125,900	125,800
Rural	75,600	73,000	75,800	78,500	75,600	78,200
Urbanization of poverty (percentage)	36.9	46.3	55.5	60.8	62.5	61.7
	Percent of poor households					
Total households	45	35	–	41	38	36
Urban Area (a)	29	25	–	35	32	30
Rural Area (b)	67	54	–	58	56	54
Rural/Urban relation (b/a)	2.3	2.2	–	1.6	1.7	1.8

Note: percent of poor households (100: Total households according to area of residence).

Source: CEPAL, 1994b, 1999.

The poor in rural areas, compared to those who are not poor, generally have worse health, since the families are more numerous and more dependent and access to health services is more limited. The availability of information on the delivery of health services and other services is very scarce. Nonetheless, from 2000 to 2005 infant mortality—one key indicator of health—was 35.4 per 1,000 live births in LAC; for males it was 38.8 per 1,000 live births and for females 31.8 per 1,000 live births. In addition, for most of the countries, those rates are considerably greater in rural than in urban areas (CEPAL, 2006ab). Infant mortality has declined gradually since 1990 in most of the countries, although it is still alarming in Haiti, at 54.1 per 1,000 live births; and Bolivia has the highest infant mortality in South America, at 45.6 per 1,000 live births.

CEPAL (2004) reports that chronic malnutrition in the region affects 15% of children under five years. In most of the countries of the region, children in rural areas, where food is produced, have the highest levels of malnutrition (Dirven, 2004). In addition, an inverse relationship has been noted between malnutrition and agricultural output. Countries with malnutrition of 0 to 10% have 400% greater per capita food production than countries with malnutrition of 10 to 20% and 320% greater per capita food production than countries with malnutrition of 20 to 65%.

Another factor behind social deterioration in the region is the lack of employment and its low quality (Dirven, 2004). The degradation of working conditions in the countryside in LAC is reflected in the low incomes of rural families and, therefore, in a persistent increase of migration from rural areas to the cities, creating mega-cities with areas of extreme poverty and greater demand, in many cases impossible to meet, for services in the main cities of LAC (Davis, 2005). The structural adjustment programs promoted and imposed by the International Monetary Fund, combined with economic liberalization, have provoked a massive exodus from the countryside to the cities (Bryceson et al., 2000). In addition, there is migration to industrialized countries, either in the region, or to Europe or the United States. Examples of this phenomenon include Mexico, Ecuador, El Salvador, Peru and Nicaragua; remittances become a very important source of income for rural and urban poor families in these countries (Comunidad Andina, 2006) (see 1.5.3).

1.5.2.2 Inequality in land tenure

Latin America and the Caribbean represent the most extensive reserve of arable land in proportion to population. The region has 576 million ha (UNEP, 2002b), equivalent to 30% of the arable land in the world and 28.5% of the total land in the region. Nonetheless, the region has the greatest inequality in land distribution in the world (Figure 1-4; Ferranti et al., 2004). Historically, the land tenure systems in LAC were based on private property, the concentration of agricultural lands in the hands of a few families and the existence of a large number of peasant families or landless workers, in what was called the *latifundia-minifundia* complex and the plantation economy (Lastarria-Cornhiel and Melmed-Sanjal, 1998). The latifundistas had vast expanses of land and those best suited for agriculture, while the small farms, or *minifundia*, survived in the marginal areas.

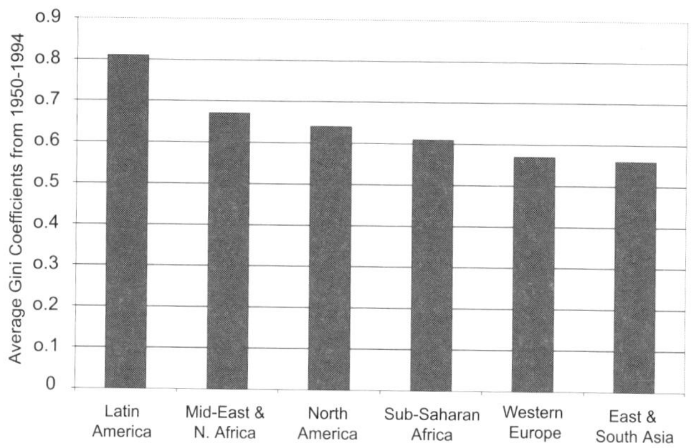

Figure 1-4. *Distribution of operational holdings of agricultural land worldwide.* Source: Author's elaboration using data from Deininger and Olinto, 2000

The agrarian reforms of the 1950s, 1960s and 1970s attempted to modify this situation of inequity by expropriating and purchasing large properties and redistributing them to peasants with little or no land, in general in the context of political and social mobilizations. Nonetheless, from an economic perspective, the agrarian reforms of this period did not succeed in reducing the levels of poverty of the rural population (Groppo, 1997). The reforms were limited in terms of the redistribution of land and allocation of land was not accompanied by supplemental measures (such as technical assistance, loans and market access) that might enable the small-scale producers to emerge from poverty.

Several decades later, the effects of the agrarian reforms on relations of production in agriculture, the development of a modern capitalist economy and the problems of poverty and equity continue to be part of the debate (Van Dam, 1999). In several countries large haciendas have given rise to commercial agriculture or agroindustry that controls the lion's share of the productive process, for both the domestic market and increasingly geared to external markets. At present, the modernization of Latin American agriculture has dramatic effects in terms of tenure, since there is a high concentration of property and agricultural production, whose main effects have been to displace small producers and peasants, leading to impoverishment, migration and social exclusion (Van Dam, 1999).

Nowadays, the forms of land tenure in the region are highly varied and complex. Nonetheless, within this heterogeneous reality, the bipolarity persists in which the *latifundium* has been replaced by the capitalist enterprise that gears its production almost exclusively to the export market, which no longer maintains economic relations with the *minifundista* peasants, who produce for their own subsistence and for the local and regional markets. At the same time, the impoverished small landowners are exposed to the constant threat of being forced to sell their land and other assets to buy foods. For the landless, access to land is generally difficult, insufficient and insecure. The systems of renting (*arriendo*) or sharecropping (*aparcería*) increasingly appear as a seasonal solution to the problems of inequity.

Most authors coincide in noting that the new land policy model being applied in Latin America uses market mechanisms instead of policy reforms. Nonetheless, several analysts consider that allowing the market to be the main land policy instrument has not resolved the problem of land redistribution, nor allowed peasants access to land; rather, it has deepened the existing inequality (Thiesenhusen, 1996; Rosset et al., 2006). Indeed, the number of small-scale producers in countries such as Brazil, Chile, Uruguay, Argentina, Bolivia, Colombia and Mexico has continued to decline, while inequality in land distribution has increased (David et al., 2001).

Another indicator of inequity is access to landed property for rural women, resulting from the specific and disadvantageous conditions in which they must face poverty (CEPAL, 1999). The liberalization of the market in land is marked by a paradox, as it favors land ownership by women, yet their ability to purchase is limited by lack of income. As a result, LAC is the region with the most unequal land distribution in the world. More than 30% of the rural poor in Latin America and the Caribbean are landless. According to studies, more than half of the households with little or no land live in extreme poverty. By way of contrast, only 10% of farmers with more than three ha of land are in a similar situation of poverty. Many other studies have confirmed that the reduction in or loss of access to the land leads directly to a loss of income and access to food (CLADEHL, 2002).

As a result of the great inequity in the distribution of land, the region is the home to many social movements that advocate the rights of the landless. These include the Movimento dos Trabalhadores Sem Terra (MST) in Brazil, which is considered the largest social movement in the region, bringing together approximately 1.5 million landless persons in 23 of Brazil's 27 states (Wolford, 2003) (see Box 1-1).

1.5.2.3 Food security and food sovereignty

Food insecurity is associated with social vulnerability and difficulty in accessing food, the origin of which is to be found in the asymmetries of development. A situation of food insecurity is reached when one does not have the means to obtain sufficient food, and is associated with poverty (Torres, 2003).

There are many different definitions of food security. In 1996 Maxwell drew up a list of 32 possible definitions (Runge et al., 2003). Nonetheless, two main considerations should be taken into account: (1) the internal capacity to increase production in the different categories of demand and (2) the country's financial possibilities for completing its food supplies (Torres, 2003). In effect, in the first, emphasis is placed on what could be called food self-sufficiency and in the second, priority is accorded to food purchases based on comparative advantages. The following present various perspectives of the debate.

The United Nations Development Program (UNDP) mentions that four criteria should be adopted: (1) coping with stress and shock; (2) economic efficiency; (3) social equity; and (4) ecological integrity. It emphasizes that the policy changes are not always those needed and that capacity-building is essential at the local level (Hall, 1998). Based on this concept of food security, the city of Belo Horizonte in Brazil developed a food security program that has been recognized internationally (see Box 1-2).

For the FAO food security exists when all people have material and economic access at all times to sufficient safe and nutritious foods to satisfy their food needs and food preferences so as to lead an active and healthy life. In 1994, the Special Program for Food Security (SPFS)[3] was begun (FAO, 2006b). In 1996, more than 180 nations participated in World Food Summit and undertook to reduce by half the number of undernourished people by the year 2015.

The United States Department of Agriculture (USDA) argues that food security for a family means access for all its members to sufficient food to be able to lead an active and healthy life. Food security includes, at a minimum: (1) the availability of adequate and safe foods and (2) the assured capacity to acquire goods by socially acceptable means.

Within the free-market paradigm of the WTO, food security has been given a different definition; it went from meaning the capacity of developing countries to produce food for their own consumption, to meaning merely access to cheap food, supplied by the developed countries or by the agroindustrial sector (Glipo, 2003). By way of contrast, the concept of food sovereignty was developed by Vía Campesina[4] as an alternative to neoliberal policies and was brought into the public debate at the World Food Summit in 1996. Since then, that concept has become a major topic of the international agrarian debate, including in the United Nations bodies. Food sovereignty was the main topic of the NGO forum held parallel to the FAO's World Food Summit in June 2002 (Vía Campesina, 1996; Desmarais, 2002).

Vía Campesina defines food sovereignty as the right of the peoples, their countries, or unions of states to define their own agrarian and food policy, without dumping with respect to third countries.

The concept includes prioritizing local agricultural production to feed the population and access for peasants and the landless to land, water, seed and credit. Hence, the need for agrarian reform and the struggle against GMOs (genetically modified organisms), for free access to seeds and to preserve water as a public good that is distributed

[3] In 1994, two years before the 1996 World Food Summit, FAO implemented the SPFS as the main program for helping its developing member states reduce hunger and malnutrition. The premise on which the design of the SPFS is based is that the productivity of small farmers in developing countries could increase considerably by introducing relatively simple, economic and sustainable technological changes (FAO, http://www.fao.org/SPFS/index_es.asp). As a result of the 1996 summit, the Rome Declaration on World Food Security was issued, with seven commitments that the participating governments would implement to enhance food security.

[4] Vía Campesina is a global movement that brings together organizations of peasants, small and medium producers, rural women, agricultural workers and indigenous communities in Asia, Africa, the Americas and Europe.

Box 1-1. The MST and land tenure in Brazil

Since the early 1980s more than one million people in Brazil have transformed their lives by gaining access to land. This has been possible thanks to a strategy of organizing and peaceful protest that has forced the government to redistribute more than eight million ha of cropland to some 350,000 families and help them develop new ways of life. These families belong to what many call the largest social movement in Latin America and the Caribbean, the Movement of Landless Rural Workers (MST: Movimento dos Trabalhadores Rurais Sem Terra).

The MST's strategy is based on forcing the government to enforce the law. For almost five centuries Brazil has been plagued by major economic inequality, in particular with respect to land tenure. Large estate owners have controlled vast rural areas with impunity, in some cases by falsifying documents and in others by recourse to violence (see figure). Much of this land is not used efficiently and has resulted in stagnant development in rural areas. To combat this problem, since the early 19th century successive governments promoted the idea that to claim legal title to property, an owner must show that the land is serving a "social function." Today this concept has been incorporated into the Brazilian Constitution.

Brazil is an emerging economy, and also the eighth largest economy in the world. Nonetheless, most Brazilians live in poverty. It has the most stark economic inequality in the world, as well as very unequal land distribution (the Gini coefficient for land distribution was 0.85 in 1994). For example, 3% of the landowners hold two-thirds of the country's arable lands. The highest levels of poverty and illiteracy are in rural areas, where the main problem is land tenure.

The MST has 1.5 million members in 23 of Brazil's 27 states. Today, there are 2,000 MST settlements and more than 80,000 additional people are currently living in camps awaiting government recognition. Cooperative farms, houses, schools for children and adults, and clinics have been built in these settlements.

According to the MST, its success is based on its ability to organize and educate. The members gain access to land, and therefore to food security for their families; in addition, many of them continue to participate in the design of a sustainable socioeconomic development model that offers specific alternatives to the model of neoliberal globalization. Some of the results of the organizational efforts of the MST with respect to production and marketing include:

- 400 associations of small-scale producers in the areas of production, marketing, and services. These include:
  - 49 farming and ranching cooperatives
  - 32 service cooperatives
  - 2 regional cooperatives for marketing
  - 3 credit unions
- 96 small- and medium-scale cooperatives for processing fruits, vegetables, dairy products, coffee, cereal grains, meat, and sugar.

These economic enterprises of the MST generate employment and salaries that directly or indirectly benefit 700 small towns in the Brazilian interior.

The leaders of the MST argue that production cannot be considered in isolation from education; accordingly, many of its programs are geared to educating its members. Results of the MST's organizing efforts with respect to education include:

- 160,000 children are studying in grades 1 through 4 in public schools located in MST settlements
- 3,900 educators paid by the local (municipal) governments are developing teaching methods specifically tailored to the MST's rural schools
- In collaboration with UNESCO and some 50 universities, the MST is developing literacy programs for some 19,000 adolescents and adults in the settlements
- In collaboration with several Brazilian universities, training is being provided to teachers, administrators of settlements and cooperatives, and nurses
- In collaboration with the government of Cuba, 48 members of the MST are studying medicine in Cuba

The MST is also promoting sustainable development. For example:

- In 1999, members of the MST developed Bionatur seeds for organic production.
- Several settlements are involved in the production of medicinal plants.
- In Pontal do Paranepanema, families from the settlements work together with environmental organizations to conserve the forest.

The MST is not free of controversy. Its critics assert that the members are mainly people from cities who ended up in worse living conditions than the urban areas they left. It is also argued that the establishment of settlements in the Amazon region has contributed to deforestation. Nonetheless, a recent survey (cited by The Economist, 2007) revealed that 94% of those living in settlements have prior agricultural experience, and 79% stated that their lives had improved as a result of having obtained land and joining the MST. The MST argues that its activities in the Amazon region are mainly in areas already deforested, particularly relatively unproductive cattle ranches.

Independent of the controversy that surrounds the MST, one cannot question the impact that this social movement has had in Brazil, or its influence on the rest of Latin America and the Caribbean. The successes and failures of this massive movement may serve as an example for the governments and social movements of the other countries of the region as they seek to solve the problems associated with the stark inequalities in land tenure in LAC.

Killing fields
Land occupations and related murders in Brazil

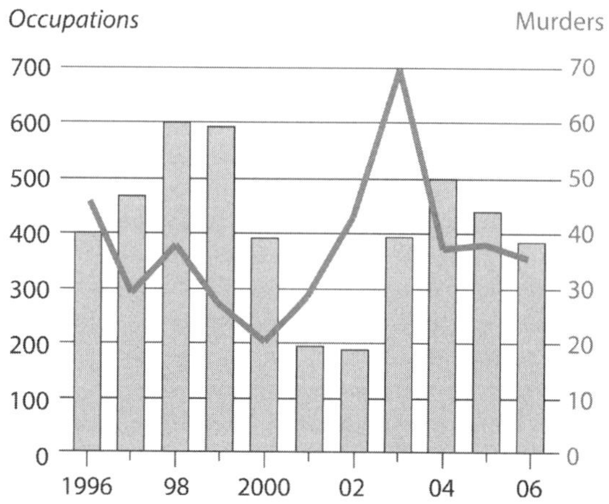

Box 1-2. Belo Horizonte: Regional food security supporting rural sustainability

In the southeast of Brazil, a few hundred km from the major cities of São Paulo and Rio de Janeiro, the municipal government of Belo Horizonte has presided over sustained improvements in nutrition and food security for its 3 million citizens for over a decade. Created in 1993, the Adjunct Municipal Secretariat of Food Security has developed programs which promote food security within the city, and which show promise as a model for improving rural livelihoods. Over the 13 years of the Secretariat's existence, millions of citizens have participated in their programs, thousands of jobs have been created, and consumption of fruits and vegetables has increased in the greater municipal area while it has decreased in other major Brazilian metropolises, and infant mortality, often attributable in large part to malnutrition, has fallen by over 41%. Indeed, the United Nations has declared Belo Horizonte a "model city" for progress that meets and in many cases exceeds the UN's Millennium Development Goals (Diário Oficial do Município, Belo Horizonte, Ano XII, Nº: 2.578, 04/01/2006).

Belo Horizonte, the capital of Brazil's Minas Gerais state, initiated its city-wide food security program in 1993 under the leadership of its then Mayor at the time, Workers' Party member Patrus Ananias de Souza. Following a period of high public attention to problems of hunger, poverty and nutrition in Brazil, Ananias held coordinating meetings between community leaders and professionals in health, education, nutrition and social assistance to create a new government office to comprehensively administer all of the city's food security-related programs. This new office, the Secretariat of Food Security "Supply" (Secretaria Municipal Adjunta de Abastecimento [SMAAB]), developed new programs and redesigned and improved old ones. In cooperation with the Secretariat of Social Assistance and with aid from the Federal government, it reinvigorated a decades-old Brazilian institution, the Popular Restaurant. Today, with 2 main facilities and several smaller "lunchrooms," the Popular Restaurant program serves over 12,000 meals each day, primarily lunches—traditionally the largest meal for Brazilians. The menus are prepared from fresh ingredients and planned by both local chefs and nutritionists. Each 1,000 calorie-plus lunch consists of rice, beans, a meat or vegetarian option, and salad or fruit, and costs the consumer one Brazilian Real (R$1 = US$0.47). (The small breakfasts and dinners at the Restaurants are R$0.25 and R$0.50, respectively.) To maintain the low cost of the meals, which is meant to promote "food with dignity," the federal and municipal governments subsidize the program to cover staff, training, and equipment costs that exceed the Restaurants' incomes. The popular high-quality, low-cost meals draw a mixed clientele: approximately 86.4% of those who eat at the restaurants are low and very-low income citizens (earning up to ~US$10,000/yr, with 34.9% of all patrons earning below US$4,000/yr), but the rest of the patrons are a mix of students and professionals from the middle- and upper-middle classes, meaning that there is little or none of the social stigma sometimes associated with assistance programs.

Like the Popular Restaurant program, the School Meals program serves meals made from fresh ingredients to all the 157,000 children in the municipal school system. Also subsidized by the federal government, the School Meals provide at least 15% of the daily nutritional requirements of the children in schools (Brazilian schoolchildren only attend school for half the day). Younger children who attend private daycares that partner with the city receive 100% of their daily nutritional requirements, and programs are underway to supplement the meals of older public schoolchildren for whom the School Lunch may be their only or primary meal. This program and the Popular Restaurants require a significant amount of food each day, especially vegetables—of which nearly 100% is provided by local farmers. Local, small and family-owned farms in Greater Belo Horizonte are primarily vegetable producers, and in cooperation with 5 municipalities in the area, SMAAB buys as much produce as possible from associations of such farms. This avoids sales through third-party intermediaries; the city receives a lower price while the small-scale farmers receive a higher income. This tactic has the added benefit of promoting rural social sustainability—especially important in a country that saw poverty and social policy push it from approximately 60% rural to 80% urban in the past 50 years. In interviews with several of the approximately 40 partner farmers, they consistently note that since joining the SMAAB program, they have seen an increase in the amount as well as the reliability of income.

In addition to selling directly to the city, the SMAAB partner farms (less than 10 ha in size) have the opportunity to participate in the "Direct from the Countryside" program. In this program, farmers are granted sales spaces throughout the city of Belo Horizonte, usually close to major thoroughfares and other highly frequented areas. Many farmers supply the Restaurants, School Meals, and other SMAAB programs, but others participate only in Direct from the Countryside or the Organic Fairs throughout the city, which have the same dual purposes of supporting local production and encouraging direct interaction between the consumers and the farmers. Such interactions have proven very valuable in other programs more familiar in the global North, such as CSAs (Community Supported Agriculture groups).

Various gains have already been realized under the Secretariat, including the astonishing decrease in infant mortality between 1993 and 2006 from 34.4 deaths per 1,000 live births to approximately 3 deaths per 1,000 live births—an achievement that surpasses the UN Millennium Development goal. This dramatic reduction has been due in no small part to cooperation with the Municipal Secretariats of Health and Social Assistance, working with their professionals and clinicians to identify at-risk children and families, and to supplement the diets of expecting and nursing mothers at little or no cost to the families. The distribution of enriched flour—wheat plus manioc, pulverized egg shells, and seeds—has been key to improving the diets of expectant and recent mothers and their young children.

Another thrust of SMAAB, and key in terms of institutional growth and sustainability, is the high importance it places on education for adult consumers and children, through school programs, community shows, average and lowest food price lists for

continued

Box 1-2. continued

consumers, workshops, cooking classes and more. These activities promote citizen ownership of the basic human right to food security (guaranteed under the UN Charter, among other international agreements) and to teach fundamental principles of nutrition to those who might not otherwise have received it. This is an especially important component in a world climate where increasing wealth is leading to obesity and nutrient-poor, high calorie diets in not just the global North, but also in other countries that are simultaneously dealing with persistent under- and mal-nutrition among their populations.

It's important to note that these are only some of the most prominent programs, and that all of the food security secretariat's programs in Belo Horizonte comprise less than 2% of the city's annual budget, at approximately US$7 million dollars per year—and even given the current level of success, there is ample opportunity to expand the comprehensiveness and size of the programs. Although SMAAB's successes are not to be taken as a direct blueprint for cities the world over, one can draw at the very least cautious hope from their example: a municipal government program cooperating across traditional health/nutrition and city/countryside boundaries, while supporting local and organic food, small-scale farmers, addressing childhood and adult malnutrition and hunger, access to food, and nutritional education, under a modest budget in a large city in the global South. From this example, we must be open to the wondrous idea that food security and small, family-farmer based rural sustainability may be mutually reinforcing, given sufficient and appropriate efforts across the many traditional borders we find between the two principles.

equitably and sustainably (Vía Campesina, 2003). The concept of food sovereignty has come about as a reaction to the definition of food security, which promotes the notion that everyone should have food, but doesn't specify where it will come from, or who will produce it, allowing control of food by large multinational companies, which may contribute to creating more dependency, poverty and marginalization. Vía Campesina also supports the concept of food as a right (see Box 1-3). The concept of food sovereignty places emphasis on local autonomy, local markets and community action. It is a process of popular resistance in the context of social movements (Grain, 2005; Niéleny, 2007).

The local space is accorded first priority because it is there that sovereignty takes on its essential meaning. It is in the spaces where the local communities create autonomy based on their own needs, beliefs and time frames. They are the custodians of thousands of years of research and creation, as a result of which their agriculture is based on biodiversity, in contrast to industrial agriculture, which fosters monoculture and only develops certain species, which are often not those grown and consumed by the local populations (Grain, 2005). Food sovereignty has a broader dimension, since it incorporates issues such as agrarian reform, territorial control, local markets, biodiversity, autonomy, cooperation, debt and health, all of which have to do with local food production. Advocates of the concept of food sovereignty argue that to attain a world without hunger one must place the communities center stage (Grain, 2005).

The Pesticide Action Network-Latin America (RAP-AL, 2007) adds that food sovereignty also has to do with the agricultural production system, since agriculture that depends on imported seed and chemical inputs does not allow for food sovereignty. This is why they support agroecological alternatives.

For civil society, food sovereignty, as a different paradigm, is needed to ensure that the developing countries can attain food security, rural employment and the goals of sustainable development. For the developing countries, food sovereignty encompasses the demand that the World Trade Organization (WTO) put an end to its control over food and agriculture. Food sovereignty basically recognizes that small farmers and landless peasants will never be able to compete in the entrepreneurial agricultural paradigm (Desmarais, 2002; Glipo, 2003; Rosset, 2006).

To the extent that food sovereignty incorporates fundamental aspects of economic equity, agrarian reform, women's rights and the rights of small farmers, it has become a broader platform for those seeking fundamental changes in the national and world order (Glipo, 2003) and represents the paradigm that is an alternative to market fundamentalism.

1.5.3 Economic context

It is generally accepted that economic growth can contribute to fighting poverty (Adelman and Morris, 1973; Dollar and Kraay, 2000). World Bank reports (2006a) indicate that for every 1% of economic growth, poverty declines by 1.25%. Nonetheless, in Latin America and the Caribbean, economic growth has not been accompanied by a significant and lasting reduction in poverty and inequality (Fajnzylber, 1990; Korzeniewicz and Smith, 2000). At the same time, poverty has a negative and very significant effect on economic growth. On average, a 10% increase in poverty reduces annual growth 1% (World Bank, 2006a).

As mentioned above, Latin America and the Caribbean is the region with the highest levels of inequality in the world (Ferranti et al., 2004). The wealthiest 10% of the population receives 48% of total income, while the poorest 10% receives only 1.6%. In the industrialized countries, the wealthiest 10% receives 29.1% of the income, while the poorest 10% receives 2.5%.

A comparison among regions within countries reveals stark differences in levels of prosperity. In 2000, the per capita income of the poorest district in Brazil was only 10% that of the wealthiest district; in the case of Mexico, per capita income in Chiapas was only 18% of per capita income in Mexico City. Regional differences account for more than 20% of inequality in Paraguay and Peru and more than 10% in the Dominican Republic and the Bolivarian Republic of Venezuela. In Bolivia, Honduras, Mexico, Paraguay and Peru, the differences in the levels of poverty between different regions is more than 40%.

Box 1-3. Food as a Human Right

The Millennium Development Goals include cutting world hunger by half by the year 2015. In the document "The Millennium Development Goals: A Latin American and Caribbean Perspective," the section on eradicating hunger in the region emphasizes food as a human right (UNDP, 2005a). This right is recognized in the International Covenant on Economic, Social and Cultural Rights, which entered into force on January 3, 1976, and to which almost all the countries of Latin America and the Caribbean are signatories.

Article 11 of the Covenant establishes as follows:

1. The States Parties to the present Covenant recognize the right of everyone to an adequate standard of living for himself and his family, including adequate food, clothing and housing, and to the continuous improvement of living conditions. The States Parties will take appropriate steps to ensure the realization of this right, recognizing to this effect the essential importance of international co-operation based on free consent.

2. The States Parties to the present Covenant, recognizing the fundamental right of everyone to be free from hunger, shall take, individually and through international co-operation, the measures, including specific programmes, which are needed:

(a) To improve methods of production, conservation and distribution of food by making full use of technical and scientific knowledge, by disseminating knowledge of the principles of nutrition and by developing or reforming agrarian systems in such a way as to achieve the most efficient development and utilization of natural resources;

(b) Taking into account the problems of both food-importing and food-exporting countries, to ensure an equitable distribution of world food supplies in relation to need."

Today, the following countries of Latin America and the Caribbean are signatories to the Covenant: Antigua and Barbuda, Argentina, Barbados, Belize, Bolivia, Brazil, Chile, Colombia, Costa Rica, Cuba, Dominica, Dominican Republic, Ecuador, El Salvador, Guatemala, Guyana, Haiti, Honduras, Jamaica, Nicaragua, Panama, Paraguay, Peru, Saint Kitts and Nevis, St. Vincent and the Grenadines, St. Lucia, Suriname, Trinidad and Tobago, Uruguay and Venezuela.

The impact of neoliberal globalization on the economy of Latin America and the Caribbean is a very controversial issue. On the one hand, some analysts argue that market-oriented reforms will eventually lead to economically sustainable growth, greater equity and a better standard of living for the population (Lustig, 1995; Sadoulet and De Janvry, 1995; Lederman et al., 2003). Nonetheless, others argue that globalization is worsening the lives of millions of Latin Americans. The statistics show that although in the 1990s (the decade of structural adjustment programs and neoliberalization) there was moderate economic growth, the number of poor by the mid-1990s was 210 million, i.e., 50 million more than the average throughout the "lost decade" of the 1980s (CEPAL, 1997; Londoño and Szekeley, 1997). On the other hand, the modest increase in economic growth has not decreased inequity in the region, which, for most countries, is still greater than prior to the 1980s (Birdsall and Londoño, 1997; Korzeniewicz and Smith, 2000).

More than an economic model, neoliberalism has been described as a mode of domination on a national and worldwide scale that stems from the restructuring of capitalist relations (Aguirre Rojas, 2005; Gilly, 2005). In the rural sector, the effects have been favorable for those who were already economically well off, but devastating for the most dispossessed; it has resulted in greater inequality and the continuation of poverty. These inequalities are expressed both among countries and among sectors within each country (Conroy et al., 1996; UNDP, 1999; Stiglitz, 2003). For example, the economic situation that the countries of the Caribbean are facing today, especially in the Lesser Antilles, is critical. The loss of the preferential treatment that had been accorded certain products of the Antilles by the European Union and which was designed to provide economic support to the former colonies will have a devastating impact on these Caribbean countries. The European Union, pressured by the World Trade Organization, will reduce the preferential price it pays for Caribbean sugar (Theodore, 2005).

In contrast with the neoliberal policies, centrist and center-left governments are drawing up proposals that point to an alternative path of inter-American economic cooperation. For example, the foreign ministers of the Caribbean countries have begun to draw up trade agreements with Mercosur and support the trade initiatives proposed by Brazil, which include technical assistance and cooperation programs in agriculture. Brazil has also offered the Caribbean countries generic drugs to fight AIDS. This is an important step, as the Caribbean is the region with the highest incidence of AIDS after sub-Saharan Africa. Recently, the Petro-Caribe agreement was signed between 13 Caribbean nations and Venezuela for obtaining Venezuelan oil. In addition, regional integration initiatives have taken place such as the "Caribbean Single Market" and the second CARICOM-Cuba meeting (Theodore, 2005).

Some countries of LAC are also putting up resistance to the negotiations of the World Trade Organization (WTO). At the WTO meeting in Cancún, Mexico, in 2003, the resistance of a coalition of Third World countries, including Brazil, Argentina and Jamaica, brought about the collapse of the negotiations. The main demands of this coalition had to do with the exclusion of agriculture from free trade agreements (Narlikar and Tussie, 2004; Rosset, 2006).

Finally, in the economic context one cannot ignore the role of family remittances. The flow of money in the form of remittances has become a major source of financing for many countries of LAC. In the last 10 years the growth in remittances has surpassed the growth of private capital investment and development assistance (Acosta et al., 2007). Although this is a trend worldwide, LAC is the region with the greatest volume of remittances in the world, with a flow of US$40 billion in 2004 and 27% of all remittances to non-industrialized countries (Acosta et al., 2007). In part, due to remittances many countries in Central America and the Caribbean have been transformed from agroexport economies to labor-exporting economies (Orozco, 2002). The volume of family remittances in LAC began to grow in the 1980s and that trend continues and is even more accentuated today. For example, remittances received in Mexico increased from US$1 billion in 1980, to US$3 billion in 1990, to US$6 billion in 2000 and by 2004 reached US$18 billion (Orozco, 2002; Acosta et al., 2007). For Haiti, in 2004 family remittances accounted for more than 50% of GDP and for Jamaica, Honduras, El Salvador, the Dominican Republic, Nicaragua and Guatemala, they accounted for 15 to 20% of GDP (Figure 1-5). In El Salvador, remittances occasionally exceed the total value of exports and in Nicaragua and the Dominican Republic they represent more than half of the value of exports (Orozco, 2002). In some countries of LAC, remittances have become a major source of support for the communities. Although very little is known about the impact of remittances on poverty, a recent study suggests that remittances contribute to economic growth of the region and to diminishing inequalities (Acosta et al., 2007).

1.5.4 Political context

In LAC, the 1980s saw the fall of the last military dictatorships and a process of democratization unfolded which, albeit with many shortcomings, provided a political opening to the most excluded sectors. In addition, in the region (with the exception of Cuba), neoliberal reforms have generated a mix of dispossessed, displaced, informal workers and migrant workers forced to survive and adapt to a new reality of unemployment or underemployment, vulnerability, precarious living conditions and hunger. The masses of dispossessed, in both the countryside and cities of LAC, are organizing new social movements that are challenging the neoliberal regimes (Aguirre Rojas, 2005). This new form of populism is expressed in the form of broad social movements that are beginning to have a major political impact in the region (Gilly, 2005; Dussel, 2007). For example, there is no doubt that the rise of the Zapatista movement in Mexico played a part in the defeat of the Partido Revolucionario Institucional (PRI), which had been in power for 79 years. In Bolivia, the indigenous movements brought an indigenous candidate to the presidency. These social-political movements without political party affiliations are changing the political landscape of the region and turning Latin America to the left.

These movements are advocating internal changes that are important in the context of this evaluation, although they do not yet have the political strength that would enable them to bring about substantial changes. Among the most important issues are: (1) recognition of the rights of indigenous nations and the growing role that indigenous organizations are playing in national politics; (2) demands for agrarian reform, especially land redistribution; (3) demands relating to access to and control and sustainable management of natural resources, including mining and energy resources and water; and (4) the insertion of the concept of food sovereignty in the national and international debate.

In Latin America, indigenous peoples live inside and outside protected areas, in tropical forests and in intertropical rural areas. Most live in marginal rural areas (Toledo, 2001). Their communities, territories/lands and natural resources continue to be subject to several pressures as well as a growing demand on the part of forces internal and external to their local communities (Kearney, 1996). This situation suggests, significantly, that the contemporary neoliberal policies of the nation-states of the region and the respective democratic regimes, among other things, (1) have not put in place or facilitated clear and coherent policies, institutions and spaces for the participation of the indigenous peoples in rural/agrarian development and in the economy and society; and (2) have not supported, in a sustained and significant fashion, the strengthening of indigenous institutions, leaders and wise people. All of this has continued perpetuating the marginalization and oppression of the region's indigenous peoples. Nonetheless, as mentioned above, the indigenous movements have strengthened significantly, becoming an important political force in some of the countries with the largest indigenous populations, such as Bolivia, Peru, Mexico, Guatemala and Ecuador (Varese, 1996; Warren and Jackson, 2003; Yashar, 2005).

1.5.5 Environmental context

1.5.5.1 General aspects of the environmental context

Latin America and the Caribbean is well known for its extraordinary biodiversity, containing five of the ten countries in the world with the highest biodiversity (Dixon et al., 2001); it has 40% of the world's plant and animal species (UNEP, 1999a). It is considered the world's leader in floristic diversity (Heywood and Watson, 1995) and in avian diversity (UNEP, 2006). While 11% of the terrestrial area of Latin America is officially under protected status (World Bank, 2006b), many protected areas exist on paper only and consequently much of the area's biodiversity is highly threatened. Almost half of the ecoregions of Latin America and the Caribbean (82 of 178) are considered critical or endangered in conservation status (Dinerstein et al., 1995). Some 873 vertebrate species in Latin America are currently estimated to be threatened with extinction and six of the twelve countries with the highest number of globally threatened bird species are found in the region (UNEP, 2002b). Unfortunately, there is little data on the extent to which arthropod species are threatened.

The Latin American region possesses 28% of the world's forest area, almost a billion ha in total (World Bank, 2005a); it contains the vast majority (68%) of the world's tropical rain forests (UNEP, 2005b). Deforestation has accelerated precipitously since 1950. It has been primarily caused by agriculture (MA, 2005a) and cattle, and more recently soybean production has been one of the major drivers for the region as a whole (Ledec, 1992; Angelsen and Kaimowitz, 2001). The overall annual deforestation rate from 2000 to 2005 in the region is estimated at 0.51% (World Bank, 2005a), but there is considerable variation across the region (Table 1-6). Historically the highest absolute amount

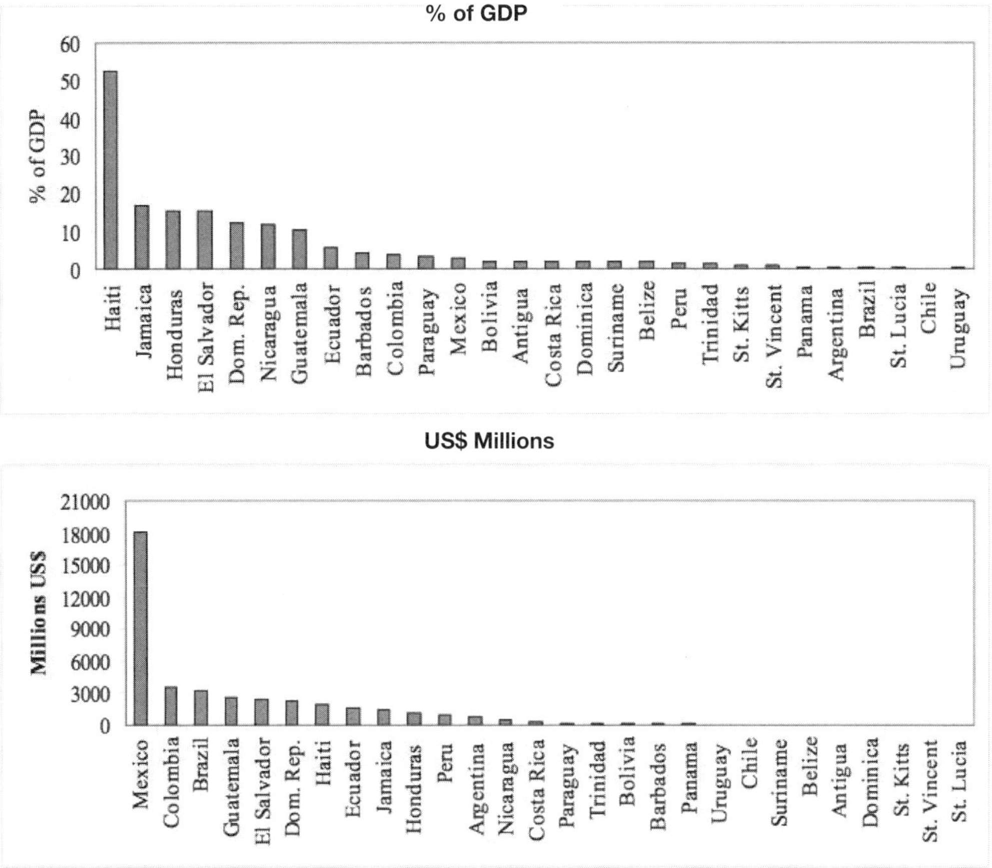

Figure 1-5. *International remittances to Latin America and the Caribbean, 2004.* Source: Acosta et al., 2007

of deforestation has occurred in South America, driven by deforestation in the Amazon; from 1981 to 1990, 6.2 million ha were stripped of forest annually in South America. However, since 2004 deforestation in the Brazilian Amazon fell by 60% due to stepped up enforcement efforts (Presidencia da República [Brazil], 2007) and lower commodity prices, namely beef and soybean and the strong Brazilian currency, which has lowered the level of land speculation (Butler, 2007). However, the growing demand for corn ethanol means that less soybean is being planted in the United States and Brazil, the biggest producer of soybean in the world, is making up the shortfall by clearing new land for soybean cultivation. Whether it will result in an increase in deforestation rates in the Brazilian Amazon or the *cerrado* remains to be seen (Butler, 2007). Soybean expansion has also affected forests in Argentina, where the rates of deforestation have increased dramatically in the last decade (Grau et al., 2005). Nevertheless, the highest rates of deforestation have consistently been found in Central America and Mexico, where deforestation in the same period reached 1.5% annually, compared to 0.7% in South America. In the Caribbean, most deforestation occurred in the 1800s and with a few exceptions (particularly the Dominican Republic), most primary moist forest suitable for agriculture had already been converted prior to the middle of the last century (Myers, 1980; Toledo, 1992). In the last decade of the 20th century, the rate of deforestation slowed throughout the region, but this slowdown was marked in South America (to 0.44% annually) and barely registered in Central America and Mexico, which still racked up 1.47% annual deforestation in that period. During this decade, forest area actually grew in the Caribbean (at 0.1% annually), driven by a rise in forested area in Cuba. It is notable that both the absolute and relative rates of deforestation in Latin America and the Caribbean during the 1980s were much higher than any other region of the world, but by the 1990s Africa had surpassed Latin America in both hectares cleared and annual deforestation rates (Barbier, 2004).

Latin America and the Caribbean are considered to have the most diverse freshwater ecosystems in the world. The region is home to one-quarter of the world's species of fish, with areas of high endemism. The Amazon in particular is noted for high freshwater fish biodiversity and tropical South America in general is a hotspot for amphibian diversity. The Caribbean and Central America are noted for their outstanding coral reefs. The Mesoamerican Reef, off the

Table 1-6. **Extent and change of forest area in Latin America.**

Subregion	Area (1,000 ha)			Annual change (1,000 ha)		Annual change rate (%)	
	1990	2000	2005	1990-2000	2000-2005	1990-2000	2000-2005
Caribbean	5,350	5,706	5,974	36	54	0.65	0.92
Central America	27,639	23,837	22,411	-380	-285	-1.47	-1.23
South America	890,818	852,796	831,540	-3,802	-4,251	-0.44	-0.50
Total Latin American and the Caribbean	923,807	882,339	859,925	-4,147	-4,483	-0.46	-0.51
World	4,077,291	3,988,610	3,952,025	-8,868	-7,317	-0.22	-0.18

Source: FAO, 2007.

Caribbean coasts of Mexico, Belize, Guatemala and Honduras, is the second longest barrier reef in the world and is one of the most diverse coral reefs in the western Atlantic. Home to over 500 fish species, 66 stony coral species and the largest population of endangered manatees in Central America, the reef is also the basis of much of the region's economy (Kramer and Kramer, 2002).

1.5.5.2 Climate change and agriculture in Latin America and the Caribbean

LAC is a very heterogeneous region in terms of climate, ecosystems and population distribution. Nonetheless, most productive activities are based on natural ecosystems and this land use interacts in a complex way with climate. Due to this complexity and the heterogeneity that characterizes the region, it is difficult to identify the effects of and vulnerability to climate change.

The Intergovernmental Panel on Climate Change (IPCC, 2007), in its latest report, forecasts a change in temperature of up to 5.8°C for this century. This climate change has the potential to create local and regional conditions that include deficits and surpluses of water in the same geographic location (Table 1-7). The potentially grave impacts that can be expected, according to the IPCC, are a considerable increase in heat waves, storms, floods, landslides and avalanches unleashed by the forecast increases in the intensity of precipitation and the rising sea level. There may be health problems in human beings, livestock and crops due to the greater incidence of pests and insects that are vectors of disease.

In addition, an increase is predicted in the sea level of up to 88 centimeters in this century, affecting (due to the intrusion of sea water in the soils subjacent to arable lands and also due to temporary and permanent flooding) approximately 30% of the agricultural regions worldwide. It is believed, in particular, that riparian and coastal settlements are at risk, but urban floods may also be a serious problem for water supply and for waste management systems that have not been designed with sufficient capacity to prevent the spread of tropical diseases. The IPCC (1997, 2001a) had already identified the following sectors as those that will be most affected by climate change in LAC: natural ecosystems (e.g., forests, wetlands, savannahs), water resources, coastal zones, agriculture and human health.

Although LAC accounts for only 4% of global emissions of greenhouse gases, the potential impacts of climate change in the region may be considerable and very costly, in both economic and social terms. In addition, the carbon emissions that result from massive deforestation in LAC have the potential to alter the carbon balance globally.

Most productive activities in LAC depend on the availability of water, such that any climate change that results in a shortening of the rainy season, greater variability of precipitation and/or greater frequency of years without rain will have extremely negative consequences for the region (IPCC, 2001a). Mexico, in particular, will be very significantly affected by drier and hotter climatic conditions as it is already suffering from very little and highly variable precipitation (Liverman and O'Brian, 1991). The Brazilian Northeast is another region highly vulnerable to drought caused by climate change. Under different climate change scenarios, global models project reductions of up to 53% in the yields in this region (Rosenzweig et al., 1993); it will be common for there to be years in which it doesn't rain and the population suffers hunger and is forced to migrate (Magalhães and Glantz, 1992).

Another effect of climate change on the productive activities of the region has to do with the effects of the Southern Oscillations, El Niño. Although there is no consensus on the effect of climate change on the El Niño phenomenon in the long term, in the short term an increase is reported in its frequency and intensity (IPCC, 2001a). In Central and South America, the relationship between El Niño and changes in precipitation is well-documented. El Niño is associated with massive fluctuations in the marine ecosystems of the western coast of South America (Ecuador, Peru and Chile), adversely affecting fishing and taking a devastating socioeconomic toll on the communities that depend on this activity (Pauly and Tsukayama, 1987; Sharp and McLain, 1993). In 2001, El Niño caused severe droughts in Central America and northern South America, with damages estimated at US$189 million, 66% of these in agriculture and affecting 600,000 people in Central America, mostly small-

Table 1-7. **Current and future temperature and precipitation, selected LAC countries/regions.**

Country	Temperature °C		Precipitation mm/day annual average	
	Current (1961-1990)	Future (2070-2099)	Current (1961-1990)	Future (2070-2099)
South Cone				
Argentina	14.65	17.89	1.63	1.66
Brazil: Amazon	26.04	30.38	5.97	5.84
Brazil: Northeast	25.58	29.46	3.58	3.52
Brazil: South	22.04	25.90	3.98	4.15
Chile	9.01	11.91	1.52	1.43
Andean Zone				
Colombia	24.31	27.81	7.25	7.44
Ecuador	22.15	25.36	5.52	6.01
Peru	19.52	23.34	4.22	4.42
Venezuela	22.44	29.17	5.33	5.31
Others				
Central America	24.23	27.76	6.51	6.18
México	20.66	24.71	2.09	1.84
Cuba	25.25	28.19	3.57	3.50

Source: Cline, 2007.

scale producers, who suffered due to the lack of food and were forced to migrate (CEPAL, 2002).

Hurricanes and tropical storms also have a devastating effect in the region. Central America and the Caribbean are the regions hardest hit by these climatic events. In these regions, 18 hurricanes and tropical storms were detected from 1960 to 2001 (Cepredenac, 2007). Hurricane Mitch, in 1998, is considered the most devastating hurricane to hit the Central American region (Pielke et al., 2003), causing total damages amounting to US$6 billion, half resulting from losses in agriculture (Ceprenedac, 2007).

It has been said that carbon dioxide has a fertilizing effect that could benefit agriculture, increasing crop yields. Nonetheless, studies in Brazil, Chile, Argentina and Uruguay, based on climate change models and crop models, predict reductions in the yields of several crops (e.g., maize, potato, soybean and wheat), even taking into consideration fertilization with carbon dioxide and moderate adaptations by producers (IPCC, 2001a).

The projected climate changes may also have a negative impact on productive activities through their effect on human health. For example, the projected increase in temperature and precipitation could expand the range of vector-transmitted diseases (e.g., malaria, dengue, leishmaniasis, Chagas' disease) and infectious diseases (e.g., cholera), making it possible for them to become established to the south of their current range and at higher elevations (WHO, 1996). Box 1-4 illustrates the relationship between changes in agriculture (which are often governed by climate changes) and the emergence of infectious diseases.

The effects of the increase in the sea level include a greater risk of flooding in the coastal zones of Central America, South America and the Caribbean and the possible loss of land area. Although the loss in land area could represent a small proportion of the national territory (except in the Caribbean), it may have a major impact in areas where large populations, tourist centers and infrastructure are located (e.g., ports) (IPCC, 2001b).

The IPCC (2001b) concluded that the alterations resulting from climate change have a high potential to negatively affect the ways of life of subsistence farmers and pastoralists who live in the high Andean planes and tropical and subtropical forests. Despite the grave socioeconomic impacts associated with climate change in the region, the governments have done very little to reduce the emissions of gases that contribute to climate change, or to implement risk management strategies and promote adaptive systems to cushion the negative effects on productive activities in the region. In Brazil, drought forecast systems have been implemented that have succeeded in reducing the negative impacts of droughts. There are also experiences in Central America involving the resistance of agroecological systems to the impacts of tropical storms (Holt-Giménez, 2002; Box 1-5).

1.5.6 Cultural context

Latin America and the Caribbean are characterized by three major cultural influences, the indigenous, the African and the European (mainly Spanish and Portuguese). The word "agriculture" emphasizes the overarching role of culture in this type of production. All the cultures, both those existing and those already lost, have affected the region's production systems to a greater or lesser extent. Nonetheless, the agriculture practiced by most small-scale producers in the region is highly influenced by the indigenous and Afrodescendant cultures.

The indigenous population of LAC accounts for about 10% of the total (IDB, 2004; Hall and Patrinos, 2005). The ethnic and cultural diversity of indigenous groups in Latin America is estimated at more than 400 ethnic groups (Deruyttere, 1997) or 800 cultural groups (Toledo, 2007), the largest percentages being in Bolivia (70%), Guatemala (47%), Ecuador (38%) and Mexico (12%). One important aspect of the relationship between agriculture and the cultures is the relationship between biodiversity and cultural diversity. In LAC, cultural diversity is highly correlated with agrobiodiversity in general. The region has two centers of origin of genetic diversity—in the territories that are today Mexico and Guatemala and Peru and Bolivia (Possey, 1999). The lands/territories of the indigenous peoples intersect/overlap to a large extent with the areas recognized as biologically megadiverse. The indigenous peoples live in 80% of the region's protected areas (Colchester and Gray, 1998). In Central America the percentage increases to 85% (Oviedo, 1999). Toledo (2003) notes that nearly 60% of the areas in central and southern Mexico recommended for protection are inhabited by indigenous peoples.

Biodiversity constitutes an irreplaceable common patrimony of humankind, the result of prolonged and ceaseless evolutionary processes, which is fundamental for socioeconomic development and for the very survival of humankind. The ethnic groups, Afrodescendant communities and peasant communities in LAC hold a large part of the cultural patrimony represented in the systems of knowledge, innovations and millenary practices of integral and sustainable management in their territories associated with biodiversity (Barrera-Bassols and Toledo, 2005). Just as biodiversity is threatened, the cultural integrity of ethnic groups is seriously threatened. Cultural erosion, the loss of land and the loss of control over their territories by these communities occur with ever greater frequency and intensity, which no doubt has a detrimental impact on the cultural patterns and appropriation of their traditional habitat.

The Green Revolution transformed the traditional agricultural culture. For thousands of years farmers, mainly women, have taken it upon themselves to select and save seeds to create, literally, thousands of "local varieties" of food crops adapted to the conditions and preferences of each place. When the Green Revolution swept across the countries of the south, the diversity that these farmers had been caring for began to weaken. Local varieties can only survive in interaction with people and disappear if not preserved and planted.

The cultures of the indigenous peoples and Euro-American societies and of the westernized/modernized societies are immersed in two profoundly different ways of knowing (epistemologies), of being (ontologies) and of relating to the world (cosmovision/world view). After more than three decades of political struggles—local, regional, national and international—the indigenous peoples have become actors known on their own terms, without mediation, or mediators, in the political arena. Their rights, albeit very slowly and still more on paper than in practice, are recognized by the United Nations (Farmers' Rights, Convention on

Box 1-4. Emergence of infectious diseases and agriculture

One of the main threats to agricultural development internationally is the emergence of diseases associated with the changes in the environment necessary for agriculture (Wilson, 2002). In Latin America and the Caribbean, the association of agricultural activities with certain diseases has been relatively little studied in comparison with other regions such as Africa and Southeast Asia (Norris, 2004). The following are four examples that illustrate the importance of this association:

1. Coffee and cutaneous leishmaniasis: Picking coffee increases the risk of infection by Leishmania parasites since it coincides with the maximum period of activity for the insect vectors of the disease (Scorza et al., 1985).
2. Irrigation and malaria: Densities of malaria vectors are much greater in irrigation canals than in bodies of water whose origin is not attributable to human activities (Zoppi de Roa et al., 2002). The density of vectors that transmit a disease tends to be linearly correlated with the risk of acquiring the disease, which is why agricultural activity increases the risk in two ways: by increasing the number of mosquitoes, and spatially, by the proximity of irrigation canals to centers of human settlement (Norris, 2004).
3. Deforestation and malaria: Agricultural development can lead to increases in temperature that facilitate the development of parasites that cause malaria in the vectors, especially when natural forests are cut down to promote agriculture (Lindblade et al., 2000). The rates of mosquito bites can be up to 278 times greater in highly deforested areas as compared to natural forest areas (Vittor et al., 2006).
4. Rural houses and Chagas' disease: One of the fundamental aspects in the epidemiology of Chagas' disease is its association with rural dwellings in precarious conditions (Rabinovich et al., 1979). In general, the more precarious the conditions of the housing units (thatched roof, clay walls) the greater the vector density and hence the greater the likelihood of acquiring the disease (Rabinovich, 1995).

The four examples presented above show the need to incorporate knowledge of infectious diseases into agricultural activities. Knowledge may have an immediate impact on agricultural practices by diminishing activities that increase the risk of acquiring disease. For example, the incidence of cutaneous leishmaniasis can be reduced by changing the hours during which coffee is picked.

Box 1-5. Measuring farmers' agroecological resistance to Hurricane Mitch in Central America. Source: Holt-Giménez, 2002, 2006.

A study using a participatory action research approach and simple field techniques found significant differences in agroecological resistance between plots on conventional and sustainable farms in Central America after Hurricane Mitch. On average, agroecological plots on sustainable farms had more topsoil, higher field moisture, more vegetation, less erosion and lower economic losses after the Hurricane than control plots on conventional farms. The differences in favor of these agroecological plots tended to increase with increasing levels of storm intensity, increasing slope and years under agroecological practices, though the patterns of resistance suggested complex interactions and thresholds. For some indicators, agroecological resistance collapsed under extreme stress.

With the help of 40 NGOs and 99 farmer-technician teams, 1,743 farmers measured key agroecological indicators on 1,804 plots paired under the same topographical conditions. These paired observations covered 360 communities of smallholders from southern Nicaragua to eastern Guatemala. The broad geographical coverage took into account the diversity of ecological conditions, a variety of practices common to sustainable agriculture in Central America, and moderate, high and extreme levels of hurricane impact. This coverage, and the massive mobilization of farmer-technician field research teams, was made possible by the existence of a widespread smallholders' network for sustainable agriculture called Movimiento Campesino a Campesino (Farmer to Farmer Movement).

Comparatively higher levels of agroecological resistance are an indication of lower vulnerability and higher sustainability. However, the effectiveness of practices appears to be bounded by a combination of steep slopes, maintenance and design of soil conservation structures, and extremely high storm intensity.

A number of methodological lessons were learned about the tradeoffs between participation and scientific rigor from the study. The ability to gather large amounts of data across wide areas had advantages, but it was learned that care must be taken to maintain the process of scientific inquiry among groups, instead of focusing solely on protocol.

After analyzing the results, agroecological and conventional farmers designed strategies for participatory, sustainable reconstruction and identified the factors driving and limiting the development of sustainable agriculture. They proposed policies for participatory sustainable reconstruction and sustainable agricultural development. Participants presented their findings in national meetings to representatives from government and international NGOs, and later distributed them publicly. Although the study was influential in reconstruction activity in villages and programs where MCAC is already present, it had negligible impact on national policies for reconstruction.

The study concludes that while the Movimiento Campesino a Campesino has successfully advanced the technical and methodological aspects of sustainable agriculture, a policy ceiling is currently limiting the generalized spread of sustainable agriculture among smallholders in Central America.

Biological Diversity, International Labour Organization [ILO] Convention 169), by financial and development organizations (World Bank, Inter-American Development Bank, USAID, European Union) and by international conservation organizations (World Wildlife Fund (WWF), World Conservation Union (IUCN), The Nature Conservancy (TNC). A number of countries of the region have adopted and ratified ILO Convention 169 on Indigenous and Tribal Peoples, which could significantly benefit indigenous peoples. Nonetheless, the states of the region, which are members of the United Nations, do not display a coherent, significant and clear will to implement, in practice, this Convention.

1.6 Recent Evolution and Current Situation of Agriculture in LAC

1.6.1 Importance of agriculture to Latin America and the Caribbean

Agriculture is much more than simply the production of economically important goods. As a source of food for human beings and animals, fiber, materials for construction and for crafts, oil and fuel, agriculture is vital for the cultures and communities that produce them and plays a critical role for the goals of sustainable development and reducing poverty and inequality. Recently special emphasis has also been placed on the role of agriculture in providing environmental services such as mitigation of the effects of climate change, regulation of the water cycle, erosion control, maintenance of habitats for wildlife and preservation of landscapes and places of religious importance. In this sense, agriculture is a multifunctional activity (Chaparro, 2000; Cahill, 2001; Dobbs and Pretty, 2004; Brunstad et al., 2005). This doesn't mean that agriculture can simultaneously satisfy all these functions, since that depends on specific contextual characteristics. Nonetheless, these multiple functions of agriculture should be taken into consideration, especially in the context of development and sustainability goals.

In the last 50 years agriculture has contributed only 10 to 12% of GDP; it has been secondary to other productive activities. Nonetheless, agriculture still represents a key sector of the Latin American economy, as it accounts for a large part (30 to 40%) of the economically active population. In those countries that lack minerals and oil, agriculture represents the main source of exports and foreign exchange. Agriculture is a relatively more important part of the economy in the Central American countries than it is for Latin America generally. While agriculture only contributed 8% of GDP in 1998 in Latin America overall (Dixon et al., 2001), in Central America in 2000 agriculture contributed from a low of 7% of GDP (in Panama) to a high of 36% (in Nicaragua). The importance of agriculture as a generator of foreign exchange is even more significant. In 2000, agricultural exports ranged from a low of 30.8% of total exports of goods in Costa Rica, to a high in Belize of 69.4% of total exports (Harvey et al., 2005). Finally, in most Latin American countries, agriculture represents a subsistence way of life for millions of people, including indigenous communities (IPCC, 1996).

Recent research has shown exhaustively that agricultural activities are diminishing in rural areas from the standpoint of the number of people involved and the income generated, while non-agricultural activities are on the rise, in particular those linked to the provision of services. For these reasons, the families that live in areas defined as rural are increasingly abandoning exclusively agricultural activities to seek out other opportunities (Da Silva, 2004; Dirven, 2004). These phenomena are responsible in part for the migrations from the countryside to the cities, but are not the sole cause. The expansion of the large transgenic monocultures in the countries of the Southern Cone is transforming the agrarian structure, increasing the concentration of land and the migration of peasants (Fearnside, 2001ab; Pengue, 2005). In addition, violence due to territorial interests are causing massive forced displacement, as in Colombia and Ecuador.

Parallel to this difficult context, fishing is also developing; it continues to be one of the key components of certain local economies in many places in Latin America, especially the Amazon region, both in terms of the value of production and in terms of employment. Bernal and Agudelo (2006) cite figures from the FAO according to which there are more than 38 million people directly engaged in fishing and fish farming on a full- or part-time basis; and the developing countries now provide 70% of the fish for human consumption. Marine fishing is also an important economic activity in LAC, generating employment and incomes; most of the fish offloaded is accounted for by the Southern Cone countries.

The current status of agriculture in LAC, in terms of production and productivity of goods and services in relation to expectations for attaining the millennium goals, is not uniform across the region. The heterogeneity in levels of agricultural knowledge is due in part to the effect of the structural reforms carried out in the region. In the last 25 years most of the countries of the region began or intensified their processes of adjustment and structural reforms, as a result of which they experienced major changes in their structure of production, productivity, competitiveness and in the profitability of various activities, including agriculture (David et al., 2001).

It should be noted that it is practically impossible to establish typologies of development models by country, as one finds the coexistence of very different and more complex situations than in the rest of the economy, given the major differences between and within the countries. The differentiation of the growth model has occurred within the countries, with repercussions both on the specially located dynamic poles and on the type of activities and actors.

1.6.2 Characteristics and trends in production in Latin America and the Caribbean

1.6.2.1 Available resources

Natural resources. Agriculture produces unprocessed agrifood products using natural resources (land, water, biodiversity) as one of the factors of production and the process may involve "cultivation" (planting, aquaculture, stockraising, forestry) or "gathering" (hunting, fishing, forestry) (Dirven, 2004). The peoples of LAC live in a territory with abundant resources in terms of land, water and biodiversity (OSAL, 2005). The water and soil, key elements in agricultural production, may or may not be considered renewable resources, depending on their degrees of cultural management. In any event, they constitute the main limitations and potential for agriculture at this level (León, 2007).

Land. Latin America and the Caribbean is the region with the largest reserves of arable lands in the world. It is estimated that 30% of the territory in LAC has agricultural potential (Gómez and Gallopin, 1995). The region had 160 million ha of land under annual and perennial crops in 1999 and another 600 million ha dedicated to grazing and pasture (Dixon et al., 2001). Nonetheless, due to the mismanagement of the soils and to the use of marginal areas for agriculture, the region has approximately 300 million ha of degraded agricultural area (FAO, 1998), while another 80 million ha of arid lands are threatened with desertification due to overgrazing, overexploitation of the vegetation for domestic uses, deforestation and the use of inappropriate irrigation methods. This represents more than 50% of the total agricultural area (including grazing areas) affected by degradation. Erosion, acidification, loss of organic matter, compaction, impoverishment of nutrients, salinization and soil contamination are a result of the intensification of agriculture through the intensive use of agrochemicals, fertilizers, and pesticides, as well as the use of inappropriate irrigation technologies and agricultural machinery (see 1.7) (UNEP, 2006).

Erosion is the main cause of land degradation in LAC and affects 14% of the territory in South America and 26% in Mesoamerica (UNEP, 1999a). This problem is especially serious in steep areas such as the Andean region (central and northern), as well as the maize and bean zone of Mesoamerica. In these areas erosion is causing low levels of production and is affecting the migration of small-scale producers to the cities or the agricultural frontier in forested areas, contributing to soil degradation there (FAO, 1998). This process is also taking place in other steep areas such as the Chiapas highlands in Mexico (Richter, 2000).

Nutrient attrition is another very serious problem that results from the intensification of agriculture and synthetic fertilizers. In South America nutrient attrition affects at least 68 million ha (Scherr and Yadav, 1997). Nutrient attrition may also be a consequence of deforestation in moist tropical zones. The conversion of forest to cropland in these areas has brought about the loss of organic matter and has accelerated erosion and the increase in the sediment load in rivers and lakes (FAO, 1998).

Chemical contamination of the soil and water, which also derives from the technologies of intensive agriculture, has been increasing in the last 30 years. Nitrification of the soil and water is directly related to the use of chemical fertilizers (UNEP, 2006); in LAC the use of fertilizers increased from less than one million tonnes in 1961 to more than 13 million tonnes in 2003 (FAOSTAT, 2005).

Water. In terms of water, the region has relatively favorable endowments compared to other areas in the developing world. It has almost half of the world's total renewable water resources and some 90% of the land area falls in the humid or sub-humid zones. While overall the region is relatively wet, there are several areas where drylands predominate, principally in northern and central Mexico and the coastal and inland valleys of Peru, Chile and Western Argentina, Northeast Brazil and the Yucatan Peninsula and the Gran Chaco area of Paraguay, Bolivia and Argentina. In total, drylands comprise some 15% of the region (FAO, 1998). Natural grasslands or savannahs, many of which are relatively dry, are found in much of Argentina, as well as in central, western and southern Brazil, Uruguay and parts of Colombia, Venezuela and Guyana. Crops occupy around 160 million ha of the region, while another 600 million ha are dedicated to pasture and grazing land (Dixon et al., 2001).

Hydrobiological resources represent another component of South America's biodiversity, with approximately 3,000 fish species. Nonetheless, very little is known of the biological cycle of the fish species dependent on the water cycle and even less of the zooplankton and phytoplankton of the continental and marine waters (Bernal and Agudelo, 2006).

Agrobiodiversity. Mesoamerica and the Andes are two major centers of origin of domesticated plants, many of which are now of global importance. Maize and beans are the most prominent of these, but the list also includes potatoes, sweet potatoes, tomatoes, cassava, chili peppers, gourds, squashes, avocado, cotton and peanuts. Wild ancestors have been discovered for some of these crops, such as maize. There is also significant genetic diversity across the region that has been developed since the introduction of non-native crops such as banana and sugar cane. With a few exceptions, the region's agrobiodiversity is not well studied.

Maize (*Zea mays*) is one of the most significant crops that originated in the Americas; it is now the most widely grown crop in the world. Due to its ability to grow under highly varied climatic conditions, it is grown in at least 164 countries worldwide (Global Crop Diversity Trust, 2007). Mexico is the center of origin and the center of diversity for maize, with more than 60 landraces and numerous local varieties, as well as the wild relatives of maize, the teosintes. Mexico provides one of the earliest examples of deliberate conservation of wild crop relatives *in situ*; the existence of teosinte was the primary reason for the creation of the Sierra de Manantlán Man and the Biosphere Reserve there in 1988 (Iltis, 1994; Meilleur and Hodgkin, 2004).

The common bean (*Phaseolus vulgaris*) appears to have been domesticated separately in Mesoamerica and in the Andean region. Wild gene pools are also concentrated in these areas. Mesoamerican cultivars dominate global production; some 60% of beans produced throughout the world are of Mesoamerican origin. Common beans are the world's most important legume food crop and are particularly important for human nutrition because of the high protein content, which is roughly double that of most cereals (Beebe et al., 2000).

Potato (*Solanum tuberosum*) was domesticated 7,000 years ago around Lake Titicaca in the Andes (Spooner et al., 2005). Potato is the most important crop for the cultures in the Andes, where over 100 varieties can be found growing within a single valley (Brush, 1992).

Relatively few animals were domesticated in the new world; only one, the turkey, has spread significantly beyond its native habitats in Mesoamerica and the present-day United States. The llama and alpaca, domesticated in the Andes, still play an important role in Andean society, as does the guinea pig, domesticated for food. The Muscovy duck was also domesticated in South America. Wild relatives of some of these animals, particularly the wild turkey and the vicuña, which is related to llamas and alpacas, are still to be found in the areas where they were domesticated (Heiser, 1990).

The agricultural genetic resources of the Latin American region are enormous. As one of only a few places where agriculture was independently invented and the center of origin of many of the world's major food crops, the area retains numerous landraces, local varieties and wild relatives of great importance to the future development of agriculture worldwide.

Economic resources. As a result of the structural adjustment processes in the context of globalization, changes have taken place in the agricultural sector in LAC that have had a differential impact on the population in three ways: (1) changes in incomes as there have been changes in wages, employment levels and the prices of goods, especially essential goods, such as food items; (2) changes in the levels and composition of public spending, especially social spending; and (3) changes in working conditions, such as type of contracting, hours and social security. The changes have included greater differentiation in the conditions of production between small and large producers and there are fewer agricultural jobs, with adverse results for many sectors due to the increase in poverty and inequality in the rural world (Da Silva, 2004).

Among the causes of the reduction in employment, Da Silva (2004) cites increases in labor productivity, relative stability of the agricultural frontier and the expansion of stock-raising and forestry, which do not require much labor. Other categories that have been expanding (such as fruit crops, vegetable crops and poultry) are using ever more contract agriculture, which is based on more capital and also reduces employment (Da Silva, 2004; Deere, 2005). According to several sources compiled by David et al. (2001), approximately 66% of the poor who live in the rural sectors—47 million people—are small-scale producers, 30% are landless rural dwellers and the remaining 4% are indigenous groups and others. Of the small-scale producers, at least 40% are farmers with little if any access to loans, technical assistance, or agricultural support services and little capacity to purchase land.

The financial sector plays a role in activities related to rural employment, favoring non-agricultural activities, which vary from country to country and depend on the ties between non-agricultural rural employment and other sectors of economic activity. In an Inter-American Development Bank (IDB) document on rural financing strategies cited by Da Silva (2004), it was recognized that the non-agricultural rural sector is an increasingly important part of the rural economy and accounts for a growing part of rural income and rural employment. Most of the document posed the need to develop financial services other than short-term loans so as to specifically increase productivity and the possibilities of expanding non-agricultural services and manufacturing and processing plants. The main conclusion of the document was that rural financial markets do not operate properly in Latin America and the Caribbean and that the underdevelopment of these financial markets has a negative impact on those investments that aim to bolster productivity, expand incomes and spur sectoral growth (Da Silva, 2004).

Technological resources. Agriculture today is experiencing major changes, leading to the rise of new scientific and technological paradigms, these are transforming the dynamics of agricultural production. These can be grouped in three major areas: the new biotechnologies, sustainable development models and the new information and communication technologies. The new biotechnologies are constituted by a set of techniques that operate at the subcellular level and make it possible to directly manipulate the genetic characteristics and process of reproduction of living beings. The main ones are: in vitro tissue cultures; molecular markers; genetic engineering, by which transgenic crops are produced (mixing genetic matter of different species); monoclonal antibodies; and bioprocesses.

These recent technological developments, especially in the field of the new biotechnologies, have created conditions that favor the private appropriation of knowledge, given their complexity, requirements for multiplication and high relative cost. This new situation has led to massive private investments in activities associated with the conservation, improvement and industrial production of biological resources and agricultural technologies, especially by transnational companies involved in the production of agricultural inputs. This is leading to a radical change in the balance between the public and private sectors. For example, 85% of current global investment in agricultural biotechnology comes from private interests. Two key controversial issues have arisen in this new context, involving intellectual property and access to genetic resources. The models of rural development in LAC have emphasized technological resources, which are capital intensive. Historically this has been one of the problems that has plagued the Green Revolution. Nonetheless, not all technological resources have to be capital intensive (Chaparro, 2000).

The second scientific and technological area includes alternative forms of agriculture, with proposals for ecological agriculture, or agroecological agriculture, as an integrated approach focusing on the sustainable management of the natural resource base (water, soil, biodiversity) and distinguished from the agriculture of the Green Revolution by its scientific, socioeconomic, political and cultural approach (León, 2007). Agroecology emphasizes technology that is knowledge-intensive, low cost and easily adaptable by small-scale producers.

Information and communication technologies constitute the third scientific and technological area that is profoundly transforming agriculture and giving rise to multiple applications with a direct impact on agricultural production and the management of natural resources. These are a set of technologies related to the processing and dissemination of information and knowledge, using Internet tools, which are important in education and for the broad and swift dissemination of the processes of globalization and its effects (Chaparro, 2000; Farah, 2004a; Farah and Pérez, 2004).

Labor. Worldwide, it is estimated that the urban population is on the way to increasing from one-third of the world population in 1975 to two-thirds in 2020. These high rates of urbanization are changing the structure of demand for food towards the consumption of processed foods with some type of value added, which fosters greater demand for non-agricultural labor (Chaparro, 2000). As a result, agricultural employment dropped in almost half of the Latin American countries, while non-agricultural rural employment continued to increase in all of them. According to data taken by CEPAL from Latin American censuses, non-agricultural rural employment climbed during the 1970s and 1980s at an average of 4.3% annually, while the economically active population in agriculture rose only 0.03% per year. In the 1990s, non-agricultural rural employment once again increased appreciably (Dirven, 2004).

The main type of non-agricultural rural employment varies across different income strata. Middle income households work mainly in non-agricultural endeavors, high-income households are mainly self-employed in non-agricultural rural activities or have small and medium enterprises that perform the same type of work, while most poor families perform agricultural wage labor that does not enable them to emerge from poverty and obtain some additional non-agricultural income from crafts or small-scale commerce (Dirven, 2004).

Working conditions (whether formal or informal; reproductive, productive, or community; remunerated or non-remunerated) have changed visibly with globalization and clearly reflect the inequalities and widening gap between rich and poor. In the processes of internationalization, work is valued on a purely mercantile basis, using the criterion that value is to be found in those things that can be bought and sold, which can be assigned a monetary value. For women, especially rural women, a considerable part of their work is not seen as economically productive, as it does not fit within the logic of the market, i.e., it takes place in the context of an economy without wages or prices and its objective is to generate products and services for household consumption (Farah, 2004ab).

The non-traditional agricultural export sector, favored by neoliberalism, has opened up salaried employment opportunities mainly for women in the rural sector. Nonetheless, these jobs are often seasonal, poorly paid and performed in precarious conditions (Deere, 2005). In the greenhouses for flowers and vegetables in Ecuador, Guatemala, Mexico and Colombia, for example, labor is mostly female and the contracts are short-term but renewed time and again. In Colombia, 80% of the flower workers are women and they generally earn the minimum wage, which covers only 45% of a family's basic needs. In Chile, Argentina and Brazil, women are contracted for seasonal positions in the production of fruit for export. Thus, for example, the employment of women in the fruit sector in Chile quadrupled from 1982 to 1992 and was concentrated in temporary jobs, such that 75% of women in the agricultural sector in Chile work under temporary contracts, harvesting fruit more than 60 hours a week during the harvest season. Of these women, one in three earns less than the minimum wage.

Market Trends. Over the last 30 years, with the accelerated pace at which the markets for Latin American products and markets worldwide, have been changing, the commercial formats of quotas and preferences have increasingly become a thing of the past. As a result, markets are fully engaged in a process of transformation in the trade arrangements between countries and between regions and a collapse in tariffs and import duties has accompanied the elimination of quotas and preferences, pointing towards more competitive global markets with a prevalence of value-added, comparative advantages, quality goods and services, as well as safe foods, traceability and biosecurity.

This transformation in the region, with tariff barriers being replaced by technical barriers, accords less importance to the volume of production in relation to factors such as efficiency and productivity. This process of abrupt change in markets has resulted not only from geopolitical changes that have produced an international dynamic in which the market approach prevails, even among countries and regions that are not on the same wavelength politically, but also from consumers themselves imposing conditions and requirements. There is a growing trend among consumers in the region towards a more conscientious, intelligent and differentiated culture of consumption with respect to the foods, cosmetics and medicinal products they consume, as well as the services they demand.

This change in the functional structure of markets has resulted in a series of challenges and opportunities for Latin American agriculture. Among these opportunities, mention can be made of the emergence of new market niches such as the organic, ecologically-sound, ethnic and functional markets, as well markets based on ethical-social considerations (for example, the fair trade market). This range of products may be produced by the small- and medium-scale producers of the region, since the volumes are not necessarily very high and what is most important is the type and denomination of origin of products. It is for that reason that many small- and medium-scale producers from countries such as Guatemala, Nicaragua, El Salvador, the Dominican Republic, Peru and Colombia have been able to become international suppliers and position themselves in markets as demanding as those of Europe, Japan and the United States. Relevant cases include coffee, cacao, banana, oriental vegetables, fruits and aromatic herbs (Salas-Casasola et al., 2006) (See Box 1-6).

The challenges posed by the markets' new structure include competitiveness, regulations and marketing strategies and structures, even in those niche markets. A large number of countries in the region are trying to access the niche

Box 1-6. Medicinal herbs and plants in the Caribbean

The Caribbean is habitat for 2.3% (7,000) of all endemic plants worldwide, and 2.9% (779) of the vertebrate species of the world, even though it accounts for only 0.15% of the earth's land mass. Hence the Caribbean is classified as one of the most important "hotspots" in the world (Myers et al., 2000). In 1988 Norman Myer defined a hotspot as a region of the earth characterized by exceptional levels of endemic species: A hotspot should be habitat for at least 1,500 species of vascular plants (the Caribbean has at least 2.3% = 7,000 plants), which represent 0.5% of the total of endemic plants in the world (as of 2000). A hotspot must also have lost at least 70% of its original endemic species. The Caribbean has met this requirement because of major deforestation, soil erosion and water pollution. In countries such as Haiti and the Dominican Republic, only 5% and 17%, respectively, of their cover remains.

The natural wealth of the Caribbean has not been economically exploited, even though one sees a trend towards the popularization of medicinal herbs and plants, reflected in the number of products available on the shelves of pharmacies, natural products and health stores, aromatherapy establishments and supermarkets (Denzil Phillips International, http://www.denzil.com/).

Currently, the Caribbean is known primarily for a small number of products derived from medicinal and aromatic herbs, despite the abundance of species. The range of products includes teas, exotic drinks made of herbs, traditional herbal remedies, nutraceutics, phytomedicines, essential oils, plant extracts such as cosmetics, condiments, tinctures, liquid extracts and functional foods. Among the best-known products are pepper, nutmeg and chili peppers. Progress has also been made in adding value; some of the better-known products include Angostura, Pickapeppa Sauce, Busha Browne's and Walkerswood.

The biggest beneficiaries of the wealth from Caribbean spices mostly devolve to the approximately 90 firms that import dry herbs from the region to markets Europe, the United States and Japan. Some 85% of the herbs are exported as dry herbs. The global market for herbs is estimated at US$12 billion, with the trade in raw extracts coming to US$8 billion.

markets, for example for oriental vegetables on the east coast of the United States, or for organic fruits in Europe and throughout the United States and Canada. This means that as quotas and tariff barriers have disappeared, the scenario offers, in the best of cases, equal conditions and, therefore, those countries that meet the technical requirements (quality, certification, traceability, biosecurity, social and environmental responsibility) will have the best opportunity to gain access to, position themselves in and stay in those markets.

LAC has a high ceiling for growing and tapping unsatisfied markets for organic and functional foods which by the year 2006 came to approximately US$40 billion. In the specific case of organic and ecologically-sound foods, the challenge is that organic agriculture requires more specialized management and the certifications are expensive for small-scale producers. This has limited the participation of these producers in the global organic market, but has also stimulated the formation of cooperative producers' organizations, which bring other secondary benefits (Bray et al., 2005) (see 1.7.1).

As for the challenge of regulations, Latin American producers and exporters have to comply not only with good agricultural and generic manufacturing practices established by Codex Alimentarius, but in addition the markets themselves have defined their protocols and quality and safety standards such as EurepGAP for the European market and USA-GAP and HACCP for the U.S. and Asian markets. These standards impose the challenge on Latin American and Caribbean agricultural producers and exporters of having to make adjustments in their production processes and physical production facilities so as to be able to comply with the markets' quality standards. Nowadays the producers in LAC who want to become inserted in the international markets are forced to adopt a culture of quality production based on continuous improvement and evolution of their products based strictly on market requirements. This process entails higher production costs and requires use of optimal methods, which at times wipes out the actual potential of many producers in the region, especially small-scale producers.

1.6.2.2 Regional trends in production

The region has a total of 2.018 billion ha, of which approximately 726 million (i.e., 36%) are under agricultural production, including seasonal crops (7.1%), permanent crops (about 1%) and pastureland (about 30%). In the last 15 years, the total agricultural area increased 4.5%, while the total covered by forest (including forest plantations) diminished 1.3%. The area under permanent crops such as cacao and coffee experienced the greatest increase in area, 10.5%, although in the last decade, with the collapse of coffee prices, the area planted in coffee diminished in almost the entire region (Calo and Wise, 2005).

The change in land use varied by region (Table 1-8). Figure 1-6 shows the increase in the total area under agricultural production by region from 1961 to 2003. The Southern Cone, the largest region in area, also saw the greatest increase in area planted. In the three decades from 1961 to 1990, the area under production increased by 27%. Although the rate of increase has diminished, since 1990 there was a 6% increase in the region; Brazil, French Guiana and Paraguay are the countries that saw the largest percentage increases. Suriname, Uruguay and Guyana have experienced almost no change since the 1990s, while Chile suffered a decline of almost 6% in the total area in agriculture.

The main change in land use in the Southern Cone has been due to the increased production of soybean (Figure 1-7), especially in Brazil and Argentina; the total area planted in soybean was almost 47 million ha in these two countries alone, which represents 8% of the total agricultural area

Table 1-8. **Land use by region.**

	Southern Cone	Andean Region	Mesoamerica (include Mexico)	The Caribbean
Terrestrial total	1,297,040	456,197	241,943	22,895
Agriculture total	450,362	133,923	128,815	13,044
% of total	34.7	29.4	53.2	57.0
Annual crops	93,842	13,263	30,736	5,327
% of total	7.2	2.9	12.7	23.3
Permanent crops	9,107	4,538	4,435	1,825‡
% of total	0.7	1.0	1.8	8.0‡
Areas with pasture	347,413	116,122	93,644	5,892‡
% of total	26.8	25.5	38.7	25.9‡
Forests and plantation forests#	675,670	255,900	72,142	4,465‡
% of total	52.1	56.1	29.8	19.6‡

Notes:
Most recent year with data on land use is 2003
Most recent year with data on land in forest and forestation is 1995.
‡ With the exception of the total terrestrial area, the data for the Caribbean does not include Aruba, The Dutch Antilles, Turks or Caicos Islands.
Source: FAOSTAT, 2005.

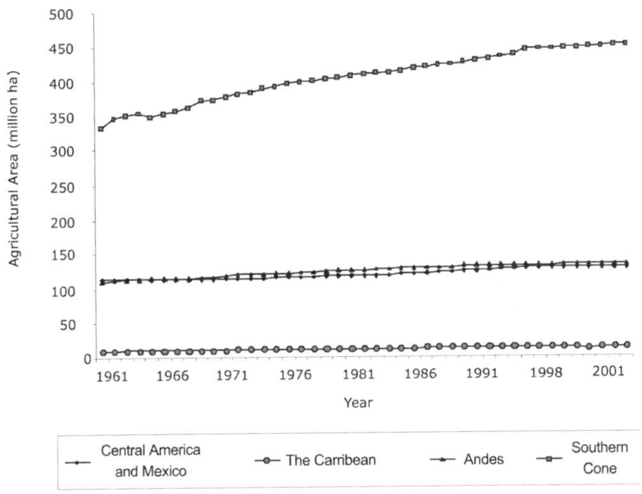

Figure 1-6. *Change in the use of land in the 4 geographic regions of Latin America and the Caribbean.* Source: Authors' elaboration from FAOSTAT data.

of the Southern Cone (including pastureland) (FAOSTAT, 2005). In Brazil, the expansion of soybean has occurred at the expense of natural vegetation and more recently of the tropical forest in the Amazon (Fearnside, 2001b), while in Argentina the increase in soybean has been at the expense of the production of milk, maize, wheat and fruit crops, as well as areas of natural vegetation such as the Yungas rain forest and the dry forest of the Chaco (Jordan, 2001; Jason, 2004; Pengue, 2005). Due to the expansion of soybean in Argentina, the rate of conversion of forest to agriculture is three to six times the global average (Jason, 2004). The expansion of this crop has also accelerated deforestation indirectly by means of the construction of railways and an extensive network of highways that attract cattle growers, mining companies and logging interests to the Amazon jungle and by displacing small-scale producers (Fearnside, 2001a) (see Box 1-7).

Another major change in this area has been the expansion of cattle-ranching in Brazil. Brazil has increased its cattle herd by 122 million animals in the last 15 years (an 83% increase) and today has 269 million animals (Figure 1-8). This expansion has also taken place at the cost of the Amazon forests. According to Giglo (2000), the expansion of cattle in Brazil (and Bolivia) was facilitated by tax incentives put in place by the governments (for example, the "Amazonas Legal" program in Brazil) and the availability of cheap labor.

The total agricultural area in Mesoamerica increased almost 9% from 1961 to 1990, but only 4% since 1990 (Figure 1-6). Though initially Belize, Costa Rica and Guatemala contributed considerably to the increase in agricultural lands in the region, since the 1990s Belize, El Salvador and Nicaragua have experienced the greatest increases (27%, 19% and 11%, respectively). Surprisingly, Honduras has been experiencing a decline in agricultural lands since the 1990s; its agricultural area has diminished almost 13%. This is mainly due to the decline in banana production, which was Honduras's main export during the first half of

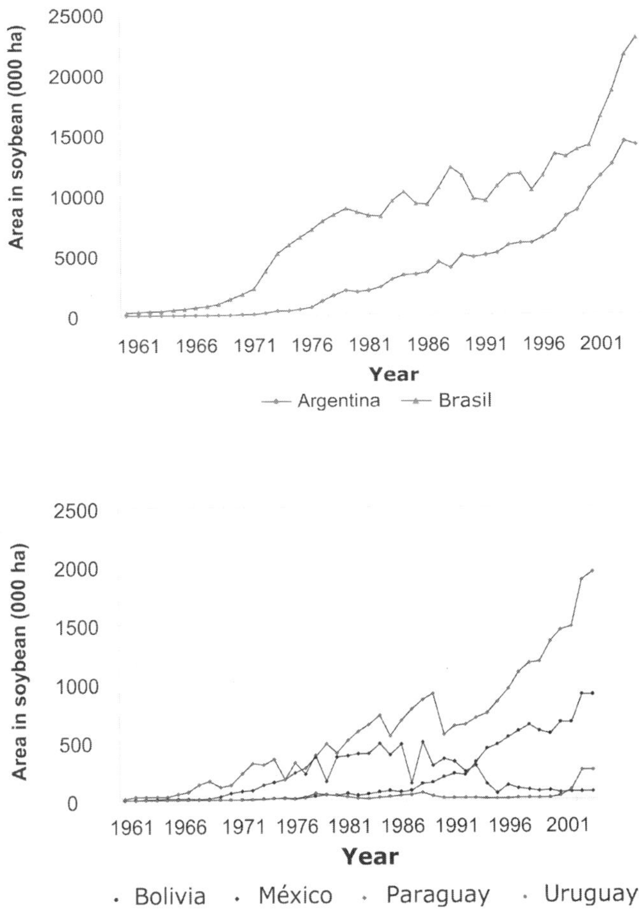

Figure 1-7. *Growth of cultivated surface in soy of LAC countries with the largest volume of production.* Source: Authors' elaboration from FAOSTAT, 2005.

the 20th century, but which began to fall as the result of a combination of diseases, labor organizing and globalization (Soluri, 2005).

The Andean region shows a similar pattern of change as Mesoamerica (Figure 1-6), with an increase in the total agricultural area of 16% from 1961 to 1990 and 4% since 1990. Ecuador is the country with the greatest change in the first three decades (65%), but it increased only 4% since 1990, whereas Peru saw an 11% increase in the same period. The other Andean countries, with the exception of Venezuela (which has seen almost no change in its total agricultural area since 1990), have seen increases of 2-5%.

The Caribbean is the region with the smallest area in LAC. This region experienced a 35% increase in the area planted; Cuba is the country that contributed most to this increase. In the first three decades of the Cuban Revolution, it expanded its agricultural area 91%, while other Caribbean countries saw decreases. Since 1990 there has been a decline in total agricultural lands of 1.3% in the Caribbean. Although most of the Caribbean countries experienced a diminution in agricultural area (including Cuba, but especially Puerto Rico, with a decline of 51%), other countries, such as Dominica, Bahamas and Saint Vincent, had rela-

Box 1-7. Transgenic soybean in Argentina

Argentina is the second leading producer of transgenic crops, with 18 million ha planted. This represents more than 5.5% of Argentina's area, larger than all of Nicaragua. The development of transgenics in Argentina is inseparable from the expansion of the soybean crop. Today Argentina plants 15 million ha of transgenic soybean, mainly Roundup® resistant (RR), producing 38.3 million tonnes (Altieri and Pengue, 2005). The low cost of the herbicide, the possibility of retaining and reusing the seed, the lower consumption of energy, the simplicity of the methods of application, and a major publicity campaign made this technological package attractive to many producers (Trigo and Cap, 2003; Qaim and Traxler, 2005; Souza, 2004). It is estimated that from 1996 to 2001, the technology of RR soybean generated profits of US$5.2 billion, 80% of it captured by the producers and the rest by the supplier corporations (Trigo et al., 2002). In 2002, soybean accounted for 20% of Argentina's export revenues.

This technology has caused major changes in the environment and in Argentina society. The economic benefits have been accompanied by social changes such as migration, concentration of landholdings and agribusinesses, and the loss of food sovereignty (Souza, 2004; Altieri and Pengue, 2005; Pengue, 2005). For example, at the same time as the production area of RR soybean tripled, some 60,000 units engaged in the production of food crops were abandoned. The replacement of traditional activities such as cattle-raising, vegetable production, fruit production, dairy production, and production of other cereal grains (maize and wheat) by the soybean crop is resulting in a lower supply of these products in the market, with the consequent rise in prices and less access for the more economically vulnerable sectors (Alteri and Pengue, 2005; Souza, 2004). From 1998 to 2002, 25% of the country's farms were lost, most of them small producers (Altieri and Pengue, 2005, 2006). From 1992 to 1999 the number of farms in the Pampas was reduced from 170,000 to 116,000, while the average size of a farm increased from 243 to 538 ha in 2003 (Pengue, 2005).

Transgenic soybean has had environmental benefits related to the practice of zero-tillage (Trigo and Cap, 2003; Qaim and Traxler, 2005). These effects are overshadowed by the dramatic increase in the use of herbicides (mainly glyphosate) (Trigo and Cap, 2003) (see Figure); the appearance of glyphosate-tolerant weeds (Papa, 2000); the increase in the use of synthetic fertilizers; the depletion of soil nutrients; the degradation of the soil structure; and the loss of habitat and biodiversity (Altieri and Pengue, 2005; Pengue, 2005). Soybean expansion has even occurred on non-farm lands, not only in the Pampas but also in susceptible and high-biodiversity ecoregions such as the Yungas, the Gran Chaco, and the Mesopotamian Forest (Pengue, 2005). Since the introduction of transgenic soybean, 5.3 million ha of non-farm lands have been converted to soybean production, and the rate of conversion of forest to agriculture is three to six times the global average (Jason, 2004).

Glyphosate, the active ingredient in Roundup ®, is a broad-spectrum herbicide classified as low (category IV) or medium (category III) toxicity. Nonetheless, there is ample evidence that glyphosate is not innocuous, as was once thought (see figure below). Most toxicological studies are done exclusively with the active ingredient (i.e. glyphosate) and not with the commercial formulations that contain the so-called inert ingredients. Roundup® contains glyphosate and the surfactant polyoxy-ethyleneamine, or POEA, which is three times more toxic than glyphosate alone (USEPA, 2002).

On the whole, transgenic soybean has been an economic success in Argentina. Nonetheless, it has not helped meet the goals of reducing hunger, poverty or inequality, nor has it helped increase sustainability in Argentina.

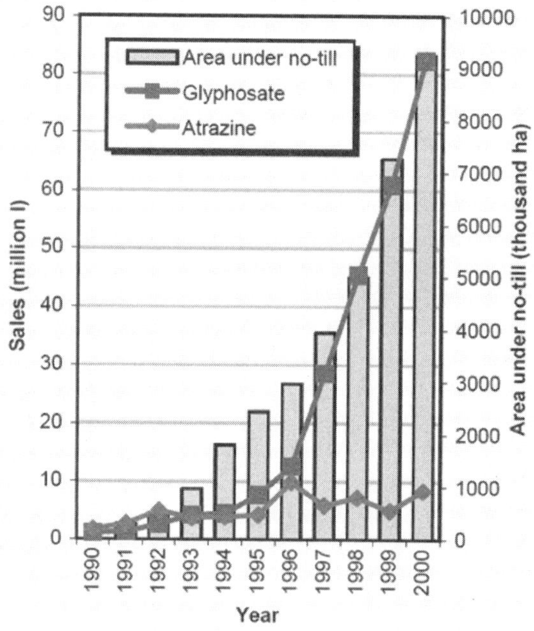

Studies that show negative effects of glyphosate or Roundup®:

- High degree of mortality in amphibians (Relyea, 2005 ab).
- Reduction in the number of aquatic species, including fish (Henry et al., 1994; Wan et al., 1985; WHO, 1994).
- Direct and indirect negative effects on beneficial soil organisms (spiders, earthworms, and others) (Hassan et al., 1988; Burst, 1990; Asteraky et al., 1992; Mohamed, 1992; Springert and Gray, 1992).
- Toxicity in nitrogen-fixing bacteria, mycorrhizal fungi and actinomycetes (all important in recycling nutrients and other ecological soil processes) (Chakravarty and Chatarpaul, 1990; Carlisle and Trevors, 1998; Estok et al., 1998).
- Stimulating effect on populations of the pathogenic fungus Fusarium, including Fusarium graminearum, which affects soybean (Levesque et al., 1987; Sanogo, 2000; Hanson and Fernández, 2003; Fernández et al., 2005).
- Synergetic effect when combined with other pesticides (Relyea, 2003).
- May accelerate the process of eutrophication of bodies of water, since it acts as a source of phosphorus (Austin et al., 1991).

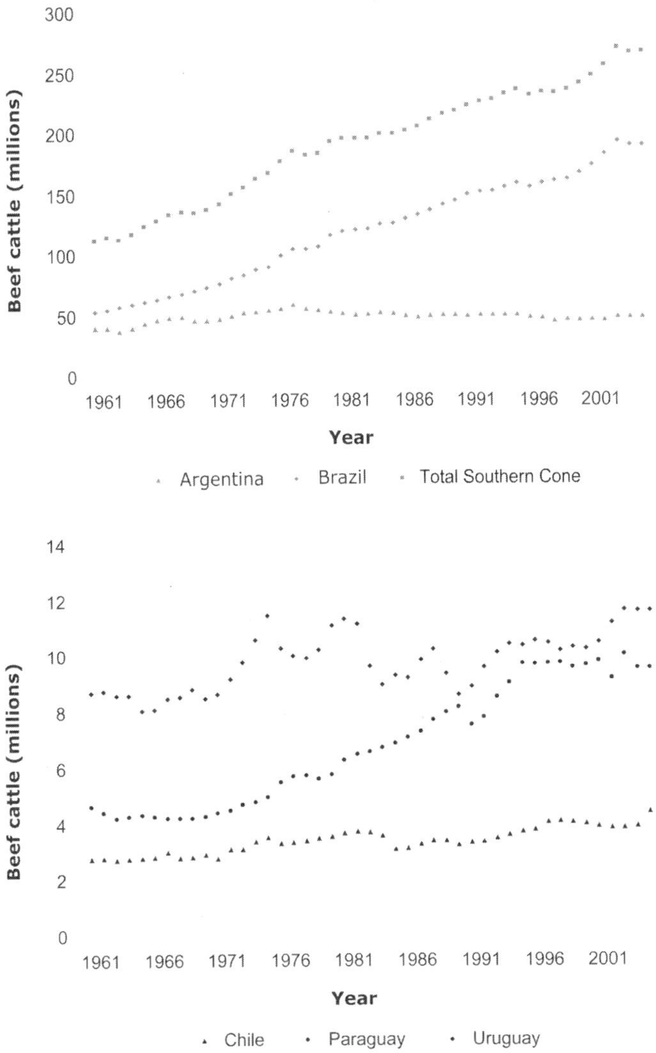

Figure 1-8. *Evolution of the number of beef cattle in Southern Cone countries.* Source: Authors' elaboration from FAOSTAT data.

tively significant increases (from 15 to 28%). One of the main trends in the English-speaking Caribbean has been the conversion of agricultural lands to urban centers and activities for tourism. Box 1-8 discusses this situation in several countries of this region.

The four subregions of LAC also differ in terms of the percentage of land that is under different uses (for example, permanent crops and pasturelands, among others). As reflected in Table 1-8, Central America (including Mexico) and the Caribbean are the two regions with the highest proportion of their territory in seasonal crops. This is related to greater population density and the predominance of the maize and bean system in Central America and sugarcane in the Caribbean. Compared to the other regions, the Caribbean also has a higher proportion of land in permanent crops. The proportion of land in pastures in the Caribbean, the Andean region and the Southern Cone fluctuates from 25 to 27%, but Mesoamerica has a higher proportion of its land in pastures (almost 40%). Finally, both the Southern Cone and the Andean region have more than 50% of their territory under forest cover, while the Caribbean and Mesoamerica have a smaller percentage (20 and 30% respectively).

In terms of products or specific groups of categories of products, there have been changes depending on the markets demands. In some products, growth has been minimal and there has even been stagnation, such as root crops and tubers, coffee, bananas, cotton and cereal grains. In contrast, there has been a jump in the production of oil-bearing crops (mainly soybean and African palm), fruits, vegetables and sugarcane.[5]

Recently sugarcane has taken on great importance given its potential for the production of ethanol. Sugarcane has the advantage of being quite efficient in the production of biomass and is a crop that can produce year-round. In the region, only Brazil has begun to make significant use of sugarcane as a raw material in the ethanol industry (Dias de Oliveira et al., 2005; Licht, 2005). It is argued that Brazil has the potential to produce enough ethanol to respond to the domestic demand for fuel if it earmarks all of its cane production to the production of ethanol, or if the area given over to this crop is doubled (in other words, if the area increases to 5.6 million ha) (Berg, 2004). Unfortunately, expanding the area of this crop has negative implications for the environment. It is estimated that sugarcane monoculture accounts for 13% of all herbicide use in all Brazil. Studies done by EMBRAPA in 2002 (cited by Altieri and Bravo, 2007) confirm the contamination of the Guaraní aquifer in the state of Sao Paulo, which is attributable mainly to the cane crop (Altieri and Bravo, 2007). The area planted in sugarcane is quickly expanding to the Cerrado region, one of the biodiversity hotspots (Myers et al., 2000) and is contributing to the destruction of this unique ecosystem, which maintains only 20% of its original vegetation (Mittermejer et al., 2000).

In addition to soybean, another oil-bearing crop that has expanded considerably in the region has been African oil palm, which has undergone expansion mainly in Central America, Ecuador and Colombia (Carrere, 2001; Buitrón, 2002; Donald, 2004). As in the case of soybean, the expansion of this crop, which is produced on large expanses as a monoculture, is threatening unique ecosystems such as the tropical forest of the Chocó in Ecuador and Colombia (Fearnside, 2001b; Donald, 2004). In Colombia, there have also been cases of violent displacements of Afrodescendant communities to grow African palm (Diocese of Quibdó, 2001).

The production of cereal grains (beans, lentils, pigeon peas and others) and root crops and tubers has remained stable in recent years, but in some cases there have been drops in production. LAC exported a total of 18.8 million tonnes of cereal grains (18% of world exports) (USDA, 2005), but almost all of this was produced by Brazil and Argentina (4 million tonnes and 14.5 million tonnes respectively). In the particular case of maize, world exports come to 74.5 million tonnes, of which only 14 million are exports from LAC, specifically Argentina, Brazil and Mexico. The maize crop and its consumption in Mexico and Central America have been affected by imports of subsidized maize

[5] Rural Development Unit of CEPAL, based on the FAO production yearbook, Rome.

Box 1-8. Land conversion from agriculture to tourism in the English speaking Caribbean

Urbanization is a phenomenon that is occurring throughout the world. As urban centers grow, agricultural land is informally converted into urban use, particularly for tourism. The English-speaking Caribbean, a chain of small islands, which is as an attractive destination for foreigners who travel thousands of miles, and spend considerable sums of money to take in the sun, sand and sea.

In the English-speaking Caribbean, the role of the State in relation to tourism has mainly been indirect. Physical incentives are common, for example, the provision of infrastructure, utilities and promotional activities accompanied by some training and development. Tourism is mainly a private sector activity, however the Bahamas, Curacao, Aruba and Grenada have put mechanisms in place to exercise control over the industry. At a Workshop on Land Policy, Administration and Management in the English speaking Caribbean held in Port of Spain Trinidad in March 2003, prominent Land Managers and Administrators throughout the Caribbean presented papers on the workshop theme. The issues common throughout the deliberations indicated the State has been delinquent in its physical planning strategies and this has resulted in unregulated control over land. For example in Tobago, local fishermen were denied access to the Pigeon Point beach; after a lengthy legal battle the State acquired the property in order to allow locals to access the beach.

The State's lack of implementation of physical planning strategies and enforcement of building codes has resulted in irresponsible development. Foreign capital channeled through ventures with local residents is the driving force behind most of the development. Once business partners have sufficient funding to undertake their venture, there is no need to interact with the Planning Authority for approval, since the change in land use patterns occurs outside of the formal process. This process results in agricultural land being converted to large tourism complexes and private holiday villas with little if any oversight. In some or almost most of the territories in the Caribbean, planning guidelines are not enforced, site development standards are breached and building designs are not always compatible with the land use classification.

Since the tourism infrastructure of the Caribbean region is mainly located on the coast, the majority of tourist facilities are located within 800 m of the high water mark. The wastes and pollution generated by the tourism industry, especially by cruise ships, are discharged into the sea. The impact of this activity is so great that in Trinidad and Tobago research on the die-off rate of fish species is being conducted by the Institute of Marine Affairs.

Jamaica has not as yet achieved the sustainable management of its land resources. Eighty percent of the island has been classified as mountainous with the remainder zoned for agriculture, commercial, industrial, mining, residential, watersheds and other uses. The State owns 22% of the land in Jamaica, and one of the challenges faced by the State is the lack of development plans and databases. Work has been initiated to address these shortcomings through legislation and policy.

In Guyana, as the boundaries of the urban center expand, agricultural land is informally converted to urban use, such as tourism. Guyana is faced with the challenge of preserving its remaining agricultural land holdings near the city in the presence of an influx of rural migrants. Guyana lacks a National Plan to address the supply of land for ecotourism as well as policies to address coastal agriculture, urban housing, and the use of land for tourist resorts. Currently physical planning is sporadic and reactionary.

In Antigua and Barbuda, the development of the tourism industry is reflected in the patterns of land use change evident on the island. Prior to 1975, the major land uses were agriculture, grazing and livestock. By 1983 the labor force in agriculture fell from 46% to 9% and by 1985, 60% of the work force was in the public or private sector with 23% in tourism. There was also a marked decline in land in agriculture from approximately 25,000 acres in 1964 to 5,500 acres in 1985. Of these 1200 acres was used for the construction of hotels and golf courses. There was a marked increase in the construction of facilities for tourists and by 1995 the acreage increased by 138%. Although this statistic signifies a large percentage of the labor force is in the tourism industry, many persons are also converting residential properties into bed and breakfast facilities thereby generating the need for additional resources. In the case of Tobago, many large agricultural estates are being subdivided into smaller parcels and converted to holiday homesteads for sale to foreigners. This trend has resulted in an astronomical increase in the price of real estate on the island and many locals are now unable to acquire a decent property at an affordable price.

In Trinidad and Tobago, in 1992 the State adopted a New Administration and Distribution Policy for Land. One of the policy goals enunciated in the document was the prevention of prime agricultural land from being converted to nonagricultural use through the institution of land use zoning. The objective was to ensure food production, food security and employment for the rural sector.

Many of the islands that comprise the English-speaking Caribbean rely heavily on the tourism industry as a significant contributor to their GDP. Some such as Trinidad and Tobago and possibly Jamaica also have additional resources e.g. oil and gas. More attention must be focused on the issue of the formulation of National Physical Development Plans with specific reference to land for tourism and the attendant site development standards. Some territories have drafted Physical Development Plans, however many are outdated or are awaiting approval from the relevant authorities. Even in instances where Plans have been accepted, implementation of the Plans has been stymied by administrative and bureaucratic challenges. It is evident that with the overexploitation of natural resources for tourism, unintended consequences can arise. If the product is no longer of interest then travelers will take their business elsewhere, hence one can not sufficiently underscore the importance of the role of physical planning in the sustainable development of the region.

from the United States and more recently by the increased use of maize to produce ethanol in the United States.

LAC is one of the most important regions in the world in livestock production. Nonetheless, beef exports are dominated by just two countries, Argentina and Brazil. Of total world beef exports, estimated at 5.72 million tonnes (USDA, 2005), Argentina and Brazil together account for 37%, with 2.14 million tonnes of beef exported between them. It is forecast that the economic take-off of Asia, mainly China and South Korea, will result in a 22% increase in demand for beef with respect to 2005 imports (USDA, 2005). As for hog production, of a total of 4.2 million tonnes sold in the international market, only 11% is produced by LAC. Once again, two countries alone account for the lion's share of these figures: Argentina (48 million tonnes) and Mexico (440 million tonnes).

Milk production in LAC is far below expectations, taking into account the proportion of arable land and pastureland in the region. The region produces only 8.96% of the milk produced in the world (FAPRI, 2006). The production of milk is concentrated in South America (Argentina, Brazil, Peru, Venezuela and Colombia). In terms of exports, the region's performs at levels below what one would expect given the world dynamic in relation to processed products. Only Argentina and Uruguay export butter, cheese and powdered milk.

The wealth of LAC's marine biomass has not been properly taken into account, as evidenced by the low levels of production of this resource. The fish supply internationally is 100.2 million tonnes, only 3.1 million tonnes of which is produced in Latin America and the Caribbean (this figure does not include Mexico).

The area in forests and timber production constitute another category with extraordinary potential. The region is one of the more forested in the world, with one-fourth of the total forests worldwide (UNEP, 2002b). The forested area comes to 834 million ha of tropical forest and some 130 million ha of other types of forest, accounting for 48% of the total. This forest cover is not evenly distributed, for Argentina, Brazil, Bolivia, Colombia, Mexico, Peru and Venezuela account for 56% of the total. There are other countries, however, with serious forest problems, such as Haiti, less than 3% of whose territory has forest cover. The forests of LAC contain 160 billion m^3 of timber, accounting for one-third of all timber in the world. In terms of exports, Brazil and Chile are the leading exporters of timber and timber products. It should be emphasized that any type of use of forest resources should take into consideration the possible environmental impacts and impacts on climate change and be done in the context of sustainable management plans. Today there are three programs for tropical timber certification that attest to the origin of the timber and whether it comes from a forest managed using certain criteria of environmental sustainability (Baharuddin, 1995). Forest resources may also be tapped by rural communities and provide an important source of income to the communities that live in forest areas. Mexico is one of the world leaders in community forest management for commercial timber production (Bray et al., 2005). The Mexican communities are attaining a balance between income-generation for the community and forest conservation.

In summary, among the main trends in the region in recent years, special mention can be made of the production of oil-bearing crops, particularly soybean, which increased considerably in Argentina, Brazil, Bolivia and Paraguay, as well as African palm in Honduras, Guatemala, Costa Rica, Ecuador and Colombia. In addition, there was an increase in the cultivation of fruits and vegetables for export, mainly in Mexico, Chile, Argentina, Brazil and Costa Rica. Another trend during the 1990s was the increase in forest products in Chile, Argentina, Uruguay and Honduras and the increase in cattle-ranching in Brazil, Mexico and Chile. In the English-speaking Caribbean there has been a transformation of agricultural lands to urban development and tourism, increasing dependence on imported foods. In many countries of the region, the increase in exports has occurred at the expense of food production for the domestic market, which has led to an increase in imports of agricultural goods (including fish and forest products, as well as agroindustrial products).

According to an extensive study by CEPAL cited by David et al. (2001), from 1979 to 2001, the region imported two times more agricultural products than it exported. Nonetheless, FAO data show that the deficit in the exports of grains and legumes is much greater for the countries of Mesoamerica and the Caribbean than for South America, although the data for South America are highly influenced by the exports of countries such as Brazil and Argentina (see Figure 1-9). This emphasis on export products also has repercussions on the food sovereignty of the countries of the region. For example, among the products with a market deficit are products essential for food in the region, such as maize, beans, rice, cereal grains, milk and other dairy products (David et al., 2001). Finally, these trends have also affected the agrarian structure of several countries in the region, since the increase in exports has taken place mainly in the most capitalized sector of agriculture (the large-scale producers tied to agroindustry and the export market) and have resulted in the displacement of small-scale producers. The CEPAL study concludes that the neoliberal reforms responsible for the changes described have accentuated the differences between those who have access to capital and market and those who do not (David et al., 2001).

Transgenic crops. Despite the controversy concerning around transgenic crops, gradually they have been adopted in LAC, with impacts perceived by some as negative and by others as positive, in relation to the goals of sustainability, poverty reduction and equity. The Southern Cone is the region with the largest production of transgenic crops, with almost 32 million ha planted in 2006 (Argentina, 18; Brazil, 11.5; Paraguay, 2; Uruguay, 0.4). Mexico, Colombia, Honduras and more recently Bolivia are also producing transgenic crops, but have less than 0.1 million ha each (James, 2006). Today, LAC produces just over one-third of the transgenic crops in the world. Most are accounted for by just three crops: herbicide-resistant soybean (Argentina, Brazil, Paraguay, Uruguay, Bolivia and Mexico), Bt maize (Argentina, Uruguay and Honduras) and Bt cotton (Argentina, Brazil, Mexico and Colombia) (Table 1-9) (James, 2006).

Transgenic crops have been an economic success story in some countries of Latin America, in particular Argentina;

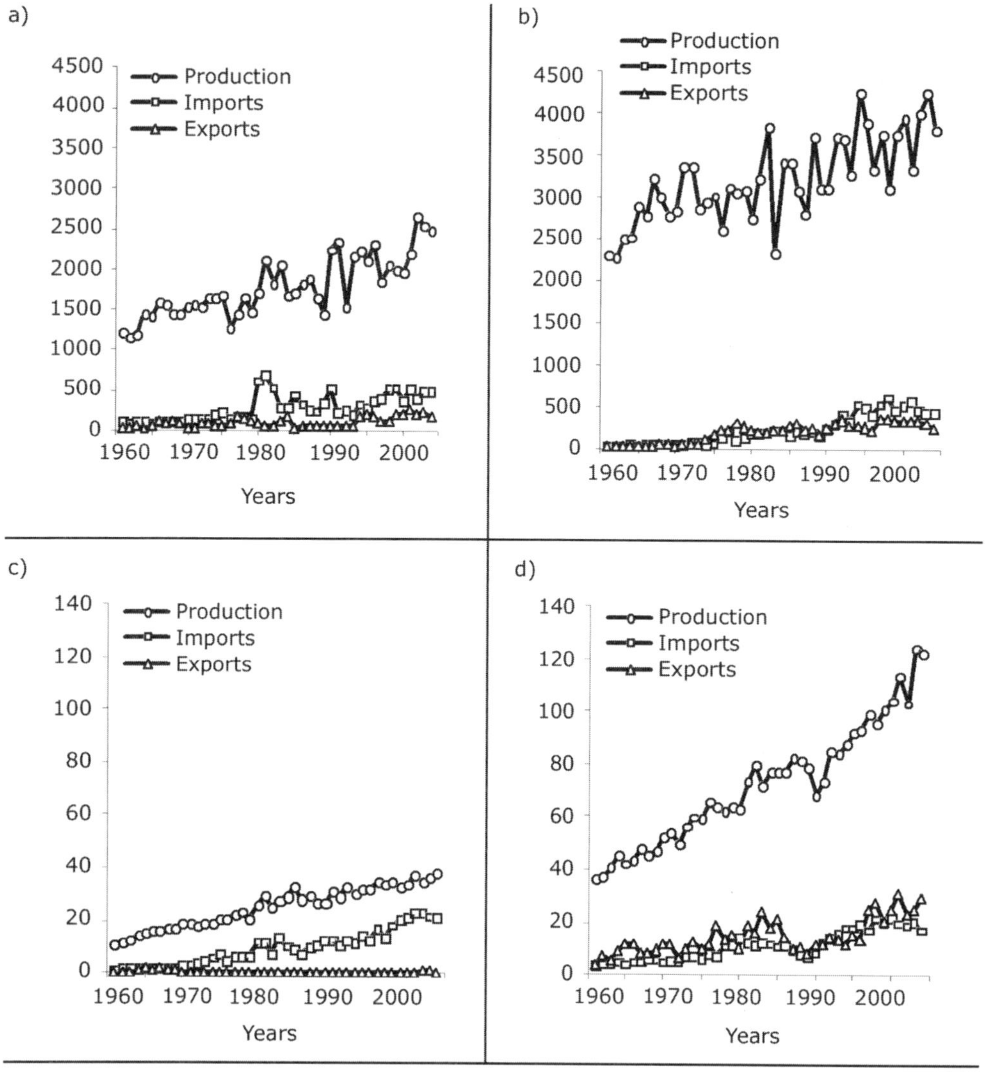

Figure 1-9. *Imports and exports of pulses and grains for Latin America and the Caribbean. Pulses data between 1961 and 2004 for countries in a) Central America and the Caribbean; b) South America; Grains data between 1961 and 2004 for countries in c) Central America and the Caribbean; d) South America.* Source: FAOSTAT, 2005.

nonetheless, thus far these benefits have been monopolized mainly by the large producers and agroindustries (see Box 1-7). Internationally, 90% of the producers who grow transgenics, i.e., 9.3 million, are small-scale producers, but they are almost all in China (6.8 million) and India (2.3 million) (Brookes and Barfoot, 2006; James, 2006). In LAC, most transgenics crops are planted in large tracts in monoculture.

Although the promoters of transgenic crops argue that this technology benefits small-scale producers and that it is a sound tool for fighting poverty and hunger in the world (Pray et al., 2002; James, 2006), there are very few empirical studies that verify these assertions for LAC. In a recent study of Roundup-resistant soybean in Argentina, Qaim and Traxler (2005) concluded that transgenic soybean was more profitable than conventional soybean and that small-scale producers benefited the most. A second study on the adoption of Bt cotton by producers in Coahuila, Mexico reached a similar conclusion (Traxler and Godoy-Avila, 2004). Both cases represent special situations. In the case of Argentina the producers do not pay for the "intellectual property rights" for the transgenic seed. Moreover, the classification of "small" includes producers of up to 100 ha with access to capital (Qaim and Traxler, 2005). In the case of Mexico, the producers pay intellectual property rights to the company Monsanto/D&PL, but they receive credit from the government to purchase the transgenic seed. In this case the benefit accrued largely due to the financial and technical support provided by the government and by the implementation of other plant protection programs (Traxler and Godoy-Avila, 2004).

The technology of transgenics has brought about major transformations in the environment and society in some

Table 1-9. **Production of transgenic crops in LAC.**

Global Ranking	Country(*)	Area (millions of hectares)	Crop
2*	Argentina	18.0	Soybean, maize, cotton
3*	Brazil	11.5	Soybean, cotton
7*	Paraguay	2.0	Soybean
9*	Uruguay	0.4	Soybean, maize
13*	México	0.1	Cotton, soybean
15	Colombia	<0.1	Cotton
18	Honduras	<0.1	Maize
Total		32.2	

*Among 14 countries that produce more than 50,000 ha of transgenic crops.

Source: James, 2006.

countries of LAC. The economic benefits have been accompanied by social changes such as the displacement of small-scale producers and the consequent migration to the cities (Pengue, 2000), the concentration of lands and agribusinesses (Verner, 2005; Altieri and Pengue 2006) and the loss of food sovereignty (Jordan, 2001; Souza, 2004; Altieri and Pengue, 2005; Verner, 2005). Moreover, environmental benefits have been reported related to the increase in area planted with reduced or zero tillage and to reduced pesticide use associated with Bt crops. For example, in Argentina, where more than half of the transgenic soybean in the region is grown, 80% of the area requires zero tillage, contributing to a reduction in the rate of soil erosion (Trigo and Cap, 2003; Qaim and Traxler, 2005). In the state of Coahuila, Mexico, where 96% of the area in cotton is planted with Bt cotton, an 80% reduction was reported in the number of applications of insecticides, although the authors recognize that not all of the reduction could be attributed to the transgenic cotton because the region also has a strong program to eradicate the boll weevil and an effective integrated pest management program (Traxler and Godoy-Avila, 2004). In general, adopting transgenic cotton appears to be highly determined by the presence of a particular pest and in many regions producers have opted to continue using the conventional seed (Traxler and Godoy-Avila, 2004; Qaim et al., 2003).

These environmental benefits of transgenics are overshadowed by other negative environmental impacts. Many scientists have expressed concern over the use of transgenic crops on a large scale considering the environmental risks, which may threaten the sustainability of agriculture (Goldberg, 1992; Paoletti and Pimentel, 1996; Rissler and Mellon, 1996; Kendall et al., 1997; Snow and Moran, 1997; Royal Society, 1998; Altieri and Rosset, 1999). For example, the widespread adoption of homogeneous transgenic varieties inevitably leads to genetic erosion and the loss of local varieties developed and used traditionally by thousands of small-scale producers (Robinson, 1996). In the case of transgenic soybean, a dramatic increase has been reported in the use of herbicides, especially glyphosate (Trigo et al., 2002; Qaim and Traxler, 2005); the evolution of resistance to glyphosate has already been reported in some weeds, limiting the possible benefit of the technology (Holt and Le Baron, 1990; Papa, 2000). The massive use of Bt crops affects other organisms and some ecological processes and can lead to resistance. For example, it has been shown that the Bt toxin may affect beneficial insects that feed on pests that eat the Bt crop (Hilbeck et al., 1998). There is also evidence that the pollen from Bt crops that is deposited on the leaves of wild plants around the areas planted in Bt crops may kill other lepidopterans that are not pests, such as the Monarch butterfly (Losey et al., 1999). There is also evidence that the Bt toxin adheres to soil colloids and lasts up to three months, having a negative impact on the populations of invertebrates that help in the decomposition of organic matter (Donnegan et al., 1995). In addition, the intensive use of Bt varieties increases the pressure of selection and generates resistance, threatening not only the future utility of these crops, but also annulling one of the most useful tools available to the organic producers for fighting pests (Pimentel et al., 1989; Mallet and Porter, 1992; Gould, 1994; Alstad and Andow, 1995).

Transgenic crops have also had a negative impact on biodiversity due to the conversion of forest areas and natural savannahs to transgenic plantations, in particular soybean. In Brazil and Argentina the expansion of transgenic soybean has affected directly and indirectly on the deforestation of unique ecosystems such as the tropical forest of the Amazon region and the Cerrado in Brazil and the Yungas forest in Argentina (Fearnside, 2001b; Montenegro et al., 2003; Pengue, 2005).

As LAC is important as a center of origin of crops of global importance, such as maize, potato and tomato, there is concern over genetic contamination should transgenic crops be introduced in the centers of origin, for example transgenic potato in Bolivia, or transgenic maize in Mexico. Indeed, there is already evidence of genetic contamination of local varieties of maize in Mexico (Chapela and Quist, 2001), although it is argued that this contamination may have been temporary (Ortiz-García et al., 2005). Also worrisome is the possible contamination by transgenics of edible crops that are given non-food uses, for example the produc-

Box 1-9. Biopharmaceutical crops and possible impacts in Mexico, center of origin of maize

Biopharmaceutical crops are plants that have been genetically modified to express substances with therapeutic properties, for example viral proteins for vaccines, hormones or antibodies (Gomez, 2001; Ellstrand, 2003; Ma, 2003). The first recombinant pharmaceutical proteins derived from plants were the human growth hormone expressed in tobacco in 1986 (Barta et al. 1986) and the human seroalbumin also from that crop, and in potato crops in 1990 (Ma et al., 2005). Twenty years later, the first drugs produced in transgenic plants are already being marketed. Although some developments use cell cultures from plants, insects, animals or microorganisms to express these molecules, others use complete plants of rice, tobacco and maize, in confined or open field crops, the latter promising lower costs. Over time, the technology has improved considerably, improving the economic feasibility of this application (Ko and Koprowski, 2005; Stewart and Knight, 2005). Of all these systems, expression in seeds has turned out to be of enormous utility for accumulating proteins in a relatively small volume; they do not degrade because the endosperm conserves the proteins without any need for low temperatures, which is a great advantage for the production, for example, of oral vaccines (Han, et al., 2006). Among cereals, maize, rice and barley are interesting alternatives; but maize has a greater annual yield, moderately high protein content in the seed, and a shorter crop cycle, which gives it greater potential protein yield per hectare overall (Stoger et al., 2005). Though maize has the disadvantage of being a cross-pollinating plant, no other cereal grain achieves such yields (Stoger et al., 2005), which makes it the most used system of expression; t holds more than 70% of the permits issued by APHIS from 1991 to 2004 (Elbeheri, 2005).

There are more than 20 firms in the US, Canada and Europe specialized in these production platforms (Huot, 2003; Colorado Institute of Public Policy, 2004). The costs are much lower than those of microbial systems (Elbeheri, 2005). The economic and technical feasibility combined with the perception of maize as an industrial raw material have resulted in it being the most widely used biopharmaceutical crop. Nonetheless, these criteria do not consider the potential risks for millions of people who have a maize-based diet. The first risk is that the grains that contain the compound may pass into the food production chain in industrial operations because it is impossible to distinguish them by sight. Careless handling in industrial processing can occur; it has already happened with Starlink maize in 2000 and with rice (USDA, 2006), although they are not biopharmaceuticals. This has happened in the US, where the rules on biosafety are well established, though they are not necessarily implemented adequately (USDA, 2005). This contamination may have a potential negative effect in the populations that consume these grains: in Mexico per capita maize consumptions varies from 285 – 480 g daily, and is the source of as much as 40% of protein intake, given its low cost (Bourges, 2002; FAO, 2006).

The potential effect may be disastrous if added to the second great risk, the risk of genetic flow. This is not a physical mix of grains, but rather the release of a pharmaceutical transgene that is inherited in the offspring, where it can endure for several generations in an open seed exchange system as one finds in Mexico (Cleveland and Soleri, 2005). The potential dangers of exposure to recombinant compounds by this means would affect practically the entire population of Mexico, particularly those that produce maize for subsistence or on a semi-commercial basis. The genetic contamination of maize could be devastating since Mexico is one of the centers of genetic diversification, and Mexican culture is tightly bound to this crop. Using maize for the production of pharmaceuticals and non-edible industrial products, which also pose health hazards, is the result of a series of decision in which Mexicans did not participate but which may directly affect them. These decisions have been made by companies and policy makers in the more technologically developed countries where lobbying has led to prohibitions on developments in animals because public opinion—which in these countries is often the driving force behind regulatory changes—considers them more similar to humans, though containing them is easier (NAS, 2002), and they have been used for a long time to produce vaccines and serums, antibodies, etc. This situation has accorded priority to production in plants worldwide, which is also cheaper. The consortia and their experts argue that there are no appreciable or verifiable risks in these crops. Even if the risks are low, which is debatable, contamination of food crops with pharmaceutical maize grains would taint the food supply of 100 million Mexicans. If maize in Mexico is contaminated by genetic flow, it would not be easy to eliminate, and it would affect 60% of the non-commercial and commercial productive units in the country, e.g., production for family consumption in Mexico, which uses 33% of the area planted in maize, and produces 37% of domestic maize production (Nadal, 2000; Brush and Chauvet, 2004). This would directly affect the safety of the food base of millions of Mexicans, not to mention the impact on megadiversity in a center of origin. Although there are methods of biological containment of trangenes such as the transformation of chloroplasts, which are inherited from the mother plant (Daniell et al., 2005), inducing the expression with substances that must be added to the crop (Han et al., 2006), and other systems of genetic containment (Mascia and Flavell, 2004), no containment system is infallible. In a case such as this, where there are possibilities of contamination, and where the consequences would be disastrous for millions of human beings, one should apply the precautionary principle.

If there is contamination, what would the potential effect be on human health?

- Plants and animals process proteins in different ways. Biopharmaceuticals may be perceived by the human body as

Box 1-9. continued

foreign substances and cause allergic reactions, including potentially deadly anaphylactic shock.

- Growth factors such as erythropoietin are active in concentrations of one billionth of a gram when injected, and could cause harm if inhaled, ingested or absorbed through the skin.
- The chemical/insecticide avidin causes vitamin deficiency and coagulation of blood, and aprotin may cause diseases of the pancreas in animals and in humans. These two chemicals are produced in transgenic maize cultivated in open fields.
- Industrial enzymes that are produced in transgenic maize (trypsin and antitrypsin) are allergens.

Can biopharmaceuticals affect the environment?

- Apritinin and other enzymes that inhibit digestion shorten the life of honeybees, while avidin directly kills or has a chronic effect on 26 insect species.
- There is no way to prevent wildlife from consuming crops that contain high concentrations of biopharmaceuticals.

These substances have not been tested for effects in the macroorganisms and microorganisms of the soil, although it is known that other proteins in transgenic crops leach from the roots and persist in the soil for months.

tion of nutraceuticals and biopharmaceuticals or non-edible industrial products that impede use of the crop for food (see Box 1-9).

On balance, despite the economic success of some transgenic crops and their swift adoption by large and medium-scale agricultural producers in some regions, thus far transgenic crops in LAC have not contributed adequately to satisfying the goals of sustainability, poverty reduction and equity. Leading social movements in Latin America and the Caribbean have openly stated their opposition to transgenic crops and in particular to intellectual property rights and genetic use restriction technology (sterile seed technology) which, they argue, threaten the rights of local producers to keep and use genetic resources (Vía Campesina, 1996; Desmarais, 2002). Despite the opposing positions on transgenics, there does appear to be consensus in the region as to the pressing need to apply and adhere to precautionary regulations in the process of generating and adopting this technology. The Cartagena Protocol on Biosafety, adopted under the Convention on Biological Diversity, is the first international agreement for the control of modern biotechnology and applies the precautionary principle to the use and transnational movement of transgenic crops (Eggers and Mackenzie, 2000). Of the countries in LAC that are growing transgenic crops, Argentina, Uruguay and Honduras have not ratified the agreement.[6]

Nanotechnology. Another component of the new technology is nanotechnology. Nanotechnology refers to the manipulation of matter on a nanometric scale (one nanometer equals one one-millionth of a meter). In LAC, the use of nanotechnologies has not yet become widespread, nor are there government initiatives in the area of research and development to produce particular applications for the region.

Nanotechnology is thought to offer society opportunities. The possible applications in agriculture include integrated pest and disease management at the molecular level, as well as technologies that improve the capacity of plants to absorb nutrients. One can already find intelligent sensors and systems on the market for applying slow-releasing inputs at the molecular level used in agriculture to fight viruses and other pathogens. There are also the so-called nanostructured catalytic materials, which bolster the efficiency of pesticides, including herbicides, possibly contributing to reduced chemical use in agriculture. Nonetheless, nanotechnology also poses major environmental and possibly health risks, as well as social, economic and ethical challenges (ETC, 2007). Nanoproducts could enter the human body or the environment and have unpredictable effects. Research studies on the impacts of nanoproducts are almost non-existent, so very little is known of the possible consequences of releasing these products in the environment. As nanoproducts are still not widely dispersed in the environment, they present an excellent opportunity to implement the precautionary principle, in order to assess potential impacts before the products are released.

Biofuels/Agrofuels. The global trend towards diminished world oil reserves plus the steadily increasing demand for fuels from non-renewable resources had induced a marked interest in the last decade (1996-2006) in identifying alternative fuel sources. In this context, major efforts have been made to optimize the use of plant biomass as an alternative renewable source for the production of bioenergy.

Traditional sources of biofuels have been used on a small scale with little technology, such as the direct fuel of firewood and manure for generating bioheat. The most widely used modern bioenergy has been microbial fermentation of manure to obtain biogas, which provides heat and electricity on rural properties. And more recently, on a larger scale are liquid biofuels, alcohol and biodiesel, obtained from crops such as sugarcane, soybean, castor-oil plant, oil palm, cassava, maize and beets, among others, more specifically called agrofuels. The possibility of producing biofuels holds out one of the great hopes in the world for reducing dependency on fossil fuels such as gasoline, gas oil and kerosene.

The Americas have traditionally held a leading place in the production of sugarcane, which has been a leading crop in the bioconversion of biomass to fuel (IEA, 2004). In LAC,

[6] http://www.biodiv.org/biosafety/signinglist.aspx?sts=rtf&ord=dt.

countries such as Argentina, Brazil, Mexico, Uruguay and Colombia produce agrofuels mainly from sugarcane and oil palm. Brazil has produced fuel alcohol since 1975; it is the leading producer of sugarcane worldwide and produces 60% of the world total of ethanol from sugar, with three million ha of sugarcane crops. In 2005, production reached a record 16.5 billion liters, two million of which were for export (Jason, 2004).

Among the advantages attributed to agrofuels as an alternative to fossil fuels that they mitigate climate change due to the reduction in gas emissions from the greenhouse effect, bring higher rural incomes for farmers and contribute to greater rural development. In Colombia, moreover, the government (in 2007) considers them an alternative to illicit crops and as a source of employment in rural areas.

Most oil-dependent countries are engaged in the discussion of biofuels today, seeing in them a viable long-term solution to the problem of regional energy insufficiency. On the other hand researchers put forth concerns because they consider large-scale production of monoculture crops for agrofuels—under the conventional/productivist system of production dependent on chemical inputs (pesticides and fertilizer) made using the fossil energy that is sought to be replaced—will have negative impacts.

The concerns are related to accelerated processes of deforestation, destruction of biodiversity, soil erosion and degradation, impacts on water and a negative balance of greenhouse gas emissions. To this situation are added the possible effects of displacement of food crops and increases in food prices, which will directly affect the food security and food sovereignty of local communities, mainly in developing countries. In Mexico, the redirection of maize crops for export to the United States to manufacture ethanol brought on a disproportionate increase in the price of maize, an essential ingredient in the tortilla, which is the main source of food and nutrition for the Mexican population. The increase in food prices is also hitting the livestock and poultry industries (Fearnside, 2001a; Bravo, 2006).

RALLT (2004) cites studies that show that producing one tonne of cereals or vegetables with modern agriculture practices requires six to ten times more energy than using sustainable farming methods. The components of modern industrial agriculture that consume the most energy are the production of nitrogen fertilizers, agricultural machinery and irrigation using pumps. These accounted for more than 90% of the energy used directly or indirectly in agriculture and all are essential to it (RALLT, 2004). In addition, the elimination of carbon-sequestering forests to open the way to these crops will further increase CO_2 emissions (Donald, 2004; Bravo, 2006).

There is also a major debate on the energy balance for making ethanol or biodiesel from some bioenergy crops. The results of the research by David Pimentel and Tad Patzek at Cornell University in the United States (Pimentel and Patzek, 2005) support the notion that the energy balance of all the crops, with current processing methods, is such that more fossil energy is spent to produce biofuels than they provide. Thus, for each unit of energy expended on fossil fuel, the return is 0.778 units of methanol from maize; 0.688 units of ethanol from switchgrass; 0.636 units of ethanol from wood; and, in the worst of the cases examined, 0.534 units of biodiesel from soybean (RALLT, 2004; Bravo, 2006).

1.6.2.3 Food chains

We understand *agrifood chain* to refer to the whole set of different movements in the process of food production that take place before, within and after agricultural production systems, linking all those involved, from the producer of inputs to the end consumer. The concept includes items whose end use is food as well as agricultural output sold to other industries. The set of all the agrifood chains, including support services, constitutes the *agribusiness* (Castro et al., 2001). The predominant model of development in the last 50 years, as already indicated, accorded priority to articulating the production systems and inputs and offered incentives for developing agroexports. The best-articulated agrifood chains in the region are for oil-bearing crops, beef, dairy products and vegetables. The opening up of Latin American markets and the need for the markets of the developed countries to expand has accelerated the economic concentration of the components of agribusiness, especially the supply of inputs and seed and marketing agrifood products, in which the multinational corporations are already the most powerful economic actors, influencing policy decisions that are restructuring agriculture generally, agrifood systems in particular and the process of technological development and technological innovations for the agricultural sector (Friedland et al., 1991; Bonanno et al., 1994; McMichael, 1994).

Although the agricultural inputs sector was already dominated by large corporations before the 1990s, that decade saw a greater rate of concentration in this sector. For example, today only 10 corporations control 84% of pesticide sales in the world. The 10 largest corporations in the seed business control 50% of seed sales worldwide and the 10 largest biotechnology companies control almost 75% of biotechnology sales, including seed for transgenic crops (ETC, 2005).

At the other end of the food chains one finds the processors, distributors and supermarkets. The penetration of transnational corporations in this sector is also proceeding by leaps and bounds in the region, even in rural areas. For example, in Argentina only seven supermarket chains control 77.5% of supermarket sales in 1999 and of these, 80% belonged to multinational chains (Carrefour, Ahold and Wal-Mart, among others). As of that date, only two national chains had survived (Gutman, 2002). In Costa Rica, supermarket chains control 50% of all food sales and the seven largest companies control 98% of supermarket sales (Alvarado and Charmel, 2002). In Chile, four companies (two national and two foreign) control 50% of the market; the milk and dairy products sector is the most heavily dominated by the supermarket chains: the five largest companies account for 80% of sales (Faiguenbaum et al., 2002). The growing control of multinational chains in the sale of foods is taking place throughout the region. As of 2003 supermarket chains controlled from 50 to 60% of all food sales in LAC, an extraordinary increase, considering that just 10 years ago they controlled 10 to 20%. Five corporations control 65% of these sales (Reardon et al., 2003).

This rapid growth and consolidation of supermarkets

has had important consequences for the structure of the markets (Gutman, 2002), for small-scale producers (Ghezán et al., 2002; Gutman, 2002; Reardon and Berdegué, 2002; Schwentesius and Gómez, 2002) and for consumers (Vorley, 2003). In Brazil, as new "retailers" with integrated operations and new rules of participation expand they are displacing small and medium rural enterprises, which were playing an important role generating employment and diversifying the ways one could make a living in the Brazilian countryside (Farina et al., 2004). In addition, the new rules imposed by the supermarkets in Brazil with respect to the beef market have ruined the small butcheries, merchants and truck drivers who were involved in this market before (Farina et al., 2004). In Chile, the growth of the large supermarket chains has taken place at the expense of traditional food outlets. From 1991 to 1995, on average 22% of these traditional outlets disappeared (Faiguenbaum et al., 2002). The same trend has been documented for Argentina, Costa Rica and Mexico (Nielsen, 1999; Alvarado and Charmel, 2002; Gutman, 2002; Schwentesius and Gómez, 2002).

The effect on small-scale producers has been equally devastating. The supermarkets are seeking a limited number of suppliers who can provide them with the volume and quantity of products they need. The supermarkets in LAC purchase 2.5 times more fresh produce (fruits and vegetables) from local producers than those which the region exports to the rest of the world (Reardon and Berdegué, 2002). With the rapid growth of supermarkets and the consolidation of that sector, local producers are increasingly subject to the rules established by a small group of transnational companies. It has been argued that for the fresh fruit and vegetables sector, the growing dominance of supermarkets may have a positive effect on producers and consumers, since the supermarkets demand a higher-quality producer (Belsevich et al., 2003). Nonetheless, these same authors conclude that the general trend is to disfavor the small- and medium-scale producers, who lack the capital and credit needed to accommodate to the new demands of the market. The negative impact on small- and medium-scale producers has been documented for several countries of the region (Alvarado and Charmel, 2002; Ghezán et al., 2002; Gutman, 2002; Schwentesius and Gómez, 2002).

It is argued that on balance the growth of supermarkets has had a positive overall impact for consumers, though there are not many studies on this (Rodríguez et al., 2002). It is assumed that supermarkets are more convenient and provide greater diversity of products along with better-quality products at a lower price. Nonetheless, as supermarket chains consolidate and the competition diminishes, these benefits will deteriorate, as with milk in some regions of the United States.

The debate continues over the impacts of the major concentration of corporations in the food sector. There is also a debate over whether the global dominance of supermarket chains is inevitable and over the possible impacts of standards and direct contracts between supermarkets and producers. Nonetheless, most of the studies in Latin America and the Caribbean indicate that this concentration and dominance in the food sales sector will have negative repercussions for small- and medium-scale producers and eventually for consumers. Although these predictions are still tentative, the evidence for this proposition continues to accumulate.

The transnational corporations continue their process of vertical and horizontal integration and continue penetrating food chains in the region. Throughout the food chain the inequality in power is greatest as between small-scale producers and the transnational corporations. To counter that inequality, some producers have organized in associations to increase their bargaining power over conditions and prices (Berdegué, 2001; Vorley, 2003). Yet Berdegué (2001) argues that these associations can only be beneficial when transaction costs are high, as in the case of milk. But when transaction costs are low, as it is in the case of grains and potatoes, the benefit of producers' associations is called into question. In the context of a globalized economy, this kind of not-very-differentiated product makes all producers worldwide compete with one another for buyers. The development of cooperatives in the context of globalization and borders open to capital poses a major challenge to small-scale producers, since transnational agribusinesses can buy their produce practically anywhere in the world.

The concentration and consolidation of these agribusiness chains have increased the gap between the prices received by food producers and the prices paid by consumers (Vorley, 2003; see Box 1-10 on soybean in Brazil). These impacts have repercussions throughout society, both rural and urban and have effects beyond the economic effects related to the displacement of small-scale producers, job losses and consumers' ability to buy food. Food is one of the pillars of any culture; how it is produced, distributed, prepared and shared with family and friends is part of what defines a culture and that pillar is quickly eroding with the expansion and concentration of transnational supermarket chains.

This imbalance in power has led the global organization, Vía Campesina, to begin a campaign to remove agriculture from the WTO based on the argument that food is different (Rosset, 2006). Consumers are playing an important role by demanding fair trade products, although they still represent an insignificant percentage of food purchases in the world. Another recent development is self-regulation in the corporate sector. Some corporations, in search of a competitive margin over their competitors, are beginning self-regulation programs with respect to social responsibility. Nonetheless, despite all the publicity, very few corporations have adopted the social responsibility agenda (Oxfam, 2004). Finally, another possible way to control the impacts of the concentration of markets is to attack it directly. Considering the rapidity with which the concentration of capital is taking place, monitoring the transnational corporations should be an urgent task (Vorley, 2003). Part of this work was done by the now-defunct United Nations Center on Transnational Corporations. In addition, the civil society sector is working on this through organizations such as Corporate Watch. Vorley (2003) argues that economic globalization makes it necessary to improve global governance on matters of monopoly and competition. Nowadays, there are no international standards for competition to regulate the activities of corporations from one continent to another. The law on competition within the WTO moves away from regulating monopolies, towards simplifying regulations across national borders to facilitate transnational

Box 1-10. Integration of the soybean food chain in Latin America: From the producers to the consumers

Only a small fraction of the soybean is consumed directly as food for humans; the rest is processed mainly to produce oil for the food industry and as high-protein tablets for animal feed.

In Brazil, it is estimated that the soybean crop employs one million persons directly and that the soybean industrial complex employs some five million people.

In the 1980s soybean production shifted from the south and southeastern regions, with small and medium producers (average 30 hectares) to the region of Mato Grosso and Goiás, including the cerrado region, with an average farm size of 1,000 hectares.

A single company, Andre Maggi, has 150,000 hectares and produces one million tons of soybean per year. The consequence of this concentration in farm size has led to an increase in rural unemployment and food insecurity, spurring migration to the cities.

The soybean market is characterized by a high degree of integration, as large corporations control the production, processing, and marketing, in both exporting and importing countries.

The four corporations that dominate soybean market, Bunge, ADM, Cargill, and Dreyfus, also process soybeans. Cargill claims to be the largest company worldwide engaged in the extraction of soybean oil. Cargill is also the largest exporter of vegetable oil and soy protein in Argentina. Dreyfus is the third leading company in terms of volume that processes vegetable oil in South America, and is the owner of and operates the giant port on the Paraná river and the giant company General Lagos crushing plant.

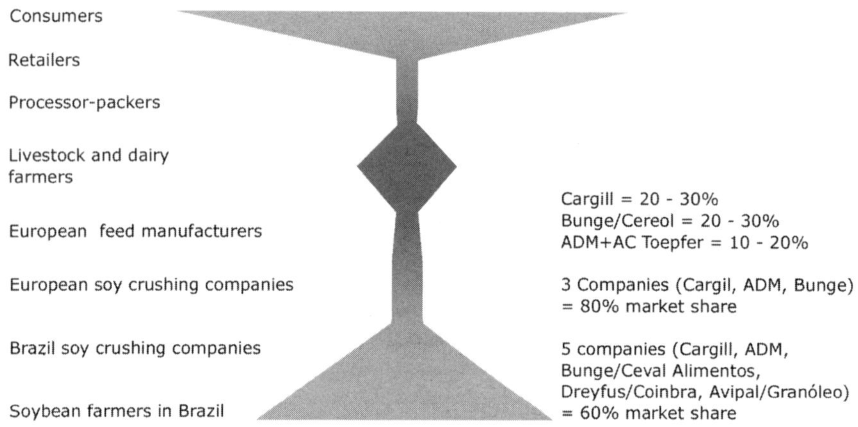

Soybean feed "Bottleneck" from Brazil to Europe. Source: Vorley, 2003

trade and access to the industrialized countries' markets for goods and services (Vorley, 2003).

1.6.2.4 Sociocultural characteristics

The agricultural sector in Latin America and the Caribbean is made up of different systems of production (traditional/indigenous, conventional/productivist and agroecological) that differ markedly from one another, depending, among other things, on working capital, quantity of assets, type of land tenure, source of income, use of labor, destination of production and especially their sociocultural characteristics. Indeed, each system is highly varied given the plurality of agricultural structures in the region. This is why, in general, family farming is marked by a wide social heterogeneity; nonetheless, it also has some characteristic sociocultural elements that distinguish it from commercial agriculture (Ahumada, 1996). For example, in family agriculture, the family lives on its farm, is at the core of all the activity and makes the decisions in the productive system and how its production is geared to meeting the needs of the family and the market; it is producer and consumer. In addition, the family is the source of labor for itself and for third persons.

There are other sociocultural aspects that determine differences within this productive system and set it further apart from commercial agriculture. The family develops socially and economically in a milieu marked by geographic isolation distinct from the urban-industrial sector. Many of its members have a common socio-historical development and families share traditions and customs that are determinant in their lives in terms of relationships and production. In this sociocultural setting tradition is the dominant institution in relationships and exchanges. In that rural milieu there is a close relationship between the degree of isolation and traditional patterns.

These aspects define more family farming of the peasant and indigenous type, where the peasants constitute a subculture, but this peasant pattern in countries such as Chile, Brazil, Argentina and Uruguay differs from that of other regions of Latin America (Peru, Guatemala, Mexico and Bolivia, among others), in which the indigenous cultural characteristic is even more determinant, in some cases to the point of having their own cultural traits (Rojas, 1986).

Another fundamental element that identifies this system socioculturally is belonging to a local community in which the networks of interpersonal relationships are essential not only for the economic strategies of the households and their members, but also for other crucial aspects of human life, such as friendship, religion, leisure and sense of belonging. The members of a peasant or indigenous community share their own sociocultural system in which beliefs and norms complement institutional and social relationships and vice versa (Durston, 2002).

In addition, in the micro, regional and national social system, the peasant occupies one of the bottom rungs on the social scale and therefore is subject to economic exploitation and social and political exclusion by the more powerful groups (Wolf, 1971), phenomena that are generally more intense when the peasants belong to ethnic groups with a history of domination by others. Moreover, peasant families have been diversifying their sources of subsistence, since scarcity of land, economic crisis and neoliberal policies have led to a situation in which this sector can no longer support itself solely from agricultural production. The response has been to seek employment off the farm (both men and women) and to migrate to the cities or industrialized countries (Deere, 2005), disarticulating rural communities and eroding the sociocultural cohesion of the rural milieu.

When subsistence family-based agriculture directs its production basically to the market, uses wage labor, has some degree of productive specialization and has assets and capacities that give it some potential for accumulation, it assumes a position of transition to commercial forms. In this transition, externally strong pressures are brought to bear that alter its traditional economic and sociocultural foundations. In this transition, some changes take place in family life, some members of the family no longer participate in the productive activity, but instead dedicate themselves to studying or working in other independent activities, there is a greater link to the urban culture and gradually the traditional rural way of life is lost (Acosta and Rodríguez Fazzone, 2005).

In contrast, the commercial agricultural system considers only the landowner as the agricultural entrepreneur and his function is primarily to organize the productive process and connect the property to the markets for inputs, financing, goods and labor. In addition, the producer and his family do not necessarily live on the property; most of their social and cultural activities are tied to the urban milieu; the enterprise uses, as the main labor force, temporary and/or permanent labor; the size of the property is an important factor behind large productive surpluses; it uses a large amount of technology; and production is for market. The further it is from the characteristics of the family agricultural system, such a system is considered more modern and commercial and less traditional (Gómez, 2000).

1.6.2.5 Knowledge

A retrospective evaluation and analysis of the current situation of the role of agricultural knowledge, science and technology in the sustainable development of Latin America and the Caribbean must acknowledge that there is a wealth in the region beyond scientific knowledge as such. One must, therefore, reconstruct the historical-cultural diversity and diversity of ways of knowing in the region and their influences on the evolution of science, as a preamble to an approach to the role, for example, of colonialism and neocolonialism, ethnicity and the ignored racial and cultural complications of the region, vis-à-vis the new and imposing paradigms such as globalization or global interdependence. In this context, it is evident that the region is broken into complexities, different bodies, memories, languages, histories, diversities and world views (Leff and Carabias, 1993; Possey, 1999; Maffi, 2001; Toledo, 2001, 2003; Toledo et al., 2001). This fragmentation, from a less uniform perspective, is considered in contrast to the assumption of a region seen from a reductionist perspective as a homogeneous mass and that advances on a symmetric front towards one or another scenario.

Recognizing the importance of historical-cultural diversity for the purposes of gauging the role of knowledge, science and technology in the development policies of the region will enable us to vindicate and value aspects such as the experience of colonialism as a present and preponderant reality in Latin America. Colonialism in its diversity of nature and time intrinsically exists in the region, not only as a territorial phenomenon, imposed and invasive, but also as a legacy, reflected in a colonial and neocolonial attitude that predominates in many Latin American countries. This colonial mentality is one of the reasons why Latin America invests less than the world average today in research and development and does not value the rich traditional/indigenous and local knowledge.

Colonialism has to date resulted in the suppression of local knowledge and wisdom for almost half a century and its legacy permeates the AKST system, restricting its creative and proactive use. The dominant AKST system has operated under the premise that scientific and technological spillover is the instrument that is going to best position the region and offer comparative advantages in today's interdependent world. Yet on the other hand, Amartya Sen (2004, 2006) suggests the contrary effect of a colonial mentality rejecting western ideas. Sen argues that rejecting the globalization of ideas and practices because of the supposed threat of westernization is a mistaken approach that has played a regressive role in the colonial and neocolonial world. As he sees it, this rejection fosters parochial trends which, given global interactions, is not only counterproductive, but can cause non-western societies to place limits on themselves and may even torpedo the valuable resources that their own cultures and wisdom represent. It should be noted that for the indigenous peoples globalization (understood as the Euro-American colonial expansion and domination) is not new. Several studies by Lumbreras (1991), Grillo (1998), Lander (2000) and Quijano (2000) illustrate how the indigenous peoples of LAC engaged in a dialogue with the colonial world.

Less in the realm of philosophy and more in that of epistemology, one can argue that LAC, even though it is a region with extraordinary resources in terms of world views, knowledge, wisdom and cultures has not taken advantage of the synergies that could be derived from the interaction between scientific knowledge and traditional/local knowledge and wisdom. This challenge puts forth, to the AKST system, another type of paradigm, as an alternative to the current dominant one, in addition to considering other structural (for example, land tenure), cultural and intercultural factors.

In terms of exclusively scientific knowledge, Latin America and the Caribbean is the region that invests the least in research and development in relation to the rest of the world. In the agricultural sector, the region invests only 0.3% of gross domestic product, whereas the rest of the world invests 0.5%. The countries that invest most in research and development in the region (Argentina, Mexico, Costa Rica, Brazil and Chile) do so at levels very far below the developing countries that are prototypes in terms of returns on research and development, such as China, India, Korea, South Africa, Singapore and Israel, among others.

Knowledge, culture and agricultural development. In LAC, the "other ecologies" (Toledo and Castillo, 1999) and their respective systems of agricultural knowledge are as diverse as the rich and diverse cultures of the region (Deruyttere, 1997; Altieri, 1999). For example, the indigenous population is made up of more than 400 ethnic groups (Deruyttere, 1997), or 800 cultural groups (Toledo, 2007).

In general, agricultural knowledge in the region is associated with the three types of agricultural production systems described in this document: the conventional/productivist system, the agroecological system and the traditional/indigenous system (including peasant agriculture). Historically, indigenous forms of agriculture (hunting, fishing, gathering, domestication and cultivation of plants and animals) not only precede the other two, but are the result of an intimate and sophisticated interaction and co-evolution with nature in general and in particular with a significant number of plants and animals (Fowler and Mooney, 1990; Valladolid 1998, 2001; Altieri, 1999; Barkin, 2005; Narby, 2007). These interactions gave way to what today are known as centers of origin of native crops (Diversity, 1991). Traditional/indigenous knowledge is very valuable for the people of the region for three reasons: First, it contributes to the cultural affirmation of the indigenous people and is useful for learning about nature and its resources, including sources of food, medicines, forage, building materials and tools, among other things (Toledo, 2005). For example, the Tzeltal of Mexico can recognize more than 1,200 plant species, whereas the P'urepecha recognize more than 900 species and the Maya of the Yucatan approximately 500 species (Toledo et al., 1985). Second, this knowledge results from the experience accumulated and shared by many men and women over thousands of years. And third, knowledge is also wisdom, as it is closely linked to the identity, values, beliefs, traditions and ideals of individuals and communities. Nonetheless, it is also important to recognize that traditional knowledge and local knowledge have weaknesses. For example, often this knowledge and wisdom is not found in books and may be lost if not transmitted from generation to generation. Traditional knowledge is also limited to a locality or region and is not easily transferable to other regions with different conditions. Finally, many natural phenomena cannot be perceived through feelings without the help of technologies, for example, microorganisms, biochemical processes and the DNA molecule (Toledo, 2005). Moreover, from the standpoint of indigenous experience, traditional/indigenous knowledge and wisdom are not necessarily limited by what one can see, hear, touch or feel. For example, anthropologist Jeremy Narby (2007) notes that a good part of the extraordinary knowledge of Amazonian plant life comes through supra-conscious/extrasensory states during ceremonies and rituals, such as those performed by the shamans of the Amazonian indigenous peoples. In his view, a process of affirmation, cultural regeneration and intercultural exchange could help recover the potential of combining the physical and the metaphysical (Narby and Huxley, 2004; Narby, 2007).

Colonial and neocolonial agriculture in the region is based on (1) the exploitation of the plants, animals, peoples and indigenous knowledge and wisdom native to the region, (2) the usurpation and violent or non-violent expropriation of lands and territories that belonged to the hundreds of indigenous peoples and (3) the exclusion of the local peasant-indigenous and agroecological knowledge and AKST systems (Crosby, 1991, 2004; Lumbreras, 1991). One might suggest that parallel to the growth of modern homogenizing agriculture, peasant-indigenous and local forms of agriculture have tended to diminish. This is summarized, for the

Table 1-10. **The reduction/disappearance of the home place: Area under the control of indigenous people of Mexico and Central America.**

Country	National area (Has)	Area under indigenous people control	
		(Ha)	%
México	195,820,000	29,399,430	15
Guatemala	10,899,000	N/A	N/A
Belice	2,296,550	N/A	N/A
Honduras	11,209,000	16,181	14
El Salvador	2,104,100	Not studied	Not studied
Nicaragua	13,000,000	5,900,000	45.3
Costa Rica	5,110,000	320,321	6.2
Panama	7,551,700	1,657,100	2.2

Source: Toledo et al., 2001.

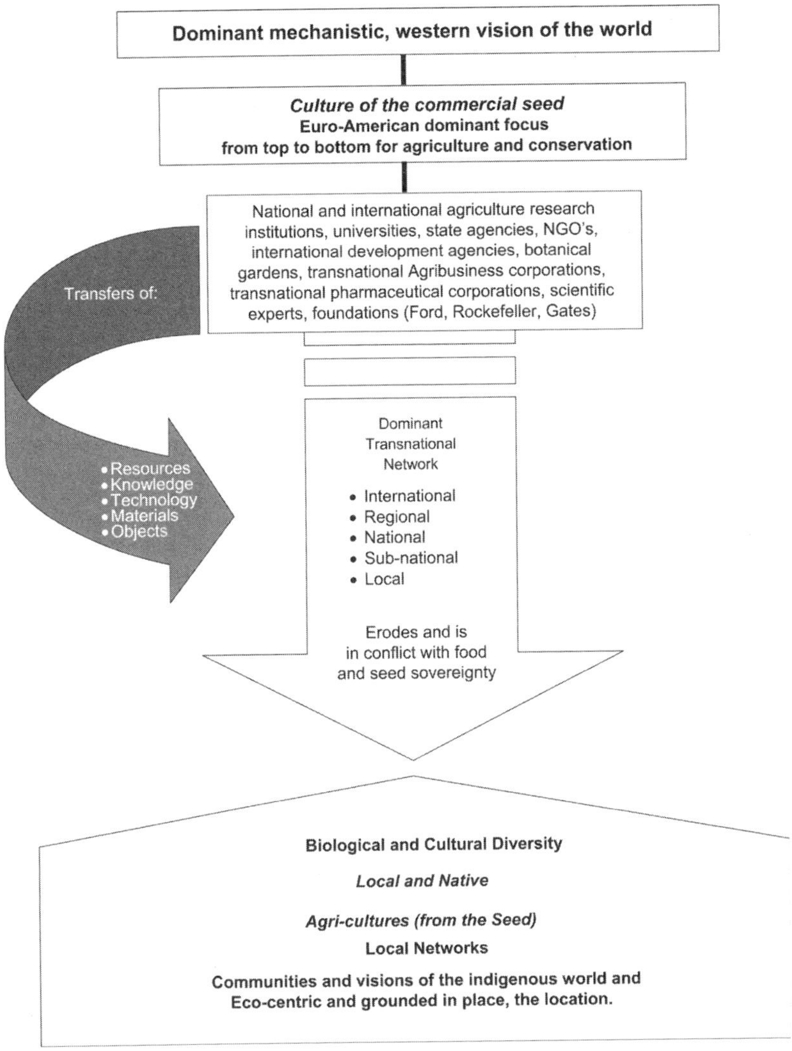

Figure 1-10. *The dominant productivist/conventional vision for agriculture and conservation, from top to bottom.* Source: Elaborated by T. Gonzales based on Pimbert, 1994; Gonzales, 1996, 1999, 2009; Escobar, 1998b, 1999

region, in the growth of the space in the face of the reduction of place, i.e., of the local world (see Table 1-10). "Place"—which is where the local, peasant-indigenous languages, cultures, rituals, knowledge and wisdom and AKST systems are, with all of life, for the last 60 years, in particular—has been eroding significantly due to the policies that accord priority to the growth of the homogenizing space related to modern single-crop agriculture (Blazer, 2004; Gonzales, 2009).

In the last 60 years modern agriculture and as a result the system of agricultural education, research and extension work was strongly emphasized by agricultural development policies. This conventional/productivist agriculture is based on the mechanistic scientific outlook that arose in western Europe (Figure 1-10). Eurocentrism,[7] in formal education generally and in agricultural education in particular, has contributed crucially to the dissemination and growing dominance of the mechanistic outlook (Rengifo, 1998; Bowers, 2002). Basic scientific knowledge on and for manipulating agriculture has been and is being generated

[7] Eurocentrism "is the imaginative and institutional context that informs contemporary scholarship, opinion and law. As a theory, it postulates the superiority of Europeans over non-Europeans. It is constructed on a set of assumptions and beliefs generally accepted without prejudices by educated Europeans and North Americans who commonly accept them as the truth, as supported by "the facts" or as "reality." A key concept behind Eurocentrism is the idea of diffusionism. Diffusionism is based on two assumptions: (1) most communities are hardly inventive and (2) a few human communities (or places, or cultures) are inventive and are, accordingly, the permanent centers of cultural change or "progress." On a global scale, this results in a world with a single center—Europe—and a periphery that surrounds it" (Battiste and Youngblood Henderson, 2000). For further thoughts, see Lander (2000), Quijano (2000).

mainly at the dominant centers that generate knowledge (international/regional research centers/institutes, universities) around the world. These centers have embraced and worked to sustain and promote the mechanistic models, theories, paradigms and world view associated with the reductionist system of conventional/productivist agricultural research and production (De Souza Silva et al., 2005). This world view and corresponding paradigms are still a key component of a transnational network made up of academic centers (Bowers, 2002; Smith, 2002; Progler, 2005; Pimbert, 2006), representatives of governments, think tanks, the business sector, international organizations and development financing agencies (Escobar, 1999; Gonzales, 2007) (Figure 1-10).

The political leadership, policy makers and civil society generally have also been permeated by the knowledge produced by the mechanistic western paradigm/world view and have become its practitioners.

A well-articulated and well-financed transnational network of scientific institutions has generated, fed into and provided feedback to the conventional/productivist system for the production of agricultural knowledge. The environmental and sustainability problems associated with the system are derived from this reductionist knowledge base. (Figure 1-11).

The agro-industrial project that emerges from the dominant AKST system proposes that the indigenous/peasant communities should modernize and progress by means of technology, machines and scientific knowledge, as well as by entering the market. This agro-industrial proposal seeks to have the agroecosystem simplified and specialized to increase labor efficiency (Toledo, 2005).

Agroecology proposes modernization by way of path different from that of agroindustry. It proposes a form of development based on respect for the environment (the Mother Earth, for the indigenous peoples), as well as the traditions, culture and history of the people. The agroecological proposal recognizes the need for scientific and technological research, yet unlike the agroindustrial proposal, it suggests a dialogue of different ways of knowing based on a respectful exchange among the researchers or technical personnel and the peasant and indigenous communities (Toledo, 2005). Ishizawa (2006) and Machaca (1996, 1998) propose a dialogue of ways of knowing from a perspective of cultural affirmation and decolonization, while at the same time suggesting the challenge posed by the world views for the dialogue.

The dominant society in general and the dominant policies and AKST system in particular, have contributed to the marginalization or exclusion of the cultures, world views, systems of knowledge, and ways of knowing and being linked to the peasant-indigenous and agroecological production systems. Several studies conclude that these two systems have a potential that has yet to be tapped or fully recognized (Altieri, 1987, 1996; Chambi and Chambi, 1995; Machaca 1996, 1998; Rosset, 1999; Toledo, 2005), or integrated to the region's AKST system. Nonetheless, agricultural movements proposing alternatives to conventional/productivist agriculture and/or decolonization and cultural affirmation suggest the potential of such alternative ways of knowing and AKST systems for making a significant contribution to attaining sustainable development objectives (Altieri, 1987, 1996; Grillo, 1998; Rengifo, 1998; Valladolid, 1998, 2001; Delgado and Ponce, 1999; Huizer, 1999; Rist et al., 1999; Toledo, 2001, 2003; Funes et al., 2001; Toledo et al., 2001; Barkin, 2005; Ishizawa, 2006; Badgley et al., 2007). This situation creates an opportunity in the region for a new, inclusive AKST policy, one which incorporates, on its own terms, the peasant-indigenous and agroecological systems of knowledge and wisdom (Leff and Carabias, 1993).

1.6.2.6 Gender aspects

The main trends associated with the neoliberal restructuring and the increase in rural poverty in LAC include greater participation of women in agriculture, both as producers and as wage workers in the agricultural sector (Deere, 2005). As the participation of men in agriculture diminishes, the role of women in agricultural production increases. Male migration is one of the main motives for the increase in women's participation in the rural economy. The expansion of non-traditional export crops, wars, violence and forced displacements are other causes of the so-called "feminization of agriculture," and with it, the feminization of poverty.

The increase in women's participation in wage labor in the agricultural sector is closely related to the expansion of non-traditional export crops (Robles, 2000; Chant and Craske, 2003; Deere, 2005). In particular, women play a predominant role in labor activities such as packing flowers (e.g., in Mexico, Ecuador and Colombia), fruits (e.g., in Mexico, Argentina, Brazil and Chile) and fresh vegetables (e.g., in Mexico, Guatemala and Brazil) for export to North America (Deere, 2005). In addition, a large proportion of women and their children (50%) provide labor in the fields where these crops are produced (Deere, 2005). The flower sector has the largest percentage of female workers of the non-traditional crops. In Mexico and Colombia it is estimated that 60 to 80% of the labor force in this sector is made up of women (Lara, 1992; Becerril, 1995; Meier, 1999). This work is mostly seasonal, lacks security and is marked by precarious working conditions and discrimination (Lara, 1995, 1998; Barndt, 2002). There is also persistent income inequality between male and female workers, as well as between white workers and those belonging to other ethnic minorities. The increase in the use of women as wage workers in agriculture is not a uniform trend throughout the region and is very much associated with non-traditional export crops. Several studies on the participation of women in wage labor show that in many countries of the region a much higher proportion of women work in the non-agricultural sector, such as in the maquiladoras, as domestic servants and in the industrial sector (Reardon et al., 2001; Katz, 2003). For example, in the Dominican Republic and Panama, 92% of economically active rural women work in the non-agricultural sector (Katz, 2003).

The literature includes a debate over whether this type of work represents greater exploitation of female labor or, to the contrary, is potentially liberating for women. In relation to this debate, Safa (1995) emphasizes the complexity and at times contradiction in the relationship between wage labor (and the discrimination, exploitation and precarious working conditions this often represents) and greater access to and control of the salary, greater purchasing power,

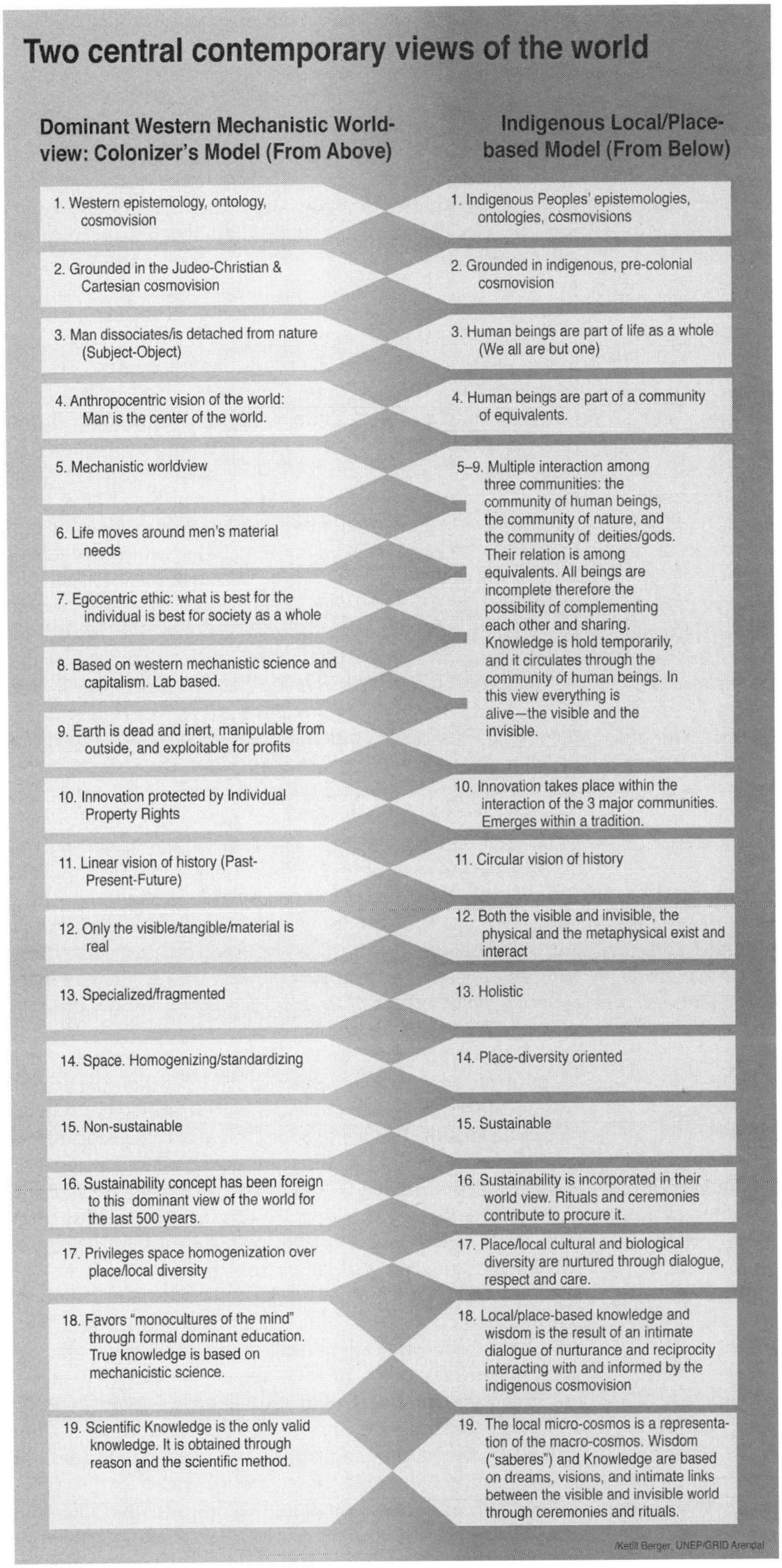

Figure 1-11. *Two contemporary views of the world.* Source: Gonzales, 1999; Gonzales et al., 1999.

changes in gender relations (which tend to favor women) and greater awareness of women's subordination to men.

The other important trend in LAC, especially in the indigenous/peasant sector of agriculture, is the incorporation of women as the main producer (Preibisch et al., 2002). This "feminization of agriculture" is occurring in some countries more than in others and is directly tied to the increase in the migration of men, the search for jobs off the farm and the diminishing viability of traditional/peasant agriculture under neoliberalism (Chiriboga et al., 1996; Preibisch et al., 2002). For example, Mexico, the country with the largest migration of men to the United States, is also one of those in which the feminization of agriculture is most evident (Robles, 2000).

The incorporation of the indigenous/peasant sector in the production of non-traditional export crops has also resulted from an intensification of the role of women in agriculture (Deere, 2005). Guatemala and Chile are the two countries where this incorporation was most successful, even if it was ephemeral (Murray, 2003). Here too there is a debate on the impact of that greater participation on women. On the one hand, studies by Dary (1991) and Blumberg (1994) conclude that the incorporation of peasant women into the production of agroexport crops had a negative impact on women because it reduced the time available for their own independent activities, reduced their power to bargain within the family and increased their dependence on men. On the other hand, the studies by Katz (2003), Hamilton et al. (2001) and Hamilton and Fischer (2003) conclude that women (in Guatemala) gained more decision-making power over productive activities.

Whether as wage workers in the agricultural sector or as producers directly, there is no doubt but that the role of women in agriculture in LAC has been expanding. This feminization of agriculture is linked to the decline of agriculture as the main economic activity of peasant families and to the greater absence of men due to migration or wage work away from the farm. As traditional agriculture becomes ever less viable, production is turning more to food security for the family and women are taking on a more important role (Deere, 2005).

1.7 Performance of Production Systems

This subchapter presents an evaluation of the three main systems of production in the region: traditional/indigenous, conventional/productivist and agroecological. This evaluation includes an assessment of the performance of these systems in terms of several indicators, such as productivity, sustainability and quality of food. In addition, this subchapter includes an assessment of the environmental, social, economic and health impacts of the three systems.

1.7.1 Productivity

Productivity is defined as an average quantity of output divided by a measure of the quantity of input. The economic concept of agricultural productivity is an evaluation of the production of a crop (i.e., yield) and its market value, so that one can estimate its profitability (i.e., profit). Agricultural economists often use a partial measure of productivity based on an area of land and/or labor. Nonetheless, for many farmers in Latin America and the Caribbean, especially those who produce for family consumption, or those who have systems using low levels of external inputs, the concept of productivity is much broader. For these producers, a productive farm is that which provides the largest amount of resources needed for the survival of the producer and his or her family. This may include foods, fuel, fiber and medicinal plants, among others. Unfortunately, there are very few studies that consider these factors; most existing statistics report only productivity per unit of land and per unit of labor.

Traditional/indigenous system. What is frequently known as *agricultura campesina* or peasant agriculture and which in this evaluation we call the traditional/indigenous system, consists of several traditional systems that predominate in many rural areas of Latin America and the Caribbean (Ortega 1986), but which are being threatened by neoliberal policies (Davis et al., 2001; Deere, 2005) among other factors. These systems, in their traditional form, have been refined over many generations and much accumulated knowledge. The marginalization and displacement of producers from their ancestral lands contributes to their being characterized as having low or moderate productivity. Nonetheless, there are traditional systems that have high productivity, in some cases higher than the conventional/productivist system (Altieri, 1999). For example, in the 1950s Sanders (1957) estimated that maize production in the *chinampas*, a traditional system in Mexico, yielded 3.5 to 6.3 tonnes per hectare. That same year, the yield of maize in the United States was 2.6 tonnes per hectare and it was not until 1965 that it reached 4 tonnes per ha (USDA, 1972, as cited in Altieri, 1999). In the 1990s the average yield of maize in LAC was only 2.56 tonnes per hectare and the countries with the highest yields were Argentina and Chile, with 4.35 and 8.49 tonnes per hectare respectively (Morris and López-Pereira, 1999). In the Amazon, traditional systems such as that of the Kayapó have yields that surpass *colonos*' yields by 200% and the yields of livestock production by 175% (Hecht, 1984).

One characteristic of the traditional systems is their high agrobiodiversity (Toledo, 2007). Multicrop systems and agroforestry systems are common in this type of agriculture (Clauson, 1985; Thrupp, 1998). In LAC, most of the subsistence crops are produced in multicrop situations. For example, it is estimated that 40% of the cassava, 60% of the maize and 80% of the beans are produced in combination with other crops (Francis, 1986). This is an important factor when comparing yields because these comparisons are normally by crop, which means that often the yield of other crops grown on the same plot is not taken into account. The multicrop systems developed by traditional and/or indigenous producers are 20 to 60% more productive (in terms of harvestable product) than monoculture systems (Beets, 1982). For example, in Mexico, 1.7 ha planted in maize in monoculture is needed to produce the same amount of food as one hectare planted in maize, squash and bean produces (Gliessman, 1998). In Brazil, multicrops of maize and bean have a 28% advantage over monocultures; under more arid conditions the multicrops of sorghum and cowpea produce 25 to 58% more than the monocultures (Altieri, 1999). The literature that shows the advantages of multicrops on yield is substantial and dates back to the 1970s (Trenbath, 1976; Beets, 1982; Francis, 1986; Vandermeer, 1989). Among the facts that have been identified as responsible for these advantages are the more efficient use of resources (nutrients and water) and the reduction in the incidence of pests and weeds (Vandermeer, 1989; Gliessman, 1998). The greatest advantages of multicropping are obtained when gramineous and leguminous species are combined, as these two plant groups tend to complement one another very well (Vandermeer 1989). Other combinations may not be as advantageous from the standpoint of yields (Vandermeer, 1989).

Producers who practice multicropping have multiple survival strategies and combine subsistence agriculture with commercial activities and wage labor (Ewell and Merrill-Sands, 1987; Barrera-Bassols and Toledo, 2005; Deere, 2005). Despite the trends towards intensification of agriculture in LAC, traditional/indigenous agriculture is still practiced by millions of producers. As of 1980 such systems of production were found in 16 million productive units and used 160 million ha, involving some 75 million people, i.e., almost two-thirds of the rural population of LAC (Ortega, 1986). In the 1980s this sector produced 41% of the food for domestic consumption and was responsible for producing 51% of the maize, 77% of the beans and 61% of the potatoes (Posner and McPherson, 1982; Altieri, 1993). Due to neoliberal policies, this sector has been weakened and it is possible that today it accounts for a lower percentage of domestic food production (David et al., 2001).

The traditional/indigenous system is also characterized by favorable rates of output per unit of energy input. For example in slash-and-burn systems (swidden agriculture), which depend on manual labor in the mountains of Mexico, estimated yields were 1,940 kg per ha, with a rate of energy efficiency (unit output per unit input) of 10:1 (Pimentel and Pimentel, 1979; Altieri, 1999). In Guatemala a similar system generated a rate of energy efficiency of 4.8:1 and when one adds fertilizer and pesticides, the yields increase (from 5 to 7 tonnes per ha), but energy efficiency drops to less than 2.5:1 (Altieri, 1999) (Figure 1-12).

Conventional/productivist agriculture. The emphasis of the conventional/productivist system has been on maximizing productivity and profit. In this regard, there is no doubt but that the conventional/productivist system has been a success for those producers who have enough capital to implement it. This system has been extending throughout the region, as the AKST system has assigned it high priority. For example, the hybrid varieties of maize developed by CIMMYT in Mexico were planted on 10.6 million ha, accounting for more than 36% of the total area planted in maize throughout the region; more than 74% was planted with some hybrid variety (Morris and López-Pereira, 1999). It's hard to know how much of this was produced under the conventional/productivist system, since many producers, who use the traditional system, also incorporate hybrid varieties.

The main objective of the Green Revolution was to increase the yields of the main food crops per unit of area. Contrary to the perception that the Green Revolution brought about a sharp increase in yields in the late 1960s, Evenson and Gollin (2003) argue that the Green Revolution has taken place in the long run, through the successive development of improved varieties. These authors divide the Green Revolution into two stages, early (1961-1980) and late (1981-2000) and argue that in the developing countries, including LAC, improved varieties contribute to a 17% increase in yields, while in a later period these varieties contributed to 50% of the increase in yields. Notwithstanding these figures, the rate of increase in yield has been diminishing in the last 10 years (Evenson and Gollin, 2003). The advocates of biotechnology argue that the only way to continue to increase yields is by the use of transgenic crops, which have been called "the new Green Revolution"

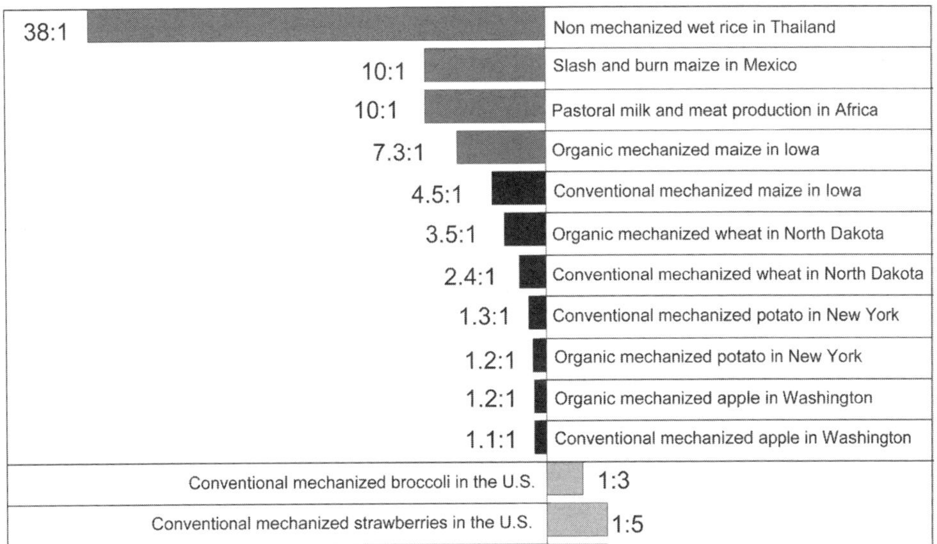

Figure 1-12. *Energetic efficiency of different production systems.* Source: Authors' elaboration based on data from: Atkins, 1979; Pimentel, 1980; Pimentel et al., 1983; Reganold, 2001.

(Smil, 2000; Trewavas, 2002). By way of contrast, the critics of conventional/productivist agriculture argue that it is possible to attain levels of production equal to those of conventional agriculture and in some cases higher, using agroecological practices and without transgenics (Pretty, 2002; Halberg et al., 2005; Badgley et al., 2007).

Based strictly on measures of yield (production per unit of area of a single crop), many economists and agronomists conclude that the conventional/productivist system has greater productivity. Nonetheless, many small-scale producers practice multicropping. Peter Rosset (1999), analyzing data from several countries, concluded that the small properties almost always produce more per unit of area than large ones. Indeed, "the inverse relationship between farm size and productivity," is widely accepted by agricultural economists, though there is a major debate over the causal mechanism (Yotopoulos and Lau, 1971; Bardhan, 1973; Sen, 1975; Berry and Cline, 1979; De Janvry, 1981; Carter, 1984; Feder, 1984; Assunção and Ghatak, 2003). The conventional/productivist system is less energy efficient than the traditional/indigenous systems and in most cases than agroecological/organic systems (Figure 1-12).

Agroecological system. This type of agriculture encompasses a wide array of systems, practices and methods that use agroecological principles to design and manage production systems. For the purpose of this evaluation we are including organic systems. Nonetheless, most agroecologists argue that organic systems are not necessarily agroecological. For example, the production of organic bananas in some parts of Central America and Ecuador, which consists of large expanses of monoculture and which are certified organic are not agroecological systems. Many small producers in LAC are adopting agroecological practices, but either because their production is not for the market or due to lack of resources to pay the certifying authorities, do not certify their production. In the last 20 years the agroecology movement has grown enormously worldwide and particularly in LAC. A recent study reports 286 projects with agroecological interventions that include 12.6 million producers on approximately 37 million ha, or the equivalent of 3% of the land in non-industrialized countries (Pretty et al., 2006). IFOAM estimates that almost 20% of all land and 28% of all farms with organic certification worldwide are in LAC (Willer and Jussefi, 2007) (Box 1-11), though this is largely due to extensive organic livestock systems, especially in Argentina, which has three million ha certified organic. Mexico is the country with the largest number of organic farms in the world, with more than 85,000 farms in organic management. It is estimated that in LAC there are some 5.8 million ha certified organic, with an annual value of US$100 million (Lernoud, 2007). Cuba is the only country in the world that is carrying out a massive conversion to organic agriculture, through the promotion of agroecological practices in both rural and urban areas (Box 1-12). In contrast to the other countries in LAC where organic production is for the export market, in Cuba organic production, with some exceptions, is not certified and is for domestic consumption.

It is frequently stated that organic agriculture, because of its lower yields, will not be able to supply enough food to feed the world. To address this question a study from the University of Michigan compiled results from almost 300 studies worldwide comparing yields of organic and conventional systems (Badgley et al., 2007). Based on the evidence the authors concluded that organic agriculture could produce enough food, on a per capita basis, to provide 2,640 to 4,380 kilocalories per person per day depending on the model used. They also found that in developing countries, where organic systems were compared to the commonly practiced agriculture, organic farms outperformed conventional practices by 57%, demonstrating that intensification using organic methods is possible.

Another study, by the University of Essex in England, carried out a census of 286 projects in 57 countries, including 45 in Latin America and the Caribbean (Pretty et al., 2003, 2006). When the yields on farmland using agroeco-

Box 1-11. Trends in organic agriculture in Latin America and the Caribbean

Organic agriculture has seen enormous growth in the last 10 years in Latin America and the Caribbean, geared mainly to the export market and focused on just a few crops, mainly coffee and bananas in Central America and the Andean region, sugar in Paraguay, and cereal grains and meat in Argentina and Uruguay. Other products are certified at low levels, such as fruits, vegetables, aromatic and medicinal herbs, and apiculture. Today there are 5.8 million hectares certified organic, and almost all the countries of the region have an organic sector, though the development of this sector has been mixed. The countries with the largest areas certified are Argentina (54%), Brazil (15%), Uruguay (13%), Bolivia (6%), and Mexico (5%). The largest share of the almost 3.9 million hectares certified in Argentina and Uruguay are lands used for extensive grazing.

Areas in organic production in Latin America and the Caribbean

In general the organic movement in LAC has grown by its own efforts and with very little government support. With the exception of Cuba, no government provides direct subsidies or economic aid for organic production. Nonetheless, in some countries the state is supporting the organic sector in several ways. For example:

- Brazil: The government announced the interagency Plan Pro Orgánico, providing incentives for research on organic production, forming associations, and stimulating the market for organic products.
- Costa Rica: Government funds for research and teaching in organic production.
- Argentina and Chile: The government export agencies support the organic producers' participation in international shows and print catalogs of organic products.
- Mexico: There is growing interest on the part of government agencies.

Box 1-12. Sustainable agriculture and food security in Cuba: Lessons for the rest of Latin America and the Caribbean

In 1989-1990, the collapse of trade relations between Cuba and the Soviet bloc plunged this small Caribbean nation into an economic and food crisis. Today Cuba has succeeded in overcoming that crisis and its experience illustrates that it is possible to feed a nation with a model based on small and medium producers, and ecological technology with low external inputs.

The Cuban agricultural system was based on the conventional/productivist model of agriculture, highly dependent on external inputs (chemical fertilizers, pesticides, oil, machinery, etc.), as well as large and inefficient state farms. With the change in the favorable terms of trade Cuba had enjoyed with the other socialist countries, there was an almost immediate 53 percent reduction in oil imports, a 50 percent reduction in imports of wheat and other cereal grains for human and animal consumption, and an 80 percent reduction in fertilizer and pesticide imports.

Suddenly, a country with high levels of inputs in its agricultural sector was submerged in a food crisis. It is estimated that as of the early 1990s, the daily average consumption of calories and proteins of the Cuban population had fallen to levels 30 percent below those of the 1980s.

Fortunately, for years Cuba had invested in the development of its human resources and had a highly educated population, as well as scientists and researchers who were mobilized to provide alternatives for the country's agricultural production and food security. The alternative model adopted rests on four pillars:

1. Agroecological technology and diversification instead of chemical inputs and homogenization. Among the practices successfully used are:
    - Diversification of production and of the farm, by intercalated crops, associated crops, multicropping, and agroforestry.
    - Biopesticides (microbial products), locally produced natural enemies, and multicropping to control pests; resistant varieties, crop rotation, and microbial antagonists to control pathogens; rotation and cover (living or dead) for weed management.
    - Biofertilizers (e.g., Azotobacter, Azosprillum), increase in populations of mycorrhyzogenic fungi, use of microorganisms that make phosphorus soluble, manure, compost, and earthworm humus, green fertilizers, natural zeolites, and minimum tillage for agroecological soil management.
    - Integration of stock-raising and crop-farming for better use of the energy byproducts generated by both sectors.
    - Use of draft animals to replace tractors, which use fossil fuels.
2. Fair prices for farmers. Cuban farmers increased production in response to the high prices farm products fetch. Through other programs and policies to bolster food security, the government is seeing to it that the population in general, and the urban population in particular, has access to food despite the high prices.
3. Redistribution of the land. The main redistribution of land in Cuba consisted of dividing up large state properties into smaller units. The arable area in the hands of the state dropped from more than 75 percent in 1992 to less than 33 percent in 1996. The small farmers and urban horticulturalists have been the most productive of all the Cuban producers under low-input conditions.
4. Major emphasis on local production, including urban agriculture. The food produced locally and regionally offers greater food security, since the population does not depend on the caprices of prices in the world economy, transportation over long distances, or the good will of other countries. Production is also more energy-efficient since so much energy is no longer consumed in transportation. Finally, in Cuba, urban and peri-urban agriculture has been an important component of the strategy of supporting local food production.

Cuba's situation is very particular and it cannot be indiscriminately applied to other countries. Nonetheless, Cuba offers us a specific example of a country that was able to transform its agriculture towards a more sustainable agriculture. The most important lesson of this example is that agroecological practices, along with fair prices for producers, agrarian reform, and local production, including urban agriculture, can make a significant contribution to food security and to improving the standard of living of both urban and rural small-scale producers.

logical or organic methods are compared, the authors found that the farms with agroecological agriculture produce the same and in most cases significantly more than those lands in conventional production. This type of agriculture is benefiting, in particular, peasants and small-scale producers. Approximately half of the producers interviewed had less than one hectare and 90% had farms with less than two hectares. The result is an increase in food consumption of the family unit and greater production, allowing the peasant/producer to consume and market a variety of products. Pretty et al. (2006) estimated an increase in food production of 79% per hectare. These results have been confirmed by other recent studies (see for example Parrott and Marsden, 2002; Pimentel et al., 2005; Halberg et al., 2006; FAO, 2007; Kilcher, 2007).

Recent studies suggest that agriculture based on agroecological principles is not only feasible for a niche market (such as products certified to be organic) but also offers a real alternative to meet food needs globally, without having to convert natural habitats to agriculture, using 30% less energy, less water and no agrochemicals (Pretty, 2002; Halberg et al., 2005; Pimentel, 2005; Badgley et al., 2007;

FAO, 2007). Yet even more important for the purposes of this evaluation, agroecological and "knowledge-intensive" agriculture offers the peasants and small-scale producers of LAC an alternative for the production not only of food, but of culture and human and social capital (Zinin et al., 2000; Pretty et al., 2003). Agroecological experiences in the region provide evidence of the potential of ecological agriculture to pull peasants out of poverty, strengthen social relations, eliminate dependency on outside inputs and knowledge and strengthen the connection with their environment. A recent report by the FAO (2007) that came out of the FAO-sponsored conference Organic Agriculture and Food Security in 2007, concludes that organic systems have a great potential to increase food access, reduce risk and build long-term investment that increase food security, all of which directly address development and sustainability goals. It also states that when total household yield and nutritional and environmental impacts are measured along with the cost-effectiveness of production, as well as energy efficiency, organic systems are superior to conventional systems.

Since the early 1990s, organic agriculture has experienced a leap in demand, which has induced a spectacular increase, representing one the areas of agriculture with the greatest commercial potential (Box 1-11).

1.7.2 Sustainability

1.7.2.1 Traditional/indigenous system

The sustainability of an agricultural system has to do with obtaining the best possible result without compromising the resource base looking to the future. The concept of sustainable agriculture integrates goals such as protecting the environment, profitability or productivity and maintenance of rural communities (Altieri, 1995). For a long time, anthropologists and ecologists have recognized the sustainability features of indigenous/traditional systems and these systems have been the basis of knowledge for the development of modern agroecology (Steward 1955; Netting, 1974; Altieri, 1995). Several specific aspects of traditional and indigenous agricultural systems tend to make them more sustainable and conducive to conserving biodiversity on and around farms. Traditional farmers have generally relied on a mosaic of fields, pasture and forests to provide the full range of their subsistence needs, which produces a variety of habitat for wild biodiversity (Altieri, 1995; McNeely and Scherr, 2003). Agricultural diversity is greater, thus providing different habitat options to biodiversity: more types of crops tend to be grown and several crops may be grown together, or intercropped. Trees are often left standing in some agricultural fields or pastures. Cultivation is usually less intensive and, in the case of the swidden agricultural systems typical of indigenous cultivation in the humid tropics in Latin America, fields are allowed to return to secondary vegetation for a considerable period after a few years of cultivation. The patchwork of land uses and in some cases use of intercropping, reduces erosion and thus sedimentation of streams and rivers. And because these farming systems use fewer or no agricultural chemicals, they also cause less pollution.

Although these traditional systems maintained and still maintain hundreds of generations of farmers, some (such as the *chinampas* in Mexico and the *camellones elevados* in Lake Titicaca in Peru and Bolivia) were not able to survive and others are in the process of disappearing due to social, economic and political pressures (Denevan 1980; Turner and Harrison, 1983; Wilken, 1987). As the crisis of rural livelihoods advances, these systems gradually disappear and with them the genetic resources and knowledge and wisdom that evolved over millennia.

1.7.2.2 Conventional/productivist system

The greatest criticism of the conventional/productivist system is that it is not environmentally sustainable. The advent of high-input agriculture has led to a simplification and homogenization of the system, which results in the loss of planned biodiversity (in other words, the diversity of crops and other productive organisms such as honey bees, fish for food and others). The reduction of planned diversity results in a diminution of the associated diversity (that is, all the other organisms that live in that agroecosystem). The loss of biodiversity has negative consequences for the sustainability of the agroecosystem, as it has a direct impact on ecological processes as well as on the environmental services provided by ecosystems (Naeem et al., 1994; Altieri, 1995; Tilman et al., 1996; Matson et al., 1997; Yachi and Loreau, 1999; Reganold et al., 2001). Some of the ecosystem services that are degraded by modern production practices are essential to the viability and sustainability of the agricultural systems themselves (McNeely and Scherr, 2002). Soil fertility is a prime example. There is increasing evidence that the rich and complex below-ground ecosystems of bacteria, fungi, protozoa, nematodes, arthropods, earthworms and other organisms play a critical role in creating and maintaining the soil conditions that are optimal for agricultural production (Buck et al., 2004). Production practices used in the conventional/productivist system, which are dependent on chemical inputs and mechanical manipulation of soils, can have devastating effects on these important but little-understood ecosystems. Erosion caused by tillage and other production practices, such as leaving bare soil exposed between planting seasons, has also gravely affected soil fertility (Buck et al., 2004).

Pollination is another key ecosystem service that can be seriously degraded in intensive agricultural landscapes. Studies in Costa Rica, Brazil and Argentina have shown that more pollinators are found in agricultural fields adjacent to forest fragments or remnants of native vegetation and that more pollen deposition actually occurs in those sites (De Marco and Monteiro Coelho, 2004; Ricketts et al., 2004; Chacoff and Marcelo, 2006). Also systems that are more diverse and harbor high levels of bee species increase pollination services (Klein et al., 2003; Steffan-Dewenter et al., 2005). Finally, it is also clear that use of agrochemicals can reduce the number of beneficial organisms available both for pollination and for control of crop pests (Buck et al., 2004).

The use of pesticides in conventional/productivist agriculture has also had a negative impact on the other beneficial fauna, such as natural enemies (predators, parasitoids and others), stimulating the evolution of resistance in pests, the resurgence of primary pests and outbreaks of secondary pests (Nicholls and Altieri, 1997). This so-called "pesticide treadmill" has caused a continuous increase in the use of pesticides in the region. The phenomenon is well-established in the scientific literature and is responsible for crop losses due to pests and diseases, which have increased notably despite the ever greater use of pesticides (Pimentel et al., 1978).

Particularly worrisome at present is the increase in weeds resistant to herbicides, mainly glyphosate, due to the establishment of herbicide-resistant or -tolerant varieties, such as Roundup-Ready soybean from Monsanto (Box 1-7). From 2000 to 2005, the number of biotypes of herbicide-resistant weeds climbed from 235 to 296 and to 178 species. All these factors combine with the vast expanses of single-crop agriculture characteristic of the conventional/productivist production system to create conditions that are unsustainable in the long run (Matson et al., 1997).

1.7.2.3 Agroecological system

The agroecological systems have emerged in response to the lack of sustainability and the environmental and health impacts of the conventional/productivist system. One of the pillars of the agroecological systems is the elimination or reduction in the use of pesticides and synthetic fertilizers; the other pillar is biodiversity. A recent study of 286 agroecological projects with small-scale producers in 57 countries of Africa, Asia and Latin America and the Caribbean found that while the average yield increased 79%, there were also increases in the efficiency of water use and the potential for carbon sequestration. Also contributing to the increase in the sustainability of the systems, the study found that 77% of the producers reported a 71% reduction in the use of pesticides. This study is significant because it covers an area of 37 million ha, which represents 3% of the area planted in the non-industrialized countries (Pretty et al., 2006). One of the strategies for managing agroecological systems is to increase biodiversity, both planned and associated (Vandermeer, 1995). The increase in biodiversity is accompanied by the restoration of ecological processes such as pollination and the predation of herbivores by natural enemies (Nicholls and Altieri, 1997). Alongside these benefits, agroecological practices may also increase the system's resistance to catastrophes, thereby bolstering its sustainability. Recently a participatory study by the Movimiento Campesino a Campesino showed that farms managed with agroecological practices were more resistant to the impacts of Hurricane Mitch in Nicaragua (Holt-Giménez, 2001) (Box 1-5).

1.7.3 Quality and food safety

Food quality and safety is understood as the guarantee that a food will not cause harm to the consumer, or in other words that it won't cause disease. The modern concept incorporates factors such as agricultural practices, genetic manipulation, the inclusion of hormones or other drugs in animals' diets (Campos, 2000) and post-harvest handling such as storage conditions and the use of unauthorized additives. The Codex Alimentarius Commission, established by the Food and Agriculture Organization (FAO) and the World Health Organization (WHO), prepares risk-based food safety standards that are used as a reference in international trade and give the countries a model for national laws (FAO, 2007).

The concept of food quality has to do with nutritional value, organoleptic properties such as appearance, color, texture and flavor and functional properties. Quality is related to characteristics that determine value or acceptability by consumers and compliance with standards that ensure that a product is safe for consumers, not contaminated, adulterated, or bearing a fraudulent presentation. Safety therefore has to do with risks associated with production and subsequent handling, processing and packaging, such as contamination with agrochemicals (pesticides and fertilizers), veterinary drugs, or unauthorized food additives; microbiological risks posed by bacteria, protozoa, parasites, viruses and fungi or their toxins (mycotoxins, aflatoxins); natural toxins present in the environment (zinc, arsenic, cyanide) or in foods themselves (solanine and histamine); and toxic industrial chemicals or radioactive waste (arsenic, cadmium, copper, lead, mercury and polychlorinated biphenyls) (FAO, 2000). Exposure to pesticide waste or other contaminants in the diet has adverse effects on the production and reproduction of animals and in human populations (Singh et al., 2007).

Although until a few years ago authorities and researchers from several countries affirmed that foods produced organically did not differ significantly in terms of food safety and nutrition from conventionally grown foods, there is more and more evidence and official recognition that organic foods contain lower amounts of residue of additives and colors, pesticides, veterinary drugs and in many cases more nitrates and other vitamins, minerals, essential fatty acids and beneficial antioxidants; and they appear to have the potential to lower the incidence of cancer, coronary heart disease, allergies and hyperactivity in children (FAO, 2000; Cleeton, 2004; Soil Association, 2005, 2007). Baker et al. (2002) performed a statistical analysis of data on pesticide residues in 94,000 food samples to describe and quantify differences between fresh fruits and vegetables from three different modes of production: conventional, integrated pest management and organic. A comparison was done of data from three programs: the Pesticide Data Program of the US Department of Agriculture; the Marketplace Surveillance Program of the California Department of Pesticide Regulation; and tests performed by Consumers Union, an independent organization. It was found that concentrations of pesticide residues in organic samples were consistently lower than in the other two categories and the greatest concentrations were found in the conventional samples, which also contain multiple pesticide residues in greater proportions.

According to Barg and Queirós (2007), in 2004 a study was carried out in Uruguay on the quality of fruits and vegetables and levels of contamination by agrochemicals, with 200 samples. Residues were detected in 72% of them; in 7% of the cases the maximum residue limits (MRLs)—established by Codex Alimentarius for individual products—were exceeded, but in many samples residues of several different pesticides were detected. Combinations of low levels of insecticides, herbicides and nitrates have proven to be toxic at levels at which the chemicals individually considered are not (Cleeton, 2004). Barg and Queirós (2007) added that the MRLs allowed are set based on the technologies available and the current economic and commercial interests and that the limits allowed today may be different in the future and from what they were in the past, thus they are not established in relation to the harm they cause to health, but have more to do with the technological packages currently available and the companies involved.

According to FAO (2000) sensory analysis studies have been performed to determine differences in the organoleptic properties of fruits and vegetables such as apples, tomatoes and carrots, in which the persons interviewed have recognized better flavor and color in organic as compared to conventional produce. In addition, it has been recognized that there are fewer losses due to fungi attacks during the storage of organic produce.

It is recognized that many developing countries have deficient food safety systems due to weak public infrastructure and incomplete or obsolete legislation that is not in line with international standards; there are even shortcomings in the developed world, when primary production is not covered. In addition, the responsibilities related to food safety and food control tend to be dispersed among several institutions and the laboratories lack the equipment and basic supplies they need, all of which is aggravated by climatic conditions. The shortcomings of the food safety systems may result in an increase in food problems and food diseases. Diarrheal diseases, for example, provoked mainly by the consumption of unhealthy food and water, take the lives of 1.8 million children each year (FAO, 2007).

Almost all chemical pesticides authorized in conventional food production are prohibited in organic production; therefore contamination may be very low in organic products. More than 500 additives are authorized in conventional foods, but only 30 additives are authorized in organic foods. It has been concluded that a predominantly organic diet reduces the amount of toxic chemicals ingested, avoids transgenics, reduces the quantity of food additives and coloring; increases the consumption of vitamins, minerals, essential fatty acids and beneficial antioxidants; and appears to have the potential to lower the incidence of cancer, coronary heart disease, allergies and hyperactivity in children (Cleeton, 2004).

The Regional Conference of Consumers of Healthy Food, held in Bogotá, Colombia, in August 2004, organized by Consumers International, Office for Latin America and the Caribbean, recognized that the use of pesticides as well as the presence of pesticide residues in foods present in the market are a major concern for the consumers' movement, since quality and safety include the primary stage of production and the processing of such products. Accordingly, emphasis was placed on the need for a comprehensive approach to ensuring safety, from production to final consumption, through sustainable agricultural production. It was emphasized that the cooperation and joint action of Consumers International with Latin American networks such as RAP-AL (Red de Acción en Plaguicidas y sus Alternativas en América Latina, Pesticide Action Network in Latin America) and MAELA (Latin American Agroecology Movement) play an essential role here. It is also crucial that strategic partnerships be strengthened with the women's movement to work on issues of food security and food sovereignty, health promotion, promoting breastfeeding and safe foods (Consumers International, 2004).

Although organic or agroecological foods are of significantly better quality than conventional ones, it cannot be said that they are totally safe. For example, one may find detectable levels of persistent organic pollutants (POPs) in organic or agroecological foods, such as DDT and other organochlorine insecticides that are no longer used because they accumulated in the soil for years.[8] Agroecological produce may also contain residues of other chemical pesticides that reach it by drift, with rain, or with contaminated waters, but also less frequently and in lower concentrations than in conventional produce (FAO, 2000; Bordeleau, 2002).

Animal manure and other organic waste such as biosolids or sludge from wastewater treatment plants, which may be used as fertilizer in ecological agriculture, may pose risks of contamination by pathogenic microorganisms that survive inadequate composting conditions (FAO, 2000).

It might be thought that organic foods pose risks of contamination with aflatoxins, a by-product of the contamination of foods with certain fungi in conditions favorable for them, given that they occur without the use of chemical fungicides. Nonetheless, it has been shown that such is not the case. Aflatoxins, which may induce cancer of the liver at very low dosages if ingested over a prolonged period of time, may be avoided by good practices in farming, post-harvest handling and storage. Studies have been reported that found that the level of aflatoxin in organic milk was lower than in conventional milk (FAO, 2000).

With regard to post-harvest handling, the vitamin C content and the dry matter are, on average, greater in organic crops and the percentage of water less, therefore they keep better than products handled with chemicals, since they are more resistant to diseases and pests (Barg and Queirós, 2007).

1.7.4 Impacts of the production systems

1.7.4.1 Environmental impacts

Agriculture general impacts. There is widespread agreement that habitat destruction and fragmentation is the major driver of biodiversity loss worldwide. While habitat destruction and fragmentation have many causes, foremost among them in terms of the area affected is agriculture (Goudie, 1990; Heywood and Watson, 1995; McNeely and Scherr, 2003; MA, 2005b). Siltation of water bodies caused by the removal of natural vegetative cover can have similarly negative effects on aquatic and marine organisms. Agriculture directly impacts aquatic biodiversity when excessive water is removed for irrigation. Production practices, such as burning cleared vegetation, can cause additional loss of biodiversity. Livestock contributes enormous amounts of methane to the world's atmosphere, which in turn contributes to climate change and impacts biodiversity (Clay, 2004). Some species introduced for agricultural purposes have become

[8] These persistent contaminants are called organic because they contain carbon in their molecule since they are manufactured from fossil fuels, but their use is prohibited in organic agriculture.

invasive and directly or indirectly caused the loss of native biodiversity as well. In short, agriculture is the human activity that has most affected the earth's environment and that has caused the most direct and indirect biodiversity loss.

Deforestation. The annual expansion in cultivated area in Latin America from 1961 to 1997 was 1.26% per year, far greater than any other region (Dixon et al., 2001). Since 1961, cultivated land has expanded by 47%, while cropping intensity has only increased by 1% (Dixon et al., 2001), meaning that most of the increase in agricultural production has been due to the expansion in cultivated area.

Expansion of the agricultural frontier in Latin America has commonly been ascribed to a set of key drivers: tax and credit policies and agricultural subsidies; agricultural colonization schemes; international and national markets; clearing for establishing land ownership; and technological factors (White et al., 2001). Frontier expansion in Latin America often starts with the cutting of logging roads into primary forest. Logging by itself deforests relatively minor areas of land. But logging roads allow colonists, usually small farmers using traditional production methods, to enter into hitherto impenetrable areas and slash and burn the forest, cultivating primarily subsistence crops for one to three years, until the soil begins to lose its fertility. Then they sell the land they have cleared to others, often large landowners, for conversion to pasture (Nations, 1992; Vandermeer and Perfecto, 2005). Cattle production is usually extensive, with low levels of inputs. Because of the characteristics of soils in tropical rain forests and grazing practices on the recently cleared land, pastures often quickly become degraded. When this happens, it can be very expensive to recuperate them and since land at the frontier is cheap, pastures are simply abandoned for newly cleared areas. In the Amazon, pastures are often abandoned within ten years and more than 50% of the area cleared is estimated to have been abandoned by the early 1990s (Hecht, 1992). Some research, however, indicates that soil fertility does not decline as markedly as widely believed and that agriculture in the Amazon may continue to be profitable over time if appropriate cultivation techniques are used (Schneider, 1995; Vosti et al., 2002).

The relative contribution of small-scale, traditional agriculture to deforestation is a matter of some dispute (Vosti et al., 2002; Sanchez et al., 2005). While small farmers using traditional cultivation methods are certainly part of the phenomenon of the expanding frontier, large-scale clearing may ultimately be responsible for a larger absolute area of deforestation (Partridge, 1989). Nevertheless, spontaneous or state-sponsored agricultural colonization, which uses the frontier as a safety valve to address the problems of land tenure, has certainly played an important role in deforestation throughout the region. In some cases, such as immigration from traditional farming areas in Guatemala to the Petén (Barraclough and Ghimire, 2000), small-scale farmers are displaced by the intensification of agriculture in the sending areas. In other cases, farmers from marginal agricultural areas move away in hope of better opportunities. This has been one reason for internal migration in Brazil and elsewhere, where farmers from the poor, drought-prone northeast of the country were among the most likely to migrate to the Amazonian agricultural frontier (Mahar, 1989; Lisansky, 1990). Typically the farming techniques that migrant farmers learned in their areas of origin are inappropriate for the fragile soils and vastly different climatic conditions of the frontier they have colonized, leading to even quicker degradation of the areas they have cleared and greater need to continually clear new areas.

The two most active agricultural frontiers in Latin America over the last few decades of the 20th century have been in the rainforests of Central America and Brazil, both areas of high biodiversity. Central America, for example, has only around 0.5% of the world's land area, but represents around 7% of the world's biodiversity. It is considered a biological hotspot and has many endemic and threatened species. Much of the original forest has already been cleared, with only 20% of the isthmus still covered in dense forest. Nevertheless, a significant swath of tropical moist broadleaf forest remains along the Atlantic Coast, stretching from southern Mexico to Panama (Dinerstein et al., 1995).

The expansion of the agricultural frontier has been linked to export cycles of commodity crops in Central America, but the ultimate use of cleared lands has been predominantly for pasture, generally using extensive systems with low levels of inputs. The total area in pasture has almost quadrupled from approximately 3.5 million ha in 1950 to over 13 million ha in 2001 (Harvey et al., 2005). Much of the cattle production was export-oriented. The decline in forest cover across the peninsula since the mid-20th century has been precipitous. Nicaragua, for example, lost 50% of its forest cover from 1963 to 1992 (Barraclough and Ghimire, 2000). The agricultural frontier has disappeared in El Salvador and Costa Rica, where most forest has already been cleared or, in the case of Costa Rica, designated as protected, but there is still an active agricultural frontier along the Atlantic Coast of the remaining countries of Central America (Harvey et al., 2005).

Government policies also provided incentives for colonization of the agricultural frontier. In both Brazil and Central America, those seeking titled land were required to show "productive" use of the land by clearing it. This has been documented as a major factor in agricultural conversion at the frontier in Costa Rica, Honduras and Panama in Central America (Barbier, 2004). Government policies that subsidized credit for certain activities have also had a big impact. In the 1960s and 1970s, Costa Rica embarked on a program of diversification of agro-exports, supported by government credits, which pushed cattle exports up to become the third largest agro-export earner (Lehnmann, 1992). By 1973, a third of the land area of Costa Rica was in pasture. State-sponsored colonization schemes, in the Guatemalan Petén, for instance, also directly added to deforestation (Barraclough and Ghimire, 2000).

Export-oriented production of commodities using conventional production systems has led to extensive clearing of native vegetation outside the rain forest in many parts of Latin America, as exemplified by the recent expansion of soybean cultivation throughout the Brazilian cerrado and the forests of Argentina. The cerrado is a mosaic of savannah and woodlands on Brazil's vast central plateau. It is one of the world's biodiversity hotspots and is home to the most diverse savannah flora in the world (UNEP, 1999a), an astonishing 44% of which is endemic (Klink and Machado, 2005). Government policies played a major role in stimulating agricultural conversion in the cerrado, as they did in the Amazon. Starting in the 1960s, government policies aimed at generating foreign exchange through the production of export crops, principally soybean, combined with a desire to populate what was perceived as a vast "empty space" in the country's interior, led to subsidized loans, the development of infrastructure and other incentives to open up the cerrado (Wood et al., 2000; Klink and Machado, 2005). As a result, by 2002 more than half the original vegetation of the cerrado had been cleared for human use (Klink and Machado, 2005), with more than 70% of the farmed area dedicated to cattle production, generally of low intensity (Wood et al., 2000). Most of the rest is dedicated to large-scale, mechanized soybean production, oriented towards the export market. Likewise, due to the expansion of soybean, Argentina now has rates of deforestation that are 3 to 6 times the world averages (Jason, 2004) (Box 1-7).

Declines in on-farm biodiversity. As an ever-increasing proportion of Latin America's land is cleared for agriculture, agricultural plots themselves and the semi-natural areas that often surround them have become more important habitats for species that are able to adapt to disturbed environments. There is evidence that use of some traditional practices leads to enhanced on-farm biodiversity, as compared to more intensive farming methods. Harvey et al. (2004) review the literature for Latin America and conclude that practices that increase the variability of habitats available on farm, such as live fences, windbreaks and isolated trees, have had a demonstrable impact on taxa such as birds and mammals. Other studies have demonstrated linkages between increased biodiversity and both organic agriculture and shaded tropical agriculture, such as shade coffee (Perfecto et al., 1996; Perfecto and Armbrecht, 2003; Buck et al., 2004). As farming systems have evolved to more technology-intensive over the last half century, many of these more sustainable practices have been abandoned (McNeely and Scherr, 2003). Consequently, the amount of wild biodiversity supported on farms has decreased over time. In his global analysis, Donald (2004) found that the increase in production of the five major commodities in the world (soybean, rice, cacao, coffee and oil palm) were achieved through an increase in the area planted as well as an increase in yield per area, both of which led to environmental degradation and a massive loss of biodiversity. These negative environmental impacts were a consequence of both habitat loss and environmental contamination due to the use of agrochemicals. Similarly, Robinson and Sutherland (2002) documented the reduction of biodiversity due to agriculture in post-war Britain. They also present evidence that the loss of biodiversity was due to both habitat loss and habitat degradation (i.e., contamination with pesticides and other agrochemicals as well as the homogenization of the farm habitat).

Impacts of freshwater ecosystems. Freshwater ecosystems are very poorly understood, but it is clear that they are highly threatened worldwide (Abell, 2002; Olson and Dinerstein, 2002; MA, 2005b). Conventional/productivist agriculture is a major source of threat to these systems. A recent assessment of Latin America's freshwater biodiversity concluded that more than 85% of freshwater biodiversity in the region is seriously threatened (Olson and Dinerstein, 2002).

Threats related to agriculture include direct habitat conversion, for example in the case of wetlands drained for agricultural use; sedimentation from the loss of riparian and catchment basin forests; and pollution and eutrophication from agrochemicals, fertilizers and fish farming. The introduction of non-native species, often as part of fish farming initiatives, is a particular problem for lakes; unintentional escapes from fish ponds into streams and rivers are also problematic (ILEC, 2005). Dams and channelizations constructed for flood control or irrigation and excessive water withdrawal, are another source of impact related to agriculture. An emerging issue with dams is the importance of environmental flows, that is, the timing and size of flows necessary for maintaining downstream ecosystems. Pollution from waste produced by processing agricultural crops also impacts freshwater biodiversity (Clay, 2004; ILEC, 2005). Finally, direct exploitation of freshwater fish for food is also an important threat.

While these problems have not been well-studied in Latin America, there is some evidence of their impact in particular places. Agostinho et al. (2005) review studies of impacts from various threats to freshwater systems in Brazil. There is evidence of reduced species diversity and alteration in community structure in freshwater bodies subject to pollution or eutrophication. Siltation caused by intensive agriculture has been documented as impacting freshwater biodiversity in the Pantanal, the Cerrado and in streams in the highly threatened Atlantic Forest, as well as the Amazon. In Chile, native lake fishes appear to have declined with the establishment of populations of rainbow trout, an exotic species, in the 1900s. With explosive growth in the Chilean aquaculture industry and Chile poised to become the worldwide leader in salmon production, there is concern about the impact of runaway salmon on native fish populations as well (Gajardo and Laikre, 2003).

Contamination and degradation of aquatic and terrestrial ecosystems. Agriculture also impacts biodiversity beyond the conversion of natural habitat. In particular, the use of agrochemicals in the conventional/productivist system results in contamination and degradation of ecosystems. Agrochemicals can harm species that utilize agricultural landscapes or nearby areas and they have a major impact on aquatic and marine biodiversity. Pesticides persist in the environment and many disperse globally as a result of drift, soil volatilization and evaporation (Kurtz, 1990). Pesticides have caused extensive contamination of the soil (Kammerbauer and Moncada, 1998), surface water and groundwater (Dalvie et al., 2003), marine and estuary sediments (Bhattacharya et al., 2003), rain (Quaghebeur et al., 2004), polar snow (Barrie et al., 1992), mammals (WWF, 2006) and even tree bark (Simonich and Hites, 1995).

Certain persistent pesticides even accumulate in human tissues and are concentrated as they pass through the links in the food chains. They are implicated in massive deaths of marine mammals (Colborn et al., 1996) and of many bird species (Goldstein et al., 1999). As a result of hormonal or endocrinal alteration, which many can cause, they are responsible for serious population losses and for the feminization of male amphibians (Hayes, 2005) and alligators (Colborn et al., 1996; Crain et al., 1997). Some halogenated pesticides, particularly methyl bromide, contribute to the destruction of the ozone layer, which protects the earth (Miller, 1996; UNEP, 1999b).

The impact of fertilizers and pesticides on the soil has been the subject of little research in LAC, yet food production ultimately depends on soil quality. This may be one of the main causes of declining crop yields and the diminution in levels of micronutrients in foods that the Green Revolution has suffered.

Another source of high levels of agricultural soil contamination is to be found in the toxic waste of pesticides, such as the packages, bottles and leftover pesticide not used. In addition, illegal and clandestine burying of obsolete or expired products has been discovered in recent years in many Latin American and Caribbean countries, such as the northern coast of Colombia. Given that the Stockholm Convention on POPs entered into force in May 2004, in several countries of LAC inventories are being taken of obsolete (prohibited or expired) pesticides, which include POPs (UNEP, 2001).

The conventional/productivist system also demands a large increase in water use, including an enormous expansion of irrigation facilities. This has reduced groundwater reserves and led to a drop in the water table in vast agricultural regions, as in Valle del Cauca in Colombia, where one finds sugarcane monoculture and the savannah of Bogotá, the main zone for the cultivation of flowers for export; wells for drawing water from the subsoil have to be dug deeper and deeper.

Coastal and marine ecosystems. The greatest impacts on marine ecosystems worldwide are caused by overfishing. Nevertheless, nutrient loading, largely due to agricultural use of fertilizers, is a major cause of degradation for coastal ecosystems (MA, 2005a).

Sedimentation caused by erosion on agricultural fields and pollution caused by agrochemicals also represent significant threats to marine ecosystems (Clay, 2004). Coral reefs, which are generally close to shore and are important repositories of the world's biodiversity, are particularly affected by these threats. Almost two-thirds of the reefs of Central America and the Caribbean are considered at risk and one-third is considered at high risk (Barker, 2002).

Aquaculture represents a relatively new but growing source of impacts on coastal ecosystems. Shrimp farming often displaces mangroves, among the most valuable and highly threatened of coastal habitats, as well as wetlands and estuaries. Shrimp production is prevalent in coastal areas throughout Mexico, Central America and the Caribbean and northern South America, especially Ecuador. In addition to outright destruction of fragile and economically valuable coastal ecosystems, shrimp farming causes considerable water pollution in coastal areas. Aquaculture was virtually nonexistent at mid-century and now represents an important economic sector in many countries and with the growth in world demand for fish, its impact on coastal ecosystems can only accelerate (Clay, 2004).

1.7.4.2 Social impacts

According to FAO (1986), the technological changes in agriculture over the last 50 years, such as the package of improved seeds, growing technologies, better irrigation and chemical fertilizers were very successful in attaining the essential objective of increasing agricultural production, crop yields and aggregate food supplies. Nonetheless, the swift modernization of agriculture and the introduction of new technologies, characteristic of the Green Revolution, had a differential impact on rural populations, depending on class and gender. The effects of modern agriculture were differentiated, depending on whether you were paid workers, growers, or consumers, from households with or without land, rich or poor, male-headed or female-headed. Moreover, there were two general trends: the rich benefited more than the poor from that technological change and men benefited more than women.

In Latin America and the Caribbean, the intensification of agriculture entailed the transformation from traditional production to production using external inputs, along with the accompanying social changes. Yet the process was carried out conservatively in the region, if we compare it with what happened in Europe, which has implied a large debt to the external banking system and the exclusion of most of the population. Agriculture saw improvements in production, exports and incomes, although poverty and rural marginality expanded, especially for thousands of small-scale producers.

However, the productive accomplishments of modern agriculture cannot be ignored; year after year millions of tonnes of food are produced, yet this is not enough to alleviate hunger and achieve food security in the region, since the poor don't have access to the food. At the same time, agrarian policies have not been able to resolve the social right to access the benefits of technology, therefore there is a growing accumulation and concentration of the wealth generated by agriculture (Rosset et al., 2000).

In addition, FAO (2000) indicates that one of the important social effects of modern agriculture has been demographic change, due to the substitution of a considerable part of the agricultural labor force by machinery, the increase in the area per worker and the consequent reduction in the number of farms, which has unleashed an intense rural exodus, also driven by the reduction in related activities (the trade in primary products, processed goods and crafts, as well as public services). This decline in the rural population has made it difficult to maintain the services (mail, schools, stores, physicians and pharmacies) and social life. The document The Millennium Development Goals: A Latin American and Caribbean Perspective identifies a lack of jobs as one of the main problems in the region (UNDP, 2005a).

Indeed, it is argued that conventional/productivist agriculture, apart from the social impacts produced by poverty and inequality, has exchanged technologies for peasants, expelling thousands of families from rural communities and devaluing everything that farmers represent for the social, economic and environmental life of the rural world. At the same time, it has generated a major increase in inequality and the continuing dismemberment and disappearance of peasant communities and with that the major loss of cultural diversity (Riechmann, 2003).

At the same time, industrial/conventional/productivist agriculture has significantly upset the land tenure of peasants and indigenous communities, since those who cannot become incorporated into this type of agriculture and are unable to compete are forced to sell their lands and seek jobs as wage workers or emigrate to the cities, which means that the concentration of landholdings in just a few hands produces greater stratification and therefore greater inequality and economic and social insecurity.

The technological changes in agriculture have resulted in a diminution of the number of small-scale producers and an increase in the number of agricultural workers. The workers employed by the agricultural enterprises have suffered deterioration of their social and working conditions: mainly low wages, unstable employment, the lack of social security and exploitation at work (Ahumada, 2000).

Giberti (2002) suggests that the impoverishment and unemployment of many agricultural producers that has been caused by the development of industrial agriculture favored the hiring of workers in unjust conditions, often disguised in pseudo-associative forms, as often happens with horticulture around large cities. This rural worker is extremely vulnerable: he or she practically lacks medical coverage and the possibility of retirement, as indicated by the tiny numbers who attain such benefits.

Another sociocultural effect has been on local knowledge and how it is disseminated. FAO (2000) suggests that since the design of the new means of production happens at research and development centers and relatively concentrated industrial and services enterprises, training for farmers and agricultural workers no longer happens directly in the countryside, but rather in public and private institutions and through technical and economic information services. In a broader perspective, the rural cultural patrimony of the past, locally developed and managed, has given way to a relatively uniform culture disseminated by the educational system and the media.

In addition, conventional/productivist agriculture has meant, for rural producers, scant participation in the choice of the technologies that have been applied, since the approach has almost always been imposed vertically, resulting in barriers to the acceptance of technology. As a result, cultural integration, specifically of local or traditional customs and knowledge, has been scant or nonexistent (Altieri, 1992).

Modern agriculture has impoverished and deteriorated the cultural aspects of how we feed ourselves. First, food customs and diversity have been lost, since numerous traditional foods have disappeared from the markets and from the rural kitchen, having been replaced by those produced by industrial agriculture and food imports. In addition, due to the whole social transformation that has taken place in the homes of peasant families, the kitchen has disappeared as the central space of the home and with it a culture whose values were quality food, sociability (*convivencia*), associated with the fact of obtaining nutrition and enjoyment of variety (Riechmann, 2003).

1.7.4.3 Impacts on health and nutrition.

Health effects of diminished biodiversity. Biodiversity is essential for nutrition and food safety and offers alternatives for improving the standard of living of communities, thus improving the overall health of human beings. Today certain communities continue using some 200 or more species in their diet, but the world trend is towards simplification, with negative consequences for health, nutritional equilibrium and food safety. Biodiversity plays a crucial role mitigating the effects of micronutrient deficiencies (iron, zinc, copper, magnesium and calcium), which weaken hundreds of millions of persons. A more diverse diet is crucial for diminishing the trend towards malnutrition and for living a healthier life (Barg and Queirós, 2007).

The loss of traditional varieties, soil degradation and contamination, the loss of biodiversity due to the establishment of large, genetically uniform expanses of single-crop agriculture and the elimination of their organic management all resulted in deficiencies in essential micronutrients and vitamins in conventional food crops. Our foods are nutritionally unbalanced, since they are fertilized generally with one to three elements (nitrogen, phosphorus and potassium), yet it is known that plants need 42 to 45 minerals to grow healthy and with this type of reductionist agriculture very few nutrients are provided to the plant (Barg and Queirós, 2007).

Statistics from the governments of the United Kingdom and the United States indicate that the levels of minerals in fruits and vegetables fell up to 76% from 1940 to 1991. By way of contrast, there is mounting evidence that organic fruits and vegetables may have a greater vitamin and mineral content (Cleeton, 2004), from 40 to 60% more (Barg and Queirós, 2007), although some recommend that additional research be done (Table 1-11) (Soil Association, 2005).

Acute and chronic toxicity due to agrochemicals. Poisonings and deaths. Pesticides account for more poisonings than any other cause worldwide. In 1990 the World Health Organization (WHO) estimated that each year three million severe cases of poisoning occur, with likely mortality of 1% (WHO, 1990), whereas others calculated 25 million poisonings that same year, estimating that an average of 3% of workers were intoxicated that year. Such figures reflect only the most severe cases and significantly underestimate unintentional poisonings due to pesticides, because they are based primarily on hospital records. Most of the rural poor do not have access to hospitals and physicians and workers in the health sector often fail to recognize and report cases of poisoning (Murray et al., 2002). In a research study on the incidence of acute intoxications due to pesticides in six Central American countries, done in the early years of this decade by PAHO, WHO, DANIDA and the ministries of health, within the project known as PlagSalud, 98% under-registration of intoxications was estimated (Murray et al., 2002; OPS, 2003).

It is estimated that 99% of the deaths occur in the countries of the South, i.e., Latin America, Africa and Asia (WHO, 1990). These data are more alarming if one considers that in Latin America, where the use of pesticides has risen the most in recent years and with it cases of poisoning, a large number of women of reproductive age and children work in agriculture, exposed to pesticides in conditions that are very dangerous in which they are highly susceptible (Nivia, 2000).

Table 1-11. **Mineral level in organically and conventionally grown foods.**

Mineral content in miliequivalent/100 grams							
Type of food	Calcium	Magnesium	Potassium	Sodium	Manganese	Iron	Copper
Lettuce							
Organic	40.5	60	99.7	8.6	60	227	69
Conventional	15.5	14.8	29.1	0	2	0	3
Tomatoes							
Organic	71	49.3	176.5	12.2	169	516	60
Conventional	16	13.1	53.7	0	1	9	3
Beans							
Organic	96	203.9	257	69.5	117	1,585	32
Conventional	47.5	46.9	84	0.8		19	5

Source: Barg y Queíroz, 2007.

Chronic intoxications. Persons subject to high levels of exposure because of their occupation may be poisoned without manifesting symptoms, which means they are not warned of the high risk they run of suffering severe intoxication and dying from a small additional exposure, which in normal conditions would not cause a critical intoxication. According to the most recent documentary research by PAN International (Pesticide Action Network), contained in its position paper on the elimination of pesticides (PAN, 2007), the main chronic effects caused by chemical pesticides include cerebral lesions and lesions of the nervous system in general, such as peripheral polyneuropathies and Parkinson's disease (Semchuk and Love, 1992; McConnell et al., 1993; Baldi, 2003; PAN Germany, 2003; Isenring, 2006); cardiovascular diseases; kidney and liver disorders; cancer (Brody and Rudel, 2003; Flower et al., 2004); genetic mutations; teratogenesis (congenital functional malformations or abnormalities) (Levario et al., 2003); endocrine or hormonal problems; reproductive problems (sterility, impotence, abortions, stillborn children, development problems in offspring) (Colborn et al., 1996; Figà-Talamanca, 2006; Bretveld et al., 2007); and suppression of the immune system. All pesticides produce chronic effects, particularly those known as persistent organic pollutants (POPs), which include DDT and other organochlorinated insecticides, which are targeted for control by the Stockholm Convention approved at the United Nations in 2001 and which entered into force in May 2004 (UNEP, 2001; UNEP, 2007).

Health effects of contamination of the environment and foods. There are growing concerns not only about the presence of pesticide residues in foods and their health effects, but also about the "cocktail effect" of multiple pesticide residues, along with food additives, hormones and antibiotics used in breeding livestock and poultry and due to the use of chemical fertilizers. Chemical fertilization in conventional agriculture results in higher levels of nitrates, which can have negative effects on health, because in certain conditions they can be converted to nitrosamines, which are carcinogenic. They may also reduce the ability of the blood to transport oxygen and pose a risk of methemoglobinemia (FAO, 2000). An effort has begun to look for multiple pesticide residues and nitrates in food samples, because the evidence suggests that when they act in combination in foods, the harmful effects may be compounded. Combinations of low levels of insecticides, herbicides and nitrates have proven toxic at levels at which the chemicals individually are not (Cleeton, 2004).

Hormonal or endocrine effects. The greatest harm from exposure to pesticides occurs during pregnancy, when toxics with endocrine effects or xenohormones limit or block the delicate natural signals that the hormonal systems of the mother and fetus send the cells and organs to guide their development. The endocrine alteration in the womb during the stage of fetal development may result in cancer, endometriosis, learning disorders, behavioral disorders, immunological and neurological disorders and other problems such as low sperm count, genital malformations and infertility. These hormonal problems may originate in fetal exposure and not manifest until puberty (Colborn et al, 1996; Figà-Talamanca, 2006; Bretveld et al., 2007). In addition, it is suggested that they may contribute to higher rates of hormone-dependent cancers such as breast and prostate cancer, in women and men occupationally exposed to pesticides. It is likely that women with breast cancer will have five to nine times more pesticide residue in their blood than those not afflicted with the disease (Bejarano, 2004; Cleeton, 2004).

Children may be particularly susceptible to pesticide residues because they consume more food and water per unit of body weight than adults and their relatively immature organs may have a limited ability to detoxify these substances. In a comparative study with children ages 2 to 4 years in Seattle, six times more pesticide residue was found in children fed conventional foods than those fed organic foods. In another comparative study in Sweden with 295 children ages 5 to 13 years from schools with different approaches to education and food, it was found that in the school with alternative approaches, in which preference is giving to organic food, there was a lesser prevalence of allergies (Cleeton, 2004).

Risks due to transgenic foods. There are many concerns about the possible effects of transgenic foods, which are prohibited in organic or agroecological foods. The potential health effects of GMOs on humans are unknown, but there are ever greater concerns because more than half of the studies that do not find negative effects on organs of laboratory animals have been done in collaboration with the industry. Other studies, done independently, relate health risks mainly in the intestinal walls, due to the transfer of transgenes to intestinal bacteria; the scientists suggest that until they are adequately researched it is best not to consume them (Cleeton, 2004).

According to statistics provided by the transgenics industry, in 2006 these crops (herbicide tolerant and insect resistant) were planted on 100.8 million ha, 12% more than in 2005 (90 million ha); global sales of these seeds reached US$6.050 billion (a 14% increase with respect to the previous year) (CropLife, 2007). Argentina was in second place in area planted after the United States, followed by Brazil in third place. Another five Latin American countries are among the 22 countries that planted transgenics in 2006, according to CropLife (2007): Paraguay (7th place), Uruguay (9th), México (13th), Colombia (15th) and Honduras (18th). The top eight countries saw growth of more than one million ha each from 2005 to 2006; geographic expansion occurred mainly in Latin America and Asia. Participation by crop in the transgenic seed market in 2006 was as follows: soybean 43.9%; maize 41%; cotton 11.9%; canola 3%; and others, 0.2% (CropLife, 2007).

1.7.4.4 Economic impacts

It is very difficult to evaluate the social and environmental costs of conventional/productivist agriculture because it is not easy to assign many values when ethical considerations come into play. For example, what value should be assigned to human life? Nonetheless, efforts have been made to try to evaluate these environmental and health costs, such as those of David Pimentel and his team of researchers at Cornell University in the United States, who have valued the costs of the public health impact of intoxications and deaths, contamination of domestic animals and cattle, loss of natural enemies and costs due to resistance to pesticides, losses of honeybees and pollination of crops, losses in fishing, crops, wild birds and contamination of groundwater.

Based on Pimentel's studies (2004), in 2004 the Pesticide Action Network—Latin America (RAP-AL) made an initial approximation of the social and environmental costs in LAC. The RAP-AL study used same methodology and data applied in the United States, yet considering that in Latin America many costs may be greater, due for example to the environmental costs stemming from the destruction of biodiversity, as the region includes some of the most biodiversity-rich countries in the world (Nivia, 2005).

To evaluate the health impacts, general approaches of the World Health Organization were used that indicate that 15% of the population of Latin America and the Caribbean lives in rural areas, with 5% poisoned, 2% hospitalized and 1% mortality (Table 1-12). With respect to the cost of human life, the 3.7 million dollar figure used by the United States EPA was based on the notion that the life of a Latin American is no less valuable than the life of a person from the United States. In this initial calculation it was estimated that there is a social and ecological debt of US$130 billion annually; as in the case of the U.S. study, the impacts on soil, loss of fertility, hormonal effects, sterility, malformations and others have yet to be calculated. In addition, although the calculations are for one year, the impact has accumulated for more than 50 years of industrial/productivist agriculture, therefore adequate economic projections remain to be done to estimate the cumulative economic impact of this type of agriculture in the region.

Historically, agriculture has been one of the largest and most important sectors receiving World Bank loans. The trend has been to capital-intensive agriculture, with growing use of chemical inputs and now genetic engineering, for export. The aggressive promotion of structural adjustment policies and rural development by the Bank favoring agricultural intensification and production for export, at the cost of smaller-scale agricultural with fewer external inputs, is the main barrier to the significant adoption of pest management plans and ecological and cultural production systems, which are called for by the Bank's new policies.

In response to the demands of civil society organizations, in December 1998 the World Bank adopted an operational policy on pesticides and pest management that requires Bank-supported projects to reduce farmers' reliance on pesticides and promote alternative integrated pest-management methods that have a sound ecological foundation. It also prohibits the use of Bank funds for the purchase of hazardous pesticides.

The Pesticide Action Network (North America) analyzed the impact on pesticide use in 107 Bank projects approved from 1999 to 2003. It showed that the Bank's policy is just on paper, because more than 90% of those projects continue to promote the use of pesticides; although they don't mention them directly, they invoke them using a different vocabulary. The Bank considers the private sector a key ally in global development, yet this collaboration tends to benefit the large corporations more than poor farmers. For example, the Bank financed more than US$250 million in pesticide sales from 1988 to 1995; from 1993 to 1995 all the contracts signed went directly to the largest pesticide companies in France, Germany, the United Kingdom, the United States and Japan. While the farmers who participated in these projects suffered the negative health effects and detrimental impact on the ecological stability of their production systems that result from pesticide use, the Bank recognized that only 1% of the projects had a complete environmental evaluation (Karen, 2004).

Table 1-12. **Estimated environmental and health costs associated to the use of pesticides in LAC.**

Effects on human health due to pesticides	Total costs (US$)
Costs for poisoning with hospitalization: 60,000 x 3 days x US$2,000/day	360,000,000
Costs of treatments of patients without hospitalization (include hospital. compensations and transportation): 3,000,000 x US$1,000	3,000,000,000
Labor lost due to poisoning: 60,000 workers x 5 days x US$80/day	24,000,000
Cancers due to pesticides: Total population 400 millions x 0.02% x US$100,000/case	8,000,000,000
Costs due to fatalities: 30,000 x US$3.7 millions (Value of a human life according to EPA)	111,000,000,000
Sub-Total	122,384,000,000
Other losses*	8,505,000,000
Total approximated environmental and health costs	130,889,000,000

*There is no data for LAC therefore the figure is that estimated from US data. These figures may underestimate the true value given the greater biodiversity in LAC.

Source: Adapted from Pimentel, 2004; Nivia, 2005.

References

Abell, R. 2002. Conservation biology for the biodiversity crisis: A freshwater follow-up. Conserv. Biol. 16:1435-1437.

Acosta L., y M. Rodriguez Fazzone. 2005. En busca de la agricultura familiar en América Latina. Oficina Regional de la FAO para América Latina y el Caribe, Santiago.

Acosta, P., C. Calderón, P. Fajnzylber, y H. López. 2007. What is the impact of international remittances on poverty and inequality in Latin America? Res. Working Pap. 4249. World Bank, Washington DC.

Adelman, I., y C. Morris. 1973. Who benefits from economic development? Economic growth and social equity in developing countries. Stanford Univ. Press, Palo Alto.

Agostinho, A.A., S.M. Thomaz, y L.C. Gomes. 2005. Conservation of the biodiversity of Brazil's inland waters. Conserv. Biol. 19: 646-652.

Aguirre Rojas, C. 2005. América Latina en la encrucijada: Los movimientos sociales y la muerte de la política moderna. Contrahistorias, México.

Ahumada, M. 1996. Estudio de la racionalidad de la economía campesina, VII Región Chile. Facultad de Ciencias Agrarias, Univ. Austral de Chile.

Ahumad, M. 2000. La innovación agrícola: cambios e innovaciones institucionales. p. 209-213. *In* FORAGRO. Memorias Reunión Agricultura con Conocimiento, IICA, INIFAP. Mexico

Alatas, S.F. 2005. Eurocentrism and the need to rethink the teaching of the social sciences. Third World Resurgence 173/174:35-37.

Alstad, D.N., y D.A. Andow. 1995. Managing the evolution of insect resistance to transgenic plants. Science 268:1894-1896.

Altieri, M.A. (ed) 1987. Agroecology: The scientific basis of alternative agriculture. Westview Press, Boulder.

Altieri, M.A. 1992. Where the rhetoric of sustainability ends, agroecology begins. CERES 134:33-39.

Altieri, M.A. 1993a. Crop protection strategies for subsistence farmers. Westview Press, Boulder.

Altieri, M.A. 1995. Agroecology: The science of sustainable agriculture. Westview Press, Boulder.

Altieri, M.A. 1996. Enfoques agroecológicos para el desarrollo de sistemas de producción sostenibles en los Andes. CIED, Perú.

Altieri, M.A. 1999. Applying agroecology to enhance the productivity of peasant farming systems in Latin America. Environ. Dev. Sustain. 1:97-217.

Altieri, M.A., y E. Bravo. 2007. The ecological and social tragedy of crop-based biofuel production in the Americas. Available at http://www.foodfirst.org/node/1662. Inst. Food Dev. Policy, Oakland.

Altieri, M.A., y W. Pengue. 2005. La soja transgénica en América Latina. Una maquinaria de hambre, deforestación y devastación ecológica. Ecol. Polít. 30:87-93.

Altieri, M.A., y W. Pengue. 2006. GM soybean: Latin America new colonizer. Seedling January:13-17.

Altieri, M.A., y P. Rosset. 1999. Strengthening the case for why biotechnology will not help the developing world: A response to McGloughlin. AgBioForum 2(3&4):226-236.

Alvarado, I., y K. Charmel. 2002. The rapid rise of supermarkets in Costa Rica: Impact on horticultural markets. Dev. Policy Rev. 20:473-485.

Angelsen, A., y D. Kaimowitz (ed) 2001. Agricultural technologies and tropical deforestation. CABI, UK.

Assunção, J.J., y M. Ghatak. 2003. Can unobserved heterogeneity in farmer ability explain the inverse relationship between farm size and productivity? Econ. Lett. 80: 189-194.

Asteraki, E.J., C.B. Hanks, y R.O. Clemens. 1992. The impact of the chemical removal of the hedge-base flora on the community structure of carabid beetles and spiders of the field and hedge bottom. J. Appl. Entomol. 113:398-406.

Austin, A.P., G.E. Harris, y W.P. Lucey. 1991. Impact of an organophosphate herbicide (Glyphosate®) on periphyton communities developed in experimental streams. Bull. Environ. Contam. Toxicol. 47:9-35.

Badgley, C., J. Moghtader, E. Quintero, E. Zakem, J.M. Chappell, K. Aviles-Vázquez et al. 2007. Organic agriculture and the global food supply. Renew. Agric. Food Syst. 22:86-108.

Baker, B., Ch. Benbrook, E. Groth III, y K. Luz . 2002. Pesticide residues in conventional, IPM-grown and organic foods: Insights from three U.S. data sets. Food Addit. Contam. 19(5):427-446. www.consumersunion.org/food/organicsumm.htm Consumers Union, New York.

Baldi, I., P. Lebailly, B. Mohamed-Brahim, L. Letenneur, J.F. Dartigues, y P. Brochard. 2003. Neurodegenerative diseases and exposure to pesticides in the elderly. Am. J. Epidemiol. 157:409-414.

Barbier, E.B. 2004. Agricultural expansion, resource booms and growth in Latin America: Implications for long-run economic development. World Dev. 32:137-157.

Barbour, I. 1993. Ethics in an age of technology: The Gifford Lectures. Vol. 2. Harper, San Francisco.

Bardhan, P.K. 1973. Size, preductivity and return to scale: an analysis of farm level data from Indian agriculture. J. Polit. Econ. 81:1370-1386.

Barg, R., y F. Queirós. 2007. Agricultura agroecológica-orgánica en el Uruguay. Principales conceptos, situación actual y desafíos. RAP-AL, Uruguay.

Barker, D. 2007. The rise and predictable fall of industrialized agriculture. Int. Forum on Globalization, San Francisco.

Barkin, D. 2005. Incorporating indigenous epistemologies into the construction of alternative strategies to globalization to promote sustainable regional resource management: The struggle for local autonomy in a multiethnic society. Univ. Autónoma Metropolitana. Xochimilco, Mexico City.

Barndt, D. 2002. Tangled routes: Women, work, and globalization on the tomato trail. Rowman and Littlefield, Boulder.

Barraclough, S., K. Land, y B. Ghimire. 2000. Agricultural expansion and tropical deforestation. Earthscan, London.

Barrera-Bassols, N., y V.M. Toledo. 2005. Ethnoecology of the Yucatec Maya: Symbolism, knowledge and management of natural resources. J. Latin Am. Geograp. 4:9-41.

Barrie, L.A., D. Gregor, B. Hargrave, R. Lake, D. Muir, R. Shearer et al. 1992. Arctic contaminants: sources, occurrence and pathways. Sci. Total Environ. 122(1-2): 1-74.

Barta, A., D. Sommergruber, K. Thompson, M.A. Hartmuth, y A.J. Matzke. 1986. The expression of a nopalinesynthase-human growth hormone chimeric gene in transformed tobacco and sunflower callus tissue. Plant Mol. Biol. 6:347-357.

Becerril, O. 1995. ¿Cómo las trabajadoras agrícolas de la flor en México hacen femenino el proceso de trabajo en el que participan? *En* S.M.L. Flores (ed) El rostro femenino del mercado de trabajo rural en América Latina. UNRISD y Nueva Sociedad, Caracas.

Beebe, S., P.W. Skroch, J. Tohme, M.C. Duque, F. Pedraza, y J. Nienhuis. 2000. Structure of genetic diversity among common bean landraces of Middle American origin based on correspondence analysis of RAPD. Crop Sci. 40:264-273.

Beets, W.C. 1982. Multiple cropping and tropical farming systems. Westview, Boulder.

Bejarano, F. 2002. La espiral del veneno. Red de Acción sobre Plaguicidas y Alternativas en México, México.

Bejarano, F. 2004. Guía ciudadana para la aplicación del Convenio de Estocolmo. RAPAM e IPEN. Red de Acción sobre Plagicidas-América Latina, México.

Belsevich, F., J.A. Berdegué, L. Flores, D. Mainville, y T. Reardon. 2003. Supermarkets and product quality and safety standards in Latin America. Am. J. Agric. Econ. 85:1147-1154.

Berdegué, J.A. 2001. Cooperating to compete: Associative peasant business firms in Chile. PhD thesis. Wageningen Univ., Netherlands.

Berg, C. 2004. World fuel ethanol analysis and outlook. F.O. Licht, Ratzeburg.

Bernal, H. y E. Agudelo. 2006. Marco general del extractivismo de los recursos naturales en América Latina y el Caribe: Con énfasis en dimensión íctica y forestal amazónica. *En* Análisis de la evolución de los sistemas de conocimiento ciencia y tecnología en América Latina y el Caribe, su efectividad e impactos. País Vasco, España.

Berry A. (ed) 1998. Poverty, economic reform, and income distribution in Latin America. Lynne Reinner, Boulder.

Berry, A. 1997. The economic income distribution threat in Latin America. LARR 32 (2):3-40.

Berry, R., and W. Cline. 1979. Agrarian structure and production in developing countries. John Hopkins Univ. Press, Baltimore.

Bhattacharya, B., S.K. Sarkar, y N. Mukherjee. 2003. Organochlorine pesticide residues in sediments of a tropical mangrove estuary, India: Implications for monitoring. Environ. Int. 29(5):587-92.

Bhouraskar, D. 2005. United Nations development aid: A study in history and politics. Academic Foundation, New Delhi.

Birdsall, N., y J.L. Lodoño. 1997. Asset inquality does matter: Lessons from Latin America. OCE Working Paper no. 344. Inter-American Dev. Bank, Washington DC.

Blazer, M. 2004. Life projects: Indigenous peoples' agency and development. *En* M. Blazer et al. (ed) The way of development. Zed Books, London.

Blumberg, R.L. 1994. Women's work, wealth and family survival strategy: The impact of Guatemala's ALCOSA agribusiness project. *En* E.N. Chow y C. White Berheide (ed) Women, the family and policy: A global perspective. State Univ. New York Press, Albany.

Bonanno, A., L. Bush, W.H. Friendland, L. Gouveia, y E. Mingione (ed) 1994. From Columbus to Con Agra: The globalization of agricultur and food. Univ. Press Kansas, Kansas City.

Bordeleau, G., I. Myers-Smith, M. Midak, y A. Szeremeta. 2002. Food quality: A comparison of organic and conventional fruits and vegetables. Ecological Agriculture. Roy. Vet. Agric. Univ., Denmark.

Bourges, H. 2002. Alimentos obsequio de México al mundo. *En* D. Alarcón-Segovia y H. Bourges (ed) La alimentación de los mexicanos, El Colegio Nacional, Mexico.

Bowers, C.A. 2002. Detrás de la apariencia. hacia la descolonización de la educación. Proyecto Andino de Tecnologías Campesinas, Lima.

Bravo, E. 2006. Biocombustibles, cultivos energeticos y soberania alimentaria: encendiendo el debate sobre biocombustibles. Accion Ecológica, Quito, Ecuador. Available at http://www.rimisp.org/boletines/bol71/doc6.zip. Centro Latinoamericano para el Desarrollo Rural, Chile.

Bray, D.B., L. Merino-Paz, y D. Barry (ed) 2005. The community forests of Mexico: Managing for sustainable landscapes. Univ. Texas Press, Austin.

Bretveld, R., M. Brouwers, M. Ebisch, y N. Roeleveld. 2007. Influence of pesticides on male fertility. Scand. J. Work Environ. Health 33(1):13-28

Brody, J.G., y R.A. Rudel. 2003. Environmental pollutants and breast cancer. Environ. Health Perspect. 111:1007-1019.

Brookes, G., y P. Barfoot. 2006. GM crops: The first ten years— global socio-economic and environmental impacts. Brief No. 36. ISAAA, Ithaca.

Brunstad, R.J., I. Gaasland, y E. Várdal. 2005. Multifunctionality of agriculture: an inquiry into the complementarity between landscape preservation and food security. Eur. Rev. Agric. Econ. 32:469-488.

Brush, S., y M. Chauvet. 2004. Evaluación de los efectos sociales y culturales asociados con la producción de maíz transgénico.

Capítulo 6 de los documentos de discusión del informe Maíz y biodiversidad: Los efectos del maíz transgénico en México. Available at http://www.cec.org/pubs_docs/documents/index.cfm?varlan=espanol&ID=1430. Comisión para la Cooperación Ambiental, Montréal.

Brush, S.B. 1992. Reconsidering the Green Revolution: diversity and stability in cradle areas of crop domestication. Hum. Ecol. 20:145-167.

Bryceson, D., C. Kay, y J. Mooij (ed). 2000. Disappearing peasantry? Rural labour in Africa, Asia and Latin America. Intermediate Tech. Publ., London.

Buck, L.E., T.A. Gavin, D.R. Lee, y N.T. Uphoff. 2004. Ecoagriculture: A review and assessment of its scientific foundations. http://www.oired.vt.edu/sanremcrsp/documents/publications/EcoAgricultureReport.pdf (verificado el 27 de abril de 2008). Cornell Univ., Ithaca.

Buitrón, R. 2002. The case of Ecuador: paradise in seven years? p. 19-25 *En* R. Carrere (ed) The bitter fruit of oil palm: dispossession and deforestation. World Rainforest Movement, Montevideo.

Bulmer-Thomas, V. 1987. The political economy of Central America since 1920. Cambridge Univ. Press, UK.

Burst, G.E. 1990. Direct and indirect effects from herbicides on the activity of carabid beetles. Pestic. Sci. 30:309-320.

Busch, L. 2000. The eclipse of morality: Science, state and market. Aldine de Gruyter, NY.

Busch, L. 2001. Implications of the change of epoch for science and technology in society and agriculture. Trabajo presentado en el Taller Towards a new institutional coherence for guiding rural research and development (RR&D) efforts in Latin America. Red Nuevo Paradigma, ISNAR, Heredia, Costa Rica, 15-20 de octubre de 2001.

Butler, R.A. 2007. Deforestation in the Amazon. Available at http://news.mongabay.com/2007/0813-amazon.html.Mongabay.com, San Francisco.

Cahill, C. 2001. The multifunctionality of agriculture: What does it mean. EuroChoices Premier Issue 2001:36-41.

Calo, M.Y., y T.A. Wise. 2005. Revaluing peasant coffee production: Organic and fair trade markets in Mexico. Global Dev. Environ. Inst., Tufts Univ., Boston.

Campos, H. 2000. Inocuidad de alimentos y negociaciones comerciales sobre productos agropecuarios. ALADI-IICA. Ponencia en Seminario–Taller: Negociaciones Int. Agricultura, Uruguay.

Capra, F. 1982. The turning point. Simon & Schuster, NY.

Cardoso E., y A. Helwege. 1992. Below the line: Poverty in Latin America. World Dev. 20(1):19-37.

Carlisle, S.M.K., y J.T. Trevors. 1998. Glyphosate in the environment. Water Air Soil Pollut. 39:409-420.

Carrere, R. 2001. Oil palm: The expansion of another destructive monoculture. p. 9-12 *En* R. Carrere (ed) The bitter fruit of oil palm: Dispossession and deforestation. World Rainforest Movement, Montevideo.

Carter, M.R. 1984. Identification of the inverse relationship between farm size and productivity: An empirical analysis of peasant agricultural production. Oxford Econ. Pap. 36:131-145.

Castells, M. 1996. The rise of the network society. Vol. I. Blackwell, Malden.

Castells, M. 1997. The power of identity. Vol. II. Blackwell, Malden.

Castells, M. 1998. End of millennium. Vol. III. Blackwell, Malden.

Castro, A.M.G. de, S.M. Lima, A. Maestrey, V. Trujillo, O. Alfaro, O. Mengo y M. Medina. 2001. La Dimensión de entorno en la construcción de la sostenibilidad institucional. Série Innovación para la Sostenibilidad Institucional. Proyecto ISNAR Nuevo Paradigma, San José.

CEPAL. 1997. La brecha de la equidad: America Latina, el Caribe, y La Cumbre Social. Naciones Unidas, Santiago.

CEPAL. 1999. El desafío de la equidad de género y de los derechos humanos en los albores del siglo xxi. Unidad mujer y desarrollo. Octava Conf. Regional sobre la Mujer de América Latina y el Caribe, 8 al 10 de febrero del 2000, Lima.

CEPAL. 2002. El impacto socioeconómico y ambiental de la sequía de 2001 en Centroamérica. Comisión Económica para América Latina y el Caribe (CEPAL). 28 de Febrero.

CEPAL. 2003. Anuario estadístico de América Latina y el Caribe 2003. División de Estadísticas y Proyecciones Económicas, Naciones Unidas, Santiago.

CEPAL. 2004a. Anuario estadístico de América Latina y el Caribe 2004. División de Estadísticas y Proyecciones Económicas. Naciones Unidas, Santiago.

CEPAL. 2004b. Boletín demográfico N° 76 de América Latina y el Caribe. Naciones Unidas, Santiago.

CEPAL. 2006a. Base de antecedentes de la División de la Estadística y Proyecciones Económicas. CEPAL, Unidad de Desarrollo Agrícola. Available at http://www.eclac.org/cgi-bin/getProd.asp?xml=/publicaciones/xml/3/28063/P28063.xml&xsl=/deype/tpl/p9f.xsl&base=/tpl/top-bottom.xslt. CAPAL, Naciones Unidas, Santiago.

CEPAL. 2006b. Anuario estadístico de América Latina y el Caribe 2006. División de Estadísticas y Proyecciones Económicas. Naciones Unidas, Santiago.

Cepredenac. 2007. Resumen de los daños causados por el Huracán Mitch. Available at http://www.cepredenac.org/04_temas/mitch/index.htm. Centro de Coordinación para la Prevención de Desastres en América Central, Ciudad de Guatemala, Guatemala.

Chacoff, N.P., A.A. Marcelo. 2006. Edge effects on flower-visiting insects in grapefruit plantations bordering premontane subtropical forest. J. Appl. Ecol. 43:18-27.

Chakravarty, P., L. Chatarpaul. 1990. Non-target effects of herbicides. I. Effect of glyphosate and hexazinone on soil microbial activity— microbial population, and in-vitro growth of ectomycorrhizal fungi. Pestic. Sci. 28:233-241.

Chambi, N., y W. Chambi. 1995. Ayllu y papas. Cosmovisión, religiosidad y agricultura en Conima, Puno. Asociación Chuyma de Apoyo Rural "Chuyma Aru," Lima.

Chant, S., y N. Craske. 2003. Gender in Latin America. Rutgers Univ. Press, New Brunswick, NJ.

Chaparro, F. 2000. La investigación agrícola internacional en un mundo globalizado. Foro Global de Investigación Agropecuaria FGIA/GFAR, Roma.

Chapela, I., D. Quist. 2001. Transgenic DNA introgressed into traditional maize landraces in Oaxaca, Mexico. Nature (London) 414:541-543.

Chiriboga, M., R. Grynspan, y L. Pérez. 1996. Mujeres del maiz. Banco Internacional de Desarrollo (BID). IICA, San Jose.

Chisholm, R. 1995. On the meaning of networks. Group and Organization Management 21:216-235.

CIA. 2008. The 2008 world factbook. Central Intelligence Agency, Washington DC.

CLADEHLT. 2002. La Tenencia de la tierra y la problemática alimentaria. Comisión Latinoamericana por los Derechos y Libertades de los Trabajadores y Pueblos, Caracas.

Clawson, D.L. 1985. Harvest security and intraspecific diversity in traditional tropical agriculture. Econ. Bot. 39:56-67.

Clay, J. 2004. World agriculture and the environment: A commodity-by-commodity guide to impacts and practices. Island Press, Washington DC.

Cleeton, J. 2004. Organic foods in relation to nutrition and helth: Key facts. *En* Coronary and diabetic care in the UK 2004. Assoc. Primary Care Groups and Trusts, UK.

Cleveland, D.A., y D. Soleri. 2005. Rethinking the risk management process for genetically engineered crop varieties in small-scale, traditionally based agriculture. Ecol. Soc. 10(1):9. http://www.ecologyandsociety.org/vol10/iss1/art9/. Resilience Alliance Publ., Waterloo.

Cline, W. R. 2007. Global warming and agriculture: Impact estimates by countries. Paterson Inst. Int. Econ., Washington DC.

Colborn, T., D. Dumanoski, y J.P. Myers. 1996. Our stolen future: are we threatening our fertility, intelligence, and survival? Penguin Books, NY.

Colchester, M., y A. Gray. 1998. Foreword. *En* A. Gray et al. (ed) From principles to practice: indigenous peoples and biodiversity conservation in Latin America. Proc. Pucallpa Conf. Pucallpa-Peru, 17-20 Mar 1997. Doc. No 87. IWGIA, Copenhagen.

Colorado Institute of Public Policy. 2004. Biopharming in Colorado: A guide to issues for making informed choices. Available at http://www.cipp.colostate.edu/pdf/bio_pharm_full.pdf. Colorado State Univ., Fort Collins.

Comunidad Andina. 2006. Estadísticas de remesas en los países de la Comunidad Andina 2005-2006. Comunidad Andina, Lima.

Conroy, M.E., D.L. Murray, y P.M. Rosset. 1996. A cautonary tale: Failed U.S. development policy in Central America. Lynne Reinner, Boulder.

Consumers International. 2004. Informe Conferencia Regional de Consumidores por una Alimentación Saludable. Bogotá, Aug 2004.

Crain, A.D., L.J. Guillette Jr, A.A. Rooney, and A.D.B. Pickford. 1997. Alterations in steroid genesis in alligators (*Alligator mississippiensis*) exposed naturally and experimentally to environmental contaminants. Environ. Health Perspect. 105:528-533.

CropLife International. 2007. Annual Report 2006-2007. CropLife, Brussels.

Crosby, A., Jr. 2004. Ecological imperialism. The biological expansion of Europe, 900-1900. Cambridge Univ. Press, UK.

Crosby, A. Jr., y W. Alfred. 1991. The biological consequences of 1492. NACLA 35:6-13.

Da Silva, J.G. 2004. Espacio rural y dimensión territorial del desarrollo en los países del Mercosur. Capítulo 4. *En* Temas Actuales y Emergentes para el Análisis Económico y la Investigación de Políticas. Vol. I. América Latina y el Caribe. Available at http://www.fao.org/docrep/006/y4940s/y4940s00.HTM. FAO, Rome.

Dalvie, M.A., E. Cairncross, A. Solomon, y L. London. 2003. Contamination of rural surface and ground water by endosulfan in farming areas of the Western Cape, South Africa. Environ. Health 2:1.

Daniell, H., S. Kumar, N. Dufourmontel. 2005. Breakthrough in chloroplast genetic engineering of agronomically important crops. Trends Biotech. 23(5):238-245.

Dary, C.F. 1991. Mujeres tradicionales y nuevos cultivos. Latin American Fac. Social Science (FLACSO), Guatemala City.

David, M.B. de A., C. Morales, y M. Rodríguez. 2001. Modernidad y heterogeneidad: Estilo de desarrollo agrícola y rural en América Latina y el Caribe. *In* B. David et al. (ed) Desarrollo Rural en América Latina y el Caribe: La Construcción de un Nuevo Modelo? Comisión Económica para América Latina y el Caribe (CEPAL) y Alfaomega, Santiago.

Davis, M. 2005. Planet slum. Verso, London.

De Janvry, A. 1981. The agrarian question and reformism in Latin America. John Hopkins Univ. Press, Baltimore.

De Marco, P. Jr., y F. Monteiro Coelho. 2004. Services by the ecosystem: Forest remnants influence agricultural cultures' pollination and production. Biodivers. Conserv. 13:1245-1255.

De Souza Silva, J., J. Cheaz, J. Santamaria-Guerra, M.A. Mato, S. Valle Lima, A.M.G. De Castro et al. 2005. La innovación de la innovación institucional: De lo universal, mecánico y neutral a lo contextual, interactivo y ético desde una perspectiva latinoamericana. Red Nuevo Paradigma, Quito, Octubre 2005.

Deere, C.D. 2005. The feminization of agriculture? Economic restructuring in rural Latin America. UN Res. Inst. Social Dev. Occas. Paper #1, Geneva.

Deininger, K., P. Olinto. 2000. Asset distribution, inequality and growth. Policy Res. Working Pap. Ser. 2375. World Bank, Washington DC.

Delgado, F.A., y B.D. Ponce B., D. 1999. Cosmovision andina para un desarrollo rural sustentable. Investigacion, Interaccion Social y Educacion Superior en Bolivia. Available at http://www.agruco.org/index.php?option=com_content&task=view&id=51&Itemid=44. Univ. Cochabamba, Bolivia.

Denevan, W.M. 1980. Latin America. P. 217-244 *In* G. Klee (ed) World systems of traditional resource management. Hasted Press, NY.

Deruyttere, A. 1997. Indigenous peoples and sustainable development. The role of the Inter-american Development Bank. Washington DC, Oct 1997, N° IND 97-101. Available at www.iadb.org/sds/IND/publication/publication_133_107_e.htm. Inter-Am. Dev. Bank, Washington DC.

Desmarais, A.A. 2002. The Via Campesina: Consolidating an international peasant and farm movement. J. Peasant Studies 29(2):91-124.

Dias de Oliveira, M.E., B.E. Vaughan, y E.J. Rykiel Jr. 2005. Ethanol as fuel: Energy, carbon dioxide balances, and ecological footprint. Bioscience 55:593-602.

Dicken, P. 1992. Global shift. Paul Chapman, London.

Dinerstein, E., D. Olson, D. Graham, A. Webster, S. Primm, M. Bookbinder et al. 1995. A conservation assessment of the terrestrial ecoregions of Latin America and the Caribbean. World Bank, Washington DC.

Diócesis de Quibdó. 2001. El precio de la tierra. PalmaDVD. Quibdó, Chocó, Colombia.

Dirven, M. 2004. Alcanzando las metas del milenio: una mirada hacia la pobreza rural y agrícola. CEPAL, Naciones Unidas, Santiago.

Dirven, M. 2004. El empleo rural no agrícola y la diversidad rural en América Latina. Revista de la CEPAL 83, Agosto 2004.

Diversity. 1991. Latin American centers of diversity. Diversity 7:1&2.

Dixon, J., A. Gulliver, and D. Gibbon. 2001. Farming systems and poverty: Improving farmers' livelihoods in a changing world. FAO and World Bank, Washington DC.

Dobbs, T.L., y J.N. Pretty. 2004. Agri-environmental stewardship schemes and "multifunctionality." Rev. Agric. Econ. 26:220-237.

Dollar, D., y A. Kraay. 2000. Growth Is good for the poor. Dev. Res. Group, World Bank, Washington DC.

Donald, P.F. 2004. Biodiversity impacts of some agricultural commodity production systems. Conserv. Biol. 18:17-37.

Donnegan, K.K., C.J. Palm, V.J. Fieland, L.A. Porteous, L.M. Ganis, D.L. Scheller, y R.J. Seidler. 1995. Changes in levels, species, and DNA fingerprints of soil micro organisms associated with cotton expressing the *Bacillus thuringiensis* var. Kurstaki endotoxin. Appl. Soil Ecol. 2:111-124.

Durston, J. 2002. El capital social campesino en la gestión del desarrollo rural, díadas, equipos, puentes y escaleras. CEPAL, Naciones Unidas, Santiago.

Dussel, E. 2007. América Latina renace la política desde el pueblo. Entrevista con el filósofo Enrique Dussel. 3 Sep 2007. Available at http://www.voltairenet.org/article151199.html#article151199. Voltairnet.org, Paris.

Eggers, B., y R. Mackenzie. 2000. The Cartagena Protocol on Biosafety. J. Int. Econ. Law 2000:525-543.

Elbeheri, A. 2005. Biopharming and the food system: Examining the potencial benefits and risks. AgBioforum 8(1):18-25. Illinois-Missouri Biotech. Alliance, Columbia.

El-Hage Scialabba, E., y C. Hattam. 2007. Organic agriculture and food availability. Int. Conf. Organic Agriculture and Food Security, 3-5 May, FAO, Rome.

Ellstrand, N.C. 2003. Going to great lengths to prevent the escape of genes that produce specialty chemicals. Plant Physiol. 132:1770-1774.

EPA. 2002. Glyphosate; Pesticide tolerances. Fed. Reg. Environ. Doc. Vol. 67 No. 188. 27 Sep 2002. US EPA, Washington DC.

Escobar, A. 1995. Encountering development: The making and unmaking of the Thirds World. Princeton Univ. Press, NJ.

Escobar, A. 1998a. La invención del Tercer Mundo: Construcción y deconstrucción del desarrollo. Editorial NORMA, Barcelona.

Escobar, A. 1998b. Whose knowledge, whose nature? Biodiversity, conservation, and the political ecology of social movements. J. Polit. Ecol. 5:53-82.

Escobar, A.1999. Biodiversity: A perspective from within. www.grain.org/seedling/?id=3. Grain, Barcelona.

Estok, D., B. Freedman, y D. Boyle. 1988. Effects of the herbicides 2,4-D, glyphosate, hexazinone, and triclopyr on the growth of three species of ectomycorrhizal fungi. Bull. Environ. Contam. Toxicol. 42:835-839.

ETC Group. 2005. Oligopoly, Inc. 2005: Concentration of corporate power. Available at http://www.etcgroup.org/upload/publication/44/01/oligopoly2005_16dec.05.pdf. ETC Group, Ottawa.

ETC Group. 2007. Gambling with Gaia. ETC Communiqué. Issue # 90. Jan 2007. Available at http://www.biodiversidadla.org/content/view/full/29901. ETC Group, Ottawa.

Evenson, R.E. y D. Gollin. 2003. Assessing the impacts of the Green Revolution, 1960 to 2000. Science 300:758-762.

Ewell, P.T., y D. Merrill-Sands. 1987. Milpa in Yucatan. A long-fallow maize system and its alternatives in the Maya peasant economy. p. 95-129. *En* B.L. Turner y S. Brush (ed) Comparative farming systems. The Guilford Press, NY.

Faiguenbaum, S., J.A. Berdegué, y T. Reardon. 2002. The rapid rise of supermarkets in Chile: Effects on dairy, vegetables, and beef chains. Dev. Policy Rev. 20(4):459-471.

Fajnzylberg, F. 1990. Unavoidable industrial restructuring in Latin America: Were we stand. Duke Univ. Press, Durham.

FAO. 1986. Mujer y la Revolución Verde. Available at http://www.fao.org/FOCUS/S/Women/green-s.htm. FAO, Roma.

FAO. 1998. Agriculture 21. Noviembre. Available at http://www.fao.org/ag/. FAO, Roma.

FAO. 2000. Food safety and quality as affected by organic farming. 22nd FAO Reg. Conf. Europe, Porto, Portugal, 24-28 Jul 2000.

FAO 2002. Organic agriculture, environment and food security. *In* N. El-Hage Scialabba and C. Hattam (ed) Environment and Natural Resources Series 4. Available at http://www.fao.org/DOCREP/005/Y4137E/Y4137E00.htm. FAO, Roma.

FAO. 2004. El estado mundial de la agricultura y la alimentación 2003-04. ¿Biotecnologías agrícolas para satisfacer las necesidades de los pobres? Available at http://www.fao.org/docrep/006/y5160s/y5160s00.htm. FAO, Roma.

FAO. 2006a. Estadísticas. Available at http://faostat.fao.org/. FAO, Roma

FAO. 2006b. Programa especial para la seguridad alimentaria. Available at http://www.pesacentroamerica.org/pesa_ca/reflexion_indicadores.htm.pdf. FAO, Roma.

FAO. 2007. Incidentes y deficiencias en los sistemas de inocuidad de los alimentos. FAO y OMS instan a todos los países a que refuercen la vigilancia. Available at www.fao.org/newsroom/eS/news/2007/1000629/index.html. FAO, Roma.

FAOSTAT. 2005. FAOSTAT on-line statistical service. Available at http://faostat.fao.org/. FAO, Roma.

FAPRI. 2006. U.S. baseline briefing book. FAPRI-UMC Rep. 01-06. Available at http://www.fapri.missouri.edu/outreach/publications/2006/FAPRI_UMC_Report_01_06.pdf. FAPRI, Columbia.

Farah, M.A. 2004a. Algunos elementos de análisis sobre el trabajo rural remunerado y no remunerado en América Latina, desde una perspectiva de género. (With English abstract) *En* Revista de Fomento Social. 59(236). ETEA, España, Octubre-Diciembre 2004.

Farah, M.A. 2004b. Globalizacion, pobreza y mujeres rurales en America Latina. En otras palabras 13-14: Enero-Diciembre 2004. Grupo Mujer y Sociedad. Univ. Nacional de Colombia, Bogotá.

Farah, M.A. y E. Pérez. 2004. Mujeres rurales y nueva ruralidad en Colombia. Cuadernos de Desarrollo Rural 51:139-160.

Farina, E.M., R. Nuñes, y G.F. Monteiro. 2004. Modelling produce procurement and merchandising strategies of traditional retailers in the face of competition with supermarkets in the São Paulo Region. 88th Seminar Eur. Assoc. Agric. Econ., Paris, 5-6 May.

Fearnside, P.M. 2001a. Land tenure issues as factors in environmental destruction in Brazilian Amazonia: The case of southern Pará. World Dev. 29:1361-1372.

Fearnside, P.M. 2001b. Soybean cultivation as a threat to the environment in Brazil. Environ. Conserv. 28:23-28.

Fearnside, P.M. 2005. Deforestation in Brazilian Amazonia: History, rates, and consequences. Conserv. Biol. 19:680-688.

Feder, G. 1984. The relation between farm size and farm productivity. J. Dev. Econ. 18:297-313.

Ferranti, David de, G.E. Perry, F. Ferreira, y M. Walton. 2004. Inequality in Latin America and the Caribbean: Breaking with history? World Bank, Washington DC.

Figà-Talamanca, I. 2006. Occupational risk factors and reproductive health of women. Occup. Med. 56:521-531.

Flower, K.B., J.A. Hoppin, C.F. Lynch, A. Blair, C. Knott, D.L. Shore, y D.P. Sandler. 2004. Cancer risk and parental pesticide application in children of Agricultural Health Study participants. Environ. Health Perspect. 112:631-635.

Fowler, C., y P. Mooney. 1990. Shattering. Food politics, and the loss of genetic diversity. The Univ. Arizona Press. Tucson.

Francis, C.A. 1986. Multiple cropping systems. MacMillan, NY.

Friedland, W., L. Busch, F. Buttel, y A. Rudy (ed) 1991. Towards a new political economy of agriculture. Westview Press, Boulder.

Friedman, H. 1993. The political economy of food: A global crisis. New Left Rev. 1993:29-57.

Funes, F., L. Garcia, M. Bourque, N. Perez y P. Rosset (ed) 2001. El movimiento cubano de agricultura orgánica. p. 15-38. *En* Transformando el campo cubano: Avances de la agricultura sostenible. Food First-ACTAF-CEAS, La Habana.

Gajardo, G., y L. Laikre. 2003. Chilean aquaculture boom is based on exotic salmon resource: A conservation paradox. Conserv. Biol. 17:1173-1174.

Gallagher, K.P. 2004. Free trade and the environment: Mexico, NAFTA and beyond. Stanford Univ. Press, Palo Alto.

Ghezán, G., M. Mateos, y L. Viteri. 2002. Impact of supermarkets and fast-food chains on horticulture supply chains in Argentina. Dev. Policy Rev. 20:389-408.

Giberti, H. 2002. Cincuenta años de evolución de la agricultura argentina. Available at http://www.inta.gov.ar/balcarce/info/documentos/econo/rural/giberti.htm. Instituto Nacional de Tecnología Agropecuaria, Balcare.

Gilly, A. 2005. The emerging "threat" of radical populism. NACLA 39(2):37-40.

Glaeser, B. (ed) 1987. The green revolution revisited: Critique and alternatives. Allen & Unwin, London.

Gliessman, S.R. 1998. Agroecology: Ecological process in sustainable agriculture. Arbor Press, Ann Arbor, MI.

Glipo, A. (ed) 2003. Acuerdo sobre agricultura y soberanía alimentaria: Perspectivas de Mesoamérica y Asia. Global Issue Pap. 3. Available at http://www.boell.de/alt/downloads/global/GIP%203%20Agricultura_span.pdf. Fundación Heinrich Böll, Berlin.

Global Crop Diversity Trust. 2007. Priority Crops. Maize. Available at http://www.croptrust.org/main/priority.php?itemid=30. Global Crop Diversity Trust, FAO, Roma.

Goldberg, R.J. 1992. Environmental concerns with the development of herbicidetolerant plants. Weed Tech. 6:647-652.

Goldschmidt, W. 1978. As you sow: Three studies in the social consequences of agribusiness. Allenheld, Osmun, NY.

Goldstein, M.I., T.E. Lacher, M.E. Zaccagnini, M.L. Parker, y M.J. Hooper. 1999. Monitoring and assessment of Swainson's Hawks in Argentina following restrictions on monocrotophos use. Ecotoxicology 8:215-224.

Gómez, S. 2000. Organizaciones rurales en América Latina (marco para su análisis). Rev. Austral de Ciencias Soc. (4):27-54.

Gómez, I. y G. Gallopín. 1995. Potencial agrícola de América Latina. *En* G. Gallopín et al. (ed) El futuro ecológico de un continente: Una visión prospectiva de la América Latina. Vol. I. Editorial de las Naciones Unidas/ Fondo de Cultura Económica, Colección Lecturas, El Trimestre Económico, México.

Gómez, L.M.A. 2001. Producción de vacunas y compuestos farmacéuticos en plantas transgénicas. Avance y Perspectiva 20:365-375.

Gonzales, T. 1996. Political ecology of peasantry, the seed, and NGOs in Latin America: A study of Mexico and Peru, 1940-1996. Ph.D. thesis. Univ. Wisconsin, Madison.

Gonzales, T. 1999. The cultures of the seed in the Peruvian Andes. pp. 193-216. *En* S.B. Brush (ed) Genes in the field: On farm conservation of crop diversity. IDRC Books, Ottawa.

Gonzales, T. 2009. Sense of place and indigenous people's biodiversity conservation in the Americas. *En* V. Nazarea y R. Rhoades (ed) Seeds of Resistance/Seeds of Hope: Crossing Borders in the Repatriation and In situ Conservation of Traditional Crops. Univ. Arizona Press, Tucson.

Gonzales, T., N. Chambi, y M. Machaca. 1999. Agricultures and cosmovision in contemporary Andes. *En* D.A. Posey (ed) Cultural and spiritual values of biodiversity. Intermediate Tech. Publ. UNEP, Nairobi.

Goodman, D., y M.J. Watts (ed) 1998. Globalizing food: Agrarian questions and global restructuring. Routledge, London.

Goodman, D., y M. Redclift (ed) 1991. Environment and development in Latin America: The politics of sustainability. Manchester Univ. Press, NY.

Goudie, A. 1990. The human impact on the natural environment. MIT Press, Cambridge, MA.

Gould, F. 1994. Potential and problems with high-dose strategies for pesticidal engineered crops. Biocontrol Sci. Tech. 4:451-461.

Grain 2005. Soberanía alimentaria y sistema alimentario mundial. Available at http://www.grain.org/biodiversidad/?id=305. GRAIN, Barcelona.

Gratius, S., y F.E. Stiftung. 2002. El proyecto del ALCA visto desde Europa (II). La Insignia, 15 de octubre 2002. Available at http://www.lainsignia.org/2002/octubre/dial_003.htm. La insignia, Madrid.

Gray, J. 2004. Una visión con futuro. Letras Libres 71:12-17.

Grillo., E. 1998. Development or cultural affirmation in the Andes? p. 124-145. *En* F. Apffel-Marglin con PRATEC (ed) The spirit of regeneration. Andean culture confronting Western notions of development. Zed Books, NY.

Groppo, P. 1997. La FAO y la reforma agraria en América Latina: Hacia una nueva visión. FAO, Roma.

Gutman, G.E. 2002. Impact of the rapid rise of supermarkets on dairy products in Argentina. Dev. Policy Rev. 20:409-427.

Halberg, N., H.F. Alrøe, M.T. Knudsen, E.S. Kristensen (ed) 2005. Global development of organic agriculture: Challenges and prospects. CABI Publ., UK.

Hall, D.O. 1998. Food security: What have sciences to offer? Available at http://www.icsu.org/Gestion/img/ICSU_DOC_DOWNLOAD/221_DD_FILE_Foof_Security.pdf. Int. Council Science, Paris.

Hall, G., y H.A. Patrino. 2005. Pueblos indígenas, pobreza y desarrollo humano en América Latina: 1994-2004. Banco Mundial, Washington DC.

Hamilton, S., y E. Fisher. 2003. Non-traditional agricultural exports in highland Guatemala: Understanding of risk and perceptions of change. Latin Am. Res. Rev. 38(3): 82-110.

Hamilton, S., L. Asturias de Barrios, y B, Tevalán. 2001. Gender and commercial agriculture in Ecuador and Guatemala. Culture Agric. 23(3):1-12.

Han, M., T. Su, Y.G. Zu, y Z.G. An. 2006. Research advances on transgenic plant vaccines. Acta Genet. Sinica 33(4):285-293.

Harvey, C.A., N.I.J. Tucker, y A.Estrada. 2004. Live fences, isolate trees, and windbreaks: Tools for conserving biodiversity in fragmented tropical landscapes. p. 261-2289. *En* G. Schroth et al. (ed) Agroforestry and biodiversity conservation in tropical landscapes. Island Press, Washington DC.

Harvey, C., F. Alpizar, M. Chacon, and R. Madrigal. 2005. Assessing linkages between agriculture and biodiversity in Central America: Historical overview and future perspectives. The Nature Conservancy, San Jose.

Hassan, S.A., F. Bigle, H. Bogonschuetz, E. Boller, J. Brun et al. 1988. Results of the forth joint pesticide testing programme carried out by the IOBC/WPRS-Working group pesticides and beneficial organisms. J. Appl. Entomol. 105:321-329.

Hayes, T. 2005. From silent spring to silent night: Sndocrine disruption, amphibian declines and environmental justice. Pestic. News 70.

Hecht, S.B. 1984. Indigenous soil management in the Amazon basin: some implications for development. p. 166-181. *In* J.O. Browder (ed) Fragile lands of Latin America. Westview Press, Boulder.

Hecht, S.B. 1992. The logics of livestock and deforestation. *In* T.E. Downing et al. Development or destruction: The conversion of tropical forest to pasture in Latin America. Westview Press, Boulder.

Heiser, C.B. Jr. 1990. Seed to Civilization: The story of food. Harvard Univ. Press, Cambridge.

Heywood, V.H., and R. Watson (ed) 1995. Global biodiversity assessment. Cambridge Univ. Press, UK.

Hilbeck, A., M. Baumgartner, P.M. Fried, and F. Bigler. 1998. Effects of transgenic Bacillus thuringiensis corn fed prey on mortality and development time of immature Chrysoperla carnea Neuroptera: Chrysopidae. Environ. Entomol. 27:460-487.

Hoffman, K., y M.A. Centeno. 2003. The lopsided continent: inequality in Latin America. Ann. Rev. Sociology 29:363-390.

Holt, J.S., y H.M. Le Baron. 1990. Significance and distribution of herbicide resistance. Weed Tech. 4:141-149.

Holt-Giménez, E. 2002. Measuring farmer's agroecological resistance after Hurricane Mitch in Nicaragua: A case study in participatory, sustainable land management impact monitoring. Agric. Ecosyst. Environ. 93:87-105.

Holt-Giménez, E. 2006. Compesino a compesino: Voices from Latin America's farmer to farmer movement for sustainable agriculture. Food First, Oakland, CA.

Huizer, G. 1999. People's Spirit of Resistance in Latin America. p. 165-176. *En* Food for thought. Ancient visions and new experiments of rural people. B. Haverkort and W. Hiemstra (ed) COMPAS, Netherlands.

Huot, M.F. 2003. Plant molecular farming: Issues and challenges for Canadian regulators, options consommateurs. Available at http://www.option-consommateurs.org/dc_pdf/pdf/molecular_farming_oc0603.pdf. Consumer, Affaire Office, Industry Canada.

IDB. 1989. IDB Notes. Inter-American Dev. Bank, Washington DC.

IDB. 2004. Operational Policy on indigenous peoples. Inter-American Dev. Bank, Washington DC.

IEA. 2004. Biofuels for transport: An international perspective. Int. Energy Agency, Paris.

IICARD. 1989. Poverty, conflict, and hope: A turning point in Central America. Int. Comm. Central Am. Reconstruction and Dev. and Duke Univ. Int. Dev. Res., Durham.

ILEC. 2005. Managing lakes and their basics for sustainable use: A report for lake basin managers and stakeholders. Int. Lake Environ. Committee Foundation, Kusatsu, Japan.

Iltis, H. 1994. New Year's card leads to newly discovered species of enormous economic potential. #103 R&D Innovator 3(6). http://www.winstonbrill.com/bril001/html/article_index/articles/101-150/article103_body.html

IPCC. 1996. Climate change 1995: Economic and social dimensions of climate change. Contribution of Working Group III to the second assessment report of the Intergovernmental Panel on Climate Change. Cambridge Univ. Press, UK.

IPCC. 1997. Impactos regionales del cambio climático: evaluación de la vulnerabilidad. Cambridge Press, UK.

IPCC. 2001a. Impactos, adaptación y vulnerabilidad. Resumen técnico. Cambridge Press, UK.

IPCC. 2001b. Latin America. Chapter 6. *In* O.F. Canciani and S. Díaz (ed) IPCC special report on the regional impacts of climate change: An assessment of vulnerability. Cambridge Press, UK. http://www.grida.no/climate/ipcc/regional/122.htm

IPCC. 2007. Climate change 2007: The physical science basis. Summary for Policymakers. IPCC and Cambridge Press, UK.

Isenring, R. 2006. Paraquat: Unacceptable health risks for users. Berna Declaration. Available at http://www.evb.ch/cm_data/EvB_Paraquat_E.pdf. PAN UK, PAN Asia and the Pacific.

Ishizawa, J. 2006. What Next? From Andean cultural affirmation to Andean affirmation of cultural diversity-learning with the communities in the Central Andes. Available at www.dhf.uu.se/whatnext/papers_public/Ishizawa-Draft-01Sep2006.pdf.

James, C. 2006. Situación global de los cultivos trangénicos/GM comercializados: 2006. Resumen Ejecutivo. Brief 35. Available at http://www.isaaa.org/resources/publications/briefs/35/executivesummary/pdf/Brief%20 35%20-%20Executive%20Summary%20 -%20Spanish.pdf. ISAAA, Ithaca, Manila.

Jason, C. 2004. World agriculture and the environment. Island Press, Washington.

Jordan, J.F. 2001. Genetic engineering, the farm crisis, and world hunger. BioScience 52:523-529.

Kaimowitz, D., and J. Smith. 2001. *In* A. Angelsen and D. Kaimowitz (ed) Agricultural technologies and tropical deforestation. CAB Int., UK.

Kammerbauer J. and Moncada J. 1998. Pesticide residue assessment in three selected agricultural production systems in the Choluteca River Basin of Honduras. Environ. Pollut. 103(2-3):171-181.

Karen, B. 2004. The persistence of pesticide dependence— A review of World Bank projects and their compliance with the World Bank's management policy, 1999-2003. Pesticide Action Network North America, San Francisco.

Katz, E. 2003. The changing role of women in the rural economies of Latin America. *In* B. Davis (ed) Food, agriculture and rural development: Current and emerging issues for economic analysis and policy research – II. Vol. I. Latin America and the Caribbean. FAO, Rome.

Kearney, M. 1996. Introduction to ethnicity and class in Latin America. Latin Am. Perspect. 23(2):5-16.

Kendall, H.W., R. Beachy, T. Eismer, F. Gould, R. Herdt, P.H. Ravon et al. 1997. Bioengineering of crops. World Bank, Washington DC.

Kilcher, L. 2007. How organic agriculture contributes to sustainable development. JARTS, Supplement 89:31-49.

Klein, A.M., I. Steffan-Dewenter, and T. Tscharntke. 2003. Fruit set of highland coffee increase with the diversity of pollinators. Proc. R. Soc. Lond. B. 270: 955-961.

Klink, C.A., and R.B. Machado. 2005. Conservation of the Brazilian cerrado. Conserv. Biol. 19:707-713.

Ko, K., and H. Koprowski. 2005. Plant biopharming of monoclonal antibodies. Virus Res. 111:93-100.

Korzeniewicz, P.R., and WC. Smith. 2000. Poverty, inequality, and growth in Latin América: Searching for the high road to globalization. Latin Am. Res. Rev. 35(3): 7-48.

Kramer, P.A., and P.R. Kramer. 2002. Ecoregional conservation planning for the Mesoamerican Caribbean reef (MACP). M. McField (ed) World Wildlife Fund, Washington DC.

Kurtz, DA. 1990. Long range transport of pesticides. Lewis Publ., MI.

Lander, E. 2000. Eurocentrism and colonialism in Latin American social thought. Nepantla: Views from South 1:519-532.

Lara, S.M. 1992. La flexibilidad del mercado de trabajo rural: Una propuesta que involucra a las mujeres.Rev. Mexicana Sociol. 54: 29-48.

Lara, S.M. 1995. La feminización del trabajo asalariado en los cultivos de exportación no tradicionales en América Latina: Efectos de una flexibilidad salvaje. *In* S.M. Lara Flores (ed) El Rostro Femenino del Mercado de Trabajo Rural en América Latina. UNRISD and Nueva Sociedad, Caracas.

Lastarria-Cornhiel, S., y J. Melmed-Sanjal. 1998. Land tenancy in Asia, Africa and Latin América: A look at the past and a view to the future. Land Tenure Center, WI.

Ledec, G. 1992. New directions for livestock policy: An environmental perspective. p. 27-65. *In* T.E. Downing et al. (ed) Development or destruction: The conversion of tropical forest to pasture in Latin America. Westview Press, Boulder.

Lederman, D., W.F. Maloney and L. Servén. 2003. Lessons from NAFTA for Latin America and Caribbean (LAC) Countries: A summary of research findings. World Bank, Washington DC.

Leff, E., y J. Carabias. 1993. Cultura y manejo sustentable de los recursos naturales. Vol. I y II. *En* Cultura y Manejo Sustentable de los Recursos Naturales. E. Leff y J. Carabias (ed) Univ. Nacional Autonoma de Mexico, México.

Lehmann, M.P. 1992. Deforestation and changing land-use patterns. p. 58-76. *In* H.K. Steen and R.P. Tucker (ed) Changing tropical forests: Historical perspectives on today's challenges in Central & South America. Proc. Conf. Forest History Society and IUFRO Forest History Group.

León S. T.E. 2007. Medio ambiente, tecnología y modelos de agricultura en Colombia. Hombre y arcilla. Inst. Estudios Ambientales, IDEA. Univ. Nacional de Colombia. ECOE Ediciones, Bogotá.

Lernoud, A.P. 2007. Organic farming in Latin América. p. 155-156. *In* H. Willer and M. Jussefi (ed) The world of organic agriculture. Statistics and emerging trends 2007. IFOAM, Bonn.

Levario-Carrillo, M., D. Amato, P. Ostroski, C. Gonzalez-Horta, Y. Corona, and L.H. Sanin. 2003. Relation between pesticide exposure and intrauterine growth retardation. Chemosphere 55(10):1421-1427.

Licht, F.O. 2005. Homegrown for the Homeland: Industry Outlook 2005 Renewable Fuels Assoc., Washington DC.

Lisansky, J. 1990. Migrants to Amazonia: Spontaneous colonization in the Brazilian frontier. Westview Press, Boulder.

Liverman, D.M. y K.L. O'Brien. 1991. Global warming and climate change in Mexico. Global Environ. Change 1(4):351-363.

Londoño, J.L., y M. Szekely. 1997. Persistent poverty and excess inequality, 1970-1995. IDB Working Pap. 357. Inter-Am. Dev. Bank, Washington DC.

Losey, J.J.E., L.S. Rayor, and M.E. Carter. 1999. Transgenic pollen harms monarch larvae. Nature 399: 214.

Lumbreras, L.G. 1991. Misguided development. NACLA 24:18-22.

Lusting, N. (ed) 1995. Coping with austerity: poverty and inequality in Latin America. Brookings Institution, Washington DC.

Ma, J.K.C. et al., 2005. Molecular farming for new drugs and vaccines. Current perpectives on the production of pharmaceuticals in transgenic plants. EMBO Rep. 6:593-599.

Ma, J.K.C. 2003. The production of recombinant pharmaceuticals proteins in plants. Nature Rev. 4:794-805.

Machaca, M. 1998. Reflections: Planning, qipa hampaq, ñawpapaq, patachay? Forests Trees People Newsl. 35:21.

Machaca, M. 1996. La crianza de la biodiversidad y la cultura Andina. p. 102-122. *En* PRATEC (ed) La cultura andina de la biodiversidad. PRATEC, Lima.

Maffi, L. 2001. On biocultural diversity: Linking language, knowledge, and the environment. Smithsonian Institution Press, Washington DC.

Magalhães, A.R., y M.H. Glantz (ed) 1992. Socioeconomic impacts of climate variations and policy responses in Brazil. UNEP, SEPLAN, Esquel Brasil Foundation, Brasilia.

Mahar, D.J. 1989. Government policies and deforestation in Brazil's Amazon region. World Bank, Washington DC.

Mallet, J., and P. Porter. 1992. Preventing insect adaptations to insect resistant crops: are seed mixtures or refugia the best strategy? Proc. Roy Soc. London B 250:165-169.

Mascia, P.N., and R.B. Flavell. 2004. Safe and acceptable strategies for producing foreign molecules in plants. Curr. Opinion Plant Biol. 7:189-195.

Matson, P.A., W.J. Parton, A.J. Power, and M. Swift. 1997. Agricultural intensification and ecosystem properties. Science 277:504-509.

Maxwell, D. 1995. Measuring food insecurity: the frequency and severity of "coping strategies". FCND Disc. Pap. 8. Disponible at http://www.ifpri.org/divs/fcnd/dp/papers/dp08.pdf. IFPRI, Washington DC.

McConnell, R., S. Henao, O. Nieto, L. Rosenstock, A. Zanaga, and C. Wesseling. 1993. Pesticides. p. 147-201. *En* Environmental epidemiology: A project for Latin America and the Caribbean. Pan Am. Center Human Ecol. Health, WHO, ECO, México.

McMichael, P. (ed) 1994. The global restructuring of agro food systems. Cornell Univ. Press, Ithaca.

McNeely, J.A. and S.J. Scherr. 2003. Ecoagriculture. Island Press, Washington DC.

Meier, V. 1999. Cut-flower production in Colombia: A major development success for women? Environ. Planning 31:273-289.

Meilleur, B. A. and T. Hodgkin. 2004. *In situ* conservation of crop wild relatives: Status and trends. Biodivers. Conserv. 13:663-684.

MA (Millennium Ecosystem Assessment). 2005a. Ecosystems and human well-being: Biodiversity Synthesis. World Resources Inst., Island Press, Washington DC.

MA (Millennium Ecosystem Assessment). 2005b. Ecosystems and human well-being: Synthesis. Island Press, Washington DC.

Miller, M. (ed) 1996. The technical and economic feasibility of replacing methyl bromide in developing countries. Friends of the Earth, Washington DC.

Mittermeier, R.A., N. Myers, C.G. Mittermeier. 2000. Hotspots: Earth's biologically richest and most endangered terrestrial ecoregions. CEMEX, Mexico City.

Mohamed, A.I., G.A. Nair, H.L. Abbas, and H.H. Kassam. 1992. Effects of pesticides on the survival, growth and oxygen consumption of *Hemilepistus reaumurei* (Isopoda: Oniscidea). Trop. Zool. 5:145-153.

Montenegro, C., M. Strada, M.C. Parmucci, I. Gasparri, and J. Bono. 2003. Mapa Forestal Provincia del Chaco. Actualización Año 2002. UNSEF, Dirección de Bosques, Secretaria de Ambiente y Desarrollo Sustenable, Buenos Aires.

Mooney, P.R. 2002. El Siglo ETC: Erosión, transformación tecnológica y concentración corporativa en el siglo XXI. Dag Hammarskjold Foundation, Winnipeg.

Morris, M.L., and M.A. López-Pereira. 1999. Impacts of maize breeding research in Latin America 1966-1997. CIMMYT, Mexico.

Murray, D., C. Wesseling, M. Keifer, M. Corriols, y S. Henao. 2002. Surveillance of pesticide-related illness in the developing world: putting the data to work. J Int. Occup. Environ. Health 8:243-248.

Murray, W. 2003. From dependency to reform and back again: The Chilean peasantry during the twentieth century. *In* T. Brass (ed), Latin American peasants. Frank Cass, London.

Myers N., R.A. Mittermeier, C.G. Mittermeier, G.A. da Fonseca, and J. Kent. 2000. Biodiversity hotspots for conservation priorities. Nature 403:853-858.

Myers, N. 1980. Conversion of tropical moist forests. National Acad. Sciences, Washington DC.

Nadal, A., 2000. The environmental and social impacts of economic liberalization on corn production in Mexico. World Wide Fund for Nature, Oxfam GB, London.

Naeem, S., L.J. Thompson, S.P. Lawier, J.H. Lawton, R.M. Woodfin. 1994. Declining biodiversity can alter the performance of ecosystems. Nature 368:734-737.

Najmanovich, D. 1995. El lenguaje de los vínculos: De la independencia absoluta a la autonomía relativa. p. 33-76. *In* E. Dabas and D. Najmanovich (ed) Redes-El lenguaje de los vínculos: Hacia la reconstrucción y el fortalecimiento de la sociedad civil. Paidos, Buenos Aires.

Narby, J., y F. Huxley (ed) 2004. Shamans through time: 500 years on the path to knowledge. J.P. Tarcher/Penguin, New York.

Narby, J. 2007. An anthropologist explores biomolecular mysticism in the Peruvian Amazon. Shaman's Drum 74:43-46.

Narlikar, A., and D. Tussie. 2004. The G20 at the Cancun Ministerial: Developing countries and their evolving coalitions in the WTO. World Dev. 27(7):947-966.

National Academy of Science (NAS). 2002. Animal biotechnology: Science-based concerns. Available at http://darwin.nap.edu/books/0309084393/html/R9.html. NAS, Washington DC.

Nations, J.D. 1992. New directions for livestock policy. *In* T.E. Downing et al. (ed) Development or Destruction: The conversion of tropical forest to Pasture in Latin America. Westview Press, Boulder.

Netting, R.M. 1974. Agrarian ecology. Ann. Rev. Anthropol. 3:21-56.

Nicholls, C.I., and M.A. Altieri. 1997. Conventional agricultural development models and the persistence of the pesticide treadmill in Latin America. Int. J. Sustain. Dev. World Ecol. 4(2).

Nielsen, A.C. 1999. Censo de comercio minorista 1997/1998. A.C. Nielsen SA, Buenos Aires.

Nivia, E. 2000. Mujeres y Plaguicidas. Una mirada a la situación actual, tendencias y riesgos de los plaguicidas. Estudio de caso en Palmira, Colombia. RAPAL-Ecofondo, Palmira.

Nivia, E. 2005. Deuda ecológica por el uso de agrotóxicos. *En* Memorias del Tribunal por la Soberanía Alimentaria: Juicio al BM y al BID, deuda social y ecológica en la agricultura ¿Cuánto nos deben el Banco Mundial y el BID? Casos y veredicto. Inst. Estudios Ecologistas del Tercer Mundo y Jubileo Sur, Quito.

Norris, D.E. 2004. Mosquito-borne diseases as a consequence of land use changes. EcoHealth 1:19-24.

Nyéleny. 2007. Declaración final del 1° Foro Mundial por la Soberanía Alimentaria. Selingué, Mali. Febrero 23 al 27 de 2007. Available at http://www.nyeleni2007.org

O'Donnell, G. and V. Tokman (ed) 1998. Poverty and inequality in Latin America. Notre Dame Press, IN.

Olson, D.M., and E. Dinerstein. 2002. The global 200: Priority ecoregions for global conservation. Ann. Missouri Bot. Garden 89:199-224.

OPS Proyecto PLAGSALUD. 2003. Plagsalud Costa Rica, una mirada al pasado, presente y futuro (agosto 1994-septiembre 2003). Ministerio de Salud de Costa Rica, San Jose.

Orozco, M. 2002. Globalization and migration: The impact of family remittances in Latin America. Latin Am. Polit. Society 44:41-66.

Ortega, E. 1986. Peasant agriculture in Latin America. Joint ECALC/FAO Agric. Div., Santiago.

Ortiz-García, S., E. Ezcurra, B. Schoel, F. Acevedo, J. Soberón, y A.A. Snow. 2005. Absence of detectable transgenes in local landraces of maize in Oaxaca, Mexico (2003-2004). PNAS 1021:12338-12343.

OSAL (Observatorio Social de América Latina). 2005. Movimientos sociales y recursos naturales en América Latina: resistencias al neoliberalismo, configuración de alternativas. http://www.bilaterals.org/article.php3?id_article=3099.

Oviedo, G. 1999. La perspectiva del WWF sobre la conservación con los Pueblos Indígenas. WWF-Fondo Mundial para la Conservacion. WWF Internacional, UK.

Oxfam. 2004. Trading away our rights: Women working in global supply chains. Available at http://www.oxfam.org/en/files/report_042008_labor.pdf. Oxfam, Boston.

PAN Germany. 2003. Paraquat exposure and Parkinson's Disease. Fact Sheet 2003. PAN, Hamburg.

PAN. 2007. Documento de posición sobre eliminación de plaguicidas. Available at www.pan-international.org. Pesticide Action Network, San Francisco.

Paoletti, M.G., and D. Pimentel. 1996. Genetic engineering in agriculture and the environment: assessing risks and benefits. BioScience 46:665-671.

Papa, J. 2000. Malezas tolerantes que pueden afectar el cultivo de soja. Nat. Inst. Agric. Res. (INTA), Santa Fe Regional Center, Extension Agency, Oliveros.

Parrott, N. and T. Marsden. 2002. The real green revolution. Green Peace Environ. Trust, London.

Partridge, W.L. 1989. The human ecology of tropical land settlement in Latin America: An overview. p. 3-19. *In* D.A. Schumann and W.L. Partridge (ed) The human ecology of tropical land settlement in Latin America. Westview Press, Boulder.

Pauly, D., y I. Tsukayama. 1987. The Peruvian anchoveta and its upwelling ecosystem: Three decades of change. *In* ICLARM Studies Rev. 15, Instituto del Mar del Peru (IMARPE), Callao, GTZ, Eschborn and ICLARM, Manila.

Pengue, W.A. 2005. Transgenic crops in Argentina: the ecological and social debt. Bull. Sci. Tech. Society 25:314-322.

Perfecto, I. y I. Armbrecht. 2003. The coffee agroecosystem in the Neotropics: Combining ecological and economic goals. p. 159-194. *En* J. Vandermeer (ed) Tropical agroecosystems. CRC Press, Boca Raton.

Perfecto, I., R. Rice, R. Greenberg, and M. Van der Voolt. 1996. Shade coffee as refuge of biodiversity. BioScience 46:589-608.

Pielke, R.A. Jr., J. Rubiera, C. Landsea, M.L.

Fernandez and R. Klein. 2003. Hurricane vulnerability in Latina America and the Caribbean: normalized damage and loss potential. Natural Hazards Rev. 4(3): 101-114.

Pimbert, M. 2006. Transforming knowledge and ways of knowing for food sovereignty. IIED, London.

Pimentel, D. 1980. Handbook of energy utilization in agriculture. CRC Press, Boca Raton.

Pimentel, D. 1984. Energy flow in agroecosystems. p. 121-132. *En* R. Lowrance et al. (ed), Agricultural ecosystems: Unifying concepts. Wiley, New York.

Pimentel, D. 1996. Green revolution agriculture and chemical hazards. The Science of the Total Environment 188, Suppl. 1, S86-S98.

Pimentel, D. 2004. Environmental and economic costs of the application of pesticides primarily in the United States. Cornell Univ., Ithaca.

Pimentel, D., y M. Pimentel. 1979. Food, energy and society. Edward Arnold, London

Pimentel, D., J. Krummel, D. Gallahan, J. Hough, A. Merrill, I. Schreiner et al. 1978. Benefits and costs of pesticide use in the U.S. food production. BioScience 28:772, 778-784.

Pimentel, D., M.S. Hunter, J.A. LaGro, R.A. Efroymson, J.C. Landers, F.T. Mervis et al. 1989. Benefits and risks of genetic engineering in agriculture. BioScience 39:606-614.

Pimentel, D., P. Hepperly, J. Hanson, D. Douds, and R. Seidel. 2005. Environmental, energetic, and economic comparisons or organic and conventional farming systems. BioScience 55:573-582.

Pimentel, D., y T.W. Patzek. 2005. Ethanol production using corn, switchgrass, and wood; biodiesel production using soybean and sunflower. Nat. Resourc. Res. 14:65-76.

Portes A., and K. Hoffman. 2003. Latin American class structures: their composition and change during the Neoliberal Era. Latin Am. Res. Rev. 38(1):41-82.

Portes, A. 1997. Neoliberalism and the sociology of development: Emerging trends and unanticipated facts. Popul. Dev. Rev. 22:229-259.

Posey, D.A. (ed) 1999. Cultural and spiritual values of biodiversity. UNEP's Global Biodiversity Assessment Volume. Cambridge Univ. Press, UK.

Posner, J.L., y M.F. McPherson. 1982. Agriculture on the steep slopes of tropical America. World Dev. 10:341-53.

Pray, C., J. Huang, R. Hu, and S. Rozelle. 2002. Five years of Bt cotton in China—the benefits continue. The Plant J. 31(4):423-430.

Preibisch, K.L., G. Rivera Herrejón, y S.L. Wiggins. 2002. Defending food security in a free-market economy: The gendered dimensions of restructuring in rural Mexico. Human Organiz. 61:68-79.

Presidencia da República [Brasil]. 2007. Objetivos de desenvolvimento do milenio: Relatório Nacional de Acompanhamento. Sept 2007. Gov. Brasil.

Pretty, J.N. 2002. Lessons from certified and non-certified organic projects in developing countries. p. 139-162. *In* N.E. Scialabba and C. Hatta (ed) Organic agriculture, environment and food security. FAO, Rome.

Pretty, J.N., A.D. Noble, D. Bossio, J. Dixon, R.E. Hine, F.W. Penning de Vries, and J. I. Morrison. 2006. Resource-conserving agriculture yields in developing countries. Environ. Sci. Tech. 40:1114-1119.

Progler, Y. 2005. White studies and the university in ruins. N-173/174. Third World Resurgence, Malaysia.

Psacharopoulos, G. 1993. Returns to investment in education: A global update. Policy Res. Ser. Working Papers. World Bank, Washington DC.

Qaim, M., E.J. Cap, y A. de Janvry. 2003. Agronomics and sustainability of transgenic cotton in Argentina. AgBio Forum 6:41-47. Available at http://www.agbioforum.org/v6n12/v6n12a10-qaim.htm.

Qaim, M., and G. Traxler. 2005. Roundup Ready soybeans in Argentina: Farm level and aggregate welfare effects. Agric. Econ. 32:73-86.

Quaghebeur, D., B. De Smet, E. De Wulf, and W. Steurbaut. 2004. Pesticides in rainwater in Flanders, Belgium: Results from the monitoring program 1997-2001. J. Environ. Monitor. 6(3):182-90.

Quijano, A. 2000 Coloniality of power, Eurocentrism and Latin America. Nepantla: Views from the South 1(3):533-580.

Rabinovich, J.E., J.A. Leal, and M.D. Feliciangeli de Pinero. 1979. Domiciliary biting frequency and blood ingestion of the Chagas's disease vector *Rhodnius prolixus* Stahl (Hemiptera: Reduviidae), in Venezuela. Trans. Roy. Soc. Trop. Med. Hyg. 73:273-282.

Rabinovich, J.E., R.E. Gurtler, J.A. Leal, and D. Feliciangeli. 1995. Density estimates of the domestic vector of Chagas disease, *Rhodnius prolixus* Stal (Hemiptera: Reduviidae) in rural houses in Venezuela. Bull. WHO 73:347-357.

RALLT (Red por una América libre de transgénicos). 2004. Alimentando al mundo bajo el cambio climático. *En* Boletín 113, nov 2004. Available at www.i-isis.org.uk/FTWUCC.php

RAP-AL. 2007. RAP-AL presente en Nyéleni 2007. Enlace 76:4-7

Reardon, T., y J.A. Berdegué. 2002. The rapid rise of supermarkets in Latin America: Challenges and opportunities for development. Dev. Policy Rev. 20(4): 371-388.

Reardon, T., C.P. Timmer, C.B. Barrett, y J. Berdegué. 2003. The rise of supermarkets in Africa, Asia, and Latin America. Am. J. Agric. Econ.85:1140-1146.

Reardon, T., J. Berdegué, y G. Escobar. 2001. Rural nonfarm employment and incomes in Latin America: Overview and policy implications. World Dev. 29:395-409.

Reganold, J.P., J.D. Glover, P.K. Andrews, and H.R. Hinman. 2001. Sustainability of three apple production systems. Nature 410:926-930.

Relyea, R.A. 2003. Growth and survival of five amphibian species exposed to combinations of pesticides. Environ. Toxicol. Chem. 23:1737-1742.

Relyea, R.A. 2005a. The impact of insecticides and herbicides on the biodiversity and productivity of aquatic communities. Ecol. Applic. 15:618-627.

Relyea, R.A. 2005b. The lethal impact of Roundup on aquatic and terrestrial amphibians. Ecol. Applications 15:1118-1124.

Rengifo, G. 1998. Education in the modern west and in the Andean culture. p. 172-192. *En* F. Apffel-Marglin with PRATEC (ed) The spirit of regeneration. Andean culture confronting western notions of development. Zed Books Ltd. New York.

Restivo, S. 1998. Modern science as a social problem. Social Problems 35(3):206-225.

Richter, M. 2000. The ecological crisis in Chiapas: A case study from Central America. Mountain Res. Dev. 20(4):332-339.

Ricketts, T.H. 2004. Tropical forest fragments enhance pollinator activity in nearby coffee crops. Conservation Biology 18:1262-1271.

Riechmann J., 2003. Cuidar la tierra. Políticas agrarias y alimentarias sostenibles para entrar en el siglo XXI. Icaria, Barcelona.

Rifkin, J. 2000. The age of access: The new culture of hypercapitalism where all is paid-for experience. Penguin Putnam, New York.

Rissler, J., and M. Mellon. 1996. The ecological risks of engineered crops. MIT Press, Cambridge.

Rist, S., J. San Martin, and N. Tapia. 1999. Andean cosmovision and self-sustained development. p. 177-190. *En* B. Haverkort and W. Hiemstra (ed) Food for thought. Ancient visions and new experiments of rural people. ETC-COMPAS, Netherlands.

Robinson, R.A. 1996. Return to resistance: Breeding crops to reduce pesticide resistance. AgAccess, Davis CA.

Robinson, R.A., y W.J. Sutherland. 2002. Post-war in arable farming and biodiversity in Great Britain. J. Appl. Ecol. 39:157-176.

Robles, R. 2000. El ajuste invisible. *En* J. Aranda et al. (ed) Tiempo de crisis, tiempo de mujeres. Univ. Autónoma Benito Juárez de Oaxaca, Mexico.

Rojas, A. 1986. La agricultura campesina y el desarrollo del sector agrícola nacional. Revista Universum, año 1, Univ. Talca, Chile.

Rölling, N. 2003. From cases to reasons: The human dimension of agricultural sustainability. Int. J. Agric. Sust. 1(1):2003a.

Rosenthal, G. 1996. On poverty and inequality

in Latin America. J. Interam. Stud. World Aff. 38(2-3):15-37.

Rosenzweig, C., M.L. Parry, G. Fischer, y K. Frohberg. 1993. Climate change and world food supply. Res. Rep. 3, Environ. Change Unit, Oxford Univ., UK.

Rosset, P.M. 1999. The multiple functions and benefits of small farm agriculture in the context of global trade negotiations. Policy Brief 4. Inst. Food Dev. Policy, Oakland.

Rosset, P.M. 2006. Food is different: Why the WTO should get out of agriculture. Zed Books, London.

Rosset, P.M., J. Collins, y F. Moore F. 2000. Lecciones de la revolución verde. Revista del Sur, julio y agosto. Institute Food Dev. Policy, Oakland.

Rosset, P.M., R. Patel, y M. Courville (ed) 2006. Promised land: Competing visions of agrarian reform. Food First, Oakland.

Royal Society. 1998. Genetically modified plants for food use. Statement 2/98. Royal Society, London.

Runge, C.F., B. Senauer, P. Pardey, y M. Rosegrant 2003. Ending hunger in our lifetime. IFPRI, Washington DC.

Sachs, W. (ed) 1992.The development dictionary: A guide to knowledge as power. Zed Books, London.

Sadoulet, E., y A. de Janvry. 1995. Poverty alleviation, income redistribution, and growth during adjustment. *En* N. Lusting (ed) Coping with austerity: Poverty and inequality in Latin America. Brookings Institution, Washington DC.

Safa, H. 1995. The myth of the male breadwinner: Women and industrialization in the Caribbean. Westview Press, Boulder.

Salas-Casasola, I., F. Boucher, y D. Requier-Desjardins. 2006. Agroindustria rural y liberalización comercial agrícola: El rol de los sistemas agroalimetarios localizados. Agroalim 11:22 Mérida.

Sanchez, P.A., C.A. Palm, S.A. Vosti, T.P. Thomich, and J. Kasyoki. 2005. Alternatives to slash and burn: Challenge and approaches of an international consortium. p. 3-37. *In* C. Palm et al. (ed) Slash and burn agriculture: The search for alternatives. Columbia Univ. Press, New York.

Sanders, W.T. 1957. Tierra y agua: A study of the ecological factors in the development of Meso-American civilizations. Ph.D.thesis, Harvard Univ., Cambridge.

Sanogo, S., X.B. Yang, and H. Scherm. 2000. Effects of herbicides on *Fusarium solani* f. sp. *glycines* and development of sudden death syndrome in glyphosate-tolerant soybean. Phytopathology 90:57-66.

Santamaría-Guerra, J. 2003. Institutional innovation for sustainable agriculture and rural resources management: Changing the rules of the game. PhD. thesis. Wageningen Univ., Netherlands.

Sarukán, J. 2003. Implicaciones del flujo genético. Proc. workshop gene flow: What does it mean for Biodiversity and Centers of Origin, Pew Initiative on Food and Biotech., Mexico.

Scherr, S.J., y S. Yadav. 1997. Land degradation in the developing world: Issues and policy options for 2020. 2020 Brief No. 44. Available at http://www.ifpri.org/2020/briefs/number44.htm. IFPRI, Washington DC.

Schneider, R.R. 1995. Government and the economy on the Amazon frontier. Environ. Pap. 11, World Bank, Washington DC.

Schwentesius, R., y M.A. Gómez. 2002. Supermarkets in Mexico: impacts on horticultural systems. Dev. Policy Rev. 20:487 502.

Scorza, J.V., L. Castillo, S. Rezzano, M. Márquez, and J.C. Márquez. 1985. El papel del cafeto en la endemicidad de la leishmaniasis cutánea en Venezuela. Bol. Dir. Malar. San. Amb. 25:82-88.

Semchuk, K., y E. Love. 1992. Parkinson's disease and exposure to agricultural work and pesticide chemicals. Neurology 42:1328.

Sen, A. 1975. Employment, technology and development. Oxford Univ. Press, London.

Sen, A. 2004. ¿Cómo importa la cultura al desarrollo? Letras Libres, No. 71, noviembre. Available at http://www.letraslibres.com/index.php?art=9972.

Sen, A. 2006. Identity and violence: The illusion of destiny. W.W. Norton, New York.

Serratos, A. 2003 La vigilancia del maíz modificado genéticamente. ¿Qué se necesita y cómo podemos manejarlo?. Proc. workshop gene flow: What does it mean for biodiversity and centers of origin, Pew Initiative on Food and Biotech., Mexico.

Sharp, G.D., y D.R. McLain. 1993. Fisheries, El Niño-Southern Oscillation and upper ocean temperature records: an eastern Pacific example. Oceanography 5(3):163-168.

Shiva, V. 1993. Monocultures of the mind: Perspectives on biodiversity and biotechnology. Third World Network and Zed Books, UK.

Simms, A., y H. Reid. 2004. Up in smoke? Latin América and the Caribbean. The threat from climate change to the environment and human development. The third report from the working group on climate change and development. New Economics Foundation, London.

Simonich, S.L., y R.A. Hites. 1995. Global distribution of persistent organochlorine compounds. Science 269:1851-1854.

Singh, N., A. Sharma, P. Dwivedi, R. Patil, and M. Kumar. 2007. Citrin and endosulfan induced teratogenic effects in Wistar rats. J. Appl Toxicol. 27(2):143-51.

Smil, V. 2000. Feeding the world— A challenge for the 21st Century. MIT Press, Cambridge.

Smith, P.H. 1999. Talons of the eagle: Dynamics of U.S.-Latin American relations. Oxford Univ. Press, UK.

Smith, R. 2002. A tapestry woven from the vicissitudes of history, place and daily life: Envisioning the challenges for indigenous peoples of Latin America in the new millennium. Ford Foundation and Oxfam America, Lima.

Snow, A.A., and P. Moran. 1997. Commercialization of transgenic plants: Potential ecological risk. BioScience 47:86-96.

Soil Association. 2005. Executive summary of the organic farming, food quality and human health report. Inform. Sheet 12/01/2005, Vers. 2. Soil Assoc., Bristol.

Soil Association. 2007. The nutritional benefits of organic milk— a review of the evidence. Inform. Sheet 01/12/2007. Vers. 4. Soil Assoc., Bristol.

Soluri, J. 2005. Banana culture: Agricultura, consumption, and environmental change in Honduras and the United Status. Univ. Texas Press, Austin.

Souza C.J. 2004. Impacto de los cultivos transgénicos en la estructura agraria y en la alimentación. Análisis de la situación en Argentina. Centro de Estudios sobre Tecnologías Apropiadas de la Argentina (CETAAR) y Red de Acción en Plagicidas y sus Alternativas para América Latina (PAP-PAL).

Spooner, D.M., K. McLean, G. Ramsay, R. Waugh, and G.J. Bryan. 2005. A single domestication for potato based on multilocus amplified fragment length polymorphism genotyping. PNAS 102:14694-14699.

Springert, J.A., R.A.J. Gray. 1992. Effect of repeated low dosage of biocides on the earthworm *Aporrectodea caliginosa* in laboratory culture. Soil Biol. Biochem. 24(12):1739-1744.

Steffan-Dewenter, I., S.G. Potts, and L. Packer. 2005. Pollinator diversity and crop pollination services are at risk. Trends Ecol. Evol. 20:651-652.

Steward, J. 1955. Theory of culture change. Univ. Illinois Press, Urbana.

Stewart, P., y A. Knight. 2005. Trends affecting the next generation of U.S. agricultural biotechnology: Politics, policy, and plant made pharmaceuticals. Tech. Forecast. Social Change 72:521-534.

Stiglitz, J. 1998. Distribution, efficiency and voice: Designing the second generation of reforms. Working Paper Summaries. Int. Conf. Asset Distribution, Poverty and Econ. Growth, Brasilia, 14-17 July 1998.

Stiglitz, J. 2003. Globalization and its discontent. W.W. Norton, New York.

Stoger, E., J.K.C. Ma, R. Fischer, and P. Christou. 2005. Sowing the seeds of success: pharmaceutical protein from plants. Curr. Opin. Biotech. 16:167-173.

Teubal, M., and J. Rodríguez. 2001. Neoliberalismo y crisis agrária. *En* N. Giarraca (ed) La protesta social en Argentina. Alianza, Buenos Aires.

The Economist. 2007. This land is anti-capitalist land. The Economist, 26 Apr

2007. Available at http://www.economist.com/printedition/displayStory.cfm?story_id=9079861&fsrc=RSS.

Theodore, S.C. 2005. Decision time for the Caribbean. NACLA 39(2):3.

Thiesenhusen, W. 1996. Trends in land tenure issues in Latin América: Experiences and recommendations for development cooperation. GTZ, Germany.

Thrupp, L.A. 1998. Cultivating diversity: Agrobiodiversity and food security. World Resources Inst., Washington DC.

Tilman, D., D. Wedin, and J. Knops. 1996. Productivity and sustainability influenced by biodiversity in grassland ecosystems. Nature 379:718-729.

Toledo, V.M. 1992. Bio-economic costs. p. 67-93. *In* T.E. Downing. Development or destruction: The conversion of tropical forest to pasture in Latin America. Westview Press, Boulder.

Toledo, V.M. 2001. Indigenous peoples and biodiversity. Encyclo. Biodivers. 3:451-463.

Toledo, V.M. 2003. Ecologia, espiritualidad y conocimiento. De la sociedad del riesgo a la sociedad sustentable. PNUMA Oficina Reg. America Latina, Univ. Iberoamericana.

Toledo, V.M. 2005. Los curadores de la tierra. Centro de Investigaciones en Ecosistemas, Morelia.

Toledo, V.M. 2007. Indigenous people and biodiversity. *In* S. Levin (ed) Encyclopedia of biodiversity. Academic Press, NY.

Toledo, V.M., y A. Castillo. 1999. La ecologia en Latinoamérica: Siete tesis para una ciencia pertinente en una region en crisis. Interciencia 24:3.

Toledo, V.M., P. Alarcón-Chaires, and P. Moguel. 2001. El atlas etnoecológico de México y Centroamérica: Fundamentos, métodos y resultados. Etnoecológica vi(8).

Tomish, T.P., P. Kilby, and B.F. Johnston. 1995. Transforming agrarian economies: Opportunities seized, opportunities missed. Cornell Univ. Press, Ithaca.

Torres, F. 2003. La visión teórica de la seguridad alimentaria como componente de la seguridad nacional. p. 15-21. *En* F. Torres (ed) Seguridad alimentaria: seguridad nacional. UNAM/IIEc, Plaza y Valdes, México.

Traxler, G., y S. Godoy-Avila. 2004. Transgenic cotton in Mexico. AgBio Forum 7(1&2):57-62. Available at http://www.agbioforum.org/v7n12/v7n12a11-traxler.htm.

Trejos, R., C. Pomadera, J. Villasuso. 2004. Políticas e instituciones para la agricultura de cara al siglo XXI. IICA, Costa Rica.

Trenbath, B.R. 1976. Plant interactions in mixed crop communities. Am. Soc. Agron. Spec. Publ. 27:129-169.

Trewavas, A. 2002. Malthus foiled again and again. Nature 418:668-670.

Trigo, E.J., y E.J. Cap. 2003. The impact of the introduction of transgenic crops in Argentinean agriculture. AgBio Forum 6(3):87-94.

Trigo, E.J., D. Chudnovsky, E. Cap, and A. Lopez. 2002. Genetically modified crops in Argentine agriculture: An open ended story. Libros del Zorzal, Buenos Aires.

Turner II, B.L. and P. Harrison (ed) 1983. Pulltrouser swamp: Ancient Maya habitat, agriculture and settlement in northern Belize. Univ. Texas Press, Austin.

UNDP. 1999. Human development report, 1999. UNDP, NY.

UNEP. 1999a. Global environment outlook 2000. UNEP, Nairobi.

UNEP. 1999b. Inventory of technical and institutional resources for promoting methyl bromide alternatives. Ozone Action Programme, UNEP, Paris.

UNDP. 2005a. The Millennium Development Goals: A Latin American and Caribbean perspective. Available at http://www.unhchr.ch/spanish/html/menu3/b/a_cescr_sp.htm. UNDP, NY.

UNDP. 2005b. Human development report. UNDP, New York.

UNEP. 2001. Stockholm Convention on Persistent Organic Pollutants (POPs). Available at http://www.pops.int/. UNEP, Geneva.

UNEP. 2002a. The sustainability of development in Latin America and the Caribbean: Challenges and opportunities. UN, Santiago.

UNEP. 2002b. Global environment outlook 3. UNEP, Nairobi.

UNEP. 2006. GEO Yearbook 2006. Available at http://www.unep.org/geo/yearbook/yb2006/. UNEP, Nairobi.

UNEP. 2007. List of signatories and parties to the Stockholm Convention. Available at www.pops.int/reports/StatusOfRatifications.aspx#notes. UNEP, Nairobi.

USDA. 2005. Circular del 5 de Julio de 2005. Available at http://www.fas.usda.gov/grain/circular/2005/07-. USDA, Washington DC.

USDA. 2006. Statement by Agriculture Secretary Mike Johanns regarding genetically engineered rice. 18 Aug2006. USDA, Washington DC.

Valdés M. 2005. La feminización de la pobreza. Un problema global. Revista Mujer y Salud N° 4-2005. RSMALC.

Valladolid, J. 1998. Andean peasant agriculture: Nurturing a diversity of life in the *Chacra*. p. 51-88. *En* F. Apffel-Marglin with PRATEC (ed) The spirit of regeneration. Andean culture confronting western notions of development. Zed Books, NY.

Valladolid, J. 2001. Andean cosmovision and the nurturing of biodiversity in the peasant chacra. p. 639-670. *En* J. Grim (ed) Indigenous traditions and ecology. The interbeing of cosmology and community. Harvard Univ. Press, Cambridge.

Van Dam, C. 1999. La tenencia de la tierra en América Latina. El estado del arte de la discusión en la región iniciativa global tierra, territorios y derechos de acceso. Unión Mundial para la Naturaleza (UICN).

Vandermeer, J.H. 1989. The ecology of intercropping. Cambridge Univ. Press, NY

Vandermeer, J.H. 1995. The ecological basis of alternative agriculture. Ann. Rev. Ecol. Systemat. 26:201-224.

Vandermeer, J.H., y I. Perfecto. 2005. Breakfast of biodiversity: The political ecology of deforestation. Inst. Food Dev. Policy, Food First, Oakland.

Varese, S. 1996. The ethnopolitics of Indian resistance in Latin America. Latin Am. Perspect. 23(2):58-71.

Verner, D. 2005. Rural poverty and labor markets in Argentina. http://siteresources.worldbank.org/INTARGENTINAINSPANISH/Resources/Argentina_Rural_Poverty_Labor_Market_062105_2.pdf. World Bank, Washington DC.

Via Campesina. 1996. The right to produce food and access to land. Position on food sovereignty presented at the World Food Summit, Nov 1996, Rome.

Vía Campesina. 2003 ¿Qué significa soberanía alimentaria? Posición de Via Campesina sobre la soberanía alimentaria. Enero 15, 2003. Available at http://viacampesina.org/main_sp/index.php?option=com_content&task=view&id=78&Itemid=27.

Vittor, A.Y., R.H. Gilman, J. Tielsch, G. Glass, T. Shields, W.S. Lozano et al. 2006. The effect of deforestation on the human-biting rate of Anopheles darlingi, the primary vector of Falciparum malaria in the Peruvian Amazon. Am. J. Trop. Med. Hyg. 74:3-11.

Vorley, B. 2003. Food, Inc.: Corporate concentration from farm to consumer. UK FoodGroup, London.

Vosti, S.A., J. Witcover, and C.L . Carpentier. 2002. Agricultural intensification by smallholders in the western Brazilian Amazon: From deforestation to sustainable land use. Res. Rep. 130. IFPRI, Washington DC.

Wallerstein, I. 1999. The end of the world as we know it: Social sciences for the twenty-first century. Univ. Minnesota Press, Minneapolis.

Wan, M.T., R.G. Watts, and D.J. Moul. 1989. Effects of different dilution water types on the acute toxicity to juvenile Pacific salmonids and rainbow trout of glyphosate and its formulated products. Bull. Environ. Contamin. Toxicol. 43:378-385.

Warren, K.B., and J.E. Jackson (ed) 2003. Indigenous movements, self-representation, and the state in Latin America. Univ. Texas Press, Austin.

White, D., F. Holmann, S. Fuijsaka, K. Reategui, and C. Lascano. 2001. Will intensifying pasture management in Latin America protect forests—or is it the other way round? *In* A. Angelsen and D. Kaimowitz (ed) Agricultural technologies and tropical deforestation. CAB Int., New York.

WHO. 1990. Public health impact of pesticides

used in agriculture. WHO and UNEP, Geneva.
WHO. 1996. Climate change and human health. WHO/EHG/96.7. WHO, Geneva.
Wilken, G.C. 1987. Integrating forest and small-scale farm systems in middle America. Agro-ecosystems 3:291-302.
Willer, H., y M. Jussefi (ed) 2007. The world of organic agriculture. Statistics and emerging trend. IFOAM, Bonn.
Wilson, M. L. 2002. Agricultural production and human disease: Evolving interactions present new challenges. p. 245-262. *In* J.H. Vandermeer (ed) Tropical agroecosystems: New directions for research. CRC Press, Boca Raton FL.
Wolf, E. 1971. Los campesinos. Nueva Colección Labor, Barcelona.
Wolford, W. 2003. Families, fields, and fighting for land: The spatial dynamics of contention in rural Brazil. Mobilization: An Int. Q. 8(2):157-172.
Wood, A., P. Stedman-Edwards, and J. Mang. 2000. The root causes of biodiversity loss. Earthscan, UK.
World Bank. 1992. Reportes de evaluación de pobreza. Banco Mundial, Washington DC.
World Bank. 2005a. The little green data book '05. World Bank, Washington DC.
World Bank. 2005b. World development report 2006. Equity and development. World Bank, Washington DC.
World Bank. 2006a. Informe sobre el desarrollo mundial-2006. Panorama general, equidad y desarrollo. Banco Mundial, Washington DC.
World Bank. 2006b. The little green data book '06. World Bank, Washington DC.
WWF. 2006. Killing them softly: Health effects in Arctic wildlife linked to chemical exposures. WWF, Washington DC.
Yashar, D.J. 2005. Contesting citizenship in Latin America: The rise of indigenous movements and the postliberal challenge. Cambridge Univ. Press, NY.
Yotopoulos, P., y L. Lau. 1971. A test for relative efficiency and application of Indian agriculture. Am. Econ. Rev. 61:94-109.
Zinin, W.J., A.C. Conçalves Jnr., V.H. Pereira, and V.P. Zonin. 2000. Agroecology as an option for income improvement in the family farming associations at "Erexim" region, Brazil. World Congress of Rural Sociology.
Zoppi de Roa E, E. Gordon, E. Montiel, L. Delgado, J. Berti y S. Ramos. 2002. Association of cyclopoid copepods with the habitat of the malaria vector Anopheles aquasalis in the peninsula of Paria, Venezuela. J. Am. Mosq. Control Assoc. 18:47-51.

2

AKST Systems in Latin America and the Caribbean: Evolution, Effectiveness and Impact

Coordinating Lead Authors:
Hugo Cetrangolo (Argentina) and Jesús Moncada (Mexico)

Lead Authors:
Hernando Bernal (Colombia), Cecilia Gelabert (Argentina), Marcelo Regúnaga (Argentina), and Mario Samper (Costa Rica)

Contributing Authors:
Antonio Castro (Brazil), David E. Williams (USA)

Review Editor:
Celia Harvey (US)

Key Messages

1. Latin America has a rich tradition of individual and institutional efforts in agricultural science, knowledge and technology (AKST). While these have made significant contributions to food security and the agro-exporting sector, they have not taken full advantage of the existing potential for agriculture-driven development. LAC's different sub-regions have a heterogeneous AKST system structure involving public, private, local, national, regional, and international institutions and organizations of varying sizes and capabilities, as well as major differences between countries and subregions. Recently, innovative alternatives have emerged for the management of relevant bodies with the participation of civil society. However, the way the systems are put together does not respond to this diversity and potentiality—which has impeded optimizing the use of the regional AKST system, and blocked its technical spill-over effects.

The needs that have been detected are the following: Strengthening AKST system institutions, particularly in the relatively less developed countries; Improving linkages and cooperation within the AKST system, including public- and private-sector users; Promoting the participation of civil society to ensure greater social oversight and moral, political, and economic support.

2. Priorities on the AKST system agenda in the past were food security, the production of agroindustrial commodities, and low-cost foods for local consumption and export. While these remain significant, the challenge today is to develop technologies, innovations, and systems aimed at addressing the environmental and social dimensions and the specific demands of indigenous, traditional, and agroecological systems. The lines of research prioritized before were directed at boosting productivity in the primary sector. Fewer efforts were made to produce technological developments geared to the competitiveness of the agrifood chains, the production of non-agricultural goods and services in rural areas, and other activities that reflected agriculture's multifunctionality.

More attention must be paid, in all three main productive areas, to social, cultural, and environmental aspects often neglected in the past. Not enough importance, moreover, has been attached to the sustainable use of the region's enormous resources with regard to biodiversity, fresh water availability, and marine resources. Not enough concern has been shown, either, for the direct impact of productive systems on water and soil resources and tree cover, or the impact of deforestation, the expansion of the agricultural frontier, and climate change.

It is to be hoped that the AKST system will manage to reconcile conflicting goals such as competitiveness, on the one hand, and environmental, economic and social sustainability on the other.

3. In response to social demands, the AKST system agenda has become more diverse and complex. In its efforts to address problems like poverty, food security, environmental degradation, deforestation, biodiversity loss, natural disasters, and global climate change, it has incorporated social, economic, and environmental considerations as well as the notion of working with all the links in production chains, from primary production to marketing. Yet few AKST system institutions can, by themselves, respond to such diverse and complex demands in a holistic manner.

Strengthening cooperation through global, regional and national networks, with proper strategic planning, execution and follow-up, is essential. Such networks should be more systemic and incorporate more broadly the various social actors. This will put to the test the solidarity and co-responsibility between countries and institutions.

4. The AKST agenda has not paid enough attention to the problems that affect the nutrition, health, and well-being of the urban and rural poor. There is a need to design, fund, and implement an agenda in favor of the poor at the global, regional and national level.

5. The AKST system has made significant agronomic contributions that have mostly benefited large producers and well-organized medium producers. Traditional, indigenous, and agroecological producers, who share a limited availability of resources and are less organized, have not benefited as much. Their equitable participation in defining the AKST agenda has not yet been achieved. There is a need to develop a participatory innovation and development system that can meet the needs of these three groups, take into account their capabilities, and help them fulfill their potential.

6. Investment in agricultural research and development (R&D) in LAC varies among countries and subregions but in all cases is lower than in industrialized nations, and even developing countries in other regions. There is a need to increase government funding of AKST systems, since for developing countries it remains the best investment.

7. In spite of AKST's contributions to agricultural production and productivity, recent decades have ironically seen a decrease in public funding. Regulations governing relevant institutions, moreover, are not conducive to research. This generates uncertainty as well as the inefficient use of resources. There is a need to provide public institutions with sufficient funding and establish mechanisms to reduce uncertainty and improve the efficient use of resources.

8. Private-sector R&D focuses on the development of appropriable technologies that have benefited from patent and intellectual property legislation. It has also played an important role in the local adaptation of technologies coming from industrialized nations. However, AKST contributions by the private sector do not meet development needs, particularly among traditional and indigenous producers. LAC needs an increase in private investment on agricultural research and development. This, in turn, entails public policies that will encourage such research. In certain countries, political, economic and institutional problems have limited policies of this nature. The hope is to achieve an appropriate balance in this area between the interests of producers and society, on one hand, and on the other a fair retribution for private investment.

9. International cooperation and NGOs have also engaged in efforts to supplement the role of government bodies in AKST, mainly in the environmental, cultural and social fields. But such efforts have been scattered, insufficient, and lacking in continuity. It will be necessary to increase such investments and promote their integration into the AKST system.

10. Several factors, external to agricultural technical development, condition AKST's potential to build more productive, sustainable, and equitable systems that contribute to food supply, food security, and poverty reduction. AKST has not been taken into consideration as much as it should have when formulating macroeconomic, commercial, and financial policies and those related to access to markets, education, and information. It will be necessary to find mechanisms to better link the AKST system with policy-makers and implementers.

11. In the region, the lack of strategic plans, and the poor participation of the AKST system in their formulation, has prevented an integral response to complex rural issues. The AKST system must be an integral part of the promotion, design, and execution of strategic plans.

12. Although society has a good perception of the AKST system, there is a certain ignorance of the importance and impact of agricultural technology, hence little social support for AKST, and adverse reactions to technology that are often baseless or negatively influenced by prejudices. Improved communication on the importance and potential positive effect of agricultural technology, based on a strategy of transparency and accountability, is a must.

13. Research institutions benefiting from public funding lack balance in their human resources, in terms of the variety of disciplines and cultures represented, and in terms of gender. Moreover, their researchers and support staff are growing older and few institutions have a program to renew their personnel. Programs must be developed that contemplate the training, updating, and diversification of scientific and technical cadres through incentives that encourage research in priority fields.

14. The AKST system has contributed to improving production and productivity (with subregional differences), but mainly within the conventional or productivist system.

15. The AKST system has not interacted sufficiently with traditional or indigenous systems, nor has it taken advantage of their capabilities and potentialities.

16. The agroecological system has emerged as an option for finding solutions to environmental, economic, and sociocultural problems. It has arisen as a result of the interaction between the AKST system and producers who share such concerns.

17. Technological development has sometimes had its environmental and social costs. The balance of agricultural, economic, social, cultural, and environmental impacts has not been studied thoroughly enough. Neither have strategies been developed to mitigate the negative effect of various technologies and production systems. There is a need to assess the results of AKST in a holistic manner, bearing in mind not only their economic and productive impact but also their environmental, social, cultural, and political implications.

2.1 Inventory, characterization and evolution of the AKST system and its interactions

Latin America has a rich tradition of individual and institutional efforts in science, technology, and knowledge regarding agriculture. They have made significant contributions to many countries in the region. LAC's different sub-regions have an abundant but heterogeneous AKST system structure, with major differences between countries involving numerous institutions and organizations—public, private, local, national, regional, and international—as well as bilateral and multilateral cooperation programs, sometimes with contrasting agendas and capabilities.

The AKST system in LAC has gradually incorporated different institutions, programs, and other cooperation mechanisms—the aim having been to provide the needed geographical and thematic coverage. It has also sought to take advantage of, coordinate, and integrate the efforts of various types of public and private stakeholders at different levels (local, national, regional, and international). As a result, it has become a complex weave of institutions, programs, and cooperation mechanisms involving (1) local and third sector organizations; (2) National Agricultural Research Institutes (NARIs), universities and other national organizations; (3) regional centers; (4) cooperative programs; (5) consortia and specialized networks; (6) international centers such as Consultative Group on International Agricultural Research (CGIAR) and Global Forum on Agricultural Research (GFAR); (7) Regional Fund for Agricultural Technology (FONTAGRO); and (8) Regional Forum for Agricultural Research and Technological Development (FORAGRO) (Figure 2-1).

2.1.1 Local and third sector organizations

The complex and intricate network of local organizations, each with its own links to and interactions with the AKST system, generates opportunities but also constraints that have expressed themselves in different ways, especially in the last three decades. There is a rich and varied experience in the creation and successful operation of civil society institutions that support publicly funded AKST system programs. In Mexico, for instance, studies have been made of "interest groups"—in this case, farmers—who have voluntarily organized themselves in Patronatos to provide moral, political, and economic support to research programs of interest, implemented in INIFAP's experimental fields (Box 2-1).

The main constraints on the interactions between NGOs

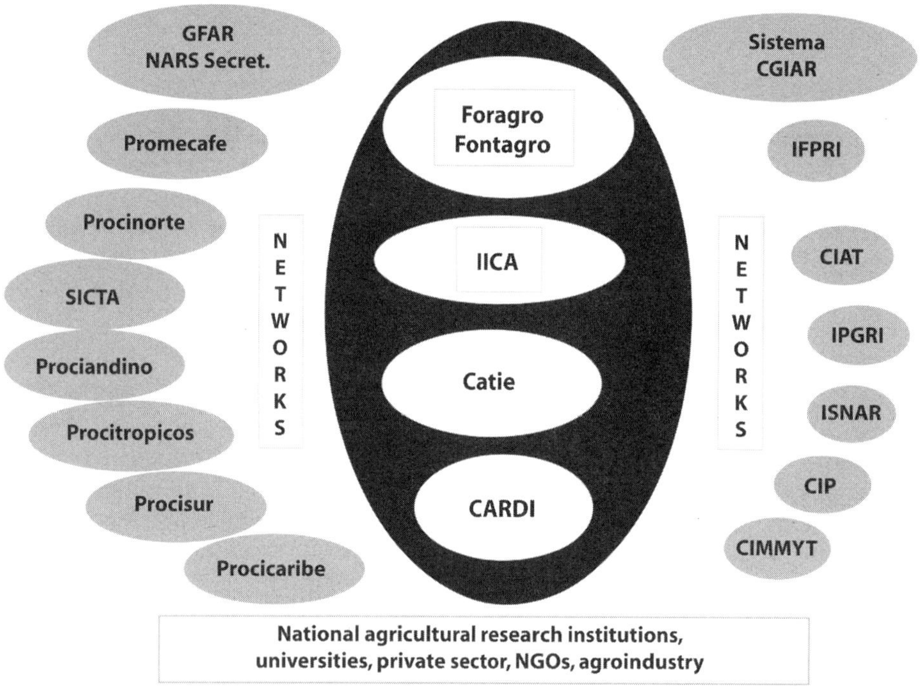

Figure 2-1. *Regional agricultural technology innovation system for the Americas.* Source: Ardila, 2006

and AKST system institutions can be attributed to regional contrasts within each country, decision-making of a political nature, and limited social participation. They also reflect a trend toward privatizing research, technical assistance, and technology transfer to small and medium-scale producers, as a result of administrative decentralization, structural adjustment, and market liberalization—all phenomena that have accelerated in the last two decades (Quiroz, 2001).

Several countries have attempted, through public policies, to develop production systems that break the cycle of exclusion and environmental degradation, and also incorporate a gender perspective and an indigenous and Afro-American worldview. However, much remains to be done to ensure the real participation of those stakeholders in decision-making at the local level (Dirven, 2001).

Rural societies are also becoming more complex. More interactions between different types of stakeholders blur the boundaries between the rural and the urban. New scenarios are emerging, created by the demands of the various actors and their respective local organizations.

With regard to the AKST system, local development processes pursued by communities, either independently or in partnership with universities, foundations, corporations, cooperatives, producers' associations, and both national and international non-governmental organizations, offer the possibility of reappraising traditional knowledge, developing greater negotiating power, improving territorial management, and strengthening claims for access to land. This is evident in various social movements such as the Zapatistas in Chiapas, Mexico; the Landless Peasants' Movement in Brazil; and the claims of the Mapuche indigenous people in Chile and Argentina—all of which have had local impact as well as regional and international repercussions on the design of a new paradigm regarding AKST at the Latin American level.

Most Latin American states have not yet resolved their agrarian problem, one that affects their respective societies, particularly local rural sector organizations. However, this phenomenon is no longer associated exclusively with the rural milieu, but has also spread to urban areas (Machado 2004).

In spite of some isolated experiences, new advances in AKST involving bioelectronics, bioinformatics, and biotechnology have not been widely adopted by local organizations or *campesino* farmers. Moreover, no reconciliation processes have emerged to take advantage of their positive aspects (Amaya and Rueda, 2004; León et al., 2004).

2.1.2 National organizations

LAC's AKST system is made up of a vast network of public, private, and third sector institutions in the various countries that have generally had a major impact, reflecting the relative importance of agriculture to the region. Within this system, the national public agricultural research institutes, generally known as NARIs (or INIAs in Spanish), have a long history—many were created more than half a century ago—and have played a significant role in generating technologies for this sector.

Just as LAC is a heterogeneous geographic area, the NARIs of the different countries also display varied characteristics. Some enjoy a high profile and receive the major share of their country's investment in agricultural science

Box 2-1. Synthesis: Assessment of the Patronatos that support AKST—Experiences in Mexico

The *Patronatos* are civil society organizations that support agricultural or livestock research in Mexico. They are led and financed to varying degrees by farmers, the main users of the products and services generated by publicly funded agricultural research institutions. They are an example of synergy between civil society and government, within what is known as "participation and/or social monitoring of innovation", which helps to ensure an appropriate correlation between the AKST System agenda and users' needs, and contributes to transparency and accountability.

The *Patronatos* offer the following advantages: they provide moral, political and economic support to specific research and technology transfer projects of interest to their members; they promote positive synergies between the federal institutions responsible for research and civil society (producers and agro-entrepreneurs) as well as the users of the products and services generated, such as improved seeds, vaccines, and technological know-how and innovations. They ensure that agricultural research projects meet the interests of the productive sector. In addition, they facilitate and promote the early and rapid adoption by farmers of innovations.

The Mexican federal government, through INIFAP, covers salaries and part of the operating and investment costs, which are complemented by the *Patronatos'* own contributions. In times of financial crisis, this helps to reduce or mitigate government budget cuts and ensure the continuity of the research projects under execution.

Although their effectiveness varies, other advantages offered by the *Patronatos* are setting research priorities based on real needs; encouraging researchers to generate results that are applicable in real agroecological and economic conditions; establishing permanent communications between researchers and farmers; enhancing the credibility and acceptance of the technology generated; taking advantage of the experience and vision of farmers; administering resources more efficiently and promptly; building consensus; diversifying the sources of financing; and reducing political influence in decision-making.

Most *Patronatos* have been established by groups of organized market-oriented farmers with medium to large-scale operations. Small subsistence-oriented farmers with few resources and little organization have not participated.

The *Patronatos'* performance has been variable, with notable examples of effectiveness, efficiency, and continuity over several decades, and also failures due to interference by federal or state governments; the use of the *Patronato* and its resources for party politics; conflicts of interest in the management of resources, and the improper use of the *Patronatos'* products (improved seeds, services, etc.) for personal benefit.

The *Patronatos'* success or failure reflects the degree of organization, education, and civic responsibility of the farmers and local officials involved, and is expressed in their solidarity on issues of community interest, as well as in joint responsibility, synergy and respect between society and the government. It would be useful to study the development, operation and performance of these institutions, since they constitute a first step in a strategy of "participatory innovation development" and are an example of "social monitoring of innovation".

and technology as well as regional investments. These include Empresa Brasileira de Pesquisa Agropecuária, or Brazilian Agricultural Research Institute (EMBRAPA) in Brazil, Instituto Nacional de Investigaciones Forestales Agrícolas y Pecuarias, or National Forestry, Agricultural and Livestock Research Institute (INIFAP) in Mexico, Instituto Nacional de Tecnologia Agropecuaria, or National Agricultural Technology Institute (INTA) in Argentina, Instituto Nacional de Investigaciones Agropecuarias (INIA) in Venezuela and Corporación Colombiana de Investigación Agropecuaria, or Colombian Agricultural Research Institute (Corpoica) in Colombia. In other countries, investment in AKST system has been limited and no significant institutional structure exists at the national level.

Parallel to the work carried out by NARIs, universities have played a significant role in basic and applied research, and some have made important contributions to the dissemination of technology in the region. In general, coordination between NARIs and universities has not been satisfactory and, except in some specific cases, is an aspect that deserves greater attention, since the capabilities of both types of institutions could be enhanced, as shown by some success stories.

Certain LAC countries also have national science and technology institutions of a more general nature, with additional centers specializing in topics related to agriculture and natural resources. These have made important contributions in some fields, mainly basic research. However, it should be noted that the lack of coordination between scientific research and technology development is a feature common to nearly all countries.

In the larger countries with political structures involving decentralized resources at the provincial or state level, the AKST system usually includes public institutions of a provincial or regional nature, often specializing in certain crops, production areas, or issues of local importance. Some of these have made important contributions to the development of specific activities; such is the case of the Obispo Colombres Experimental Station, in Argentina's Tucuman Province, with regard to sugarcane production and other products of local interest.

In most LAC Countries, the public AKST system developed vigorously in its initial stages and made substantive contributions during the 1960s, 1970s, and part of the 1980s. However, the situation changed in the last two decades, when their relative importance and contributions declined with regard to conventional/productivist agriculture vis-à-vis the private sector. This has resulted from two simultaneous processes: (1) a gradual decline in the importance and, in many cases, in the competencies of the state, which has led to reductions in the budgets allocated to AKST, and in certain cases to the closure or merger of institutions specialized in this field; and (2) economic, social and technological processes, particularly in the Southern Cone, that have affected the agricultural sector in recent decades, particularly the scale and concentration of production. Both processes have placed greater emphasis on appropriable technologies directed at increasing productivity, with the private sector playing a key role in generating and adapting technology, mainly in fields related to plant and animal genetics, chemical fertilizers, health products, and agricultural machinery.

The scale of the R&D investments needed to obtain technology products consistent with growing demands for competitiveness in modern agriculture means that many R&D efforts are beyond the scope of national science and technology (S&T) bodies. In many cases such initiatives can only be undertaken by global technology firms, which obtain benefits through the sale of inputs and capital goods, and income from royalties for developments protected by intellectual-property rights.

In some countries, private mechanisms for generating and disseminating technology have eclipsed the work of public institutions, whose efforts have focused on addressing the needs of small- and medium-sized farmers—groups that are seldom of interest to firms that supply inputs, particularly when the potential customers are not able to purchase them in significant quantities.

Beyond the role of the private companies specialized in generating innovations and technology for the agricultural sector, private or public-private partnerships based on production chains have emerged in recent years that, in some countries of the region, implement research programs on topics they themselves have identified. Such innovative, albeit incipient, activities are carried out in close association with science and technology institutions and universities, and are good examples of identifying demands and engaging in planning and coordination to resolve technological problems.

Many significant advances in technology have been achieved by "catching up" with technologies generated in developed countries and adapting them to local or regional conditions in different countries. This has led to some very competitive developments in certain crops and regions—especially in temperate zones—with relatively little effort or investment in science and technology at the national level, by simply adapting the technology of other countries with similar agroecological conditions. However, it should be noted that certain LAC countries with fewer resources, particularly those in tropical and subtropical zones, have been unable to address specific local needs due to the lack of basic and applied research, and because they have not developed sufficient capacity in the field.[9]

National public institutions has focused R&D mainly on the most relevant ways of improving farmers' livelihoods and incomes, while social and environmental aspects have traditionally received less attention. It is only in the last two decades that these issues have become more important in NARIs' activities.

The region's public AKST system has also placed greater emphasis on generating "hard" production technologies than on "soft" organizational technologies, due to the characteristics of its own member institutions. This has hindered their linkages with production models—a situation aggravated by the fact that technology products are often generated from the supply side, without considering the needs and capabilities of their recipients. As a result, support is growing for a line of thought that holds that the management of technological development should involve a greater participation by end users.

Demand-side requirements are becoming increasingly important in determining the types of technologies needed. Consumers and more concentrated distribution channels require new services like product traceability, certifications of origin and processes, respect for the environment, and "natural" products. This, in turn, has placed new demands on the AKST system.

Given that technology is both an economic and a social good, and given the negative social and economic trends in many Latin American and Caribbean countries in recent years, public AKST institutions have begun to incorporate social issues, such as subsistence agriculture and urban agriculture, in their agendas. However, S&T institutions are still a long way from being able to respond to specific demands in terms of developing appropriate technologies for the most disadvantaged sectors.

In some countries, extension and technology transfer systems have undergone major changes in the last two decades as a result of public institutions assigning greater importance to social issues and to small farmers due to the aforementioned emergence of the private sector as the main provider of appropriable technologies to larger producers, toward whom agricultural extension and technology transfer is generally directed. For specific types of farmers, independent professionals—both agronomists and veterinarians—are an important factor in technological development.

It should be noted that in some cases there is an important *spillover* effect, with the technology used by larger producers being adopted by small farmers, especially when they are not prevented from doing so by economic or cultural constraints.

2.1.3 Regional organizations, international centers and other regional cooperation mechanisms

LAC has had a long experience—more than half a century—of regional cooperation between countries and institutions on agricultural research and education. The existence of common problems in different regional and sub-regional spheres and in some fields of interest, as well as the constraints encountered in attempting to develop significant independent agricultural research programs, especially in the smaller countries, led to the implementation of various initiatives. In some cases, these efforts were consolidated in new regional institutional structures: In others, they resulted in joint or cooperative research projects and programs

[9] It should be noted that in developed countries, technology for temperate zone crops is more readily available than for tropical ones; consequently, there are fewer possibilities of using foreign technology and adapting it to the tropical climate of LAC countries.

and a growing exchange of knowledge[10] among the region's national institutes, and between these and various regional and international institutions.

Some regional organizations are of long standing and in some countries even predate the creation of the national institutes (NARIs). One example is Inter-American Institute of Agricultural Sciences, currently known as the Inter-American Institute for Cooperation on Agriculture (IICA), an institution created in 1942 in Turrialba, Costa Rica, where an experimental station and postgraduate education center was established that subsequently led to the creation of Tropical Agriculture Research and Higher Education Center (CATIE) in 1973. In that year, the research and training activities were separated from more comprehensive efforts of hemispheric scope undertaken by IICA, which established its headquarters in the canton of Coronado, also in Costa Rica but in the outskirts of the country's capital.

Also In the mid-1970s, the twelve members of the Caribbean Community (CARICOM), a trade and integration initiative, created Caribbean Agricultural Research and Development Institute (CARDI) with the aim of strengthening agricultural research and development activities and supporting the agricultural sectors of member countries. These functions had previously been carried out by a regional Research Center, created in 1955 by the English-speaking Caribbean countries to meet the growing and increasingly complex challenges of agriculture.

In addition to the sub-regional centers mentioned above, in the 1970s and 1980s the NARIs and other public and private institutions of LAC countries gradually established cooperative agricultural research programs (known as PROCIs), which have grown notably and continue to function today. These programs evolved, from initial exchanges of knowledge among participating institutions, to the execution of joint research activities and the implementation of regional research projects and informal training efforts. Nowadays there are various cooperative programs for several topics and for all the sub-regions of the Americas.[11] The majority of these initiatives received support from IICA and the IDB during their initial stages. Such cooperative mechanisms, which do not require new institutional structures, have had a positive impact in promoting technological development in the countries involved, as shown by various impact assessments.

There are also *consortia and specialized networks* for different topics, products, and sub-regions that have received support from FAO's national and regional offices and other international institutions. Some of the most important include the regional Cooperative Potato Program; the regional Cooperative Program on Beans for Central America, Mexico and the Caribbean; the regional Maize Program, coordinated by the International Maize and Wheat Improvement Center (CIMMYT); the Latin American Agricultural Conservation Network; the Consortium for the Sustainable Development of the Andean Ecoregion; the International Network of Farming Systems Research Methodology; the Technical Cooperation Network on Plant Biotechnology; and various cooperative research programs funded by the United States Agency for International Development (USAID) and administered by US universities.

LAC's institutional AKST system also has two other types of components, implemented in the 1990s in an effort to complete the region's institutional architecture and fill some of the gaps observed in its functioning: FONTAGRO and FORAGRO.

The Regional Fund for Agricultural Technology (FONTAGRO) is a consortium created to promote strategic agricultural research of regional scope with direct participation by LAC countries in setting priorities and funding research projects. It was established by a group of countries of the region[12] with sponsorship from IDB, IICA, the Rockefeller Foundation, and Canada's International Development Research Center (IDRC). Its purpose its to improve the competitiveness of the agricultural sector, ensure the sustainable management of natural resources, and work to reduce poverty through the development of technologies that qualify as international public goods. It should do this by facilitating the exchange of scientific knowledge within the region and with other regions of the world.

The goal is to establish an endowment fund of 200 million dollars and use the annual dividends to provide sustained non-reimbursable financing for regional strategic research projects. Project funding is allocated through a competitive mechanism based on projects' coherence with the Fund's objectives and on technical, economic, environmental and institutional criteria established for the priority research areas defined in the Medium Term 2005-2010 Plan. The design and execution of the proposals is undertaken by different organizations in the Fund's member countries (research institutes, universities, foundations, private groups),

[10] Known generally as "spillover"

[11] The Cooperative Research and Technology Transfer Program for the Northern Region, involving Canada, Mexico, and the U.S. (PROCINORTE); the Caribbean Agricultural Science and Technology Networking System for the CARDI countries plus Suriname (PROCICARIBE); the Central American Cooperative Program for the Improvement of Crops and Animals (PCCMCA); the regional Cooperative Program for the Technological Development and Modernization of Coffee Cultivation in Central America and the Dominican Republic (PROMECAFE); the Central American Agricultural Technology Integration System, involving the Central American countries and Panama (SICTA); the Cooperative Research and Technology Transfer Program for the Andean Subregion, which includes Bolivia, Peru, Ecuador, Colombia and Venezuela (PROCIANDINO); the Cooperative Research and Technology Transfer Program for the South American Tropics, covering Brazil and the countries of the Amazon Basin—Colombia, Ecuador, Guyana, Peru, Suriname and Venezuela (PROCITROPICOS); and the Cooperative Program for the Development of Agricultural Technology in the Southern Cone, which includes Argentina, Bolivia, Brazil, Chile, Paraguay, and Uruguay (PROCISUR).

[12] In 2000, its members included Argentina, Bolivia, Chile, Colombia, Costa Rica, Ecuador, Nicaragua, Panama, Paraguay, Peru, Dominican Republic, Uruguay, Venezuela and the International Development Research Center (IDRC). www.fontagro.org.

together with regional and international research centers, in association with national technology development organizations.

Taking into account the growing importance of operating in knowledge networks, FORAGRO is a mechanism designed to facilitate discussion and support the definition of a regional agricultural technology research and development agenda. FORAGRO's general objective is to contribute to the consolidation of the Agricultural Technology Innovation System for the Americas by facilitating dialogue, coordination, and strategic alliances between the stakeholders that comprise national, regional, and international technology research and development systems. In 1997, the Inter-American Board of Agriculture (IABA) decided to support the Forum's creation and asked IICA to set up its Technical Secretariat. In May 1998, FORAGRO held its first meeting. The Forum includes a wide range of members: national public and private agricultural research institutions, national science and technology councils, university education centers and private sector organizations, producers' associations, NGOs, public and private foundations that implement or promote technological innovation, sub-regional cooperative research programs, regional networks, CATIE and CARDI centers, CGIAR Centers located in the Americas, as well as FONTAGRO and IICA, which acts as the Forum's Technical Secretariat.[13] Although FORAGRO does not have official representation in CGIAR, it plays an important role in the design of that body's overall strategy by providing regional inputs for determining its priorities at the global level.

Finally, the regional Technology Research and Development Center of the Americas is supported by the international centers of CGIAR, the main global agricultural research network. Three of these centers are located in the LAC Region: CIMMYT, headquartered in Mexico; International Center for Tropical Agriculture (CIAT), based in Colombia; and the International Potato Center (CIP), headquartered in Peru. The region also receives support from the network of international research centers for different activities and products with headquarters in other countries, including those specializing in policy (International Food Policy Research Institute—IFPRI), plant genetic resources (International Plant Genetics Resources Institute, now known as Bioversity International—IPGRI), livestock production (International Livestock Research Institute—ILRI), and forestry and agroforestry (Center for International Forestry Research—CIFOR, and the International Center for Research in Agroforestry—ICRAF). All these institutes carry out activities in LAC and in some cases have offices in several countries in the region (Box 2-2).

In brief, we can say that the present AKST system in LAC consists of a complex web of institutions, programs and other cooperation mechanisms created over time with the aim of ensuring sufficient spatial and thematic coverage, and taking advantage of potential contributions from public and private stakeholders at the different levels (local, national, regional and international) (Figure 2-1).

Nevertheless, various authors have noted that the lack of inter-institutional links has been a major weakness of AKST systems in LAC (Níckel, 1989; Eckboir et al., 2003; Parellada and Eckboir, 2003; Piñeiro et al., 2003).

In the Amazon region, the evolution of the institutional complex has been based on integrating its important contribution at the global level to the respective national economies, and reinforcing national sovereignty in the face of the possible internationalization of tropical rainforests (Walschburger 1992; Chaves de Brito, 2001; Becker, 2005). In this subregion, the key problem is the lack of an autonomous research corps and hence of regional capacity in science and technology for the agricultural sector (Aragón, 2001, 2005; Sicsú and Lima, 2001; Perez-Garcia and Domingue, 2004; Becker, 2005).

The advance of democracy and subsequent economic liberalization at the end of the 1980s and beginning of the 1990s redefined and energized the roles and functions of the State—all this in the context of an environmental crisis that has encouraged new ideas within the framework of sustainable development. Special reference must be made to the U.N. Conference on Environment and Development (UNCED), or Earth Summit, which was held in Rio in 1992 and promoted the development of AKST systems both by governments and non-governmental organizations.

In the 21st Century, a new AKST agenda is emerging in the region. It involves, for instance, South-South cooperation for eco-development and sustainable water management in the Amazon basin (Aragón, 1998), the Initiative for the Integration of Regional Infrastructure in South America and the United States Agency for International Development's Amazon Basin Conservation Initiative.

2.1.4 Institutional and administrative constraints in national AKST systems

Although LAC's national AKST systems vary greatly in size, organizational structure, effectiveness, and level of support, and have very different characteristics stemming from their institutional, cultural and political context, a study identified a number of common problems affecting these institutions (Nickel, 1996). The most outstanding include limited inter-institutional cooperation (Table 2-1a), lack and poor allocation of resources (Table 2-1b), organizational and management weaknesses (Table 2-1c) and labor-related weakness (Table 2-1d).

National AKST leaders in LAC have acknowledged the existence of these problems and several efforts have been made to correct them, often through externally financed projects. ISNAR, for instance, sent specialists to various countries to assess their institutional situation and offer advice on the best measures to improve organizational structure and administration and management procedures. It also devised tools for research management and made them available to institutions through publications and training programs. This has led to a significant improvement in the effectiveness and efficiency of some national institutions. But many problems persist because certain institutions continue to operate in a policy and cultural environment that is not conducive to the changes required.

[13] FORAGRO implements biannual plans based on the interaction between the agreed political-institutional lines of action and the priority technical lines of action, consisting of 11 major research topics adopted for hemispheric cooperation (www.iicanet.org/foragro).

Box 2-2. Examples of linkages between the Consultative Group on International Agricultural Research (CGIAR) and civil society in Latin America

The scientists who work at the 15 CGIAR centers collaborate closely with a broad spectrum of civil society groups. These include farmers, producers' associations, and community organizations. Participatory research is a way of ensuring that the results of CGIAR's research efforts rapidly reach small farmers with limited resources so they can use them to improve their quality of life and livelihoods. The examples described below offer a brief synthesis of the participatory research projects currently under implementation and other programs that foster important linkages with civil society.

Local Agricultural Research Committees (CIALs). In these committees, coordinated by the International Center for Tropical Agriculture (CIAT), farmers express their views on the development and evaluation of agricultural technologies. Researchers benefit from the feedback provided by farmers. Farmers, in turn, are encouraged to evaluate new options for increasing agricultural productivity and improving the management of natural resources. Currently, 249 local committees are active in eight Latin American countries. The benefits of this initiative range from increased local capacity in formal research methods and improved local planning and management skills to a greater availability of improved seed, not to mention food security. For example, in Cauca, Colombia, over 80% of farmers from the village of Pescador have adopted a bean variety recommended by the local committee. CIAT has estimated a 78% rate of return on investments to implement the CIALs approach (www.ciat.cgiar.org).

Learning partnerships for agribusiness development in Latin America. CIAT, in association with CARE, Catholic Relief Services and other institutions, is creating "learning partnerships" in Central America. These innovative partnerships are made up of research and development organizations that jointly design and implement strategies and interventions aimed at building local capacity in specific geographical areas. Members of these partnerships, including farmers, jointly analyze the strategies to determine which ones work. The lessons learned are applied and generate new learning cycles. In Nicaragua, thanks to this participatory learning process, an agribusinesses initiative that began in one municipality is now being applied in 10 others (www.ciat.cgiar.org).

Combating bacterial wilt in the Andean region, CIP scientists have developed an inexpensive detection kit that can be used in an organized seed system to eliminate infected potato seed before it reaches farmers' fields. Although crop rotation can help eliminate the pathogen from the potato fields, the recommended method—abandoning potato cultivation for a few years—is not an economically or socially viable option for thousands of poor farmers who depend on the tuber for their income and nutrition. With CIP's participation, farmer/researcher groups have identified a promising solution that enables farmers working in highly infested soils to sanitize their fields in 9-17 months by planting three successive non-solanaceous horticultural crops with high market value (e.g., onion, leek, or cabbage), or two successive food crops such as lupine, sweet potato, or *arracacha* (an Andean root crop) after the potato harvest. Using this method, farmers were able to recover their fields for potato production in a short time—and also managed to triple their potato yields (www.cipotato.org).

CIMMYT and the Agricultural Research and Experimentation Board (Patronato) of the State of Sonora. In the Yaqui Valley in Sonora, located in Northwestern Mexico, a group of private farmers and the *Patronato* have donated a new sprinkler and drip irrigation system to CIMMYT that will help scientists avoid water wastage and better manage this valuable resource in a dry zone. The system will directly benefit farmers in the Yaqui Valley who produce wheat, maize, and other crops. *Patronato* leaders work on a voluntary basis and make sure that the organization only invests in research efforts aimed at minimizing the obstacles to agricultural production (www.cimmyt.org).

Self Help International, an NGO based in the United States, is promoting quality maize with high protein content in Nicaragua. This new and more nutritious variety of maize, developed by CIMMYT, is helping to reduce malnutrition in a community located in the southern tip of Lake Nicaragua (near Costa Rica) that has the second highest maternal mortality rate in the world. After Hurricane Mitch, Self Help International, in collaboration with CGIAR, established an innovative seed bank program, giving farmers a bag of seed to be paid back later with two bags of seed that in turn would be distributed to other farmers, allowing them to benefit from the new technology. By December 2002, more than 7,000 farmers were planting the new maize seed (www.cimmyt.org).

Consortium for the Sustainable Development of the Andean Eco-region (CONDESAN). The consortium works with the Water and Food Challenge Program for Andean Region Watersheds. CONDESAN provides support to this program by creating links between research networks, and providing its infrastructure and experience, in order to contribute to the efficient execution of research activities. By combining the program with other regional initiatives, CONDESAN prevents duplication of efforts while promoting complementary aspects and fostering synergies. The main purpose of this collaborative effort is to promote an eco-regional approach to meet development challenges in the Andean region.

Conserving agricultural biodiversity. Cassava, maize, beans, potato, and sweet potato are Latin America's leading crops. The Center for Advanced Research and Studies of the National Polytechnic Institute (CINVESTAV) brings together the main national research programs and the CGIAR centers in order to promote conservation activities throughout the region. The International Plant Genetic Resources Institute (IPGRI), for example, has implemented an international cooperation project in nine countries to strengthen basic science for *in situ* conservation of cultivated plants and to incorporate agricultural biodiversity into agricultural development strategies. Similarly, the Latin American and Caribbean Consortium to Support Cassava Research and Development (CLAYUCA) works to increase cassava production and expand marketing opportunities for poor farmers throughout Latin America (www.ipgri.cgiar.org).

Table 2-1. **Problems common to NARIs in LAC (Most outstanding examples).**

(a) Limited inter-institutional collaboration
Mutual antagonism and lack of cooperation between the institutions in charge of agricultural research and universities and university faculties involved in agriculture.
Ineffectual links between plant and livestock research.
Insufficient use of socioeconomic disciplines.
Research on production and on-farm systems relegated to isolated, separate organizational units. Insufficient support for the concept that researchers should carry out research on-farm from a production systems perspective.
Ineffective linkages between research and extension activities.
Limited interaction and linkages between public and private bodies engaged in agricultural research.
Insufficient participation of producers in the definition of research agendas and the evaluation of results.
Dispersion of agricultural research over a large number of ministries and other agencies.
Excessive intra-institutional fractioning, with researchers and other team members spread over too many small experimental stations or scientific fields, leading to the lack of a critical mass for the efficient use of infrastructure and proper supervision, tutoring, and collaboration.
(b) Resource problems
Severe lack of resources.
Allocation of resources by crop, system, product or research area that do not reflect national priorities and the needs of producers.
Diluted distribution of scarce resources among a large number of crops or research areas without the necessary setting of priorities.
Inappropriate balance of resources (the greatest percentage of budgets is assigned to paying for salaries, leaving insufficient resources for operations).
A resource allocation process that is too centralized.
Excessive dependence on resources from externally financed projects for the acquisition of equipment and vehicles.
Inefficient use of costly equipment and specialized infrastructure due to their dispersion and fractioning, aggravated by ineffectual linkages.
Budgetary allocation guided more by experimental station than by research area.
Budgets are more a compilation of "requests" than tools for the effective allocation of resources.
(c) Organizational and managerial weaknesses
Hierarchical organizational structures and attitudes instead of "flatter" structures linked to a more collegiate management style more conducive to scientific innovation.
Insufficient delegation of authority.
Purchase of inputs that is subject to complex and slow bureaucratic procedures.
Lack of management information (information systems).
Inappropriate procedures for the preparation and revision of budgets.
Inadequate research planning and follow-up.
(d) Organizational and personnel weaknesses
Hierarchical organizational structures and attitudes, instead of "flatter" structures linked to a more collegiate management style that is more conducive to scientific innovation.
Insufficient delegation of authority.
Lack or weaknesses in assessing individual performance.
Promotion based on seniority rather than on merit.
Heads of institutions or units not chosen on the basis on their administrative performance.
Inadequate training regarding leadership, administration, and management.
Lack of incentives.
Lack of flexibility in civil service regulations regarding the administration of human resources dedicated to agricultural research.

Source: Nickel, 1996.

In order to overcome these problems, a variety of semiautonomous institutions have been established, based on the assumption that they would be free from political influence in such fields as hiring and would enjoy greater flexibility in such areas as their administrative regulations.

Often, however, the institutional changes proposed could not be implemented, o were only done so partially. When examining the reasons, one or more of these factors seem to have played a role: (1) the Ministry of Agriculture or its equivalent agency would not renounce control of the AKST body; (2) the new human resource policies were not all that different from those applied in Ministry departments; (3) administrative procedures and financial controls remained too complex.

Human resource issues cannot be attributed to the quality of researchers, who are often cited as among the most capable and productive scientists in the field, but rather to the working atmosphere and the resources available to those centers. It should also be noted that simply improving salaries to attract and retain competent personnel does not automatically increase productivity nor the quality of research unless, at the same time, more attention is paid to the processes whereby staff is hired, evaluated, and provided with incentives.

Sometimes, particularly in traditional government systems, annual salary increases and promotions are based on seniority, not on productivity. The reason such systems were adopted was to discourage "favoritism." This is undoubtedly a consideration. However, it has become a crutch for a majority of the personnel of these institutions, aggravated in some countries by the existence of labor laws that make it almost impossible to sanction or fire unproductive employees. Productivity is thus rarely valued or rewarded, a severe weakness of some national institutions that, unless corrected, will condemn them to mediocrity.

In addition, LAC's oldest publicly funded research institutions such as EMBRAPA, INTA, INIA and INIFAP of Brazil, Argentina, Chile and Mexico respectively are faced with a problem of ageing researchers and support staff. Few of these institutions have adopted plans to renew or replace human resources due for retirement. In some countries, such as Mexico, this has resulted from a government policy of "indiscriminately downsizing the state apparatus"—an issue that merits critical assessment with a view to designing rational, efficient and effective policies.

Few AKST institutions have programs for training their scientific and technical staff and keeping them up to speed on current developments in their field, nor do they offer incentives to attract talented young people into cutting-edge research in new, highly promising fields like biotechnology or nanotechnology. Even less attention has been paid to other fields of knowledge—economic, social, anthropological—that are not so new or popular, but are very valuable when it comes to explaining and encouraging individual and collective attitudes and actions in order to generate and implement innovations leading to productive, sustainable and equitable development.

The abovementioned challenges justify efforts to promote a greater and more effective interaction between research centers and advanced training and education institutions, and to promote their participation in projects of interest to their respective countries and societies involving what is known as Participatory Innovation Development.

In the administrative field, it is clear that senior managers of AKST institutions feel more comfortable with bureaucratic procedures than with more flexible systems for administering financial resources and purchasing inputs, since the former protect them from being accused of mismanagement. Safeguards or controls are necessary to prevent abuses, but it is also essential to adopt more flexible and effective administration and financing systems. This is particularly crucial in AKST system institutions, where significant delays in making funds available, or in purchasing equipment and inputs, can negatively affect the effectiveness of research.

However, either because of the nature of their legal constitutions or because of subsequent administrative decisions by the Central Government, most NARIs have operated within the administrative restrictions and political interference that characterize Latin America's public sector (Bisang, 2003).

Piñeiro, (2003) cites Argentina's National Agricultural Technology Institute (INTA) as an example of the progressive erosion of their autonomy (Piñeiro et al., 2003). Created in 1958, INTA's charter granted it financial and administrative autarchy. However, over the years, the political authorities gradually curtailed this independence, converting it *de facto* into an institution with the same restrictions as the rest of the central administration. (Recently, this situation was reversed when INTA recovered its budgetary autonomy.)

A similar situation occurred with Mexico's National Agricultural Research Institute (NARI), which was widely recognized for its effectiveness, efficiency and productivity. Legally, it was a deconcentrated body of the central administration; from the beginning it was endowed with a trust fund that allowed for flexible and timely financing and operational autonomy. This mechanism was canceled in 1982, as part of a general government instruction to cancel public trust funds, and thereafter the Institute became subject to the regulations of the central administration, which were not very suitable to research functions. However, national public research centers like INIFAP currently enjoy a trust fund that contributes to the flexible and timely financing of their research activities.

At present, the effectiveness and relevance of AKST system institutions is in doubt. The lack of consistent political support, the ensuing weakness and randomness of public funding, and institutional "obsolescence" in the face of the growing complexity of science and extraordinary changes in the economic context, all call for AKST institutions in LAC countries to embrace modernization (Piñeiro and Trigo, 1983), including modifications to their management processes and their links with users.

To be more efficient and effective, changes to AKST system institutions must be approved, implemented, and audited. External political pressures must ensure that these changes are approved by higher-level government authorities. That will not be easy. This external political pressure may be exerted more naturally and efficiently by society through the social oversight of stakeholders, who will ensure that AKST institutions implement the approved changes. In other words, the advancement of AKST systems in LAC depends in large measure on their capacity to monitor the risks and opportunities posed by their external context and their capacity to communicate with their users and obtain their feedback.

Such a legal framework would allow for a responsive and flexible management style, essential for achieving greater efficiency—including salary levels and promotion system for scientific personnel, flexible recruitment policies, links and associations with the private sector, royalty contracts, and/or a share in income derived from intellectual property. Examples of this trend in the region include Chile's INIA and the Colombian Agrarian Research Corporation (CORPOICA) (Piñeiro, 2003). In response to this problem, Mexican lawmakers took the initiative of creating a new definition for public research institutions.

There has also been a growing tendency among NARIs to include representatives of leading private sector trade organizations on their governing bodies at the national and regional levels. Argentina's INTA has enjoyed a long history in this regard; half the members of its Board of Directors have been representatives of producers' organizations since it was established in 1956. Among the more interesting examples of this trend one can mention Uruguay's INIA, CORPOICA, and INIFAP. However, sometimes the composition or actions of the governing body could be improved, as in the case of INIFAP in Mexico (Piñeiro et al. 2003).

2.1.5 The evolution of the AKST system

Technology generation in LAC dates back to pre-Columbian times. Notable contributions have been made throughout history, for instance in the Andes and the Amazon basin.[14] Towards the end of the nineteenth century and the beginning of the twentieth century, the AKST system was institutionalized; that is to say, the first stage of organized agricultural research began in universities or specialized national institutions sponsored by the state. In those early stages, these institutions were organized into departments, that is, by branches of knowledge. Their researchers interacted very little with each other, and their sphere of action was the Experimental Station.

In the second half of the twentieth century, farming system research was incorporated, forcing researchers to interact directly with the rural milieu. From the relatively simple environment of the Experimental Station, the move was made to the more complex and multifaceted context of farms and production systems, leading to an acknowledgement of the need for interdisciplinary work. By working with "cooperant producers", researchers adopted an informal but highly effective role as extension workers that were broadly appreciated by producers. Some LAC countries have pursued such a researcher/extension-worker strategy as an effective means for the transfer of technology.[15]

In many LAC countries, however, extension services have not been integrated with agricultural research efforts, often separate agencies of Agriculture Ministries. The question of how to improve the effectiveness and efficiency of technical outreach and technology transfer has been, and remains, a highly significant and relevant issue.

Some LAC countries have pursued a participatory strategy involving farmers and extension researchers as an effective means of experimentation and transfer of technology (Piñeiro et al. 2003). These participatory systems have not only become important in technology transfer and training projects with low-income farmers and women but are also being used for such purposes as the genetic improvement of plants or the characterization and management of natural resources (Araya and Hernández, 2006).

One objective of participatory research programs is to take advantage of farmers' knowledge, which obviously implies identifying their needs, their preferences, and the reasons for what they do. Although society recognizes farmers' role in managing and improving germplasm, there is little agreement on how to appraise the role of farming communities—traditional, indigenous and agroecological systems, not conventional agricultural systems—and their potential contribution to formal systems of genetic improvement.

Technical cooperation can only grow and develop if potential barriers of mistrust are discussed and addressed ethically. The key issue here is to ensure that plant breeders—both producers and scientists—have access to germplasm.

In some LAC countries, we have witnessed over the past two decades a trend toward taking advantage more integrally of existing research institutions, mostly state-sponsored, and considering them part of a research and technology transfer system whose challenge is to promote a networking synergy based on interinstitutional complementarity.

The design, establishment, and operation of more efficient and effective AKST systems is at different stages of development in LAC countries, going from rhetorical discourse to efforts aimed at responding to specific demands from society. In the institutional discourse, it is often said that institutions have evolved from a supply-driven model to a demand-driven model. However, the weakness of AKST systems in most LAC countries has limited their capacity to develop interinstitutional links, as reflected in a limited number of partnership-based projects.

A new current of thought proposes that the greatest challenge is to shift from existing AKST systems to Participatory Innovation and Development (PID) systems that focus on specific production chains or commodities. Another conception, wider and more inclusive, involves the application of such systems to watersheds as the natural spaces or territories in which one or more production chains op-

[14] Among other significant innovations that have been documented as part of Inca civilization, one can mention drainage systems, as well as anthropic soils and other recent archaeological findings in the greater Amazonian basin. The diversity of genetic resources to be found in Peru is an achievement of its indigenous peoples who, over at least 10,000 years, domesticated native plants, selected them, and adapted them to ecological niches of varying altitudes. Thanks to this, and to the domestication of various species of fauna, Peru is one of the richest world centers of genetic resources, having domesticated 182 species of plants and five species of animals.

[15] For instance, in Mexico, INIFAP formally established Cattle Ranchers Groups for Technology Validation and Transfer, with initially promising results (Piñeiro *et al.* 2003), as well as Experimental Farmers for vegetable production. In both cases, small-scale producers were targetted.

erate and interact with each other and with the broader environment.

These developments have brought about new requirements regarding the attitudes and communication processes needed to facilitate dialogue and linkages between, on the one hand, those who generate technological knowledge and innovation and, on the other, those responsible for other links or factors indispensable to the development, productivity, and competitiveness of the production chain or watershed—suppliers, producers, traders, and financiers, as well as officials in charge of infrastructure, public policies, and institutions, and those in charge of information and communication mechanisms aimed at enhancing participatory development.

It is also necessary to improve the efficacy and efficiency of universities and other existing research, development, and technology transfer institutions. This calls for the creation of formal and informal mechanisms for interaction, including service contracts between such institutions and private sector users. In that respect, special programs and mechanisms have already been established to promote and facilitate linkages between agricultural research bodies and farmers.[16]

For the past several decades, moreover, private enterprise has become actively involved in the AKST system and has assumed an increasingly important role in the development of certain innovations (such as genetic products, machinery, and agrochemicals) and their dissemination among producers through the sale of inputs or services. As a result, public research institutions find themselves in the dilemma of either (1) competing, (2) withdrawing from the field and focusing their efforts on developing other innovations, or (3) attempting to cooperate on joint strategies. In other words, public AKST institutions face the challenge but also the opportunity of working with private AKST institutions on projects of mutual interest. This decision has strategic political implications that must be considered. It will test governments' vision and their willingness to generate new game rules, or standards, for public-private partnerships, in the interest of safeguarding the interests of society.

Another challenge facing AKST institutions in LAC is to take advantage of the enormous potential offered by new fields of knowledge such as biotechnology and nanotechnology, which are being incorporated at a different pace by the countries of the region.[17]

Although such developments may offer interesting alternatives related to people's well-being and quality of life, the level of investment required, together with patent- and copyright issues, could become insurmountable obstacles to taking advantage of their potential to benefit the region's poor. New developments are being used mostly by industry and the service sector, where users have purchasing power and the interests of investors are protected by intellectual property rights and patents. One of the greatest challenges facing small- and medium-sized countries in LAC is to review, update, and reinforce mechanisms and processes for regional cooperation in this area. Tables 2.2 and 2.3 summarize the factors that condition AKST's potential to develop more productive, sustainable, and equitable systems. They also summarize AKST's most significant impacts in Latin America.

2.1.6 Interactions between organizations and knowledge networks

In the early 1950s, formal national research organizations would transfer their technological innovations through public extension services and private agents. They did so with varying degrees of success, depending on the type of crop, type of producer, or agroecological area. The interaction between science, on one hand, and local technology and knowledge systems on the other, tended to be one way, frequently leading to the latter being undervalued.

Starting roughly around the 1980s, and varying from country to country, a reappraisal was made of the relations between organizations and knowledge networks. Two reasons accounted for this: the need to provide agile and innovative responses to the changing environment; and the redefinition of the role of public and private actors in agricultural research and technological innovation.

Although the ways in which networks have developed in the different countries display major differences, some important changes that have occurred in the last 25 years can be identified across the board:

In many countries, the relative importance of government investment in agricultural research declined, although it continued in the universities, increasingly relying on resources from the productive sector.

The role of extension services has been redefined for budgetary reasons and due to the restructuring of the state's role in agriculture. As a result, some extension tasks have been privatized and different types of civil society associations and organizations have intervened more actively in the provision of technical support.

In general, private or non-governmental actors have taken a more active role in the generation, validation, and transfer of agricultural technology, partly on the initiative of agroindustrial firms and providers of seeds and inputs, but

[16] For example, INTA in Argentina has implemented a technology transfer program, while Brazil's EMBRAPA and Chile's INIA have special programs in their regional centers. In Mexico, INIFAP has established the Cattle Ranchers' Technology Validation and Transfer Groups, the Experimental Farmers, and the MOCAT groups. For its part, civil society has created the Patronatos and the *Produce* Foundations to support agricultural and livestock research. In Bolivia, SIBTA has moved toward a model in which a good deal of technological innovations is carried out by private foundations that obtain financial support from the Government's budget.

[17] For example, biotechnology is not limited to the world of genetic engineering (DNA). There have been other agronomic efforts in this field, focused on integrated pest and disease management or the integrated management of agroecosystems. Biotechnology includes knowledge and management of soil microorganisms, different types of compost, green manures, forage crops, multiple-crop systems, biocultures, rhizosphere microbial cultures, efficient microorganisms, and bacteria that promote growth in plants and induce systemic resistance. These are just some examples that expand the horizons of biotechnology, and should be given equal consideration in government financing policies (León et al., 2004).

Table 2-2. **Factors that condition AKST's potential for developing more productive, sustainable and equitable systems.**

Type of factor	Description
Political	Lack of linkages between AKST systems and public policies—macroeconomic, commercial, financial, environmental, and related to access to markets, education, and information. Lack of policies to promote and support AKST. Lack of vision on the strategic role of the sector. Regulatory insecurity.
Institutional	Lack of cooperation in national, regional and international AKST networks. Lack of strategic plans and AKST participation in the same. Ageing of scientists and technicians and lack of human-resource policies within the system. Lack of balance in human resources with regard to interdisciplinary, intercultural, and gender issues. Lack of linkages between research and technology transfer.
Economic	Reduction of public investment in AKST. Insufficient private investment in AKST.
Social	Lack of acknowledgement of the importance and impact of AKST among the general population (reflected in little public investment in AKST). Lack of participation of social actors in defining the agenda and management of AKST Systems.

Source: Authors' elaboration

Table 2-3. **Most important impacts of AKST Systems in Latin America.**

Aspects	Positive Impacts	Negative Impacts and Risks
Productive	• Improvement in crop and animal production yields by surface and water quantity, mainly in conventional production systems. • Development of new varieties and races that are resistant to diseases and adapted to different agro-ecological conditions. • Development of safer and higher-quality products. • Generation of new agricultural technologies.	• Loss of agro-biodiversity. • Loss of soil fertility. • Loss of productive systems' resilience. • Negative impacts on health due to lack of hygiene and on-the-job safety.
Economic	• Reduction in production costs. • Reduction in food prices, particularly for basic food items. • Increase in the income and profits of conventional farmers. • Increase in countries' GDP and exports. • Access to new markets for traditional, indigenous, and agro-ecological farmers.	• Reductions in employment. • Migration. • Concentration of profits. • Lower incomes for traditional or indigenous farmers.
Ecological	• Soil and water conservation in some production systems. • Generation of less polluting agrochemicals.	• Loss of agro-biodiversity and wildlife biodiversity. • Contamination of water and soils by agrochemicals. • Contributions to climate change.
Social	• Improvements in the social conditions of conventional and agro-ecological producers.	• Little impact on the social conditions of traditional and indigenous producers. • A devaluation of local knowledge.

Source: Authors' elaboration.

also due to a greater role by local and international NGOs and producers' associations themselves.

There has been a revaluation of farmers' own knowledge of agroecosystems and production systems better suited to local conditions. This has coincided with agroecological studies that examine comprehensively the complexity of these systems from a scientific perspective.

Our understanding of the interfaces between local technological knowledge systems and the scientific-technical system has improved with experiences in cooperation or joint experimentation. Studies have begun on both the constructive and negative interactions between formal and informal networks for the dissemination of agricultural knowledge.

Formal research networks are beginning to transcend the national sphere through joint efforts at the international level, although this remains incipient.

The development of such interactions differs greatly, especially between relatively small countries and larger ones where the size of the agricultural sector itself, and public and private investment, have made it possible to establish institutions with more significant human and financial resources and their work has developed on a larger scale and with a more long-term projection, as in the case of Brazil, Argentina, Mexico, Colombia, and Venezuela.

In Central America, by contrast, the economic problems and policies of the 1980s, together with structural adjustment and state reform, led to a weakening of public agricultural research institutions and their links with international organizations and local universities, where a good part of formal agricultural and livestock studies continued to be carried out. Some undergraduate and postgraduate education centers with international projection, however, continued to promote concerted research efforts and served to link researchers within and outside their respective countries—such as the Tropical Agriculture Research and Higher Education Center (CATIE), the Zamorano Pan-American Agricultural School, in Honduras, and the Escuela Agricola de la Region del Tropico Humedo (EARTH University).

At the same time, the "Farmer to Farmer" movement and analogous experiences supported by producers' organizations and non-governmental cooperation agencies encouraged smallholder (*campesino*) experimentation, reconfigured the relations between technicians, scientists and farmers, and promoted alternative technological approaches in pursuit of a greater agroecological and social sustainability.

In the 1990s, efforts began to develop more participatory relations between public and private stakeholders engaged in producing and transferring technological knowledge. Such efforts involved exploring more participatory forms of research and extension, setting agendas through consultations and negotiations, and testing different forms of participation by farmers and their organizations in the various phases of the research process as well as in the assessment and dissemination of results. Different positions have been taken on the effectiveness of these activities, albeit at a very preliminary stage. But it is clear that consensus mechanisms are required in public-private agricultural and livestock research that may take a variety of forms and follow different paths.

2.1.7 Society's perception of AKST systems

The *public perception of science and technology* may be defined as a set of factors that have to do with the general public's understanding, knowledge, and attitudes towards scientific and technological activities (Albornoz et al., 2003). It is important to note that society has a positive perception of science in general, and technology in particular. This attitude is associated with the notions of modernity that prevailed over recent decades. However, negative views of technology, usually associated with concerns over environmental and social crises, also exist. (Albornoz et al., 2003; Piñeiro et al., 2003; Casanovas, 2006).

The lack of response to environmental problems linked to agricultural production techniques—like the contamination of water, soil and food with agrochemicals, the loss of biodiversity, and the clear-cutting of forests to expand the agricultural frontier—has often provoked determinist postures among certain sectors of society, especially social movements and NGOs linked to the rural sector. Much of the debate around these issues is based on a lack of information, or incomplete or biased information. This underscores the importance of promoting an effective liaison with the mass media (Albornoz et al., 2003).

A greater participation by society in the social oversight of AKST system institutions, both in terms of their work agenda and their performance, is also needed—among other reasons, to provide moral and political support through "positive external political pressure" on AKST system institutions, as well as on the Government itself. (SECYT, 1997; Polino et al., 2003) *(Box 2-3.)*

2.2 Research Approach, Agenda, and Processes

2.2.1 The AKST system agenda

From 1945 onwards, the AKST system agenda in Latin America and the Caribbean (LAC) had a strong biological orientation and was driven by agricultural export activities based on the premises of modernization and import substitution (Dixon and Gibbob, 2001; Ballarin, 2002; Kalmanovitz and López, 2006; Méndez, 2006). These lent special weight to economies of scale.

The current agenda and processes for generating knowledge and technological innovation in AKST institutions in LAC have become more diverse and complex. Nowadays, AKST system institutions are expected to address issues related to all the links in the agricultural production chain.

At the national level, AKST institutions face growing challenges in their efforts to address a wide range of diverse research agendas. These are aimed at generating:

1. Technological innovations for specific production systems of strategic interest to a particular country and/or watershed;
2. Innovations to explore and develop new agricultural products with high export value;
3. Innovations aimed at benefiting the poor and designed to meet their needs.

The design, application and financing of some of these research agendas has been, is, and will remain the responsibility of the state, since the goal is to generate public goods

Box 2-3. Civil society's role in supporting the AKST System—the case of the *Produce* Foundations, Mexico

According to a recent assessment (Ekboir et al. 2006), the *Produce* Foundations have been a highly significant institutional innovation in Mexico. In their ten-year history, the *Produce* Foundations have promoted links between the federal and state political authorities, on the one hand, and rural production sectors on the other, to support the transformation of public research organizations and influence the design and implementation of agricultural policies, including scientific, technological, and innovation policies for the rural milieu. New channels of interaction have also opened up between federal and state authorities, on the one hand, and groups of commercial agricultural producers on the other.

Mostly, these impacts did not originate in the activities for which the Foundations were established—that is, administering competitive funds for agricultural research and extension—but on actions the Foundations themselves started to engage in as they evolved.

The growth of the Foundations was made possible by the presence of a group of highly motivated and innovative individuals (Ekboir et al. 2006). They did not work only for the Foundations but also for the federal government and several state governments. Acknowledging the central role of such individuals is crucial for the design of policies and programs. Frequently a great deal of attention is paid to building organizations and regulations, while their effectiveness often depends on the people who are involved in the administration and operation of those organizations (Ekboir et al. 2006).

The Foundations have had a significant impact because they have developed effective learning mechanisms. Initially, research priorities and the selection of projects to be financed were determined in an *ad hoc* manner. Currently, the Foundations use structured methods to identify priorities and have adopted a clear division of tasks between the state levels, on the one hand, and regional and national levels on the other. They have also established new contractual mechanisms to transfer resources to researchers and providers of agricultural services.

By contrast, aspects related to extension have not received sufficient attention and until now remain one of the weaker aspects of the Foundations' work. For this reason, extension services are another area of opportunity.

According to Ekboir et al., 2006, the future recognition of the *Produce* Foundations will largely depend on their capacity to continue offering valuable elements for the consolidation of the agricultural innovation system and for the transformation of agricultural research organizations into more efficient and effective institutions in generating or identifying products and services to support innovation in the production processes.

Diversifying their funding sources and encouraging increased contributions of resources from state governments and from the users themselves for innovation projects of mutual interest is another short-term challenge facing the *Produce* Foundations.

for society as a whole but mainly for the poorest sectors. Due to their implications, other efforts regarding the AKST system agenda, such as the development of new agricultural products with high export value, will have to be financed mainly by the private sector. However, government support should not ruled out, given the interest by any country in improving its balance of trade.

A wide range of issues, such as postharvest handling, food safety, nutraceuticals, and organic products, also form part of society's new and growing demands. For this reason, it is said that today's AKST system agenda is driven more by consumers than producers.

Such considerations, together with a growing environmental awareness, means that some sectors of society expect AKST institutions to address and reconcile seemingly conflicting objectives, like productivity and environmental sustainability (Moncada and Muñoz, 1999).

Countries also face the challenge of responding to subregional AKST agendas (in Central America, the Caribbean, the Southern Cone, and the Andean countries) that are directed at generating knowledge and technological innovations and providing relevant subregional public goods for local application in fields such as:

- Climate change
- Diseases
- Biodiversity
- Water availability and quality
- Land degradation
- Management of persistent organic residues
- Air pollution

Traditional government institutions have little capacity to meet such a broad array of demands. As a result, others have begun to emerge. They specialize in specific areas, such as postharvest handling, food quality and safety, and certain promising cutting-edge fields such as biotechnology and genetic engineering.

We are just beginning to witness the emergence of institutions in a front-line scientific field—nanotechnology. As what might be considered an unprecedented preventive action, governments, industry and the world's research organizations have started to study ways to take advantage of its potential benefits while minimizing its potential risks. However, despite commitments to that effect, many opportunities have been missed to establish cooperative research programs.

The following question, however, remains unanswered: who will finance research projects aimed at using the potential of nanotechnology in areas of interest to the poor, such as health, nutrition, or energy?

Reducing poverty has been a secondary concern for the AKST system agenda in LAC. The primary goal has been to boost productivity in order to increase the food supply and reduce food prices. Implementing a research agenda aimed at helping the poor has been discussed by Hazell and Haddad (2001). More recently, in 2005, the International Food Policy Research Institute organized a meeting to explore poverty-related issues that might be of interest for public-private financing of pro-poor research projects.

Particularly noteworthy are certain research initiatives regarding the poorest social groups (see Box 2-4). The initiative by the Mexico-based International Maize and Wheat Improvement Center (CIMMYT) to promote the use of QPM (Quality Protein Maize) in several Central American and South American countries could be mentioned. Another example is INIFAP's adaptation of genetic material produced by CIMMYT to areas with a high concentration of poverty in the states of Oaxaca and Guerrero. In combination with the National Institute of Nutrition, INIFAP has already gathered statistical evidence to show the nutrition benefits offered by these types of maize to indigenous children in Oaxaca.

2.2.2 Clients of the AKST system

Different socioeconomic segments strive to determine the focus of research in relation to their own needs and aspirations. Assessments have been carried out of the importance to the public agricultural R&D sector of a variety of economic-social segments as target groups or beneficiaries of research in the field.

Castro et al. (2005) analyzed the situation in six Latin American countries (Brazil, Cuba, Mexico, Panama, Venezuela and Peru). Their study revealed general agreement among researchers in the region regarding the relatively low importance of social segments such as subsistence farmers and small family producers vis-à-vis medium- and large-scale producers. This work offers at best a partial perspective—only researchers were consulted. It therefore does not reflect the points of view of other sectors of society. Trigo and Kaimowitz's research (1994) on Latin America and the Caribbean, however, confirms that the benefits derived from the agricultural research undertaken by NARIs were mainly directed towards the larger, market-oriented farmers located in favorable ecological zones (Trigo and Kaimowitz, 1994).

This view of agricultural research is much more closely linked to economic development and agribusiness, and less to the social development of underprivileged segments like subsistence farmers and indigenous communities in agroecosystems (Trigo and Kaimowitz, 1994; Castro et al., 2005; Santamaría et al., 2005).

A study by Castro et al. (2005) also found that nongovernmental organizations were considered of little importance as agricultural research clients in Venezuela and Peru; of medium importance in Panama, Mexico and Brazil; and of high importance only in Cuba—even though Trigo and Kaimowitz (1994) noted the importance of NGOs with regard to the development of sustainable technologies, which involves highlighting local demands difficult to identify through the traditional approach to technology transfer.

Decentralizing research activities through the training of local non-governmental organizations, extension agencies, and farmers, in order to carry out simple adaptive research, would appear to be in order (Trigo and Kaimowitz, 1994). New priority clients also mentioned in studies on the subject include public policymakers and agroindustry. This takes into account recent advances in scholars' concept of agricultural research as not only directed at rural producers but at society as a whole—in this case, represented by consumers.

The greater importance of agroindustry as a client suggests a conception of agricultural research as linked to production chains and the development of processes technology capable of adding value to primary agricultural production, as well as competitiveness to those chains. This concept, more recent in the region, replaces the view of agricultural research as linked exclusively to primary production that prevailed until the 1980s. Trends governing demand imply greater specialization and a call for technology products aimed at a broader typology of producers (Lindarte, 1990; Trigo and Kaimowitz, 1994; Castro et al., 2005).

Finally, a notion emerged in the 1990s that attaches greater importance to clients such as policymakers, input providers, wholesalers, and retailers: It suggests a more politically influenced organization of research and a search for partners to resolve the shortage of financial resources (Trigo and Kaimowitz, 1994; Cetrángolo, 1996; Castro et al., 2005) (Table 2-4).

Historically, agricultural research organizations have found it difficult to determine the focus of research for each socioeconomic segment, involving as it does many complex dimensions—political, scientific, technological, environmental, economic, and administrative. To make matters worse, scientific progress has been uneven throughout the region (Castro et al., 2005).

While knowledge regarding the demands of medium- and large producers is ample, research organizations know little about the demands of other segments, such as subsistence farmers, indigenous communities, and small family farmers linked to production chains, and do not much value them.

2.2.3 Research styles

Research activities may be geared to different purposes. These purposes are commonly associated with the different types of research: basic, applied, adaptive, and strategic.

Studies that assess current research efforts by the public and private sectors regarding agricultural research of each type show that organizations involved in these activities are strongly oriented toward applied research, followed by adaptive research. Strategic research is the least important at present, but will become more important in the future, along with basic research.

During the 1950s, the dominant approach was adaptive research, based on the belief among policymakers that sufficient technology existed for the modernization of agriculture. This view prompted the establishment of agricultural extension systems in nearly all Latin American countries (Trigo and Kaimowitz, 1994).

The role of the private sector was limited to supplying seeds and agrochemicals. The food processing industry was still in its early stages, strongly dependent on public sector support. Except in the case of a few export products, private research was virtually non-existent. (Malan, 1984; Moura, 1990).

An analysis of historical trends suggests a gradual decline of applied and adaptive research in the public sector in favor of increasing efforts in basic and strategic research (Castro et al. 2005).

The development of biotechnology has prompted a change of emphasis towards basic research, which is evident in the growing importance of laboratory work with regard to fieldwork. Greater importance is attached to research institutions involved in basic science. For their part, Trigo and Kaimowitz (1994) note the importance of restrictions in

Box 2-4. A pro-poor AKST system agenda for LAC

Reducing poverty and its negative impacts has been of secondary importance to the AKST System agenda in LAC. The primary goal has been to boost productivity in order to increase the food supply and reduce food prices—and to increase the productivity of agricultural, forestry, fishery, and aquaculture export commodities.

Agricultural research policies often do not mention poverty relief among their specific goals. The incentives system for researchers does not encourage their interest in this issue (Gunasena 2003). A current and growing challenge facing governments, public AKST System organizations and civil society is to define, sponsor, and execute a research agenda to help the poor—with their active participation, It would be aimed at developing products and services accessible to poor populations whose use may serve to decrease or mitigate the negative effects of poverty.

Does AKST have the potential to generate knowledge and innovations that will contribute to reduce or mitigate the negative effects poverty on nutrition, health, energy use, and the degradation of natural resources? These are factors that influence the development of human capital, in terms of health, life expectancy, education, empowerment, organization, recreation, development, and well-being.

According to Nickel (1989), "Obviously, agricultural research *per se* cannot solve all social problems and inequalities." However, as he suggests, "Research policies and strategies may be designed in such a way as to direct the benefits toward relieving poverty." It is also possible to "develop technologies that will give a comparative advantage to farmers with limited resources and to poor consumers."

Both Nickel (1989) and Gunasena (2003) agree that a pro-poor research agenda should focus on product-systems of interest to the poor, and on the zones where they are concentrated such as barren highlands, the semiarid tropics, and marginal lands. Although these areas are extensive, their limited ecophysical conditions mean that the poor will not benefit unless research is focused on the natural resources available in the region they inhabit. Research should be designed to find ways of helping the poor to emerge from poverty.

The technologies most likely to succeed in these marginal areas are those associated with mixed livestock and agroforestry production systems, with improvements in deferred grazing, cover crops, etc., which are more in tune with the agroecological farming system (Gunasena 2003).

Science and technology policies to support the poor should promote the development of plots or farms in ways that do not require them to purchase more external inputs. A challenge facing AKST is to develop technologies that require little capital and low energy and can be used by small farmers with few resources. (Dialo, 2005; Pretty and Hine 2001).

A pro-poor AKST System agenda should aim to optimize integrated pest control and promote strategies to increase the organic matter content in the soil, improve the efficiency of fertilizers through biological nitrogen fixation, or develop technological innovations to conserve genetic resources (FAO 2005).

In short, according to Gunasena (2003), "The second green revolution—for poor peasant farmers on marginal lands—should not be a copy of the first. It should seek environmental sustainability [and] low-cost inputs and better yields on small plots, and should reduce risks to a minimum. It should focus less on crops and more on systems, and on finding ways to diversify production and use the different resources available."

Biotechnology and the poor. New developments in molecular biology offer opportunities for researching and resolving problems that affect developing countries, such as the increase in water scarcity. The development of drought-tolerant and salt-tolerant crops would be of value, as would genetic improvement to develop tolerance or resistance to pests and diseases.

However, it is unlikely that biotechnology and nanotechnology's potential will be used to solve these problems. Substantial investments would be required in laboratories, equipment, and highly specialized human resources, as well as financial resources to pay for royalties for access to and use of patented genes and processes. Small farmers with few resources—the potential users of such innovations, products, and services—have very limited purchasing power. Because biotechnology research is mainly concentrated in the private sector, large biotechnology companies focus on crops and livestock products that enjoy a large market. The users of these biotechnology products and innovations are large-scale producers with significant purchasing power.

Accordingly, basic research aimed at understanding the mechanisms and problems that affect crops grown by small farmers in developing countries will not receive financial backing. For this reason, it is essential that the international community create a trust fund to finance the use of frontier knowledge and advanced methodologies to address major problems affecting the poor in developing countries.

Financing a pro-poor agenda will test the solidarity between the public and private sectors, both at the country level and at the regional level, for instance in Central America and the Caribbean, throughout the entire region, and globally. And the primary responsibility for generating public goods (products and services) and making these available falls on governments.

Table 2-4. **Evolution of the AKST agenda in Latin America and the Caribbean over the last 50 years.**

AKST Dimension	Until the 1980s	Currently
Main AKST objectives	• Increasing production and productivity. • Increasing the food supply.	• Increasing production and productivity. • Increasing the food supply. • Ensuring food security. Conserving natural resources and providing environmental services. • Alleviating poverty. • Mitigating the impact of climate change and natural disasters. • Incorporating local knowledge.
Issues researched	• Production aspects: genetic improvement, fertilization and soil management, pest and disease management and control, agricultural machinery, animal and plant health.	• Production aspects: genetic improvement, fertilization and soil management, waste management, pest and disease management and control, agricultural machinery, animal and plant health. • Biotechnology and biosafety. • Postharvest treatment. • Environmental services valuation. • Agro-biodiversity and wildlife biodiversity conservation. • Impact of production on natural resources (water, soil, biodiversity). • Value added to the production chain. • Socioeconomic and anthropological issues. • Environmental-, ecological-, and natural- resource economics issues.
Technological tools used	• Animal and plant genetic improvement. • Crop and livestock technologies. • Soil management and conservation. • Water management and conservation.	• Advanced animal and plant genetic improvement. • Biotechnology and genetic engineering. • Crop and livestock technologies. • Precision farming methods. • Soil management and conservation. • Water management and conservation. • Information and communications technology. • Participatory methods. • Nanotechnology. • Aquaculture.
Dimensions assessed	• Agronomic.	• Agronomic. • Environmental and ecological. • Social. • Anthropological. • Economic (environmental and ecological). • Cultural. • Policy-related.
Main focus of AKST research	• Personal consumption and food self-sufficiency. • Agro-exports of commodities and other products.	• Personal consumption and food self-sufficiency. • Agro-exports of commodities and other products (including fruit, garden greens, and handicrafts). • Products with value added. • Non-agricultural products and services. • Biofuels.
Main AKST customers	• Conventional producers (in medium- to large-scale organizations). • Agro-industries.	• Conventional producers. • Agro-industries. • Agro-ecological producers. • Traditional/indigenous producers. • Women farmers.
Sectors included	• The primary sector.	• The primary sector. • The secondary sector and other stages of productive and service chains. • The non-rural sector.

continued

Table 2-4. **continued.**

AKST Dimension	Until the 1980s	Currently
Sectors included	• The primary sector.	• The primary sector. • The secondary sector and other stages of productive and service chains. • The non-rural sector.
Places where AKST activities take place	• Experimental stations.	• Experimental stations. • Demonstration farms. • Producers' farms and small farms. • Watersheds. • Non-rural milieus.
Legal nature of AKST institutions	• Centralized. • Mainly public, with a high degree of autonomy. • With little participation from NGOs.	• Decentralized. • Para-statal. • Public corporations run according to private law. • Public research centers. • Greater participation of the private sector in appropriable technologies. • Greater participation of small-scale producer NGOs.
Participation of civil society	• Low.	• Growing: moderate to high.
Valuation and incorporation of local knowledge in AKST	• Low.	• Growing.

the free flow of information, with a greater exclusion of research results from the public domain given their increased market value.

The private sector plays an active role in developing biotechnologies. Its interest grew with the advent of deregulation, economic liberalization, regional economic integration processes, and the growing recognition of intellectual property rights related to genetic material and other agricultural inputs (Piñeiro and Trigo, 1983; Trigo and Kaimowitz, 1994). This will have major implications for the region stemming from the wide dissemination of new biotechnologies, increased use of intellectual protection mechanisms, and support to regional industries, and will affect the interactions between the different public research institutions.

With regard to strategic research initiatives, according to Trigo and Kaimowitz (1994), efforts that do not have short-term commercial application require direct participation by the public sector. At present, according to Castro et al. (2005), strategic research only represents about 10% of public research in the six countries analyzed.

2.2.4 Priority research processes

Castro et al. (2005) point to the high historical importance of research on factors that affect production efficiency and, at the same time, the low importance assigned to research approaches more focused on scientific topics and social and environmental aspects. This shows that agricultural research finds itself at a crossroads, where the well-trodden paths towards the search for efficiency in production that have sustained research in the last fifty years have been exhausted but new paths are not yet known and research organizations do not have sufficient capacity to pursue them.

To identify the technology demands of users and define their research priorities accordingly, the national institutes have taken several steps, among the most outstanding ones decentralizing and regionalizing their activities. To this end, they have taken advantage of their experimental stations located in different areas of each country, which tend to specialize in specific commodities according to local characteristics. (Piñeiro et al., 2003)

It has also been pointed out (Castro et al., 2005) that the selection of priority lines of research requires:

- A strategic institutional planning mechanism to help develop a prospective approach to long-term needs that can provide a framework and nourish discussion by the scientists themselves regarding the relative importance and likelihood of success of various lines of research;
- Institutional mechanisms to facilitate effective linkages with technology users and ensure that these users can exert the necessary social oversight over decisions regarding priorities and resource allocation; and
- A financial structure to align research initiatives with the needs that have been identified.

However, national AKST system institutes are implementing these types of mechanisms to varying degrees and at different paces (Castro et al., 2005).

2.2.5 Monitoring and assessment of institutional performance regarding AKST

The follow-up and assessment of institutional performance has not been sufficiently attended to by most AKST institutions in LAC. In general, assessment occurs as an isolated action that is seldom used to improve organizational performance due, among other reasons, to a lack of the information needed to identify structural, organizational, or administrative and managerial problems.

Efforts to assess the results achieved by S&T institutions overall, and not just specific projects, only began in the 1980s and 1990s, and the issue has still not been addressed with the dynamism, energy, and depth needed to ensure a better use of resources and improve the planning and general efficiency of these bodies. The complexity and scale of NARIs has produced vertical organizations with many hierarchical levels and a bureaucratic management style, because they were established to respond to the problems of every region in the country, leading to highly complex institutions both from the organizational point of view and in terms of the quantity, variety, and heterogeneity of the topics to be researched. (Piñeiro et al., 2003).

Recent literature emphasizes the need for research institutions to adopt decentralized management styles with a horizontal organizational structure that promotes discussion and consensus-building among peers. In pursuit of this type of organizational structure and management style, two complementary paths have been followed (Piñeiro et al., 2003). The first has sought to develop a highly decentralized organizational structure in which different units enjoy a high level of operational autonomy, a model exemplified by American universities. The second approach, inspired by the reforms introduced in Great Britain, Australia and New Zealand, has been to create relatively small bodies with specific mandates, highly focused on regions, products, or scientific topics.

The main challenges facing AKST system institutions in LAC are to: (1) identify and measure all outputs, emphasizing productivity in terms of the products and services generated for clients/users; (2) address crucial management issues and constraints; (3) create consensus and a sense of ownership; (4) improve internal and external transparency; and (5) strengthen knowledge of the institution's strengths, weaknesses, and constraints. (Peterson et al., 2003).

The region's AKST institutions can improve their performance by assessing periodically, and critically, the relevance and quality of their research through the peer review system accepted by the international scientific community. It is also useful to review the modern and practical concept of assessment, which has progressed "from the notion of finding weaknesses and culprits, to an approach where the assessment is at the service of users, with an emphasis on learning to improve organizational and institutional performance" (McKay and Horton, 2003).

2.2.6 Knowledge, science and technology from an agroecological perspective

Starting in the 1970s, alternative production models have been developed with a view to reducing the use of pesticides in agricultural production. This has led to a variety of practices, among them integrated pest management (IPM), integrated crop management (ICP), and agroecological pest management (Burlet and Speedy, 1998).

In the early 1980s, an agroecological alternative to the commercial agricultural system began to develop. This alternative is based on a systemic approach to managing agricultural production that identifies the ecological, social, economic, cultural, and geopolitical dimensions related to the management and use of natural resources, revaluing the exchange between local know-how and scientific knowledge (Sevilla and González, 1995; Sevilla and Woodgate, 2002; Bernal, 2006). Other sustainable management approaches have emerged, such as agroforestry, integrated soil management, and integrated watershed management.

The agroecological approach has been adopted by producers' organizations, public research institutions, universities, and non-governmental organizations. The most prominent include the Latin American Consortium for Agroecology and Development (CLADES), based in Chile, the Masters Program in Ecological Agriculture of the Tropical Agriculture Research and Higher Education Center (CATIE) in Costa Rica, and the Masters Program in Agroecology of the University of Caldas, Colombia. Leading NGOs in the field include the Ecological Agriculture Network and the Agroecological Movement of Latin America and the Caribbean (MAELA), an open, pluralistic and diverse movement involved in research, development, training and promotional activities that brings together over 65 institutions.

2.3 Financial Resources and Administration of the AKST system

2.3.1 Development and impact of investment in AKST

In Latin America, total investment in agricultural R&D in 2000 came to 2.6 million dollars; of these, 2.5 million (95.2%) were public investments (Pardey and Beintema, 2006). Most studies carried out in the region, as in other regions, show extremely high rates of return on investment in agricultural research and development (Alston et al., 2000; Ávila et al., 2002, Días Ávila et al., 2006) (See Table 2-5).

Despite this, from the mid-1980s onward, and especially during the nineties, public investment in agricultural research and development declined. As a result of fiscal and public debt problems, most countries in the region implemented profound reforms in their macroeconomic, commercial, sectoral, and overall public investment policies, aimed at limiting state intervention and reducing public spending and deficits. These policies restricted agricultural credit, making it more expensive, and reduced the budgets allocated to investment in rural infrastructure and those aimed at agricultural research and extension and other programs and services to support rural development.[18]

This less favorable context of macroeconomic and sectoral policies was reflected in lower growth rates for agricultural production in LAC countries—both in terms of cultivated area and average productivity—for the period 1982-2001, compared with those recorded for the period 1962-1981 (Table 2-6). As the authors note, average growth of production for the main agricultural commodities was 3.05% annually in the 1960s and 1970s, and fell to 1.98% in the last two decades. But there were significant differences in the growth patterns of the different subregions. In the Andean countries, Central America, and the Caribbean,

[18] These policy changes to support agriculture in LAC also coincided with the start of a review of subsidies and food self-sufficiency policies in developed countries, especially the Common Agricultural Policy.

Table 2-5. **Impact assessments of agricultural research in different LAC countries.**

Authors	Countries	Products/Levels	Rates of Return* (%)
Ayer & Schuh (1972)	Brazil	Cotton	77
Fonseca (1976)	Brazil	Coffee	23-26
Moricochi (1980)	Brazil	Citrus	28-78
Ávila (1981)	Brazil	Irrigated rice	87-119
Cruz & Ávila (1983)	Brazil	Aggregate	20 -38
Roessing (1984)	Brazil	Soy	45-62
Monteiro (1985)	Brazil	Cacao	61-79
Barbosa, Cruz & Ávila (1988)	Brazil	Aggregate	34-41
Teixeira *et al.* (1989)	Brazil	Aggregate	43
Gonçalves, Souza & Rezende (1989)	Brazil	Rice	85-95
Evenson & Ávila (1995)	Brazil	Wheat Soy Maize Rice	40 58 37 40
Oliveira & Santos (1997)	Brazil	Aggregate	24
Almeida, Ávila & Wetzel (2000)	Brazil	Soy	69
Almeida & Yokoyama (2001)	Brazil	Rice	93-115
Barletta (1971)	Mexico	Wheat Potato Maize Other crops	74-104 69 26-59 54-82
Himes (1972)	Peru	Maize	65
Ardila (1973)	Colombia	Rice	58
Montes (1973)	Colombia	Soy	79
Peña (1976)	Colombia	Potato	68
Scobie & Posada (1977)	Colombia	Rice	87
Pazols (1981)	Chile	Rice	16-94
Yrrarazaval R. 91982)	Chile	Wheat Maize	21-28 36-34
Martinez (1983)	Panama	Maize	47-325
Norton (1987)	Peru	Beans Maize Potato Rice Wheat	14-24 10-31 22-48 17-44 18-36
Mendoza (1987)	Ecuador	Potato Rice Soy Palm oil	28 44 17 32
Scobie (1988)	Honduras	Fruits, nuts Other crops	16-93 17-76
Cordomi (1989)(**)	Argentina	Aggregate	41
Echeverria (1989)	Uruguay	Rice	52
Evenson & Cruz (1989b)	PROCISUR Southern Cone Region	Wheat Maize Soy	110 191 179
Ruiz de Londoño (1990)	Peru / Colombia	Beans	15-29
Traxler (1990)	Mexico	Wheat	22-24

Table 2-5. **continued**

Authors	Countries	Products/Levels	Rates of Return* (%)
Pino (1991)	Ecuador	Wheat Potato Maize Beans	29 29 3 5
Palomino & Echeverria (1991)	Ecuador	Rice	34
Taxler (1992)	Mexico	Wheat	15-23
Cruz & Ávila (1992)	Andean Region	Aggregate	245
Vivas, Zuluaga & Castro (1992)	Colombia	Sugarcane	13
Racines (1992)	Ecuador	Palm oil Soy	32 35
Palomino & Norton (1992)	Ecuador	Flint Maize	54
Byerlee (1994)	Latin America / Caribbean Mexico	Wheat Wheat	81 53
Cap (1994)	Argentina	Beef cattle Milk Maize Potato Wheat Other crops	74 55 77 69 67 54-59
Macagno (1994)	Argentina	Maize Wheat Other crops	47 32 34
Pena (1994)	Argentina	Potato	53-61
Romano, Bermeo & Torregrosa (1994)	Colombia	Sorghum	70
Byerlee (1995)	Latin America	Wheat	82
Fonseca (1996)	Peru	Potato	26
Ortiz (1996)	Peru	Potato	30
Farfan (1999)	Colombia	Coffee	21-31
Manzano (1999)	Ecuador	Rice	58
Amores (1999)	Ecuador	Cacao	31

*Internal rates of return, except in the cases indicated with (**) which are estimates of the marginal internal rates of return.

Source: Adapted from Días Ávila et al., 2006.

growth rates declined. By contrast, growth rates increased in the Southern Cone countries, influenced mainly by increases in the productivity of the land both for crops and for livestock.

Moreover, the restrictions imposed on public budgets for AKST in the last few decades have come precisely at a time when LAC's producers have faced growing pressure to improve their productivity in order to compete at the international level in the context of free trade policies—those stemming from unilateral reforms implemented by the countries of the region, as well as those resulting from multilateral trade negotiations in GATT and the WTO, those corresponding to the different sub-regional integration initiatives (CARICOM, CAN, MERCOSUR, NAFTA), and a growing number of bilateral agreements signed by some countries, especially Mexico and Chile.

It is also important to emphasize that the decline in public investment in the AKST system in LAC has coincided with new demands, associated with sustainable rural development, that have traditionally been assigned a low priority in the agendas of the region's institutions.

The most important of these demands are: (1) conservation of natural resources and the environment; (2) conservation and sustainable use of genetic and biodiversity resources; (c) the development of human resources and social capital as strategic assets for competitiveness and progress; (3) the empowerment of civil society; (4) proper attention to aspects related to gender and ethnicity; (5) the incorporation of new leading-edge technologies that require substantive changes in institutional structure and organization, such as biotechnology, genetic engineering, nanotechnology, telecommunications, and computer science; (6) emerging new topics or issues that may have significant impacts on production and on future food demand, e.g., biofuels; and (7) new demands linked to such issues as product differentiation and value added.

Table 2-6. **Growth Rates of Agricultural Production in Different Regions of LAC during the Period 1962-2001 (annual %).**

Regions	Crops			Livestock			Average Growth		
	1962/1981	1982/2001	Average	1962/1981	1982/2001	Average	1962/1981	1982/2001	Average
Southern Cone	2.79	2.98	2.89	1.74	2.95	2.34	2.27	2.96	2.62
Andean	2.43	2.65	2.54	3.95	2.92	3.44	3.19	2.79	2.99
Central America	3.60	1.32	2.46	4.35	2.84	3.59	3.97	2.08	3.03
Caribbean	1.20	-0.71	0.24	2.78	0.77	1.78	1.99	0.03	1.01
Averages	2.55	1.57	2.06	3.56	2.38	2.97	3.05	1.98	2.51

Source: Días Ávila et al., 2006.

In short, the political, fiscal and institutional crisis of the State in most LAC countries over the last two decades and the resultant reforms in macroeconomic, trade, and sectoral policies—including cuts in public investment in research and development—have created a less favorable context for promoting sustained growth in the value of agrifood production and a decline in the system's capacity to address traditional demands. And this comes at a time when the new context calls for a change in Latin America and the Caribbean's NARIs, in their institutional strategies, structure, and management models, so that they can fit into the global AKST system (Machado, 2004; Martínez, 2006).

2.3.2 AKST funding amounts, trends and consequences

Ardila (2006) underscores that public investment in agricultural research and development in most LAC countries was always low compared to international standards. It is a situation that has worsened in recent decades. Thus, while the ratio of research spending to GDP for the period 1970-75 in industrialized countries was around 2.5%, the average in LAC was 0.65%; and that ratio fell to 0.5% in 1975-85 and to a range of between 0.1 and 0.4% in 1985-95.

According to Hertford et al., (2005), in the mid 1990s—the last date for which global figures can be compared internationally—a total of US$21.7 billion were spent worldwide on agricultural R&D. LAC countries spent US$1.95 billion (at 1993 international prices) or close to 8.8% of the world total. This was nearly double what those countries spent in 1976. However, there were great disparities. More than half the investment in agricultural research corresponded to Brazil. If Mexico is added, both countries accounted for nearly two-thirds of the region's total. Other three countries spent over US$100 million annually. However, a significant number of countries spent US$16 million or less, resulting in a serious erosion and decline in the installed capacity of specialized institutions. Moreover, these have not been replaced by equivalent investments in the private sector.

When one measures overall expenditure in agricultural research as a proportion of the share of GNP that corresponds to agriculture, in the mid 1990s in LAC the average was 1.12%, almost twice as much as was spent in 1976 (Table 2-6). However, great disparities persisted, from barely 0.13 in Guatemala to more than 1.7 in Brazil and Uruguay. These coefficients of agricultural research intensity in Brazil and Uruguay are far superior to those of most countries in the region, albeit far inferior to those recorded in industrialized countries, which on average spent 2.62% on such activities. Although funding from non-governmental organizations (mainly commodity producer organizations) doubled from 1976 to 1996, this increase started out from a very small base and undoubtedly continues to be insufficient to increase the poor intensity coefficients in the region.

Other private research has not been able to reduce the gap. While in rich countries approximately half of all agricultural research is carried out by private firms, by the late 1990s, in LAC, total expenditures by the private sector in agricultural R&D amounted to no more that 4.4% of total expenditures,[19] and with extreme asymmetries, since most of the private investment was carried out in Brazil. In Honduras, private research accounted for 7% of total agricultural R&D. In Panama, the figure reached 46%. Regardless, most private technologies used in the region are based on research carried out in industrialized nations.

Even in those countries where public investment in agricultural R&D increased in the first half of the 1990s, recovery was fragile. Investment was greatest in Brazil and Colombia, but suffered cutbacks in the second half of the decade. In the region's smallest countries, research activity has experienced no growth whatsoever, revealing an asymmetry between richer and smaller countries that left the latter lagging behind.

At present, only a handful of countries—Brazil, Mexico, Argentina, Colombia, and Venezuela—can boast of important organizations that have kept up significant levels of investment.

2.3.3 Consequences of reduced financing

In LAC, when analyzing the 1981-2002 period (Figure 2-2 and Table 2-7), a negative evolution in public research can be detected vis-à-vis industrialized nations. In the least devel-

[19] R&D investments are measured on the basis of where they are carried out, regardless of where the company's headquarters may be located.

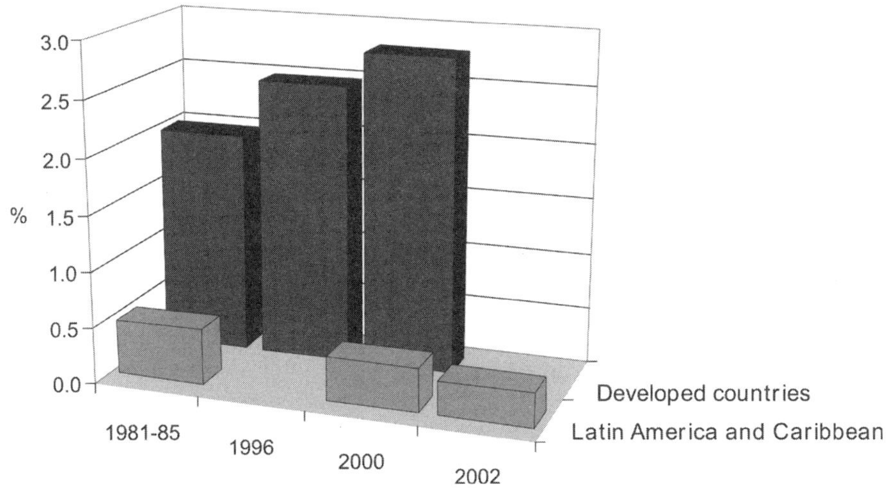

Figure 2-2. *Evolution of the intensity of public agricultural research in Latin America and the Caribbean compared to developed countries.* Source: Ardila, 2006.

oped countries, the lack of public investment in agricultural research is a significant threat if one considers the growing demand for knowledge to ensure the sustained growth of food production—something that can only be secured by innovation and increases in soil and water productivity. It should be noted that in many of these countries the availability of agricultural land per capita will tend to fall over the coming decades, making it likely that food production will not meet local demand. Not only will the balance of trade be affected; the population with the lowest income levels will have to pay more for food. Recent increases in international maize prices are indicative of this phenomenon.

Even in the five countries in the region that can boast of relatively strong public research institutions, the decrease in government funding has significantly affected their productivity. By degrading the ratio between operational and personnel costs, they have reduced their efficiency and the possibilities of carrying out the institutional transformation that contextual changes in recent decades call for.

Among other effects, this situation has led to the implementation of a variety of agreements between public institutions and the private sector for the development of certain technologies appropriable by private firms. The lack of government funding has altered the focus of NARI research. It is currently guided by the contributions and demands of companies, particularly those that specialize in providing agricultural inputs, although it also extends to groups of producers, agroindustries, and other components of society.

This entails a reconceptualization of NARIs to incorporate new management schemes that contemplate strategic planning aimed at forging alliances and cooperation

Table 2.7 **World and LAC: Indicators of public and private R&D activities around 1995.**

(a) Expenditure in agricultural research and development (millions of dollars at 1993 rates)

	Developing World		Developed World	World Total
	LAC	TOTAL		
Public	1,947	11,469	10,215	21,684
Private	91	672	10,829	11,511
Total	2,038	12,141	21,044	33,194

(b) Intensity ratio of agricultural research (percentage)

	Developing World		Developed World	World Total
	LAC	TOTAL		
Public	0.98	0.62	2.64	1.04
Private	0.01	0.04	2.80	0.61
Total	0.99	0.66	5.43	1.65

Source: Pardey and Beintema, 2001.

mechanisms at the national and international level with the various public and private AKST players—that is, the building of research networks (Lindarte, 1997; Salles-Filho et al., 1997) without ignoring the demands of the sector as a whole, which in most countries features small producers.

2.3.4 Changes in approaches to mobilizing resources

In the early stages, public funding for NARIs normally came from the national government budget. The main exception to this rule was Argentina's National Agricultural Technology Institute (INTA), whose charter allowed it to receive a direct percentage of revenues from leading agricultural exports. More recently, Uruguay's INIA began to receive a percentage of revenues from agricultural exports, complemented with an equal sum from the national budget. Since the end of 2002, Argentina's INTA has received a percentage of the earnings from imports coming from outside the MERCOSUR trade alliance (Piñeiro et al., 2003).

The limited experience of these funding arrangements suggests that it is advantageous for NARIs to have an independent financing system in which funds are assigned for specific purposes. This provides security regarding the sums that can be spent and their availability in the course of the fiscal year. Both elements are essential to proper planning. They also encourage a careful use of available resources since, if unused, they remain at the disposal of the institution.

Governments have tended to assign AKST funds as part of their overall budgets. A total annual amount has been generally allocated, divided into partial, normally monthly, payments. However, this periodicity has often not been observed, especially when it comes to operating costs, which are sometimes disbursed in random fashion. This allocation is supposed to cover: salaries, operating expenses, maintenance of infrastructure and equipment, and investment.

The aforementioned trend of declining governmental support for AKST institutions confronts them with shrinking and untimely budgets that reduce their effectiveness and efficiency. They are forced to cover, first of all, their payroll, for which they must use part of the resources earmarked for operations, maintenance, and investment. It is common to find ratios of 90:10:0 regarding salaries, operations and maintenance, and investment. Experts consider that this ratio should be 50:35:15.

Consequently, AKST system institutions have been forced to seek external resources to reduce their budget deficits. This has led them to diversify their funding sources through a variety of projects. It has also led them to identify and approach other financial agents they may turn to (multilateral banks, regional research funds, international cooperation), which are not necessarily a solution for AKST institutions confronting a budget deficit and a reduced capacity to cover their essential payroll, operational, and maintenance expenses.

Recently, national AKST system institutes have made major efforts to adapt to the new conditions. In general, they have solved their budgetary problems. In some cases, they have even managed to improve their finances significantly. As a result, changes are evident in their financial structure and composition, and many now generate their own resources through the sale of non-essential assets and technological services and solutions.

Similarly, these organizations are taking their first steps to harness the benefits derived from the intellectual property of some of their own technology packages. This has implied developing new regulatory frameworks on issues such as intellectual property legislation for seeds, genes, and other appropriable innovations that encourage private investment in agricultural R&D, as well as laws to properly regulate the appropriation of benefits in the case of joint initiatives between public institutions and private firms (based on the notion of public goods and private goods).

Finally, it is important to note that the debt crisis of the 1980s and the effects of globalization have forced governments to rethink the administration of science and technology. In developed nations, direct government contributions have been reduced and new mechanisms have been introduced to finance innovation activities, such as competitive funds for research, contracts for the development of specific products, the purchase of new products by the public sector, subsidies for innovation activities in companies, and the formation of public-private consortia (Branscomb et al., 1999; Huffman and Just, 1999; Echeverría and Alvaro, 2000).

These new mechanisms have not replaced the traditional financing mechanisms, but instead have complemented them. Although experts agree that funding for public research institutions should combine fixed budget allocations with variable appropriations (Huffman and Just, 1999; Echeverría and Alvaro, 2000; Huffman and Evenson, 2003), developing countries have given almost exclusive priority to the use of competitive funds.

Gil and Carney (1999) mention that competitive funds can be an efficient mechanism if there is sufficient research capacity in the country. However, the experience of some of the larger research systems of developing countries (including Brazil and India) shows that these conditions are not always met.

Competitive funds have been used in LAC by the World Bank and the Inter-American Development Bank as part of loans to support AKST. In Mexico, competitive funds are the preferred mechanism for allocating public resources for research and innovation. The *Produce* Foundations used these funds from the outset, though their implementation gradually evolved as they gained more experience. However, efforts to identify more effective mechanisms have been slow, in the absence of studies to assess these experiences.

Given the limited AKST institutional capacity in some LAC subregions, it is essential to promote inter-institutional projects to complement and utilize the comparative advantages of each institution. A financing mechanism using competitive funds shared by two or more institutions engaged in cooperative projects is a more effective and efficient strategy. In Mexico, the *Produce* Foundations have used the mechanism of competitive funds through public bids but give preference to inter-institutional projects.

The financing system using shared funds has proven to be a powerful instrument for: (1) guiding research based on pre-established priorities, so that it is possible to link the demands or needs of users with research activities; (2) enhancing the definition of project objectives and methodology, thereby helping to achieve the expected results; and (3) facilitating the development of monitoring and evaluation mechanisms for research activities.

Experience suggests that financing research through competitive funds is extremely useful (Bisang, 2003; Piñeiro et al., 2003). However, this form of financing should be complementary to institutional financing, given that each fund sets its own priorities and has its own mechanisms for resource allocation, follow-up, and monitoring. For institutions that finance part of their research projects through competitive funds, this entails increased administrative costs, since several control and monitoring systems must be applied, each following the rules of the specific fund. Similarly, the fact that special resources granted for research are subject to different criteria from those of the institution that receives them tends to alter previously established research priorities and creates asymmetries in the flow of information between researchers and those who are cognizant of the available resources.

One complementary financing mechanism, independent from national budget allocations, is to levy rates or charges on the first-time sale of specific products. This method is used extensively in Australia, and also in Colombia through the so-called Parafiscal Funds, but it is not common in LAC. In both cases, the funds received are channeled to private corporations administered by governing councils made up of representatives of the public sector and producers' associations linked to the specific product, and the resources can only be used to support research and the promotion of exports.

Some AKST system institutions have succeeded in generating income through the sale of technological services not directly linked to their research activities, such as soil analysis, agrochemical tests, and other types of studies (www.inifap.gob.mx). However, these cases are only justified to the extent that there is surplus capacity and the income can help finance research activities; aside from exceptional situations, it would be advisable to use that surplus capacity for research, to avoid sidetracking institutions from their specific goals.

2.3.5 Support institutions

It is important to mention the foundations that have emerged as an initiative of NARIs themselves, created to raise funds to sponsor research and technology-transfer projects. Some of these foundations even execute their own projects, or do so through NARIs and universities.

In Argentina, for example, INTA participated in the creation of a foundation called ArgenInta and set up a technological liaison unit for this purpose. It has also established a company to strengthen links with the private sector.

In Mexico, in order to support specific research projects related to agrifood or agroindustrial chains, INIFAP promoted the establishment of the Mexican Foundation for Agricultural and Forestry Research (FUMIAF A.C.), comprising the leading agribusiness and agroindustrial entrepreneurs.

At the regional level, countries are being encouraged to cooperate on AKST system projects of mutual interest. To support this strategy, FONTAGRO was created as a consortium to promote strategic agricultural research of regional interest with the direct participation of Latin American and Caribbean countries in setting priorities and financing research projects.

2.4 Responses of the AKST systems to Changes in the Most Influential Contextual Variables

2.4.1 Water

Since the 1950s, knowledge, science and technology efforts related to water in LAC have focused on finding ways to promote its rational and sustainable management, particularly in areas of water scarcity, as well as carrying out inventories, systematizing hydrological and hydro-biological resources, and trying to reverse unsustainable processes like the pollution caused by domestic waste water (IDEAM, 2002). However, it is essential to consolidate a science and technology system that addresses the demands of the 21st century (UNESCO, 2006).

Historically, research on water has focused on such issues as its role as a factor in agricultural production and on irrigation systems, the introduction of drought-tolerant materials, and the adaptation of species to saline and sodic soils.

In the case of smallholders and indigenous and Afro-American farmers, some AKST strategies have managed to achieve a positive impact in situations of limited—or in extreme cases, no—water availability (through drip irrigation, microaspersion, or gravity irrigation systems), aspects that were emphasized in integrated rural development programs until the end of the 1980s.

In the 1990s, field capacity irrigation through remote sensing began to be implemented, making advances possible in the knowledge of water resources regarding such issues as consumptive use, soil field capacity, water sources, wetlands, and pest and disease control (Vörösmarty et al., 2005, cited in UNESCO, 2006).

Another AKST advance for areas with permanent or seasonal water limitation is the production of biological inputs (biofertilizers, mycorrhizae) that potentiate and capitalize on soil dynamics, expanding the horizons of knowledge regarding soil biology.

The current agenda is revaluing the small irrigation systems used in extensive areas around the world, and especially in LAC (Palerm and Martinez, 1997). This reverses the historical tendency to ignore the role played by local communities in territorial water management, leading to a central strategy to regulate consumption and promote a rational use of the resource that is essential for its sustainability (Aguilera, 2002).

In urban and semi-urban contexts, most of the research focuses on aspects related to the efficient management of water resources and the decontamination of water sources. Semi-dry rivers, exhausted or salinized aquifers, sedimented lakes, high levels of organic material, the presence of heavy metals, and the disappearance of wetlands are only part of the current panorama (Fundación Ecología and Desarrollo, 2006).

An important area of AKST research is the contamination of water with heavy metals produced by activities like crop-spraying to combat illegal crops and the exploitation of hydrocarbons and minerals such as gold, which creates ecological imbalances and has adverse effects on human health. Another adverse factor that threatens water resources is oil spills (Aragón, 2002).

Climate change has also forced a shift in the direction of research, partly in response to the El Niño phenomenon and its effects on the spatial and temporal distribution of water. This has affected weather patterns, with increasingly frequent reports of extreme events related to maximum and minimum water flows and changes in ocean currents (Obasi, 2000; IDEAM et al., 2001; MMA and IDEAM, 2002). Networking has been an important factor in mitigating the impact and designing policies at the regional and global level through bilateral and multilateral cooperation.

One of the most recent trends in water use planning centers around advanced research centers and water treatment laboratories. Outstanding examples include the Network for Water Management in Agriculture, Irrigation and Ferti-irrigation *(Red para la Gestión del Agua en Agricultura, Riego and Fertirriego)*; the Ibero-American Water Quality Laboratories Network *(Red Iberoamericana de Laboratorios de Calidad de Agua)*; and the project known as "Indicators and Appropriate Technologies for the Sustainable Use of Water in Ibero-America's Drylands" *(Indicadores and Tecnologías Apropiadas de uso sostenible del agua en las tierras secas de Iberoamérica)* (Fernández, n.d.).

Activities include the desalinization of seawater to extract potable water, the use of water as a source of energy (either from hydrogen or kinetic energy from water and tides), the study of ground waters and their decontamination, geothermy, and research on the estuaries of large Latin American rivers like the Amazon, the Río de la Plata, and the Orinoco. Major efforts and progress have also been made in the field of limnology. These new strategies increase our knowledge base and—with the help of case studies, best practices, partnerships between organizations, and the exchange of experiences—constitute essential actions to enhance the capabilities of national statistics institutes and their management of water resources (UNESCO, 2006).

2.4.2 Biodiversity

LAC is an exceptionally rich territory in terms of agrobiodiversity because it spans important cultural centers for domestication and agriculture: Meso-America, Amazonia, and the Andean region. Approximately 10,000 years ago, the original settlers domesticated scores of native species, originating agriculture in the New World and leading to the rise of highly developed pre-Hispanic civilizations involving extensive empires based on the success of autochthonous agriculture, its genetic and agronomic diversification, and its broad geographical diffusion.

The inter- and infra-specific diversity of these native crops constitutes a rich heritage of genetic resources and an enormous comparative advantage, since this agrobiodiversity contains the elements (unique genes) that are essential for plant genetic improvement and the long-term sustainability of agriculture.

However, in spite of the enormous value of genetic resources in the region, the institutional and political capability of most countries is too weak to conserve such assets properly and use them rationally.

The conservation of genetic resources is achieved through two different but complementary strategies: *ex situ* (in germplasm banks) and *in situ*. In LAC, germplasm banks are typically associated with public agricultural research institutions and agronomic improvement programs. Germplasm collections conserved *ex situ* at these banks are well documented and catalogued, with information regarding their place of origin, agronomic characteristics, and other information that can facilitate their direct use by farmers, in improvement programs as a source for desirable characteristics, or for their eventual repatriation to the communities of origin should they have been lost for any reason and there is a desire to bring them back.

Advantages of *ex situ* conservation include the assurance provided by banks that the materials will survive, their availability for research and improvement, and comparative studies of different strains to test, for instance, for resistance to a given pest or disease. Disadvantages of this strategy include the cost of the facilities and technical staff needed to regenerate, characterize, and document the conserved materials, and the fact that samples are relatively small with regard to the genetic diversity found in wild populations. In addition, the process of evolution—of natural selection—pretty much stops while the materials are stored in the bank, where they are regenerated no more frequently than 5, 20, or more years in between.

In situ conservation refers to preserving various species or varieties in their natural field conditions in the places where they developed their particular characteristics. In the case of domesticated plants, *in situ* conservation is carried out "on-farm", in the fields of the farmers who have traditionally grown these crops or varieties. For the *in situ* conservation of wild plants (such as the wild relatives of common crops), efforts are made to preserve the ecosystems where the natural populations of such species are to be found, whether in national parks, protected areas, or other ecosystems that have not been intervened. The advantage of *in situ* conservation is that evolutionary processes continue, thanks to large populations of individuals with wide genetic variability. The disadvantages of this strategy include the difficulties of monitoring and protecting wild or cultivated populations in remote areas, the relative lack of documentation and characterization of the genetic materials, and the logistical difficulties of accessing those materials easily to apply them to research or genetic improvement.

Neither *in situ* nor *ex situ* conservation by themself are enough to safeguard the survival and integrity of genetic resources in the long terms. Each strategy has its strengths and weaknesses, which makes it necessary to rely on both mechanisms (*in situ* and *ex situ*) so that they can function together in an integral strategy known as "complementary conservation". Thus, if for some reason farmers lose their seed in the field they may reclaim it from the bank, while if due to some accident a bank loses some of its materials it will know where to go to once again collect them in the field and restore them to their germplasm collection. It may also make sense to encourage the exchange of seeds among farmers in the same region, or even different regions and countries. An AKST challenge would be to improve national institutional and technical infrastructure for safeguarding and making good use of the agrobiodiversity (genetic resources) that make up the heritage of each country.

The Convention on Biological Diversity (1992) acknowledged the sovereignty of each country over the genetic resources to be found within its borders. But with sovereignty comes the responsibility of conserving those unique and irreplaceable natural resources, not only for the welfare and agricultural development of the country but also for humanity as a whole, which must rely on them to feed future generations.

At the national level, this responsibility implies every government's duty to invest in its national agricultural research institutions so they have the basic resources needed to compile, maintain, characterize, and utilize their genetic resources, both native and imported, to meet the needs of their people and confront the problems of national, regional, and global agriculture. At the regional and international level, it would be advisable for all countries to become affiliated with the multilateral system for accessing and sharing the benefits associated with vegetable resources through FAO's 2004 International Treaty on Plant Genetic Resources for Food and Agriculture.

2.4.3 Soils

AKST system advances regarding soils have gone through several historical and mutually interrelated stages in LAC that have made it possible to advance and systematize knowledge about edapho-biodiversity. Before the 1960s, regional research focused on aspects of taxonomy, fertility, and valuation for cadastral purposes. Then there was a turn toward fertility, management, and conservation studies. During the 1980s, experts introduced research at the watershed level for land use management purposes, with the subsequent development of Landscape Ecology Theory (LET), leading to ecological-economic zoning. In the 1990s, research regarding plant nutrition moved toward the impact of applying fertilizers and pesticides to the soil, their effects on microbial biomass, and their dynamics. At present a great deal of work is being carried out in soil biology based on molecular techniques and working with DNA and RNA to inventory mezzo-organisms and microorganisms. Another field of activity relates to ethnotaxonomies and traditional soil-management techniques, an outstanding example being the case of the Pacha Mama, or Mother Earth, ritual in the Andes.

2.4.4 The social variable

From the 1950s until the end of the 1970s, AKST systems directed their efforts at boosting agricultural productivity in response to the need to produce more food at a lower cost. This was accomplished through the development of technology packages that, due to their characteristics, achieved their best results in large landholdings but provided few benefits to poor farmers with lower levels of organization, or to Afro-American and indigenous communities (Piñeiro and Trigo, 1983).

The need to respond effectively to local demands, mainly from farmers who benefited the least from the technology transfer models that characterized the agricultural modernization phase described in the previous section, led to the first attempts to regionalize AKST (Piñeiro and Florentino, 1977; Trigo et al., 1982). This reflects a changing perception of the role and effects of technology on the economic organization of society (Valdés et al., 1979; Gilbert et al., 1980; Norman, 1980; Trigo et al., 1981).

Later, in the 1980s and especially from the nineties onward, the social changes that occurred as a result of urban growth required the agricultural sector to develop new technologies associated with more advanced linkages of the production chain such as postharvest handling and storage, improving the quality of the final product and the strengthening the industrialization of agricultural producers. To respond to these new demands, AKST system institutes began to rethink their objectives. However, according to Lindarte (1997), NARIs and extension services have not achieved significant results in this respect, possibly due to constraints in the development model, the interests that govern institutional structures, or a lack of conceptual clarity regarding the direction and implementation of the necessary changes.

Lindarte (1997) also emphasizes the importance of incorporating different stakeholders involved in the process of technology generation. This is evident in the growing involvement of private sector representatives and those from producers' organizations, foundations, and NGOs in national research institutes, and also in the development of technology transfer programs such as *Cambio Rural,* implemented by INTA in Argentina, and other experiences carried out by EMBRAPA in Brazil and INIA in Chile (Cetrángolo, 1992). The limitations of this new approach are mostly due to the lack of new and appropriate forms of social and cultural integration (Lindarte, 1997).

2.4.5 Policies

The performance of AKST systems, the focus of research and, in particular, the incorporation of innovations, are conditioned by the general public policy context, and are not only limited to specific aspects of AKST. In most LAC countries, the relatively high contribution of agriculture to GNP and employment generation in the second half of the 20th century pushed production, rural development, and food self-sufficiency policies toward the top of the agendas of governments, cooperation programs and international development agencies. From the 1950s to the 1980s, these agendas contemplated a broad range of rural development policies and programs with active participation by governments in financing production and the physical infrastructure needed to support both production and marketing. Governments also implemented policies on land-use and irrigation, intervened in commodity and input markets, introduced measures to protect agricultural trade (through the application of tariffs and other quantitative limits on imports), and implemented initiatives to support research and development.

During that period, public policies emphasized the generation and transfer of technology, strengthening the human and financial resources of specialized public institutions and paving the way for the creation of NARIs. In some countries, particularly the larger ones, the activities undertaken by these institutions and the favorable policy context played a significant role in boosting productivity and agricultural production in the 1960s, 1970s and 1980s. However, they did not have a similar affect on reducing rural poverty, nor did they pay much attention to the conservation of natural resources and the environment.

Ample evidence suggests that the sustained and sustainable growth of agricultural production and, in consequence, its positive impacts on the development of rural communities and on the economy as a whole, depends in great measure on the systematic incorporation of innovations, since the current possibilities of increasing the cultivated area are fairly limited. Although there are still opportunities to expand the agricultural frontier in some LAC countries, the main way to increase the growth of the food supply and farmer income is by increasing the productivity of the land. Similarly, most of the studies carried out in LAC, and in other regions, show that the rates of return on investment in agricultural research and development are extremely high (Alston et al., 2000; Ávila et al., 2002) (Table 2-8).

Despite the points mentioned above, starting in the mid-1980s and especially during the 1990s public investment in agricultural research and development declined in LAC. As a result of their fiscal and public debt problems, most countries in the region implemented profound reforms in their macroeconomic, trade, sectoral and public investment policies with the aim of limiting state intervention and reducing public spending. These policies also restricted agricultural credit, making it more expensive, and reduced the budgets allocated to investments in rural infrastructure, and those corresponding to agricultural research and extension and other programs and services to support rural development.

This less favorable context of macroeconomic and sectoral policies was reflected in lower growth rates of agricultural production in LAC countries—both in terms of the cultivated area and average productivity—for the period 1982-2001, compared with those recorded for the period 1962-1981. The average growth of production for the main agricultural commodities was 3.05% annually in the 1960s and 1970s, and fell to 1.98% in the last two decades. But there are significant differences in the growth patterns of the different LAC subregions. In the Andean countries, Central America and the Caribbean, growth rates declined. By contrast, growth rates increased in the Southern Cone countries, influenced mainly by increases in the productivity of the land both for crops and livestock.

When analyzing public investment in agricultural research and development in most LAC countries, it can be seen that it was always low compared with international standards, but the situation has worsened in recent decades. Thus, while research spending for the period 1970-75 in industrialized countries amounted to 2.5% of GDP, the average for LAC was 0.65%; and it fell to 0.5% during the period 1975-85, and to a range of 0.10 to 0.40% during the period 1985-95 (Ardila, 1997).

The aforementioned reductions in public investment in agricultural research have not been homogeneous throughout the region. At present only a few countries (Brazil, Mexico, Argentina, Colombia and Venezuela) can boast of large organizations that have maintained significant levels of investment. Hertford (2004) underscores that in the mid 1990s more than half the investment in agricultural research corresponded to Brazil. If Mexico is added, both countries accounted for nearly two-thirds of the region's total. Only the other three countries mentioned spent over US$100 million annually each. In most countries, instead, public investment was very low, and in recent years fell to such extremes that it has given rise to a serious erosion and decline in the installed capacity of official specialized institutions. Moreover, these have not been replaced by equivalent investments in the private sector.[20]

In the least developed countries, the lack of public investment in agricultural research constitutes a major threat, in terms of responding to a growing demand for knowledge to ensure the sustained growth of food production, which should essentially be based on innovation and on increased productivity of the land. In many of these countries, the availability of farmland per capita will tend to fall in the coming decades, leading to a high probability that they will be unable to produce enough food to be self-sufficient. This will not only have negative repercussions on their balance of trade, but will also result in higher food prices for the poorest segments of the population, who depend to a large extent on personal consumption.

Even in the five LAC countries that have relatively strong public research institutions, the decline in public funding has had a significant effect on their productivity. In most of these institutions the ratio between operating costs and personnel costs has deteriorated, thereby reducing their efficiency and the possibilities of implementing the necessary institutional changes required by the broader contextual transformations that have occurred in last two decades. This has implied, among other things, implementing different types of agreements between public institutions and the private sector to develop technologies that can be appropriated by companies. The lack of public resources has shifted the focus of research in NARIs, which is now conditioned by the contributions and demands of companies, mainly suppliers of agricultural inputs. But it also affects producers, agroindustries and other social organizations.

These changes in the public policy context call for the establishment of a new institutional framework that goes beyond that of the traditional public AKST system institutions. In other words, it is necessary to redefine the roles and scope of the public and private spheres, with regulatory frameworks that allow for effective links between both sectors. Among other aspects, this implies rethinking the NARIs, with the aim of incorporating new management systems that contemplate strategic planning for the implementation of partnerships and cooperation mechanisms at the national and international level with different public and private stakeholders of the AKST system. In other words, a high priority should be given to the formation of research networks (Lindarte, 1997; Salles-Filho et al., 1997).

The restrictions imposed on public budgets for AKST in recent decades have come precisely at a time when LAC's producers have faced growing pressure to improve their productivity in order to compete at the international level—all this in the context of free trade policies stemming from the reforms implemented by the countries of the region, as well as those resulting from the multilateral trade negotiations in GATT and the WTO, those corresponding to the different sub-regional integration initiatives (CARICOM, CAN, MERCOSUR, NAFTA) and a growing number of bilateral agreements signed by some of the countries, especially Mexico and Chile. The agenda of future or imminent multilateral

[20] It should be noted that in LAC private investment in AKST is even less significant than that of the public sector.

Table 2-8. **Global investment in research & development in selected countries (in billions of international 2000 dollars and in percentage).***

Selected countries/regions	Amount 1995	Total 2000	Participation 1995	Total (%) 2000
Developed Countries (23)	461.4	574.0	82.1	78.5
USA	196.4	263.0	35.0	36.0
Japan	90.0	99.5	16.0	13.6
Developing countries (141)	100.3	157.0	17.9	21.5
Asia Pacific (26)				
China	19.5	48.2	3.5	6.6
India	11.7	20.7	2.1	2.8
LAC (32)	17.2	21.2	3.1	2.9
Brazil	9.8	12.4	1.7	1.7
World Total	561.6	730.9	100	100

*Local currency converted to international dollars using the Purchasing Power Parity index (PPP).

Source: Pardey and Beintema, 2006.

and regional trade negotiations, including those that Central America is launching into with Europe and those that being explored with Asian countries, is copious and will produce new challenges in terms of improving the competitiveness of agriculture in the region.

2.4.6 Markets

Urbanization and globalization processes in LAC and worldwide, together with increases in per capita income, have had a major impact on creating demand for different types of goods, and also on the characteristics of the products and services demanded by consumers. The last few decades have brought changes in consumption patterns and new requirements associated with changing consumer preferences in terms of health, food safety, food quality and certification, which are being incorporated into national regulations and the international agreements that regulate world food trade.

The growing demand for differentiated products, with more services and value added, plus other characteristics such as the environmental and cultural nature of products, identification of origin and processes, and so on, imply modifications to the traditional demand for innovations from the AKST system. It is not enough to have an approach centered on the product, the producer, or the use of technologies to increase productivity and the food supply; every day brings more demands, but also new opportunities to build competitiveness through value added, based on a proper understanding of demand and the supply of products and services that are aligned with consumer preferences.

In this respect, reference should be made of the many organizations dedicated to Fair Trade, a movement that began in the mid 1980s. Its purpose is to treat rural producers of goods and services in poor countries fairly. This entails offering fair compensation for these products, to cover production and labor costs. It also leads to a revaluation of the work carried out by indigenous peoples, Afro-Americans, and other ethnic minorities, and discourages slave labor and child labor. It makes it possible to secure long-term contracts that ensure a steady source of income and reduce market uncertainty. It also encourages the improved management and conservation of biodiversity and the environment, and provides support for producers to acquire the knowledge and skills needed to become better at business and marketing, and even increases their self esteem. Products marketed under this scheme vary in their characteristics and points of origin. Countries that stand out include Guatemala, Honduras, Nicaragua, Ecuador, Bolivia, Brazil, Panama, Peru, Colombia, Mexico, Chile, and Costa Rica.

Until now, most AKST system institutions have not assigned a high priority to these aspects, or to the different links of the agrifood chains. Moreover, they do not have the necessary technical and human resources. These new challenges will become more critical in the coming decades. It is clear that, in future, the AKST system will be unable to limit its activities to the traditional supply-side approach to technological innovation. A high priority will have to be given to identifying and responding to demand, and to developing new ways of organizing the production and marketing of agrifood products (organizational innovations), so as to effectively meet new consumer demands.

2.5 Effectiveness and Impact of the AKST System

2.5.1 On production systems

2.5.1.1 The traditional indigenous and campesino systems

Traditional indigenous and campesino production systems have historically been considered by the AKST system an obstacle to development. Its social actors have suffered from a low political and organizational profile, and it has been addressed in a marginal and reductionist way, ignoring the complex dynamics of production in the rural milieu (Armiño, 2002; Macías, 2002; Santamaría et al., 2005; Martínez et al., 2006; OAC and IICA, 2006; Raigoza et al., 2006).

In the last two decades, the traditional *campesino* and Afro-American farming systems and the indigenous production systems in LAC have started moving into alternative trade spaces, producing organic and ethnic products, free of transgenic material, with denomination of origin, as well as raw materials for multinationals, among others. They sometimes use advanced technology and marketing strategies (online communications, networks of farmers and consumers of ecological products, dietetic products, and natural pharmaceuticals and cosmetics). Recently, there has also been a move towards the service sector with the adoption of multi-activity systems (hiking trails, horse-riding, photography, environmental education, and ecological or alternative tourism (Toledo, 1980; Naredo, 2006) that respond to the new concerns of international agendas with regard to forests, water, biodiversity, desertification, wetlands, a gender perspective, intellectual property rights, the precautionary principle, cyber-agriculture, fourth generation rights, and the exchange of know-how, among other issues.

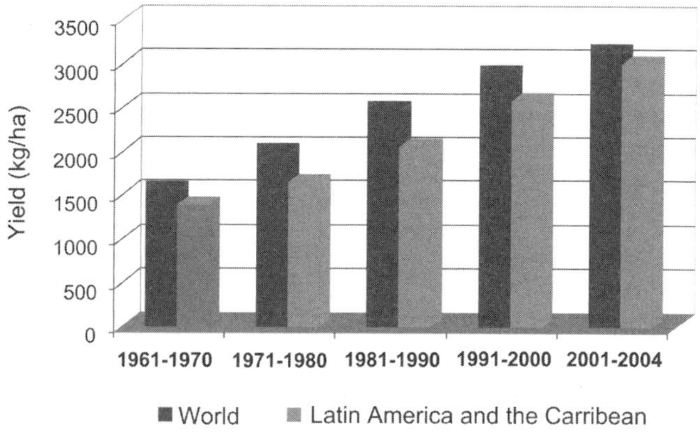

Figure 2-3. *Trends in the median yields of food crops in LAC and the world, 1961-2004.* Source: Ardila, 2006

2.5.1.2 The agroecological production system

The Agroecological Production System emerged as an approach at odds with the practices and philosophy of conventional production systems. The AKST system framework is increasingly seeking to revalue traditional knowledge or know-how based on local research and "farmer to farmer" extension, with participatory research mechanisms, *in situ* protection of agrobiodiversity, and the study of collective forms of social action (Sevilla and Woodgate, 2002). These changes in the traditional, indigenous, and agroecological production systems have provided new ways of generating, adapting, and transferring AKST system services at different scales and intensities from the spheres of governments, nongovernmental institutions, and cooperation agencies.

In efforts related to the study of production systems, geographic information system (GIS) platforms have provided AKST systems with important support and are an essential tool for the identification, delimitation, and management of territories (Echeverri and Alvaro, 2000; Ofen, 2006). The preparation of biodiversity inventories; the assessment of population dynamics, efficient water management, and renewable energy sources (especially biofuels); the monitoring of pests and diseases; the assessment of CO_2 sinks; the survey of aquifers and ground waters; the mapping of current and potential soil uses; and modeling, are just some of the activities undertaken within the AKST context in LAC that involve GIS.

2.5.1.3 The conventional system

The AKST system has had a significant impact on the productivity of agricultural units in recent decades. Starting in the 1980s, one can detect an increase in yields that continues to this day (Figure 2-3). Most of this growth has been the result of incorporating new technologies, mostly improved seeds, crop protection, and fertilizers. The increase in the production of certain crops, and the resultant increase in the food supply, brought with it a decrease in the price of agricultural products.

In spite of this increase in yields, it should be noted that they have been lower than those secured in industrialized nations. Perhaps this difference has been influenced directly or indirectly by the agricultural subsidies prevalent there, which facilitate a greater adoption of new technology. But countries in East and Southeast Asia have also enjoyed a faster rate of growth than in LAC, where the rate of growth has been diminishing in the last five years.

2.5.2 On the advancement of knowledge and innovation systems

Biotechnology, nanotechnology, and information technology are fields of scientific knowledge from which innumerable new technologies are derived. Advances in biology and information science are considered the most influential scientific foundations for agricultural research in the last decade.

Although some authors already note a decline in its rate of progress (Oliver, 2000), information science is indicated as one of the most influential branches of science in research organizations. It is possible that many organizations have not yet been able to take full advantage of the potential provided by this progress.

Nanotechnology is another branch of science that could have a major impact on generating other cutting-edge technologies in coming years. In 2004, it is estimated that worldwide investment in this area was in the order of 3.7 billion dollars (Roco, 2004).

Various constraints, however, have slowed the pace of development in biotechnology and the information sciences in developing countries, especially limited financial resources, lack of information, inadequate research infrastructure, and limited access to technology. In addition, there are groups that are ideologically opposed to biotechnology and its possible impacts on biodiversity and the environment as well as its implications for food security (Castro et al., 2006).

Commercial biotechnology in the region has focused mainly on the transfer of genes to make crops resistant to herbicides and protect them from several types of insects and pathogens that affect commercial commodities, especially soy, maize, and potato. A typical example is the case of RR Soy seeds in Argentina which, according to Regúnaga et al., (2003), is the most dynamic example of large-scale adoption of technology innovation in world agriculture. The authors note that in a period of five years, RR soy accounted for 95% of the total soy crops planted in the country; it was adopted by farmers because of the lower complexity of the production system and the reduction in prices per unit.

Most countries of the region still face an unresolved conflict between supporters of biotechnology and its products (mainly those associated with public and private agricultural research institutions) and stakeholders linked to NGOs and other social and political movements who oppose the spread of genetically modified organisms. This has curtailed the use and even the production of biotechnology innovations in certain countries.

In the aforementioned study by Castro et al., (2005), basic and applied research in nanotechnology was deemed as of the lowest strategic importance; in recent years, the advances and impacts of these new frontiers of knowledge were assessed to be of medium to low significance in the region. For biotechnology, the assessment figures were slightly higher, but did not exceed the category of medium importance. An interpretation of this result reaffirms the point made previously regarding the slow rate of uptake in the use and production of biotechnological innovations in LAC.

It should also be noted that innovation not only had an impact of the productivity of agricultural units but has also enabled the development of many inputs and productive management technologies that are environmentally friendly, like crop rotation, biological inocula, and natural fertilizers.

With regard to the regulatory bias of science and technology, there are asymmetries between the knowledge of users, producers, and generators of innovation. In LAC we repeatedly find that new technologies are beyond the reach of the very populations for whom they were generated, for a variety of reasons. This problem, in turn, is connected to another issue mentioned in the studies, i.e., the isolation of the various innovation systems due to lack of participation and linkages between all the actors involved in the innovation process, which generates a regulatory bias (Arocena and Sutz, 1999).

Regarding the notion of an innovation system as a political objective, data gathered through several recent surveys on industrial innovation in different countries indicate that national spending on innovation is fairly low. For this reason, private companies carry out internal R&D activities, even though these may be of an informal character (Arocena and Sutz, 2002).

If we analyze the particular case of innovation systems in MERCOSUR, these respond to the region's current economic situation. In this context, it should be emphasized that numerous transnational corporations based in MERCOSUR delegate innovation activities to their parent companies. Although we observe a growing trend regarding cooperation for research purposes, the technological divide between Latin American countries and industrialized nations is still very wide. Hence much of the innovative technology in the region comes from technological advances that arrive to LAC through inputs, mostly seeds and agrochemicals, produced and distributed by multinationals.

According to Lundvall (1985), innovation stems from a convergence of technical opportunities and user demand, which suggests the importance of citizens' participation in research processes—an issue that should be considered by AKST system institutions in the design of innovation systems. It is also important to consider the systemic nature of innovation, taking into account all related processes and their interdependence.

2.5.3 On consumers

There were, as of 2000, approximately 520 million consumers in Latin America and the Caribbean. According to figures from the Department of Economic and Social Affairs of the United Nations Secretariat, disseminated in the studies World Population Prospects: The 2002 Revision and World Urbanization Prospects: The 2001 Revision (Perez, 2005), this population grew significantly since 1985, by around 120 million people (they were 401 million in 1985, 441 million in 1990, and 481 million in 1995). These consumers, located both in urban and rural areas, represent a plethora of demands for goods and services.

Consumer-oriented processes have traditionally had little influence. However, even in cases where end consumers were not the main priority of research, they have indirectly benefited from the other priorities that have been set, that led for example to significant reductions in food prices. Over the period in question, for instance, the population benefited from decreases in the prices of basic foods of almost 70%. This occurred due to a decrease in production costs due to increases in productivity obtained as a result of agricultural research efforts and innovation processes. Consequently, end consumers benefited even though research priorities were more concerned with farm performance and productivity (Figure 2-4).

Consumer segmentation leads to the generation of supply-side production alternatives. Over time, these develop into different knowledge, science and agricultural technology initiatives. In the case of the rural sector, this translates into, and is materialized in, agricultural innovation and technology transfer processes (Jacobs 1991; Funtowicz and Ravetz, 2000).

At the same time, advances achieved by agricultural science and technology have sometimes been questioned, as in the case of genetically modified organisms (GMOs) or practices that are believed to cause undesirable effects such a climate change or soil contamination and erosion (Beca, 1988; Sartori and Mazzoleni, 2005; Duarte et al., 2006).

As part of this analysis, it is important to emphasize that new spaces for discussion and feedback are emerging between the so-called "responsible consumers" sector and producers, as part of a general policy to ensure compliance with standards and principles related to intellectual property, certification mechanisms, fair trade strategies, denominations of origin, and ecolabelling.

2.5.4 Social aspects

The modernization of Latin America's agricultural sector sharpened the contradictions between the modern and traditional sectors. On the one hand, it led to poverty for the social groups who were displaced towards large urban centers and border zones or who joined the transborder migratory flows. At the same time, it produced environmental impacts and caused the large-scale destruction of natural resources and the erosion of traditional knowledge.

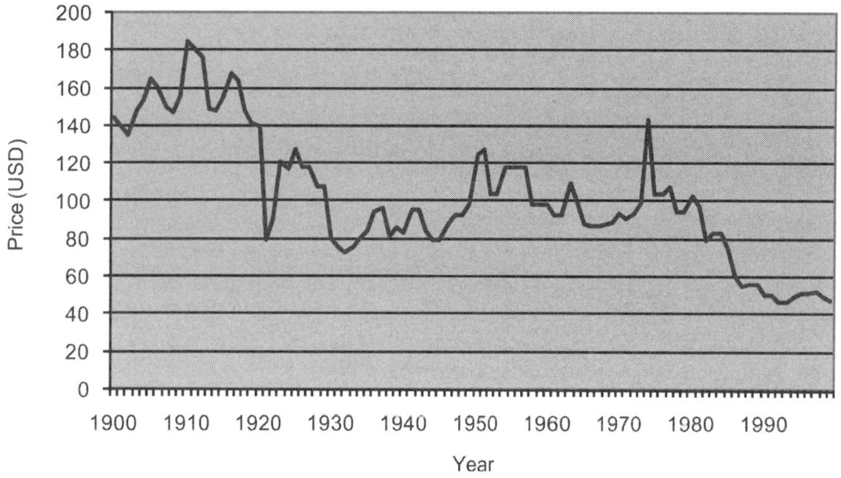

Figure 2-4. *Historical trends of average prices of primary commodities (Weighted average prices in real dollars).* Source: Authors' elaboration based on World Bank, 2000.

With regard to the gender dimension, it is clear that the modernization of the agricultural sector provoked changes in labor relations both for men and women. Rural women have a greater presence in the production chains of fresh and processed foods and in other agricultural export products. However, their working conditions remain precarious (Farah, 2004) except in the case of exporting firms that have been certified internationally.

In general terms, public policy in Latin American countries has prioritized economic growth as a strategy for overcoming poverty in all its manifestations. This economicist vision has ignored the complexity of the situation of rural populations, failing to consider that poverty is multidimensional and cannot be resolved with one-dimensional strategies (Sen, 2000).

2.5.5 On the competitiveness of chains and conglomerates, and on territorial development

The AKST system has had a significant impact on the competitiveness of production chains over the period analyzed. The region's growing agricultural output has largely been the result of the technological development promoted by the AKST system (Regúnaga et al., 2003). This has occurred despite the fact that, as previously mentioned, the system did not begin to address production chains as a whole until the middle of the 20th century, focusing before that on specific projects due to the region's considerable technological backwardness.

For several decades, research efforts pursued productivity without taking into consideration the social aspects of a given territory. The populations historically and culturally linked to these territories were not adequately inserted into the technological changes underway, often not only for cultural reasons but also for economic and financial ones. This lack of a holistic vision of the system has produced negative externalities such as social exclusion and the degradation of natural resources (Molina, 1980; Trucco, 2004).

Although agricultural R&D began to be implemented through individual projects a few decades ago, it was not until the end of the 1990s that strategies were developed to address the requirements of the production chain as a whole. An example is Argentina's Multi-annual National Science and Technology Plan (SECYT, 1997), which used the concept of the production chain to design its technology policy and worked with this unit of analysis in pursuit of the greater competitiveness of the whole.

In recent years, the development and expansion of the concept of agribusiness (Davis and Goldberg, 1957) and the implications of the new institutional economy for the competitiveness of production chains (Zylbersztajn, 2001) have introduced an institutional and organizational framework that has improved the productivity and competitiveness of chains and conglomerates.

This new vision of agribusiness is encouraging discussion on ways of ensuring a more harmonious and balanced development of production chains and their stakeholders. The concept, however, is being incorporated mainly in the more competitive chains, leaving aside the weaker ones or those whose stakeholders have fewer opportunities to make them heard.

Consequently, this new way of integrating technological development with institutional aspects has limited importance for the communities linked to a territory, since there is less interest, knowledge, or efforts on the part of the AKST system to improve their conditions of relative development.

In this regard, non-governmental organizations committed to social and territorial development, as well as certain specific institutions, plays an important role in promoting better conditions for local populations with respect for their culture (Feito, 2005).

References

Aguilera, K.F. 2002. Hacia una nueva economía del agua: cuestiones fundamentales. Ponencia presentada en el III Congreso Ibérico sobre Economía y Gestión del Agua. Sevilla. Noviembre.

Albornoz, M., L. Vaccarezza, C. Polino, y M.E. Fazio. 2003. Resultados de la encuesta de percepción pública, cultura científica y participación ciudadana. Available at http://www.redhucyt.oas.org/ricyt/interior/biblioteca/Percepcion.pdf. Documento de trabajo N° 9. RIYCT/CYTED-OIE.15 de Mayo, 2003.

Allegri, M. 2002. Partnership of producer and goverment financing to reform agricultural research in Uruguay. *In* D. Byerlee and R. Echeverria (eds) Agricultural research policy in an era of privatization. CABI Publ., UK.

Alston, J.M., C. Cheng Kang, M.C. Marra, P.G. Pardey, y T.J. Wyatt. 2000. A Meta-analysis of rates of return to agricultural R&D: Ex pede herculem? Res. Rep. 113. IFPRI, Washington DC.

Amaya, P.J.P., y M.E. Rueda. 2004. Ciencia, tecnología e innovación en el sector agropecuario (CTIA). Documento preparado para el Proyecto de Apoyo a la Transición de la Agricultura Ministerio de Agricultura y Desarrollo Rural.

Aragón, L. 2005. Building regional capacity for sustainable use of tropical rainforest in Latin America: Experience and challenges of the South-South Cooperation. Int. Symposium Tropical forest in a changing global context. UNESCO, Paris.

Aragón, L. 2002. Uso da agua. Há futuro para o desenvolvimento sustentável na amazônia ?. Futuro da Amazonia. Secretaria de Tecnologia Industrial (STI). Ministério do Desenvolvimento, da Indústria e Comércio Exterior. Available at http://www.desenvolvimento.gov.br/sitio/sti/publicacoes/futAmaDilOportunidades/futAmaDilOportunidades.php.

Aragón, L. 2001. Ciencia e educação superior na Amazonia: Dasafios e oportunidades de cooperaçao internacional. UNAMAZ/NAREA, Belén.

Aragón, L. 1998. Amazônia 21: Uma Agenda para um Mundo Sustentable. UNAMAZ/SCA, Brasilia.

Araya-Villalobos, R., y J.C. Hernández-Fonseca. 2006. Mejora genética participativa de la variedad criolla de fríjol "Sacapobres". Agronomía Mesoamericana 17(3):347-355.

Ardila, J. 2006. Retos y oportunidades para el agro Latinoamericano: Hacia una visión compartida. IICA, San José.

Armiño Pérez, K. de. 2002. Seguridad alimentarla: implicaciones para las políticas nacionales e internacionales. p. 63-83. *En* Desarrollo y cooperación en zonas rurales de América Latina y África para adentrarse en el bosque. Univ. País Vasco, España.

Arocena, R., y J. Sutz. 2002. Sistemas de innovación y países en desarrollo. SUDESCA Res. Pap. No. 30, Dep. Business Studies, Aalborg University, Denmark. Versión inglesa: Innovation systems and developing countries. DRUID (Danish Research Unit for Industrial Dynamics) Working Paper No. 02-05 (2002). Versión revisada y ampliada de un trabajo preparado para el Seminario Internacional de SUDESCA, Estrategias para el Desarrollo Sostenible en Centroamérica, El Salvador, mayo del 2001.

Arocena, R., y J. Sutz. 1999. Mirando los sistemas nacionales de Innovación desde el Sur. Trabajo presentado en la Conferencia Sistemas Nacionales de Innovación, Dinámica Industrial y Políticas de Innovación. Danish Res. Unit on Industrial Dynamics (DRUID) en Rebild, Dinamarca, 9-12 de junio de 1999.

Avila, J.A., J.L. Rebolledo, V.H. Santoyo Cortez, and R. Cardenas. 2002. Indicadores de desempeño en programas de fomento para la innvocación en el sector agropecuario mexicano. Available at http://www2.ricyt.org/docs/VII_Congreso/DIA_24/SALA_B/14_00/Aguilar_Avila.pdf.

Ballarin, P. 2002. Consideraciones técnicas: biotecnológicas y agricultura ecológica. p. 103-119. *En* G. Ramírez de Haro et al., (ed) Desarrollo y cooperación en zonas rurales de América Latina y África para adentrarse en el bosque. Coordinadores: Instituto de Estudios sobre Desarrollo y Cooperación Internacional -Hegoa. Univ. País Vasco, España.

Banco Mundial. 2000. Global commodity markets. A comprehensive review and price forecast. Enero N° 20306. Banco Mundial, Washington DC.

Beca, E. 1988. Comentarios y perspectivas del Seminario sobre educación rural básica. Educación rural básica. Grupo de Investigaciones Agrarias (Santiago, CL). Academia de Humanismo Cristiano.

Beck, U. 1998. La sociedad del riesgo. Hacia una nueva modernidad. Editorial Paidos Ibérica, Barcelona.

Becker, B. 2005. Geopolítica da Amazônia. Estud. av. 19(53).

Bernal, Z.H. 2006. El agua de la Amazonia: recurso estratégico para la comunidad sudamericana de naciones (casa) en el siglo XXI .X Jornadas de Economía Crítica: Alternativas al capitalismo. Barcelona. Available at http://www.ucm.es/info/ec/jec10/index.htm

Bisang, R. 2003. Apertura economica, innovacion y estructura productiva: la aplicacion de biotecnologia en la produccion agricola pampeana argentina. Desarrollo Económico 43:413-442.

Branscomb, L.M., F. Kodama, y R. Florida. 1999. Industrializing knowledge. MIT Press, Cambridge.

Burlet, J., y A.W. Speedy. 1998. Investigación agroforestal. Perspectivas globales. FAO, Rome.

Casanovas, E.2006. Problemática de los fertilizantes en Venezuela. Univ. Central de Venezuela, Instituto de Edafología, Maracay AP 4579, Aragua.

Castro, A.M.G. de y S.M.V. Lima López. 2006 O Futuro do melhoramento genético e do mercado de sementes no Brasil. EMBRAPA, Brasilia.

Castro, A.M.G. de, S.M.V. Lima; J. de S. Silva, A. Maestrey, y J.I. Gastón. 2005. Proyecto quo vadis: El futuro de la investigación agrícola y la innovación institucional en América Latina y el Caribe. Red Nuevo Paradigma, IFPRI, Quito.

Cetrángolo, H. 1996. Cordinación vertical del negocio agroalimentario desde la producción primaria. Tesis de Maestría, Centro de Estudios Avanzados UBA.

Chaves de Brito, D. 2001. A Modernizaçao da Superficie: Estado e dsenvolvimento na Amazônia. UFPA/NAEA/PDTU, Belen.

Davies, J.H., y R.A. Goldberg. 1957. A concept of agribusiness. Harvard Business School, Cambridge.

Dirven, M. 2001. Entre el ideario y la realidad: capital social y desarrollo agrícola, algunos apuntes para la reflexión. División de desarrollo Politico y Social Comisión Económica para América Latina y el Caribe.

Dixon, J.G., y A. Gibbob. 2001. Compendio de sistemas de producción y pobreza. Cómo mejorar los medios de subsistencia de los pequeños agricultores en un mundo cambiante. M. Hall (ed) Available at http://www.fao.org/docrep/004/ac349s/AC349s00.htm#TopOfPage. FAO, Banco Mundial, Washington DC.

Duarte O.L., A. Roa, y A. Atkinson. 2006. La bioprospección como mecanismo de cooperación para la construcción de capacidades endógenas en ciencia y tecnología y análisis de las capacidades de Colombia para adelantar procesos de bioprospección. en Eje 1 de Sociedad y producción de conocimiento. (CDRom), Memorias Esocite 2006. VI Jornadas Latinoamericanas de Estudios Sociales de la Ciencia y la Tecnología. Observatorio Colombiano de Ciencia y Tecnología, Universidad Nacional de Colombia, Universidad de los Andes, Universidad Central, Instituto Colombiano de Antropología y Cultura, Pensar

(Instituciones Organizadoras). Bogotá, Colombia. 19, 20 y 21 de Abril.

Echeverri, J., y J. Álvaro. 2000. Reflexiones sobre el concepto de territorio y ordenamiento territorial indígena. p. 173-180. *En* Territorialidad indígena y ordenamiento en la Amazónia. J. José Vieco et al. (ed) Instituto Amazónico de Investigaciones. Univ. Nacional de Colombia. Programa COAMA. Editorial Unibiblos, Santa Fe, Bogotá.

Ekboir, J.M., G. Dutrénit, G. Martínez, V.A. Torres Vargas, y A. Vera-Cruz. 2006. Las Fundaciones Produce a diez años de su creación: pensando en el futuro. Informe Final de Evaluación. IFPRI. Coordinadora Nacional de las Fundaciones Produce. (COFUPRO A.C.), México D.F.

Ekboir, J., J.A. Espinosa, J.J. Espinoza, G. Moctezuma, y A. Tapia. 2003. Análisis del sistema mexicano de investigación agropecuaria. CIMMYT, México.

Farah, M, 2004. Algunos elementos de análisis sobre el trabajo rural remunerado y no remunerado en América Latina, desde una perspectiva de género. Revista de Fomento Social. No. 236 Volumen 59. Octubre-Diciembre 2004. ETEA. España.

Feito, C. 2005. Políticas de Intervención y desarrollo rural: el caso de la Horticultura Bonaerense. 1° Jornadas de Antropología Rural. Tucumán Argentina

Fernandez, C.A. (sin fechas). Situación del agua potable en América Latina. Centro de Estudios Transdisciplinares del agua. Ciencia y Tecnología para el Desarrollo (CYTED). Cooperación Iberoamericana. Available at http://www.cnea.edu.ar/xxi/ambiental/agua-pura/presentaciones-encuentro/Present%20Dra%20Fernandez%20Cirelli%20Situacion%20Agua%20Potable.pdf.

Fundación Ecología y Desarrollo. 2006. Guía práctica de tecnología ahorradora de agua para viviendas y servicios públicos. Zaragoza, España.

Funtowicz, O., y S. Raventz. 2000. La ciencia postnormal. Ciencia con la gente. Icaria editores, Barcelona.

Gilbert, E.H., D.W. Norman, F.E. Winch.1980. Farming Systems Research: A Critical Appraisal. Florida, Estados Unidos. University of Florida. 1° ED. p135.

Gill, G.J y Carney D. 1999. Fondos Competitivos en Tecnología Agrícola en países en desarrollo. Natural Resources Perspectivas. N° 41. ODI, UK.

Gunasena, H.P.M. 2003. Food and poverty: Technologies for poverty alleviation. South Asia Conference on Technologies for Poverty Reduction, New Delhi.

Hertford, R., P. Gpardey y R.S. Wood. 2004. Panorama Estratégico del Sector Agropecuario en América Latina y el Caribe: Perspectivas de investigación y desarrollo. Mimeo, IPFRY.

Huffman, W. y R.E. Just. 1999. Agricultural research: Benefits and beneficiaries of alternative funding mechanisms. Rev. Agric. Econ. 21:2-18.

Huffman, W. y R. Evenson. 2003. Econometric evidence on agricultural total factor productivity determinants: Impact of funding sources. Iowa State Univ.

Jacobs, M. 1991. La economía verde. Medio ambiente, desarrollo sostenible y la política del futuro. Economía crítica 12. Icaria Editores, Barcelona.

Hazell, P., y L. Haddad. 2001. La investigación agrícola y la reducción de la pobreza. Resumen 2020 No. 70 Una visión de la alimentación, la agricultura y el medio ambiente en el año 2020. IFPRI, Washington DC.

IDEAM (Instituto de Hidrología, Meteorología y Estudios Ambientales). 2002. Una aproximación a estado de la gestión ambiental de las ciudades en Colombia.

IDEAM, MMA y PNUD. 2001. Colombia primera comunicación nacional ante la Convención Marco de las Naciones Unidas sobre Cambio Climático. Servi Grafics. Colombia. Available at http://www.ideam.gov.co/publica/cambioclimatico/PrimeraComunicacionColombia.pdf. IDEAM, Bogotá.

Kalmanovitz, S., E. López y E. Encizo. 2006. La agricultura en Colombia en el siglo XX. Fondo de Cultura Económica. Banco de la República, Bogota.

León, S., Espinos, T. Alzate., y J. Alonzo. 2004. Consideraciones ambientales del sector agropecuario y el medio rural colombianos. Ministerio de Agricultura y Desarrollo Rural (MADR), el Programa de las Naciones Unidas para el Desarrollo (PNUD) y el Banco Mundial, Bogotá.

Lindarte, E. 1997. ¿Es necesario anticipar el futuro?: Segunda parte. Venezuela. SIHCA.

Machado, C.A. 2004. La cuestión rural, un fantasma que nos persigue. Revista Foro. Ediciones Foro Nacional por Colombia. Número 50. Junio. Bogotá.

Macías, A. 2002. El sistema agroalimentario mundial: implicaciones para el mundo rural. p. 39-62. *En* Desarrollo y cooperación en zonas rurales de América Latina y África para adentrarse en el bosque. Coordinadores: Gonzalo Ramírez de Haro, Antonio Rodríguez -Carmona, Alfredo Macías. Instituto de Estudios sobre Desarrollo y Cooperación Internacional -Hegoa. Univ. País Vasco, España.

Mackay, R. y D. Horton. 2003. Desarrollo de capacidades en planificación, seguimiento y evaluación. Resultado de una evaluación. ISNAR, La Haya, NL.

Malan, P.S. 1984. Relações econômicasinternacionais do Brasil. HGCB 4:53-106.

Martínez, M. 2006. Desenvolvimiento y tecnología. *En* Eje 1 de Sociedad y producción de conocimiento. CD Rom. Memorias Esocite 2006. VI Jornadas Latinoamericanas de Estudios Sociales de la Ciencia y la Tecnología. Observatorio Colombiano de Ciencia y Tecnología, Univ. Nacional de Colombia, Univ. Andes, Univ. Central, Instituto Colombiano de Antropología y Cultura, Pensar, Bogota.

Méndez Sastoque, M.J. 2006. Pluriactividad rural y modelos de crecimiento económico en el contexto Latinoamericano. Debates Agrorurales. *En* Desarrollo rural: conceptos, estrategias y métodos. I. Tobasura, E. Castaño Ramírez (ed). Univ. Caldas, Manizales.

MMA (Ministerio de Medio Ambiente), IDEAM (Instituto de Hidrología, Meteorología y Estudios Ambientales). 2002. Lineamientos de Política para el Manejo Integral del Agua. República de Colombia. Available at http://www.ideam.gov.co/apc-aa/img_upload/467567db4678d7b443628f8bc215f32d/Lineamientos_Agua.pdf.

Molina, J. 1980. Una nueva conquista del Desierto. ED. El Ateneo, Buenos Aires.

Moncada de la Fuente, J., y F. Muñoz. 1999. Respuestas a nuevas demandas tecnológicas investigación en agroindustria y en el manejo de recursos naturales. Informe de Evaluación. Proyecto INIAs-BID-ISNAR. Convenio de Cooperación Técnica ATN/SF-5210-RG. ISNAR, La Haya.

Moura, G.1990. O alinhamento sem recompensa: A política externa do governo Dutra. CPDOC/FGV, Rio de Janeiro.

Roco, M.C. 2004. The US National Nanotechnology Initiative after 3 years (2001-2003). J. Nanoparticle Res. 6:1-10.

Nickel, J.L. 1989. Research management for development: Open letter to a new agricultural research director. IICA, San José.

Nickel, J. 1996. A global agricultural research system for the 21st century. *In* C. Bonte-Friedheim and K. Sheridan (eds) The globalization of science: The place of agricultural research. ISNAR, The Netherlands.

Norman, D.W. 1980. Farming systems approach. relevancy for the small farmer. MSU Rur. Dev. Pap. No. 5. Michigan State Univ., East Lansing.

OAC y IICA. 2006. Comportamiento del empleo generado por las cadenas agroproductivas en Colombia (1990-2006). Memo agrocadenas no. 13. OAC, Bogotà. Available at http://www.agrocadenas.gov.co/novedades/memo_agrocadenas13.pdf.

Obasi, G.O.P. 2000. Agua para el siglo. Día Mundial del Agua. Organización Metereológica Mundial XXI. Available at http://www.unesco.org/science/waterday2000/wmo_message_spanish.htm

Ofen, K. 2006. La nueva política geográfica indígena y negra en América Latina. p. 37-64. *En* Región espacio y territorio en Colombia. Univ. Nacional de Colombia. Unibiblos, Bogotá

Oliver, R.W. 2000. The coming biotech age: The business of bio-materials. McGraw Hill, NY.

Palerm Viqueira, J., y T. Martinéz Saldaña. 1997. Introducción: la investigación sobre pequeño riego en México. *En* T. Martínez Saldaña y J. Palerm Viqueira (ed) Antología sobre pequeño riego. Colegio de Postgraduados México.

Pardey, P., y N. Beintema, 2006. Agricultural research: A growing global divide? IFPRI, Washington DC.

Parellada, G.H., y J. Ekboir. 2003. Análisis del sistema argentino de investigación agropecuaria. CIMMYT, México.

Perez Garcia, M. y S.R. Dominguez. 2004 Reservas extractivas ¿Alternativas para la conservación de especies forestales?. Maderas y Bosques 10(2):55-69.

Peterson, W., G. Gijsbers, M. Wilks. 2003. A new approach to assessing organizational performance in agricultural research organizations. ISNAR, The Hague.

Piñeiro, M., R. Martinez Nogueira y E. Trigo. 2003. Informe Final. Evaluación institucional externa del INIFAP. Comisión Externa de Evaluación. SAGARPA. IICA, México.

Piñeiro, M. y E. Trigo. 1983. Hacia una interpretación del cambio tecnológico en América Latina.

Piñeiro, M. y R. Florentino. 1977. La generación y transferencia tecnológica agropecuaria. Notas sobre la funcionalidad de los Centro Nacionales de Investigación. Documentos PORTAL. N° 6 IICA, Bogotá.

Polino, C., M.E. Fazio y L. Vaccarezza. 2003. Medir la percepción publica de la ciencia en los paises iberoamericanos. Aproximación a una propuesta conceptual. Revista Iberoamericana de Ciencia, Tecnología, Sociedad e Innovación. N° 5 Enero-Abril.

Pretty, J. and R. Hine. 2001. Reducing food poverty with sustainable agriculture: A summary of new evidence. Centre for Environ. Society, Univ. Essex, UK

Quiroz. A.J. 2001. Agricultura y reformas macroeconómicas en la década de los años noventa. p. 104-141. *En* R.G. Echeverría (ed) Desarrollo de las economías rurales. Banco Interamericano de Desarrollo. Washington DC.

Raigoza, B.E., J. Roncancio, y J. Valencia 2006. La seguridad alimentaria en los países de la Comunidad Andina. Debates agrorurales. p. 127-176. *En* I. Tobasura Acuña, E. Castaño Ramírez (ed) Desarrollo rural: conceptos, estrategias y métodos. Univ. Caldas. Manizales, Colombia.

Regúnaga, M., S. Fernández, y G. Opacak. 2003. El impacto de los cultivos genéticamente modificados en la agricultura argentina. Univ. Buenos Aires.

Santamaría, G., H. Guerra, V. Guillen, e I. Ruiz De León. 2005. Escenarios futuros para la tecnociencia y la innovación agropecuaria y forestal en Panamá. Instituto de Investigaciones Agropecuario de Panamá. Red Nuevo Paradigma. Impresora Continental, S.A. Panamá.

Sartori, G., y G. Mazzoleni. 2005. La tierra explota superpoblación y desarrollo. Editorial punto de lectura. Madrid.

SECYT. 1997. Plan Nacional Plurianual de Ciencia y Tecnología, 1998-2000. SECYT, Buenos Aires.

Sen, A. 2000. Desarrollo y libertad. Editorial Planeta, Barcelona.

Sevilla, G.E., y G. Woodgate. 2002. Desarrollo rural sostenible: de la agricultura industrial a la agroecología. p. 77-96. *En* M. Redclift y G Woodgate (ed) Sociología del medio ambiente una perspectiva internacional. McGraw Hill Interamericana España.

Sevilla, G.E., y M. González De Molina. 1995. El concepto de transición en el pensamiento Marxista: reflexiones desde la agroecología. Trabajo elaborado para su discusión en el Curso de Posgrado de Agroecología y Desarrollo Rural Sostenible, Sesión IV: Transición Agroecológica. Programa de Doctorado: Agroecología, Campesinado e Historia de la Universidad de Córdoba.

Sicsú, A.P. y J.P.R. Lima. 2001. Regionalizaçao das politicas de C&T: concepçao, açoes e propostas tendo en conta o caso do Nordeste. Parcerias Estrategicas. CGEE 13:23-41.

Toledo, V.M. 1980. La ecología del modo campesino de producción. Antropología y Marxismo 3:35-55.

Trucco, V. 2004. 16 millones de hectáreas solo por siembra directa en Argentina. CropLife Latín América, San Jose.

Trigo, E., y D. Kaimowitz. 1994. Investigación agrícola y transferencia de tecnología en América Latina en los años noventa. Cabernos de Ciência & Tecnología, Brasília 11:99-126.

Trigo, E; M. Piñeiro, y J.F. Sábato. 1982. La cuestión tecnológica y la organización de la investigación agropecuaria en América Latina. IICA, San Jose.

Trigo, E., M. Piñeiro, J. Chapman. 1981. Assigning priorities to agricultural research: A critical evaluation of the use of programs by product-line and production systems [Latin America]. IICA, San José.

UNESCO. 2006. Mejorar el conocimiento y las capacidades. 2° Informe de las Naciones Unidas sobre el desarrollo de los recursos hídricos en el mundo: El agua, una responsabilidad compartida. Programa Mundial de Evaluación de los recursos hídricos. Capítulo 13. Available at http://www.unesco.org/water/wwap/wwdr2/pdf/wwdr2_ch_13_es.pdf. UNESCO, Paris.

UNESCO. 2006. Evaluación de los recursos hídricos. Elaboración del balance hídrico integrado por cuencas hidrográficas. Documento Técnico del PHI-LAC N°4. UNESCO, Paris.

Valdés, A., S.G. McDonald, y J. Dillon. 1979. Economics and the design of small-farmer technology. Int. Conf. Economic analysis in the design of new technology for small farmers. Iowa State Univ. Press, Ames.

Vorosmarty, C.J., E.M. Douglas, P.A. Green, C. Revenga. 2005. Geospatial indicators of emerging water stress: An application to África. Ambio, NY.

Walschburger, A.C. 1992. La internacionalización de las selvas húmedas tropicales y de la amazonia en especial. p. 354-385. CEGA, Santafé de Bogotá.

Zylbersztajn, D. 2001. Economía de las organizaciones: Un análisis contractual de la Firma. Material de estudio de la Maestría del Programa de Agronegocios y Alimentos, Facultad de Agronomía, UBA.

3

Agricultural Knowledge and Technology in Latin America and the Caribbean: Plausible Scenarios for Sustainable Development

Coordinating Lead Authors:
Hector D. Ginzo (Argentina), Suzana M. Valle Lima (Brazil)

Lead Authors:
Salisha Bellamy (Trinidad and Tobago), Antonio Maria Gomes de Castro (Brazil), Falguni Guharay (Nicaragua), Roberto Saldaña (Mexico)

Review Editor:
Javier Souza Casadinho (Argentina)

Key Messages

1. By building five scenarios—Global Orchestration (GO), Order from Strength (OS), Life as it is, Adapting Mosaic (AM), and TechnoGarden (TG)—future alternatives are provided to answer the question: "How can we reduce hunger and poverty, improve rural livelihoods, and facilitate equitable, environmentally, socially, and economically sustainable development through the generation, access to, and use of agricultural knowledge, science, and technology?"

2. These scenarios present different challenges that require complex adjustments in order to ensure the successful performance of AKST systems and productive systems. The scenarios show us that in the real world of Latin America and the Caribbean, it is not feasible to think in terms of simple technological solutions or global solutions to respond to the growing complexity and vulnerability of these systems.

3. In most of the scenarios, the AKST systems have favorable social and environmental repercussions for society as a whole. Science generates innovation and helps improve competitiveness and production efficiency, and the quality of the products in terms of safety, diversity, bromatological quality, and nutritional value for all social groups (including the most vulnerable ones, depending on the scenario), and reduces the impact of agricultural activities on the environment.

4. The existence of trade barriers of different types would increase the cost of agricultural activity and threaten the sustainability of small farms, and it would create specific demand for AKST systems. The scenarios assume different types of barriers, which would expand over time, as a result of difficulties stemming from various factors—environmental, economic, and biological—even in the scenarios depicting a highly integrated and economically open world (GO and TG). These barriers, which could lead to the loss of important markets and a reduced capacity for economic insertion on markets suitable for small-scale agricultural producers, would be eliminated with good policies and management capacity. The barriers would in turn generate demand for AKST systems to create mechanisms and protocols that would allow for adequate compliance with international laws and rules pertaining primarily to the most vulnerable productive systems.

5. The scenarios assume institutional changes of varying intensity in the region. In some scenarios, the changes would accompany the current development model, which shows trends towards greater stability and consistency among social development, environmental, food, innovation, and biosafety policies, and greater capacity to manage these policies (except for Order from Strength). But deep-seated institutional changes—such as changes in the paradigms of agriculture itself, and consequently in the AKST system and in the expansion of power of various interest groups—would be required to introduce and implement successfully the Adapting Mosaic.

6. Losses in productivity of productive systems in response to variations in the contextual factors vary in the different scenarios. Rising temperatures, the manifestation of extreme weather events, and an increase in diseases, pests, and contamination of foods are contextual factors that have a differential impact on production systems in the different scenarios. More specifically, the greatest losses would occur in scenarios that emphasize trade or the ones that predict a limited capacity to prevent and eliminate or reduce epidemics (the case of Order from Strength).

7. Agribusiness in LAC would diversify and expand differentially, and small-scale producers would face challenges. In some scenarios, new uses would be added for existing or new commodities. In various scenarios, the participation of a limited group of countries in markets of differentiated products would develop. These markets would require substantial inputs of knowledge and technology (in the case of differentiated products) or production on a large-scale (in the case of commodities). Small-scale producers in Latin America and the Caribbean would be challenged to meet these requirements.

8. In some scenarios, there would be important interdisciplinary advances in formal knowledge, especially in relation to facilitating technologies—such as biotechnology and nanotechnology—and ecology. In others, there would be a high degree of integration between these technologies and other knowledge, such as agroecology and traditional knowledge. In GO and TG, there would be integration between materials engineering, food technology and biology, for instance, either to expand basic knowledge, or to generate new technologies capable of increasing quality and efficiency or reducing production costs. International progress in scientific and technological knowledge, which would demand large amounts of resources, should be followed by AKST systems in LAC, to prevent their knowledge from becoming obsolete and the consequent loss of relevance for the region. In view of the current situation of AKST investment in LAC, which is not only limited but is also extremely heterogeneous, these technological and scientific changes would pose important threats to the region's systems.

9. Traditional knowledge would be increasingly valued and incorporated into certain scenarios (AM, TG). Barriers, pests, diseases, and climate change would create needs for solutions using local knowledge, and its integration would be facilitated by institutional changes in these scenarios. In the other scenarios (GO, OS, and Life as it is), the integration of traditional knowledge would occur only occasionally, due to commercial interests and defective institutional structural arrangements.

10. In some of the scenarios (GO, OS, Life as it is) advances in formal knowledge and technological development linked to productive chains would remain in the hands of large transnational corporations. In other words, many countries in the region could lose the capacity to independently generate knowledge, which is the most important factor of development in the contemporary world.

The scenarios indicate that the option of using local knowledge is not sufficient to meet the demand for food, nutrition, health, and environmental development in an increasingly complex world. This would pose a serious threat to the region.

11. Scientific activity in LAC would change in the scenarios, both in terms of relevant actors (public or private sector, NGOs, and transnationals) and in terms of the sources of resources. In some scenarios, such as GO, OS, and TG, the role of the public sector in generating knowledge and technology would be reduced, and private stakeholders would play a more active role. Since the public sector is the one that has historically been responsible for guaranteeing a similar capacity for access to knowledge and technology to the most vulnerable social groups—while the private sector has not had this function (although it may engage in acts of corporate social responsibility), and NGOs do not really have the capacity to perform it—the generation of knowledge and technology to equalize adverse economic, social, and cultural conditions would not be guaranteed in these scenarios.

12. The scenarios indicate that agricultural knowledge, and science and technology applied to agriculture are necessary but not sufficient to help in achieving the purposes of the IAASTD, namely, to reduce hunger and poverty, and ensure sustainable development and food security. AKST systems are not sufficient in and of themselves, because other factors, such as governance, legal and regulatory institutions, international trade practices, and the like, are fundamental and more inclusive than science and technology in actually achieving sustainable development, which leads to a real reduction in hunger and the eradication of poverty. Based on the results of the analysis of these scenarios, in the subsequent chapters specific innovation policies oriented to achieving these objectives are described, in addition to sustainable development policies for vulnerable groups, to supplement the action of the AKST systems.

3.1 Objectives of the Chapter

This purpose of this chapter is to help answer the following question: "How can we reduce hunger and poverty, improve rural livelihoods, and facilitate equitable, environmentally, socially, and economically sustainable development through the generation of, access to, and use of agricultural knowledge, science, and technology?"

With specific reference to Latin America and the Caribbean, these future alternatives for the development of this region can be used to propose nonprescriptive recommendations as to how science and technology can best contribute to this development.[21]

To meet this objective, the chapter presents five scenarios on development of agriculture (sensu lato), agricultural production systems, and the knowledge, science and technology associated with them. The scenarios described are: (1) *Global Orchestration;* (2) *Order from Strength;* (3) *Adapting Mosaic*; (4) *TechnoGarden*; and (5) *Life as it is*.

The first four scenarios follow the Millennium Scenarios (Carpenter et al., 2005), and take the same name and broader macro-context or major premises used to analyze the relationships among the variables of the context closest to Latin America and the Caribbean and the variables that define the agricultural knowledge, science and technology systems and agricultural production systems in the region. The fifth scenario was designed as a continuation into the future of these systems, with their influences and interaction, as they are today. In other words, it portrays a world based on the premise that the future is similar to the past, whereas the other scenarios use the present as a point of departure to explore future alternatives (that are not a mere continuation of the present). Therefore, the fifth scenario is what is usually called a "trend scenario" or "business as usual."

Why use these scenarios?

The future is full of uncertainties for medium- and long-term policy makers, who need to understand what their worlds will look like in five to ten years from now, for decision-making purposes. In these times of extensive and speedy global intercommunications, the social, political, and economic contexts of societies change, and they are in turn modified with surprising speed. The task of understanding how these changes can alter the future and our societies is thus a difficult one and involves a great deal of uncertainty.

Building scenarios is a methodology used to help understand the future and, consequently to support decision-making on current policies and strategies. The scenarios are not linked to rigid mathematical formulas, unchangeable over time, but instead they offer a probable vision of the future and of the nature of complex phenomena (such as those considered in this paper) and of how that situation is arrived at on the basis of the present and a behavioral model of various types of social, economic, environmental and technological phenomena, among others, and their interaction. The scenarios make it possible to manage the uncertainty which necessarily characterizes the future, by creating *plausible* futures, or descriptions of what may occur in future, depending on the premises regarding selection of social stakeholders in relation to different macrovariables.

This vision of plausible futures is clearly subjective, but it is based on a critical analysis of existing information on the past and present and on methodologies—the scenarios—that lead to a systematic understanding of the future, or, better said, *futures*. The future *could be like this*, if it is not *like that*. This "*could be*" is reasonably credible here and now.

3.2 Conceptual Framework

Some concepts are fundamental for building the scenarios presented in this chapter. These concepts include the following.

The concept of the future. In reality, the future is something that does not exist and cannot be attained, because when you think that you have arrived at the future, in truth it is actually the present. Thus, when one studies the future, what is studied are the images or perceptions that can influence present activities of persons or of the organization that

[21] Proposals to this end are presented in Chapters 4 and 5.

is interested in them. The concept of the future is related to several basic dimensions: (1) Time, the perception and measurement of which, in some societies, is related to the cycles of nature and natural phenomena that are repeated. This dimension leads to a concept of the future as a natural sequence of the past and present; (2) Advances in knowledge and technology. This dimension brings a perspective of evolution and change for contemporary societies, which is different from the previous idea of the future as a continuation of the past. It implies a turbulent atmosphere, in constant transformation, in which studies of the future become more difficult and at the same time more necessary.

Moreover, it is important to consider present influences, or the relationships among the phenomena that influence the present, as well as the possible emergence of new influences. Thus, to be able to understand the future, the current influences on the present must be understood, but account must also be taken of possible emerging events. This last consideration implies a degree of uncertainty, for the future or futures, to the extent that it expands the horizon of time in which the future is analyzed.

The concepts of present influences and future uncertainty are combined in the concept of the future adopted in this chapter. According to this concept, *the future is the result of the interaction between historical trends and the occurrence of hypothetical events.*

A prospective view is an attempt to understand the future that considers the dynamics of various types of influences, including scientific-technological, social, economic and environmental factors, which act on social systems over time, in order to build plausible alternative futures on the basis of this analysis.

The systemic approach. In systems theory, the whole, or the system, is the product of its interactive parts, which must be understood and known as they relate to the operation of the whole. Among the conceptual frameworks of the systemic approach, the concepts of *system, limits, hierarchy,* and *systemic model* are the ones that are most useful for the prospective studies presented in this chapter.

A system is a series of interactive parts or components of interest to the researcher, according to Milsun's definition in Jones (1970). What are the systems of interest in the case of this chapter?

The very question underlying this entire evaluation provides the clues for identifying these systems. The question refers to systems of agricultural knowledge, science, and technology and also systems in which sustainable development occurs, especially in the rural environment. The question also makes specific reference to the relations among these systems, in referring to the contributions of one to the results of another.

What are the limits of the systems to be analyzed? In this chapter, the limits are defined as follows:

- For knowledge, science, and technology systems (AKST), they include the so-called systems of traditional and local knowledge, i.e., the "dynamic body of knowledge and practices accumulated by traditional communities and by agricultural production systems, based on their interaction with nature and their agricultural activities." They also include formal systems of science and technology, or, more specifically, research and development (R&D) designed to generate technology and know-how for agricultural production systems;
- For systems where there is sustainable development, the premise contained in the question underlying this evaluation is that they must be agricultural production systems, because the contribution of R&D to sustainable development implied in the question can only occur on the basis of its action on those systems.

Moreover, these two systems not only interact with each other, but are also subject to the influence of other larger, more embracing systems, the system that could be referred to as the macro-context or, more simply, the context, which involves all of the different types of influences that are not generated in the R&D systems and in agricultural production systems.

The complexity of the systems is simplified in the models that represent them. A general model to represent the question on which this evaluation and chapter are based can be found in Figure 3-1.

3.3 Methodology

The first stage in the scenario building process is to prepare a model that represents the relations among the systems of interest (the R&D systems, the agricultural production systems, and their context), in more detail than what is presented in Figure 3-1. Although consideration should be given to the model presented in Figure 3-1, it is too general to guide construction of the scenarios.

Thus we worked on the basis of a recently constructed model and variables for another study of the future. This study was undertaken in an attempt to understand the changes in the context of R&D systems that would affect the development of these systems over a period of ten years (around 2015) in six countries in Latin America (Castro et al., 2005; Lima et al, 2005; Santamaría et al, 2005).

Consequently, for the variables described by R&D systems and their context, the same variables used in that study were considered for this analysis. For the variables that describe agricultural production systems, a process of collective creation and bibliographic review made it possible to identify the relevant variables for those systems. All the variables considered in this chapter are presented in Table 3-1.

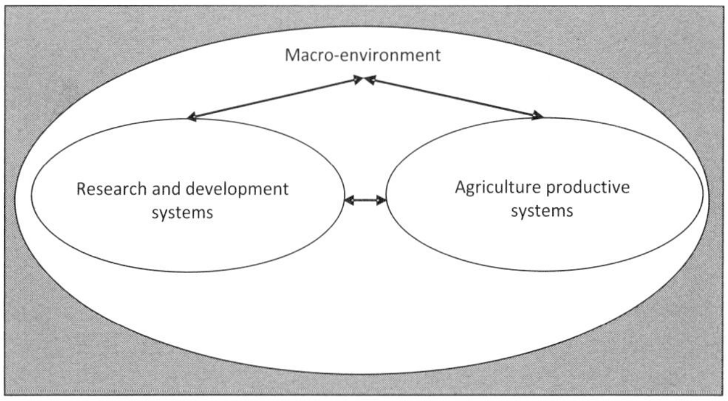

Figure 3-1. *The relationships among relevant systems in AKST.*
Source: Authors' elaboration

Table 3-1. **Definition of structures and variables included in the model.**

Structure	Variable	Variable's definition
	Macro-context	
Barriers to International trade	Non-tariff barriers based on social concerns	It is the body of official regulations and directives based on social indicators devised for the supply chain, which restrict the trade of agricultural products and services.
	Monitoring protocols and regulations for assuring product traceability and quality certification	It is the body of official regulations and directives relevant to the quality of food and products along the supply chain, with the aim of ensuring their safety to both domestic and foreign customers.
	Non-tariff barriers based on environmental concerns	It is the body of official regulations and directives based on environmental indicators devised for the supply chain, which restrict the trade of agricultural products and services.
	Tariff barriers	It is the body of official regulations and directives intended to protect the trade of domestic agri-business products from external competition.
Competitiveness of agricultural business	Competitiveness of agricultural business	Ability of LAC agri-business to displace similar products and services from markets by offering products and services at prices and qualities demanded by consumers.
	Differentiation of innovative products	Products with increased added value from the use of R&D processes and marketing.
	Access to markets of innovative differentiated products	Placement of innovative products with increased added value in international markets
	Information and Communication Technology (ICT) as a tool for facilitating commercial transactions in agri-business	Use of diverse electronic communication resources to reduce transaction costs in the trade of agri-business products.
	Commodities' costs	Production and transaction costs of commodities in agri-business.

Table 3-1. **continued.**

Structure	Variable	Variable's definition
Demands from final consumers	Demands from final consumers	Demands for diversification of agri-business products from several segments of final consumers.
	Demand for healthy and safe foods	Public's interest in foods harmless to health and nutraceuticals (foods with medicinal effects).
	Consumer information	Free access to product information as suited to the needs of final consumers.
Climate change	Climate change	The effects of increased frequency and intensity of climate phenomena driven by temperature, rainfall, wind, etc. on agricultural activities.
Epidemics/food contamination	Diseases, pests and food contamination	Occasional outbreaks of diseases, pests and/or diverse kinds of food contamination in different countries and regions.
	Advances in biology and biotechnology	Research in biology and biotechnology moves steadily on the discovery of fundamental biological knowledge.
Advances in knowledge	Advances in information technology	Information and Communication Technology (ICT)'s progress on novel modes for the communication and flow of information.
	Advances in nanotechnology	Progress on fundamental nanotechnological knowledge.
Traditional/indigenous knowledge	Traditional/indigenous knowledge	Dynamic body of knowledge and practices accumulated by traditional/indigenous communities and agricultural production systems as a result from the interaction between the latter and both nature and agricultural practice.
Social monitoring of innovation	Public perception of S&T (Science and Technology)	Public trust on the results and conclusions from scientific and technological activities.
	Social monitoring of innovation	Involvement of social actors on the aims, planning, implementation, results and impacts of S&T activities.
Governance	Governance	A wide and inclusive social compact buttresses the stability of social, economic, environmental and innovation policies in LAC.

continued

Table 3-1. **continued.**

Structure	Variable	Variable's definition
Policies for development	Integration of policies for innovation and social development	Development is facilitated by the integration of national, sub-national entity and sectorial policies.
	Proposal and implementation of agricultural policies	The ability to devise agricultural policies together with the existence of organizations and institutions prepared to implement them.
	Biosecurity policies	Policies for reducing the intrinsic risks of foods and agriculture (environmental risks included) These are policies for the security of food, health and the life of plants and animals.
	Social development policies	Policies for facilitating the access of vulnerable rural and urban populations to education, credit, health and housing.
	Incentive policies for research	Policies for the development of science and technology.
Management of regulations and standards	Regulations and standards	The mechanisms that (a) regulate intellectual property rights for the results of scientific research, including the production of living organisms (cultivars) and (b) set the provisions for trading agricultural products in LAC and other world regions.
	Implementation of regulations and standards	Set of actions addressing the implementation and monitoring of regulations and directives governing agricultural S&T and agricultural products.
Education of PS actors	Education of PS actors	Degree of schooling of productive systems' actors.
Urban food security	Access to food security	Ability of urban consumers (particularly the poor ones) for regularly purchasing food in enough quantities for ensuring their well-being.
	Access to food security	Ability of urban consumers (particularly the poor ones) for regularly purchasing healthy food in the sense of low risks as to biological contamination, allergenic potential and pollution.
Social inequality	Social inequality	Relative access to employment, food security, education and health of different social groups—like, e.g., small family farmers, subsistence farmers, large farmers, wage earners—involved in agricultural production activities.
Environmental sustainability of agriculture	Environmental sustainability of agriculture	Ability of the agro-ecosystem to keep its productive functionality in future times.

Table 3-1. **continued.**

Structure	Variable	Variable's definition
	R+D systems	
Focus of research	Focus of research	Strategic orientation of the objectives and results from R+D activities to social groups attended by them.
	Harmony between R+D organizations and their social environment	Congruence between the mission, objectives and products from R+D organizations and the needs and expectations of their clients, users, beneficiaries and other pertinent stakeholders.
	Priorized activities	Strategic choice of topics/problems for developing projects and project portfolios in R+D organizations.
	Demands for research	The need of knowledge and technology to take advantage of opportunities or to remove checks on the performance of agricultural production systems.
	Survey of future demands for research	Systematic evaluation of likely demands for research in the future.
Incorporation of formal knowledge	Incorporation of formal knowledge	Incorporation of advances in formal knowledge to the R+D process.
Incorporation of traditional/indigenous knowledge	Incorporation of traditional/ indigenous knowledge	Incorporation of traditional/ indigenous knowledge and practices to the formal process of knowledge and technology production.
Availability of resources for R+D	Alternative resources for funding R+D	Alternative non-fiscal sources of R+D funding.
	Funding for R&D production	Funding necessary for producing the technologies and knowlege demanded by the clients/users of R&D
	Infrastructure for the production of R+D	Facilities and equipment necessary for the production of knowledge and technologies demanded by the clients/users of R+D.Infrastructure for the production of R+D

continued

Table 3-1. **continued.**

Structure	Variable	Variable's definition
Performance of R+D systems	Products and services generated by R+D	Portfolio of products and services generated by R+D organizations for their clients.
	Effectiveness of R+D	Products are delivered according to consumer, client, and whole society needs.
	Efficiency of R+D	Ability of R+D organizations to generate lowest-cost products and services.
	System for project planning, monitoring and evaluation (PME) in R+D organizations	The systematic process for setting objectives/goals, procuring and distributing resources, implementing projects and programs, and adjusting the implementation and evaluation of projects and final services obtained from R+D organizations.
Management of R+D	Project portfolio	Collection of projects intended to solve a large national or regional strategic problem.
	Projects	Management tool with goals clearly defined by: the nature of a problem; a particular request; particular favourable conditions for meeting some goals; or the interest of groups that seek translating ideas into concrete results in a prestablished period of time and at a known cost.
	Management of research teams	Mechanisms of planning, monitoring, evaluation and organization of R&D work.
	Multidisciplinary approach	Interaction, synergism and interfacing among diverse fields of knowledge.
	Reward systems	Processes for valuing or approving (or both) the results of research work in R+D organizations by means of both material and immaterial rewards.
Relative spaces of public and private R&D	Relative spaces of public and private R&D	Fields covered by each of public and private research organizations
	Public-private alliances	Agreements between public and private organizations with the aim of complementing resources for projects in which there is a common research interest.
	Competition between agricultural R+D organizations	Strategies of public and private R+D organizations to predominate in markets for agricultural industry technologies.
	Privatization of the R+D system	Complete transfer of public R+D infrastructure and activities to the national or international private sector.

Table 3-1. **continued.**

Structure	Variable	Variable's definition
Social involvement in the management of R+D	Social involvement in the management of R+D	Social groups are involved in the decision making and implementation of R+D activities.
Proper technologies for agricultural activities	Proper technologies for agricultural activities	The degree with which the technologies generated by R+D systems support sustainable development and also are suitable to the culture, resources and conditions of the agricultural production systems.
Agricultural production systems		
Incorporation of knowledge to productive systems	Support to the incorporation of knowledge	Operation of mechanisms for giving technical assistance (public or private) to productive systems for adopting appropriate technologies.
	Incorporation of knowledge to productive systems	Choice and adoption of appropriate technologies by productive systems.
	Integration of production chains	Degree of connectivity with and participation of productive systems in established production chains.
Attended markets	Attended markets	These are the markets agricultural production systems send their produce.
Social organization of vulnerable production systems	Social organization of vulnerable production systems	It is a mechanism for attaining economies of scale in production, negotiation capacity, and improvements in the management and trade of agricultural productive systems goods and services.
	Social movements focalized on the most vulnerable production systems	Social mobilization as an instrument for accessing resources and empowering production systems.
Availability of resources for agriculture	Availability of resources for agriculture	Access of production systems to credit, land, water and knowledge.
Performance of agricultural productive systems	Efficiency	Relationship between costs of production and returns in productive systems.
	Quality of products and processes	Sustainability of agricultural products and processes, and the degree of agreement between them and consumer needs.
	Products, subproducts and waste	Characteristics of the products, sub products and waste in regard to their effect on the environment.
Rent inequality in agriculture	Rent inequality in agriculture	Relative access to rent by diverse social groups involved in agricultural production, like family farmers, salaried employees, subsistence farmers, large producers, etc.).

Source: Authors' elaboration.

Next, the relationships among these variables were studied with the help of a crossed impact matrix. This matrix makes it possible to analyze the direct relationships between each pair of variables in terms of intensity, type, and direction of the interaction. Based on that analysis, the model of relations shown in Figure 3-2 was built.

On the basis of this model, a selection was made of the variables considered as the critical factors for understanding the future in the scenarios. These variables are: the demands for and focus or focal point of the R&D; technologies adapted to the agricultural production systems; incorporation of knowledge into agricultural production systems; available resources for agricultural production systems, performance of agricultural production systems; income inequality; social inequality; urban food security; and, environmental sustainability in agriculture. These last four critical factors describe the results of the interactions between the context and the two (R&D and production) systems of interest. For each of the critical factors, submodels were prepared, that show the direct relationships with other variables based on the model presented in Figure 3-2. Examples of submodels for the four macrovariables of results (income inequality, social inequality, urban food security, and environmental sustainability in agriculture) are shown in Figures 3-3 to 3-6.

The scenarios were designed on the basis of these models, using the morphological analysis matrix tool. It takes into account the plausible situation of the variables for the time horizon under analysis. Then, the situation—considered as the hypothetical future development of each variable—is linked to the themes of the five scenarios: (1) *Global Orchestration*; (2) *Order from Strength*; (3) *Adapting Mosaic*; (4) *TechnoGarden*; and (5) *Life as it is*.

The first four scenarios follow the Millennium Scenarios (Carpenter et al., 2005), and take the same name and the broader macro-context or the main premises used to analyze the relationships between the variables of the context closest to Latin America and the Caribbean and the variables that define the agricultural knowledge, science and technology systems and the agricultural production systems in the region. In these scenarios, the interaction of two macrovariables (integration among countries and action related to environmental services) defines the major forces that determine the entire scenario. Table 3-2 presents these premises, both for the themes taken from the millennium scenarios and for the "business as usual" scenario.

The link between themes and descriptions of situations resulted in the matrix of scenarios and in the first version of the scenarios themselves for two time periods: 2007-

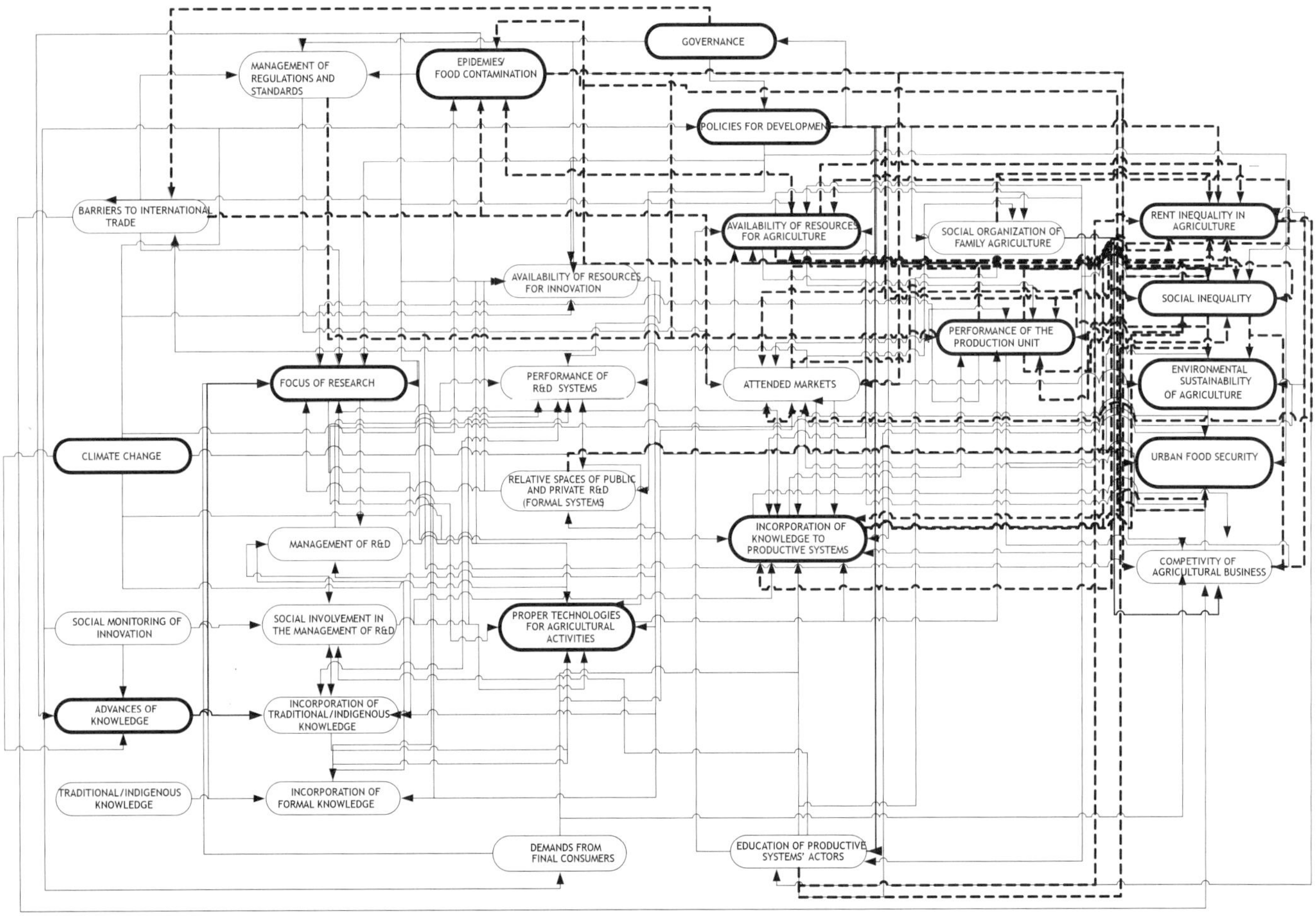

Figure 3-2. *Model for the relationships among contextual variables, R&D systems and agricultural productive systems.* Source: Authors' elaboration.

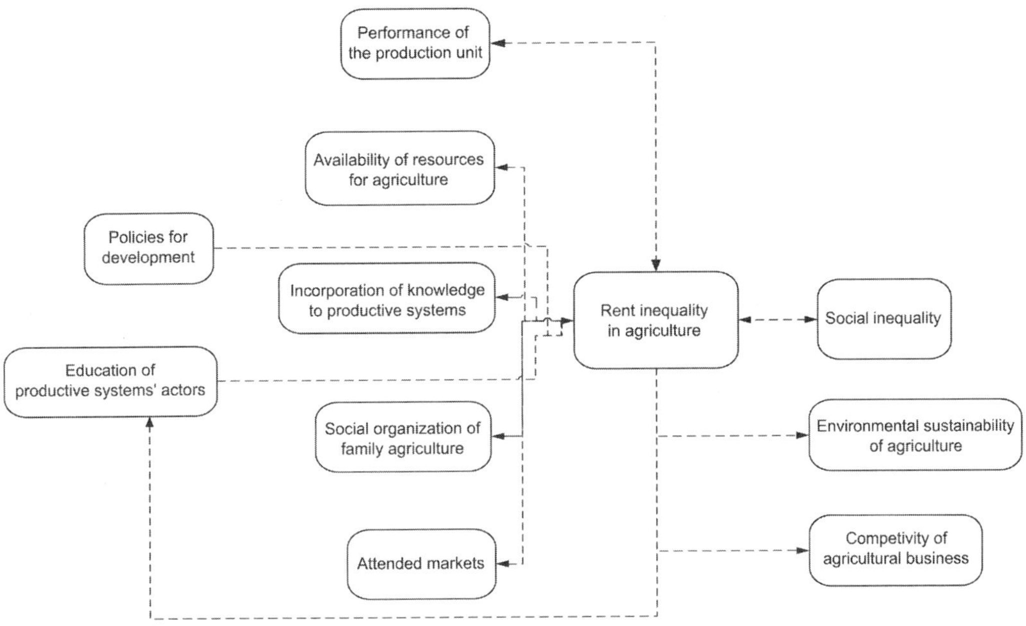

Figure 3-3. *Sub-model for the critical factor: inequality of rent in agriculture.* Source: Authors' elaboration.

2015 and 2016-2030. The authors revised these scenarios to obtain a working paper, that was submitted to around 50 specialists from Colombia and Brazil for validation on the following themes: climate change and environmental sustainability; governance and development policies; advances in know-how (biotechnology and nanotechnology); epidemics, pests, and contamination of food; economic and social development; and, traditional knowledge (appreciation of it and its inclusion in R&D).

The validation process entailed an evaluation of the plausibility of each description of these variables in the different scenarios and time periods, using a ten-point scale,

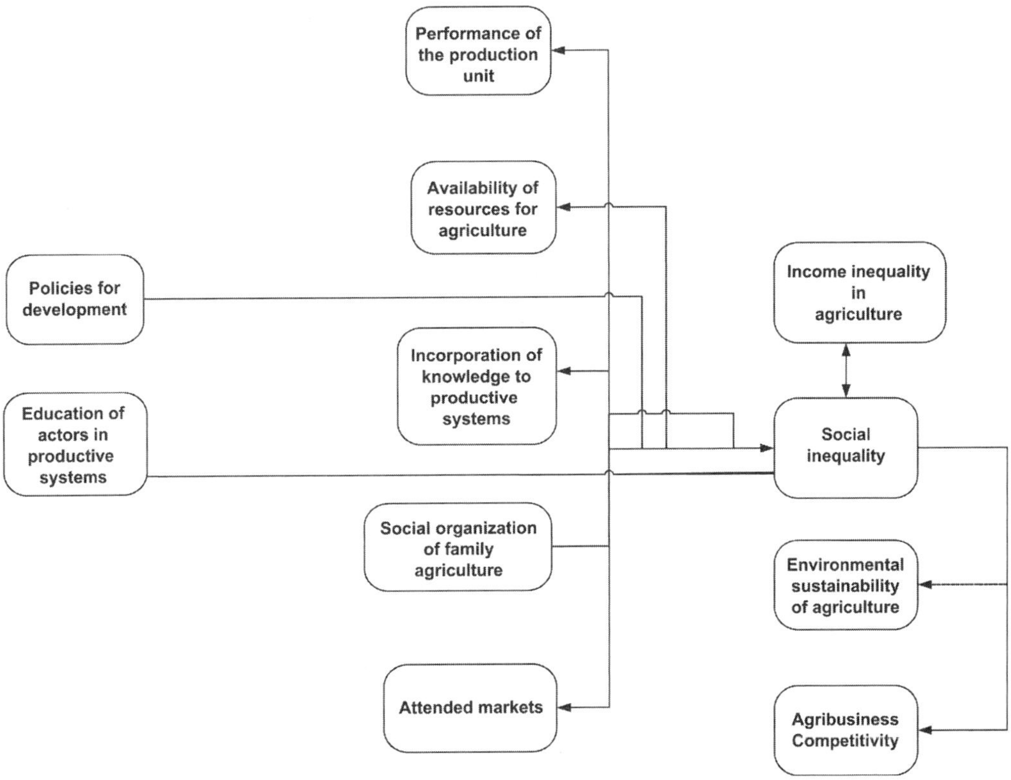

Figure 3-4. *Sub-model for the critical factor: social inequality.* Source: Authors' elaboration.

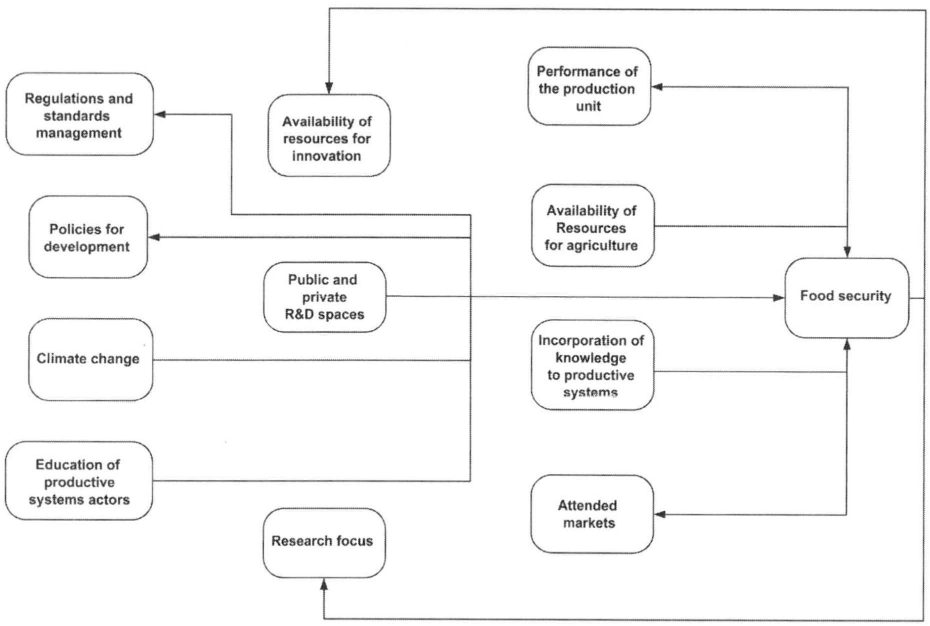

Figure 3-5. *Sub-model for the critical factor: urban food security.* Source: Authors' elaboration.

with "1" representing the point of least plausibility and "10" total plausibility. For ratings of less than 5, the specialists were asked to indicate (1) a reason justifying the assessment or rating, and (2) a suggestion for improving the plausibility of the description.

The scenarios were adjusted on the basis of that evaluation and also on the basis of comments and suggestions by other external reviewers. These adjusted scenarios are presented below.

3.4 Scenarios: AKST and Sustainable Development in LAC in the Future (2007-2030)

Table 3-3 presents the current situation of the indicators selected for the variables considered in this study of the future. Based on this table it is possible to identify that there are countries at present that are more or less vulnerable in relation to these indicators. Vulnerability is defined as "*the weak capacity of an individual or group response to risks and uncertainty . . . ; a predisposition to a drop in well-be-*

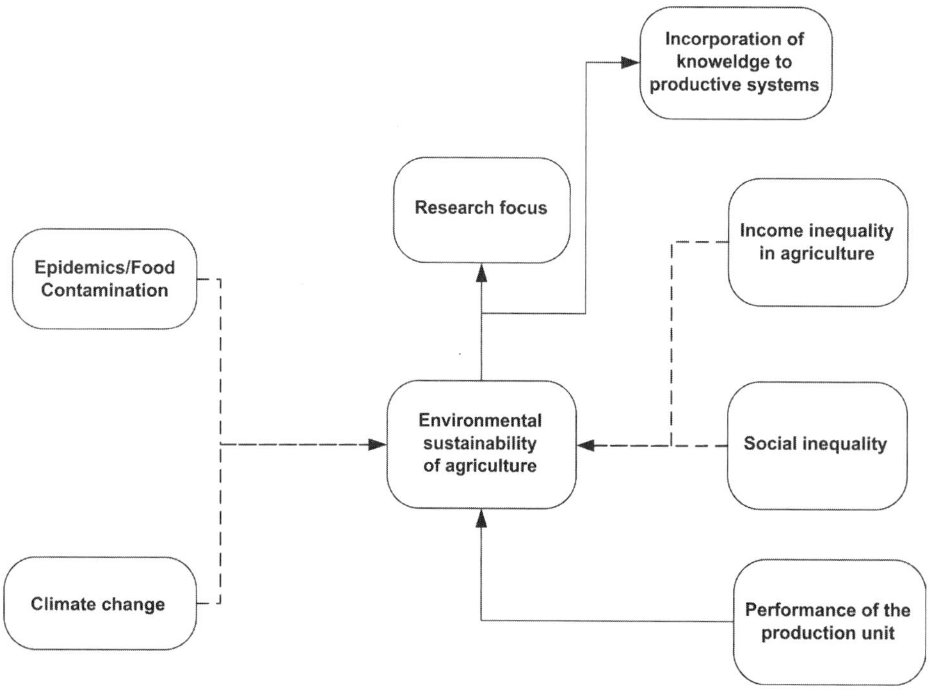

Figure 3-6. *Sub-model for the critical factor: Environmental sustainability of agriculture.* Source: Authors' elaboration.

Table 3-2. **Subjects used for scenario building.**

Approach in relation to the management of environmental services	Approach in relation to governability and economic development		
	Globalized	Mixed	Regionalized
Reactive	Global orchestration	—	Order from strength
Mixed	—	Life as it is	—
Proactive	TechnoGarden	—	Adapting Mosaic

Source: Authors' elaboration.

ing, based on a configuration of negative attributes to achieve material and symbolic returns . . . ; a negative predisposition to overcome adverse conditions." (Filgueira and Peri, 2004). All of the countries are presented with greater or less vulnerability, depending on the indicator/variable considered.

The scenarios built on the basis of the variables indicated are presented below. A summarized version of the scenarios, referring to all the variables used in their construction, is presented in Table 3-4.

3.4.1 Global Orchestration

3.4.1.1 2007-2015

3.4.1.1.1 Context of the AKST systems and agricultural production

The world and LAC are shifting toward the absence of barriers—except for health barriers—to international trade in agricultural products. This increases competition among countries, which fight for market shares on the basis of prices of differentiated products. The LAC countries already established in commodities markets (including Argentina, Brazil, Chile, Colombia, Ecuador, and Mexico) endeavor with some success to gain a place on the most dynamic markets—United States, China, and India—and on the market for differentiated products.

Throughout the world, the diversity of consumer demand for differentiated foods increases, on the basis of flavor, appearance, nutritional value, nutraceutical properties, bromatological quality, or another such factor. In many countries, consumers demand quality certification for processed foods, referring to such matters as the absence of agrotoxins, child labor, genetically modified organizations, and animal suffering. There is also a rising demand for traceable food products. In LAC, the growing education of the people and increased availability of information lead to greater exigency on the part of consumers, although consumers are more concerned over damage to their health than over aspects related to environmental protection.

During this period of time, in most of the region the frequency or severity of epidemics is not on the rise, due to the incentive to implement good management practices in production systems, the development of appropriate research for prevention and management of epidemics and the search for safe foods, and the development of capacity and regional cooperation to prevent new epidemics.

In some parts of the region, there are major changes in land use patterns. For instance, large tracts of land are used for single crops for production of biofuels, which may encourage the manifestation of new epidemics. Similarly, in areas already highly affected by early manifestations of climate change, such as floods, droughts, heat waves, and the like, and in zones where no adaptation policies have been planned, conditions are ripe for the proliferation of epidemics or emergence of new pests.

The temperature is rising at the rate of 0.22C-0.24C per decade, and so the frequency of extreme phenomena increases. Their effects are relevant and range widely for the agriculture and production systems in the region, due primarily to the equally widely ranging capacity of adaptation and mitigation. Decision-makers and societies in general, especially in LAC, do not show much concern over these climate changes.

Some countries establish social development, innovation, environmental, and biosecurity and biosafety policies that are coherent and consistent with the major economic development objectives. Consequently, those countries increasingly improve their capacity to manage these policies. Other countries in the region still have relatively ill-defined and short-sighted policies, in addition to a weak management structure. However, as a rule, the governance situation improves considerably up to the end of the period.

Education is considered an essential factor for improving the trade competitiveness of countries. The growing generation of wealth allows governments to make large investments in formal education, from basic to graduate education. The countries with a smaller economic capacity still try to provide at least good primary and secondary education for their citizens.

Education of the stakeholders in production systems is also provided by private educational institutions, along with public schools. The former gradually improve the quality of their results. Some major agricultural enterprises also cooperate in educating stakeholders in production systems, even on a graduate level, in various countries.

The most developed countries of the region make major investments to develop new technologies, such as nanotechnology, and also biotechnology and information technology. Few LAC countries have the capacity to achieve major advances in knowledge of agricultural systems and agriculture, not to mention new technologies.

Both in other regions and in the LAC in general, the value of traditional knowledge is not recognized, yet some large private enterprises seek this knowledge to create new products, such as pharmaceutics or plant-based insecticides, to be used intensively by agricultural production systems.

Table 3-3. **Selected indicators: current situation of variables.**

Variable	Current situation	Source
Context variables for AKST systems and agricultural production systems		
Tariff barriers	In LAC there are lower import tariffs, and no subsidies to exports and production of goods compared to the both the World and rich countries.	Anderson y Valenzuela, 2006
Non-tariff barriers	Agricultural exports are the most likely to be penalized with non-tariff barriers. This effect is less in LAC than in the Middle East, North Africa, Europe, USA, Canada and Japan. Agricultural exports are the most likely to be penalized with non-tariff barriers. This effect is less in LAC than in the Middle East, North Africa, Europe, USA, Canada and Japan.	Bora et al., 2002
Market competitiveness	Agricultural products: net-exporting countries: Argentina, Bolivia, Brazil, Chile, Colombia, Ecuador, Paraguay, Uruguay, Costa Rica, Guatemala, Honduras, Nicaragua; net-importing countries: Peru, the Bolivarian Republic of Venezuela, El Salvador, Mexico, Panama, Cuba, the Dominican Republic, Haiti, Jamaica and Trinidad and Tobago. Foods: net-exporting countries: Argentina, Bolivia, Brazil, Paraguay, Uruguay, Nicaragua; net-importing countries of foods and agricultural products: Peru, the Bolivarian Republic of Venezuela, El Salvador, Mexico, Panama, Cuba, the Dominican Republic, Haiti, Jamaica and Trinidad and Tobago.	de Ferranti et al., 2005
Demands from final consumers	Consumers increasingly demand better quality in foods. According to Renard (1999) quality—in its manifold dimensions and meanings—is the factor that binds together consumers, wholesalers, industry and farm production.	Renard, 1999
Epidemics/food contamination	In developed countries many episodes of transboundary diseases have been recorded since the 1980s.	Jaffee et al., 2005
	In LAC the foot-and-mouth disease and the avian flu are epizootics of much concern because of their impacts on important sources of work and earnings for rural communities. The capacity to quickly and effectively react to transboundary diseases' outbreaks would expose institutional weaknesses in many LAC countries as well as agencies responsible for monitoring, prevention and sanitary control of those kinds of diseases. The diverse agricultural production methods in use decrease the effectiveness of international monitoring and harmonization of public programmes for preventing and fighting transboundary diseases.	CEPAL, 2006
	In regards to avian flu a team from the Inter-American Development Bank assessed the integration of agricultural and health measures before an outbreak of that kind of disease. It is shown that the degree of that integration is greater in the Southern Cone than in other regions of LAC (Central America, Andean countries and Latin Caribbean). The countries in the Southern Cone show some differences in the degree of integration among themselves. In regard to poultry consumption, it accounts for 35% of meat consumption in LAC, 42% in Central America and 45% in the Latin Caribbean. These relatively high percentages point to the existence of a food insecurity risk in the event of an outbreak of avian flu.	Schneider et al., 2007
	If the following three indicators—i.e., units for veterinary practice, available personnel, and economic resources—were taken in account for combating foot-and-mouth disease on an area basis in South America, each of Bolivia, Chile, Guyana and Peru has two out of those three indicators with lower values than in the rest of the continent. Brazil, Ecuador, Paraguay and Uruguay are the less vulnerable countries.	PANAFTOSA, 2006
Climate change	Parameters related to agriculture. Severe environmental restrictions to dry land farming in LAC 1961-1990: Central America and the Caribbean, 51% (mostly arid lands); South America, 61.9% (poor soils). Lands without restrictions: 10% of LAC. Average yield potential 1961-1990 (Mtons/year): Central America, 101; South America, 543; developed countries, 0.002815.	Fischer et al., 2005

Table 3-3. **continued.**

Variable	Current situation	Source
Governance and policies in LAC	Political stability. Positive values for Chile, Costa Rica, Uruguay, Cuba and the Dominican Republic; negative values for the rest of the countries, and particularly small values for Haiti, the Bolivarian Republic of Venezuela, Bolivia, Ecuador, Colombia, Guatemala and Peru. Government effectiveness: Positive values for Chile, Trinidad and Tobago, Costa Rica, Uruguay, and Panamá; negative values for the rest of the countries, and smaller values for Haiti, Ecuador, Cuba, the Bolivarian Republic of Venezuela, Paraguay and Bolivia. Regulatory Quality: Positive values for Colombia, Brasil, Peru, El Salvador, Panamá, Uruguay, México, Costa Rica, Trinidad and Tobago and Chile; negative values for the rest of the countries, and smaller values for Cuba, Haiti, the Bolivarian Republic of Venezuela, and Ecuador. For all three indicators, positive values were given to Uruguay, Costa Rica, and Chile, and negative ones to the Bolivarian Republic of Venezuela, Ecuador, Paraguay, Argentina, Bolivia, Honduras, Guayana, and Nicaragua.	Kaufmann et al., 2006
	Education: Education quality is assessed by the average number of students with mathematical skills in three education levels: basic, primary and secondary. There is a correlation between the quality values measured in urban and in rural students, but in no case average values are greater in the rural students. Country-wise Cuba shows high skill scores in both student populations (greater than 90%); Brasil, Chile and Argentina reach 80% skill in urban students; the Bolivarian Republic of Venezuela Paraguay, México, and Colombia show skill values ranging from 50% to 70% for both urban and rural students. The rest of the countries—Perú, Bolivia, Honduras, and the Dominican Republic—show values below 60% for both the urban and rural students.	de Ferranti et al., 2005
Advances in formal knowledge	The private sector invests annually more than US$1.5 billion in biotechnology in a large part of developed countries; public organizations doing agricultural research in developing countries invest US$100-150 million per year; the CGIAR centers invest about US$25 million per year; and the Rockefeller Foundation and other non-profit organizations annually invest about US$40-50 million.	Byerlee and Fischer, 2000
	Brazil, Argentina, Mexico and Chile are the LAC countries with more firms, publications and patents in biotechnology.	Niosi y Reid, 2007
	The largest investments in nanotechnology in 2004 were made in Europe (US$1.32 billion), North America (US$1.28 billion) and Asia (US$1.16 billion); in LAC biotechnology as a whole received US$16.2 million from only three countries: Mexico (61.7%), Brazil (35.8%) and Argentina (2.5%).	Simonis y Schilthuizen, 2006
Traditional/indigenous knowledge	This knowledge is in steady progress. The following features distinguishes it from occidental scientific knowledge: (1) it is verbally recorded and transmitted; (2) it is nourished by observation and experience; (3) its cosmology is rooted in the view that Nature is instilled with spirituality; (4) it is intuitive; (5) it is qualitative; (6) it is based on data generated by its users and (7) it is grounded in a social context which sees the world through multiple social and spiritual relationships among all forms of life.	Dutfield, 2001
	The intellectual property of traditional knowledge of biodiversity, phyto-genetic resources, and products derived from natural principles found in wild species by indigenous communities and peoples is still an unsettled question.	WIPO, 2001
		Variables of AKST systems
Focus of research	There presently are three processes highly relevant for R+D in LAC and running in six countries (Brazil, Cuba, Mexico, Panamá, Peru and the Bolivarian Republic of Venezuela). Those are concerned with high productivity; increase of resistance to pests and diseases; and biological control of pests and diseases. The management of water quality and use; survey and conservation in situ and ex situ of germplasm; and management, zoning and conservationist agriculture are the best assessed environmental subjects in LAC. The applications of biotechnology, livestock and plant production were considered of greatest relevance.	Castro et al., 2005; Lima et al., 2005; Santamaría G. et al., 2005; Ramirez-Gastón R. et al., 2007; Saldaña et al., 2006

continued

Table 3-3. continued.

Variable	Current situation	Source
Focalized social segments	In most countries in LAC—except Cuba—R+D is better informed on the supply-chain segments represented by big and medium producers, agri-business, wholesalers, and retailers than on subsistence producers and indigenous communities.	Castro et al., 2005; Lima et al., 2005; Santamaría G. et al., 2005; Ramirez-Gastón R. et al., 2007; Saldaña et al., 2006
Capacity in R+D	There is a "specialization index" that equals "1" for the case of all researchers with completed tertiary (university) education, and is "3" for all researchers with a doctorate. For Brazil, the Bolivarian Republic of Venezuela and Trinidad and Tobago, the value of the index is "2"; for Costa Rica, Bolivia and Colombia, the average index is above "1.5" and for the rest of the countries, it is above "1". Countries with the lowest formation level (most of researchers with a licentiate degree) are Ecuador, Paraguay and Uruguay. There are no data for Cuba.	RICYT, 2007.
Investment in agricultural R+D	The countries which invest more in terms of average GDP (1990-2004) are Brazil (0.9%), Cuba, Chile (about 0.6%), Argentina, Mexico, and Panamá (about 0.4%); the rest of the countries invest less than 0.3%, and some below 0.1% (Ecuador, El Salvador, Honduras, Jamaica, Nicaragua and Paraguay).	RICYT, 2007
Performance	Technologies that because of their relevance are presently considered "leading technologies" for most of countries are those addressing the following changes in agricultural production systems: (a) Increase in agricultural and silvicultural productivities; (b) reduction of agricultural and silvicultural production costs; (c) improvement of product quality in production chains; (d) food security; and (e) improvement process quality in agricultural and silvicultural production chains. These technologies are more suitable for medium and big producers, but less so for agri-business.	Castro et al., 2005; Lima et al., 2005; Santamaría G. et al., 2005; Ramirez-Gastón R. et al., 2007; Saldaña et al., 2006
Relative spaces of public and private R+D	In Latin America a scenario is emerging such that the private sector is becoming keener to invest in R+D activities, particularly in the improvement of cultivars of crops like corn (and increasingly soybean), which would readily produce profits. In Brazil it is also observed a growing participation of the private sector—the national one mostly—in R+D.	Castro et al., 2005; Lima et al., 2005; Castro et al., 2006
	There are evidences that in Argentina the transnational private sector invests in biotechnology about six times the amount invested by the public sector.	Varela y Bisang, 2006
		Variables for agricultural production systems
Incorporation of knowledge to agriculture	The countries which invest more in terms of average GDP (1990-2004) are Brazil (0.9%), Cuba, Chile (about 0.6%), Argentina, Mexico, and Panamá (about 0.4%); the rest of the countries invest less than 0.3%, and some below 0.1% (Ecuador, El Salvador, Honduras, Jamaica, Nicaragua and Paraguay).	RICYT, 2007
Resources for agriculture	Expenses per rural inhabitant (1991-2001). >US$1,000: Uruguay; >US$150 & <US$300: Mexico, Argentina, Brazil, and Chile; >US$75 & <US$150: Panamá, Nicaragua, Costa Rica, Dominican Republic, and the Bolivarian Republic of Venezuela; <US$75: Honduras, Guatemala, El Salvador, Paraguay, Jamaica, Peru, Ecuador, Bolivia, and Colombia.	de Ferranti et al., 2005
	Agricultural and rural public expenses as percent of agricultural GDP. Average for 1990-2001 was 12.8%. Countries where those expenses were: (a) above the average: Uruguay, Panama, Dominican Republic, Mexico, Nicaragua and Chile; (b) equal to the average: Guatemala and Honduras and (c) below the average: Bolivia, Ecuador, Costa Rica, Jamaica, Peru, Brazil, the Bolivarian Republic of Venezuela, Argentina, Paraguay and Colombia.	Kjöllerström, 2004

Table 3-3. continued.

Variable	Current situation	Source
Performance of agricultural systems in LAC[1]	Agricultural GDP (USD million in 1995) for 2002. (a) Greater than 60,000: Brazil; (b) 10,000-20,000: Mexico, Argentina and Colombia: (c) 5,000-9,999: Peru and Chile and (d) 400-4,999: Ecuador, the Bolivarian Republic of Venezuela, Guatemala, Cuba, Paraguay, Dominican Republic, Costar Rica, Uruguay, El Salvador, Bolivia, Honduras, Nicaragua, Panamá, Haiti.	RLC-FAO, 2004
	Share of agricultural GDP of total GDP (%) in 2002. (a) Greater than 40%: Guyana; (b) 20%-39%: Nicaragua, Paraguay, Ecuador, Belize and Guatemala; (c) 10%-19%: Honduras, Haiti, Dominica, Bolivia, Colombia, Suriname, Dominican Republic, El Salvador and Costa Rica and (d) Lower than 10%: Saint Lucia, Peru, Grenada, Brazil, Uruguay, Panama, Jamaica, Chile, Argentina, the Bolivarian Republic of Venezuela, Cuba, Barbados, Mexico and Trinidad and Tobago.	RLC-FAO, 2004
	Interactions between the agricultural production and the AKST systems	
Rent	Rent per capita. More than US$9,655: Argentina, Brazil, Chile, Uruguay, the Bolivarian Republic of Venezuela, Costa Rica, Mexico, Panama and Trinidad and Tobago; US$875-3,125: Bolivia, Colombia, Ecuador, Paraguay, Peru, El Salvador, Guatemala, Honduras, Dominican Republic, and Jamaica; Less than US$875: Haiti.	World Bank 2003
Rent inequality	Between 1998 and 2005 the difference between the most rich and the most poor—an indicator of social inequality—in some LAC countries shrank between 8% and 23%; those countries were Argentina, Brazil, Ecuador, El Salvador, Mexico, Panama, Paraguay, Peru and the Bolivarian Republic of Venezuela. The relatively large difference in percent values was due to an increased participation of the lowest four population deciles as well as a decrease in the participation of the richest population decile. Chile and Costa Rica did not show any change in that indicator. Colombia, Honduras, the Dominican Republic, and Uruguay instead showed increases not greater than 13%. The value of the Gini Index confirms the emerging trend to an improvement in wealth distribution. Brazil, El Salvador, Paraguay and Peru showed a substantial decrease (4% to 7%) in the value of that index: however, Honduras showed a marked increase in the value of the Gini Index. During the longer period 1990-2005, in Uruguay and Panama urban wealth distributivity markedly increase, as attested by a decrease of about 8% in the Gini Index. Honduras followed the same path, with a decrease of 4% in the value of that index. On the other hand, urban areas in Ecuador and metropolitan Asunción in Paraguay yielded a 10% increase in the value of the Gini Index, which amounts to a sizeable increase in the concentration of wealth. The index also decreased from 4% to 7% for Argentina (Great Buenos Aires area), Costa Rica and the Bolivarian Republic of Venezuela. In 2005 Bolivia, Brazil, Honduras and Colombia showed relatively large values (ranging from 0.584 to 0.614) of the Gini Index. The lowest value of that range (0.584) was greater than the upper value of the range 0.526-0.579 obtained for Nicaragua, the Dominican Republic, Chile, Guatemala, Paraguay, México and Argentina. Inequality (as measured by the Gini Index) was still less (0,470-0,513) for Ecuador, Peru, Panama, El Salvador, the Bolivarian Republic of Venezuela and Costa Rica. Uruguay was the only country with a low inequality level: Gini Index of 0,451.	CEPAL, 2006
Social development	Concern with meeting people's basic needs (e.g., assistance to education premises, sanitation, electricity, drinking water, five or more years of schooling, dwelling, avoidance of overcrowding, etc.) as measured by an index running from 0% to 100%. High (equal to or greater than 70%): Panama, Argentina, Chile, Costa Rica, Uruguay and Brazil; medium (50%-69%): Mexico, Ecuador, Colombia, the Bolivarian Republic of Venezuela and Guatemala; below average (25%-49%): El Salvador, Paraguay, Peru, Bolivia, Nicaragua and Honduras.	CEPAL, 2005a

continued

Table 3-3. **continued.**

Variable	Current situation	Source
Food security	During 1979-2000, daily consumption increased about 10 kcal per capita in Peru, Ecuador, Honduras, Colombia and Brasil, but it decreased or did not change in Haiti, Argentina, Panama, Nicaragua, Guatemala, Cuba y the Bolivarian Republic of Venezuela.	Morón et al., 2005
	Proportion of undernourished population. Greater than 35%: Haiti (improving); 20%-34%: Bolivia (improving), the Dominican Republic, Nicaragua, Honduras (stable), Panama and Guatemala (deteriorating); 10%-19%: Peru (reached the Millenium Goal), Jamaica, Colombia, Paraguay, El Salvador, Trinidad and Tobago and the Bolivarian Republic of Venezuela (improving); 5%-9%: Brasil and México (improving); 2%-4%: Cuba, Chile, Ecuador (reached the Millenium Goal), Uruguay and Costa Rica.	RLC-FAO, 2006
Food sustainability	The most serious environmental problems in LAC are: land and forest degradation, deforestation, losses of habitat and biodiversity, pollution/contamination of fresh-water sources, marine coasts and the atmosphere. The amount of global rainfall is enough, but it is unevenly distributed; agriculture is strongly dependent on irrigation in many areas; there has been a marked increase in livestock production and many areas are under water stress. There has been a striking increment in both crop and livestock production. The latter exerts a strong pressure on forest lands, even when the rate of increase in lands under agriculture has decreased. There is a noticeable trend towards soil degradation and contamination because of the intensive use of agri-chemicals, fertilizers and pesticides, salinization and deforestation. Misuse has led to soil degradation in arid, semiarid, subhumid, and dry regions. In the 1990s important advances were made in LAC towards institution-building for environmental management, the creation of a legal framework and specific legislation directed to natural resources and the limitation of polluting/contaminant emissions, and the implementation of tools like environmental impact assessments. Despite differences among countries, total environmental expenses (i.e., public and private) did not go beyond 1% of GDP, and rarely beyond 3% of total public expenses. The degree of deforestation is very large. Deforestation is mainly due to the conversion of forested lands to other uses, like agriculture, livestock production, urban expansion, road and railway construction, and mining. Other causes of deforestation, which are very important in some areas but are much less widespread than the ones referred to above are the harvest of firewood for either household or industrial use and the intensive exploitation of some particular tree species. Fires may also result in large forest losses. Conventional silvicultural approaches to forest management and use that do not take into account the complexity of the forest ecosystem, its multiple environmental services and its benefits for the communities inhabiting them still are the preferred ones in LAC. Nevertheless there currently is a trend in most of the countries in the region to prepare national forestry plans with the idea contributing to the sustainable development of a country. Eight countries in the region are classed as mega-diverse: Bolivia, Brazil, Colombia, Costa Rica, Ecuador, Mexico, Peru y the Bolivarian Republic of Venezuela. The conservation of biodiversity is considered to be extremely important agriculture and food security. A wide variety of plants and animals make the basis of agricultural biodiversity. However, just 14 mammal and avian species altogether make up 90% of the food from animal sources people eat. And only four plant species—wheat, corn, rice and potato—provide half of the energy humankind gets from plants. Latin America is the origin of many crops species relevant for human nutrition, like corn, beans, potato, sweet potato, tomato, cacao, cassava, peanuts and pineapple.	CEPAL, 2005b

Table 3-3. **continued.**

Variable	Current situation	Source
Food sustainability (continued)	In the last 100 years three quarters of agricultural crops' diversity has been lost; this represents a serious threat to both agriculture and food production. Forest cover (1990-2000). Increased: Uruguay and Cuba; invariable: the Dominican Republic and Chile; decreased: (in decreasing order) Guyana, Bolivia, Colombia, Peru, Brazil, the Bolivarian Republic of Venezuela, Paraguay, Costa Rica, Argentina, Trinidad and Tobago, Honduras, Mexico, Ecuador (reduction less than 10%), Jamaica, Panama, Guatemala, Nicaragua (reduction between 10% and 30%), El Salvador and Haití (reduction between 30% and 50%).	CEPAL, 2005b
Population and poverty	In LAC there are 432.8 million people, of which 24.2% is rural population. There were 170.7 million employed people in 2005. Most of the urban employed (93.9%) perform non-agricultural activities, and about three-fifths (58.8%) of the rural employed are engaged in agricultural activities.	CEPAL, 2005a
	The rural population has in general relatively decreased (as a fraction of total population) in most of the LAC countries along the decade 1990-2001, except in Costa Rica, Ecuador, Guatemala, Honduras, Nicaragua, Panama, Paraguay and Peru. During 1994-2000, urban poverty has decreased in most of the countries, except in Argentina, Colombia, Ecuador, Guatemala and Nicaragua. In this same period, rural poverty decreased or remained stable, except in Guatemala, Honduras, Nicaragua and Paraguay; in Peru rural poverty increased.	de Ferranti et al., 2005
	Advances in poverty reduction in LAC (1998-2005). Large (10%-20%): Ecuador, Mexico and the Bolivarian Republic of Venezuela; intermediate (5%-10%): Colombia and Honduras; small (1%-4.9%): %): Brazil, El Salvador and Chile; Increase in poverty: Argentina, Bolivia, Costa Rica, Panama, Peru and the Dominican Republic. Paraguay did not change its poverty level.	CEPAL, 2005a

See also competitivity indicators in the same tab.

Source: Authors' elaboration.

3.4.1.1.2 AKST systems

At the start of this period, the public research and development organizations define as priority technologies ones that permit: (1) an increase in agricultural productivity; (2) a reduction in production costs; (3) an improvement in the quality of agricultural products; (4) an increase in food security; (5) an improvement in the quality of processes in production chains; (6) an improvement in the income of agricultural producers; (7) an increase in competitiveness of production chains; (8) generation of exportable surpluses; (9) an improvement in the nutritional profile of the urban and rural populations; (10) environmental sustainability of agricultural systems; (11) development of mechanisms and conditions for the preferential production of farm goods and services with a high value added; and (12) an expansion of the portfolio of basic agricultural products, including nonfood products. This last priority makes it possible to create an important autonomy of nonrenewable energy sources by developing biofuels, such as ethanol, biodiesel, biogas, and the like, particularly in countries such as Brazil, Mexico and Argentina.

In terms of the social groups to which R&D is oriented, they include first and foremost conventional large and medium-sized producers, and extend to end consumers, agroindustry, and policymakers, and, finally, in last place, traders and merchants. Indigenous communities and subsistence farmers are not very relevant for R&D organizations.

The capacity to incorporate advances in formal knowledge into the creation of new technology varies in LAC. In most of the countries, there is a small capacity to generate such technology, and so efforts focus on the adaptation or import of technology, when possible. Argentina, Brazil, and Mexico have large investments in biotechnology which, together with equally large investments in nanotechnology, allow them to achieve some progress in applying these sciences to agriculture. Traditional knowledge is taken into account only in isolated initiatives.

Some LAC countries make an effort to set aside resources for public agricultural R&D. There are also resources available from many international sources linked to countries, communities of countries, and international institutions.

The private system is the largest investor in research for economically profitable production, and endeavors to expand its portfolio of products. In a few instances this effort is shared with the public sector.

In LAC countries with more institutionalized public R&D structures, work objectives are differentiated between the public and private sectors. This differentiation is driven by the economic profit of the investment of private companies in AKST, which is promoted by knowledge protection laws.

Most R&D systems work with the following agricultural products on a priority basis: grains, vegetables and spices, tropical fruits, and beef and fish products. Other

Table 3-4. **Brief description of the states of component variables in each scenario.**

Variables	Global Orchestration	Order from Strength	Life as it is	Adapting Mosaic	TechnoGarden
Barriers to international trade	Trade barriers are removed, but sanitary and phyto-sanitary barriers are retained	Trade barriers and subsidies proliferate, particularly those intended to prevent bioterrorism	Trade barriers and subsidies proliferate, particularly those intended to prevent bioterrorism	Trade barriers are set, together with environmentally friendly tariffs and subsidies	There are trade barriers at the outset, but by 2030 only sanitary and phyto/sanitary barriers are retained
Epidemics/food contamination	At the beginning of the period there is an increase in disease and pest outbreaks, whose frequency and intensity increase steadily through 2030, when their control becomes regional. High risk of food contamination	The frequency and intensity of diseases and pest outbreaks increase. Low risk of food contamination because of strict bromatological control of food to avoid bioterrorism attacks	The frequency and intensity of diseases and pest outbreaks increase. High risk of food contamination	The frequency and intensity of diseases and pest outbreaks increase at the beginning of the scenario, but they decrease towards the end of it (2030) Decreasing risk of food contamination	The frequency and intensity of diseases and pest outbreaks increase at the beginning of the scenario, but they decrease towards the end of it (2030) At this time previously unknown pests and diseases come into the fore. Decreasing risk of food contamination
Competitiveness of agricultural business	High. LAC countries are embedded in markets for basic and differentiated products	Low, due to slower development. LAC countries only compete in markets for basic products	High. There is an increased competition for embedding into markets for differentiated products	Low. The competitiveness of LAC countries slacks off. Local markets become more relevant than international ones	High competitiveness increases because production costs are decreased and differentiated products are preferentially manufactured
Demands from final consumers	Consumer demands become more diversified. There is a stronger request for information on the origin and quality of products	Rich countries demand diversified products; poor countries demand cheap products	Consumer demands become more diversified. There is a definite demand for cheaper products	Consumers preferentially demand local products manufactured with due care for their environmental impacts along the production chain and waste disposal	In general, consumer demands become increasingly more diversified.
Climate change	Mean temperature and the frequency of extreme events increase. Society is not fully aware of climate change impacts. By 2030 countries fully cooperate to implement global mitigation and adaptation programs.	Mean temperature and the frequency of extreme events increase. Society is not fully aware of climate change impacts. Countries do not show mitigation and adaptation capabilities.	Mean temperature and the frequency of extreme events increase. Society is not fully aware of climate change impacts. Countries show variable mitigation and adaptation capabilities	Mean temperature and the frequency of extreme events increase. Society becomes fully aware of climate change impacts. Countries increase their mitigation and adaptation capabilities	Mean temperature and the frequency of extreme events decrease. Society is fully aware of climate change impacts. Countries have well developed mitigation and adaptation capabilities

Table 3-4. **continued.**

Variables	Global Orchestration	Order from Strength	Life as it is	Adapting Mosaic	TechnoGarden
Governance	Governance reasonably improves but not uniformly across the region. By 2030 problems derived from the biophysical and social environments become serious.	There is a marked worsening of governance	Governance ranges from mediocre to bad	Governance progressively improves across the region until becoming optimal just in some countries by 2030.	Governance is optimal across the region by 2030
Policies for development	Policies are not even across LAC, but show a clear trend to becoming uniform	Widely divergent policies across LAC at the beginning, but they become more uniform by 2030, because of the pressure exerted by countries endowed with (comparatively) abundant resources	Widely divergent policies across LAC, but generally addressing biosecurity issues. Scarce resources are allocated to social policies	Policies are improved and made more consistent across LAC, with emphasis on the development of traditional knowledge and the conservation of the environment and biodiversity	Policies are improved and made more consistent across LAC, with focus education, traditional knowledge and the environment and biodiversity
Management of regulations and standards	It vastly improves throughout the region	There is an improvement in management because countries endowed with (comparatively) abundant resources press for it, progress is slow	It does not substantially change, because of lack of consistency across LAC countries	It fastly improves throughout, but not a the same pace across the region	It vastly improves throughout the region. Quality standards and certification processes become universal across the region
Education of productive system-actors	Strong public and private investments in education	Scarce public and private investments in education	Scarce investments in education at the beginning, but social demand for education makes the private sector to get involved into its improvement by 2030	Scarce investments in education at the beginning; however, resources are substantially increased by 2030	There is a remarkable increase in investments, particularly in private education. This even reaches the most vulnerable population
Social monitoring of innovation	In general, public in LAC has trust in the outcomes of innovation	There is some public distrust of innovation, because its stewardship is in the hands of social elites	The public sector leads in innovation but as it progressively becomes under funded, the space thus relinquished is taken up by the private sector	The social control of innovation becomes the norm, and the focus of research is mostly aimed to solve environmental problems.	There is a growing public trust on the outcomes of innovation

continued

Table 3-4. **continued.**

Variables	Global Orchestration	Order from Strength	Life as it is	Adapting Mosaic	TechnoGarden
Advances in knowledge	Large investments in R+D are made, particularly in the richer LAC countries	There is a growing gap in R+D activities between the richer and the poorer countries. LAC imports R+D products	Rich LAC-countries make large investments in R+D, which makes its development very uneven across the region. However, the region becomes a leader in some some fields of R+D	Investments in R+D are mostly directed to environmental sustainability and biodiversity conservation	R+D advances at great strides, but close to 2030 there is a growing social concern for the environmental impacts of many engineered production systems
Traditional/ indigenous knowledge	There are few advances, because this knowledge is not valued as such	Almost nil; it is not highly rated by governments, because they are wary of it	Slow advances. There is not much incorporation of it to formal knowledge	There is a growing acknowledgement of the epistemological value traditional/ indigenous knowledge and the consequent furtherance of its application	There is a growing acknowledgement of the epistemological value traditional/ indigenous knowledge and the consequent furtherance of its application
Focus of research	Improvement of the competitivity of agricultural products and the production of biomass for making biofuels. The needs of indigenous communities and subsistence farmers are not taken into account	Food innocuousness (biosecurity) and economic efficiency of agricultural production	At the beginning, it is food production and its economic efficiency. At the end (2030) the focus is on the most dynamic food production chains, particularly in the larger countries in LAC	For all social groups, the environmental sustainability of production systems, eco-labeling of foods, and mitigation and adaptation to climate change	For all social groups, the competitivity and environmental sustainability of production systems, their adaptation to climate change and the valuation of environmental and ecosystem services, and biodiversity.
Incorporation of formal knowledge	Some countries in LAC strive to keep their capacity for integrating knowledge into new technologies	The capacity to integrate knowledge is rather restricted, and shows a large variability across countries because it depends on national circumstances	The capacity to integrate knowledge is constrained by meager resources, and shows a large variability across countries because it depends on national circumstances	It is conditioned to the putative effects of the incorporated knowledge on the environment and biodiversity	Very intense across the region
Incorporation of traditional/ indigenous knowledge	Just isolated initiatives in this regard	None	Fortuitous	Growing	Growing

Table 3-4. **continued.**

Variables	Global Orchestration	Order from Strength	Life as it is	Adapting Mosaic	TechnoGarden
Availability of resources for R+D	They are irregularly distributed across the region. A large part of resources are obtained from international sources	Their amount is substantially reduced because of decreasing national investments in R+D, which are partly compensated by international sources	Their amount is substantially reduced because of decreasing national investments in R+D, but differing among countries because of national circumstances. Qualified people leave the R+D system.	They are substantially increased, but not enough. They are mostly channelled to R+D on environmental sustainability and biodiversity	There are enough across the whole region
Management of R+D	It becomes more complex and also better appreciated by society	There is a loss in management capacity	There is a loss in management capacity	It is much appreciated by society	It is much appreciated by society
Social involvement in the management of R+D	Growing participation	Scarce participation	Scarce participation	Large and very active participation	Growing participation
Performance of R+D systems	Systems are effective and focalized on the market	Systems are efficient but not relevant because protectionism impairs international trade	Systems are effective and focalized on the market	Systems are not very efficient, but they are effective in regard to the environment and biodiversity	Systems are highly effective and efficient: they are focalized on the environment and biodiversity
Relative spaces of public and private R+D	Public-private alliances are made with transnational corporations on strictly commercial terms	Transnational corporations perform R+D activities for profit. Public R+D provides input for private R+D activities and for satisfying social needs	Transnational corporations perform R+D activities for profit. Public R+D provides input for private R+D activities and for satisfying social needs	Public R+D institutions prevail over private R+D, but they collaborate with each other. R+D is strongly focused on environmental sustainability and biodiversity conservation	Public R+D institutions either collaborate through commercial alliances or compete with each other.
Proper technologies for agricultural activities	Technologies are aimed to intensified agriculture: they are not suitable for vulnerable productive systems	Technologies are aimed to the production of few common products. When specific technologies are needed, they are imported	Technologies are aimed to the production of few common products. When specific technologies are needed, they are imported	Technologies are aimed to satisfying the demands from productive systems and are closely adapted to local conditions	Social participation in technology development results in products very much adapted to user's needs
Incorporation of knowledge to productive systems	Is high; it is substantiated through inputs and practices	It is limited; it is substantiated through commercial enterprises	It is limited; it is substantiated through commercial enterprises	It is high and particularly focused on environmental protection and the development of local innovations	It is high, unevenly distributed across the region and mostly commercially oriented

continued

Table 3-4. **continued.**

Variables	Global Orchestration	Order from Strength	Life as it is	Adapting Mosaic	TechnoGarden
Attended markets	Large productive systems serve domestic and foreign markets. Most of the small productive systems remain isolated from those markets, except when they can gain niche markets	These are restricted. Exports are generally restricted and a few countries serve niche markets. The domestic market is little developed	These are restricted. Exports are generally restricted and a few countries serve niche markets. The domestic market is well developed	Mostly local markets. These are served with sustainably produced products of good nutritional value	Both domestic and foreign markets, with sustainably produced products of good nutritional value
Social organization of vulnerable production systems	Complete; it is real through production centers or cooperatives	Restricted; it is replaced by assistentialism	Restricted. It is replaced by assistentialism, but its materialization is encouraged by NGOs concerned with the environment, biodiversity and traditional/ indigenous knowledge	Complete and strongly localist, but restricted by scarcity of resources	It is connected to production centers and aimed to product qualification
Availability of resources for agriculture	Sufficient. Natural resources are easily got at; knowledge is increasingly available	Large in rich countries; resources are easily accessed in poor countries. There is some degradation of natural resources, and a restricted access to knowledge	Large in rich countries; resources are easily accessed in poor countries. There is some degradation of natural resources, and free access to knowledge	The use of natural resources is constrained by environmental concerns. There are scarce economic resources. There is free access to available resources and knowledge	The use of natural resources is constrained by environmental concerns. There are ample economic resources, and free access to available resources and knowledge
Performance of agricultural productive systems	Greater efficiency and production quality in big firms. Performance is highly variable among small systems: efficiency and production quality are low, and the most vulnerable emigrate	Better efficiency and production quality in big firms. Performance is highly variable among small systems: efficiency and production quality are low, and the most vulnerable emigrate	Better efficiency and production quality in big firms. Performance is highly variable among small systems: efficiency and production quality are low, and the most vulnerable emigrate Niches of agro-ecological production stay put	Productive processes and their products are more sound and friendly with the environment. Problems emerge in regard to the production of enough quantities of food.	Large efficiency and production quality is attained in all productive systems because they are fully integrated as supply chains
Rent inequality in agriculture	Increases, but unevenly across LAC	Increases because most investments are made by transnational firms and those are not of a social kind	Increases, but unevenly across LAC	Unevenly decreases; deracination of farmers increases inequality	Decreases across LAC

Table 3-4. **continued.**

Variables	Global Orchestration	Order from Strength	Life as it is	Adapting Mosaic	TechnoGarden
Social inequality	High. Most of the population does not have ample access to education, health and home. There is a statistical reduction in inequality brought about by internal migration from rural to urban areas	High. Most of the population does not have ample access to education, health and home.	High. Most of the population does not have ample access to education, health and home.	Small	Generally small, but large variability across LAC
Food security	Uneven across LAC, particularly in countries with few resources for ensuring food quality	Food offer is insufficient; low quality foods	Food offer is insufficient; low quality foods	High but food quality is sub-standard	High food security and food quality
Environmental sustainability	Low	Low; particularly in the poorer countries in LAC	Low	High and stable	High, but unstable

Source: Authors' elaboration.

countries focus on apiculture and development of other species of livestock, medicinal plants, and cosmetics.

The technologies generated by the public and private R&D systems are oriented more towards intensive agriculture, large and medium-sized agricultural producers and agroindustry. A few of these technologies incorporate aspects related to environmental protection and conservation, mainly in countries such as Brazil, Peru, Ecuador, and Mexico, with a high degree of biodiversity and threats to it or in countries that have semiarid or arid regions. The technologies generated do not take into account the most vulnerable social groups, such as small-scale producers, subsistence farmers or indigenous communities.

3.4.1.1.3 Agricultural production systems

Conditions are favorable to incorporate more know-how in agriculture, due to greater investments in education, the availability of resources for agricultural activities, and the openness of borders and markets, as well as support by companies themselves. Know-how is basically incorporated in two ways: one is by promoting new inputs to improve productivity; and the other is by implementing and verifying a series of practices designed to ensure compliance with quality standards.

The large production systems supply the external market with commodities, but they also provide differentiated products to a broad internal LAC market. A considerable proportion of small-scale producers become part of major production chains, such as the poultry chain, which is efficiently coordinated, even though it is highly fragmented. Other small-scale producers manage to participate in market openings in their own country or in wealthier countries. A vast majority of vulnerable producers and subsistence farmers, however, remain isolated.

The opening of markets and borders creates a good climate for investment in agriculture. Access to natural resources, such as water and soil, is not a problem except for the most vulnerable production systems. Access to knowledge increases.

Large agricultural corporations that apply modern production and management methods operate with great efficiency and have high-quality products and processes. Consequently, they are more competitive on markets. Small-scale producers that participate in major chains are also generally successful. The ones that participate more independently in market openings in some cases do not perform well. Efficiency is critical for them.

Nevertheless, a good part of the small production units leave the business, because they cannot meet quality requirements, such as traceability, safety, etc., imposed by marketing and consumer systems, due to the fact that technologies adapted to their conditions are relatively unavailable and to the effects of climate change, which, although incipient, are not depreciable.

3.4.1.1.4 Result of interaction between the systems

National and transnational companies consolidate their control over the supply chains and markets served. Some of the production units, with better ecological and economic conditions, manage to become organized within these chains and markets and thus improve their profitability.

For some countries, however, food imports compete with local food production systems, with a catastrophic effect on small- and medium-sized production units. Displaced producers abandon agricultural activity and shift to providing small, nonspecialized services, either in the same rural areas or in nearby urban settlements. All of this exacerbates inequality in agricultural income, but this varies among the different LAC countries.

By the end of the period, there is still a considerable degree of social inequality, which is seen in differences in access to employment, food security, education, and health on the part of various social groups, such as large agricultural producers, small family producers, agricultural wage-earners, and subsistence farmers. For some of the vulnerable groups at the start of the period—small family producers and wage-earners—the unequal access is considerably reduced. This result is a continuation of a trend initiated in the last decade of the 20th century, which was also strengthened by the more widespread prosperity of that period. The situation is also heterogeneous in the case of LAC countries. In a small number of countries, thanks to public policies and to the management capacity of food regulations and standards, the urban poor also regularly have access to adequate quantities of healthy food.

For countries highly dependent on food imports and with a more reduced per capita income, the prices of these products increase, creating urban food security problems.

In the less developed countries in the region where economic efficiency is low, environmental sustainability is not a concern for production systems, except in some highly local, traditional, or indigenous production systems. Deforestation, intensive use of fertilizers and herbicides, expansion of arable land into natural ecosystems, and the consequent loss of biodiversity and neglect of soil fertility and water quality continue. In a few countries there are plans to guarantee greater productivity with environmentally friendly technologies.

3.4.1.2 2016-2030

3.4.1.2.1 Context of AKST systems and agricultural production

Trade barriers are still absent, with the exception of health-related ones. The trend towards intense competition among countries increases during this period. The race to develop new agricultural products incorporating a high degree of technology is stepped up, so that commodities lose the relative importance they had in world trade. The vast majority of markets consume products with little value added, which are frequently synthetically created in laboratories or generated by microorganisms. In many cases, commodities are only the raw material used to obtain these products. Some commodities are the principal sustenance of a few LAC communities, which preserve their identity and rituals.

In addition to concerns over quality and safety of foods prevalent in the previous period, now consumers—virtually without exception, since the entire world population is more highly educated than at the beginning of the century—demand information on genetic manipulation and nanotechnological methods incorporated in the foods. Regulations on these matters and procedures for evaluation of foods or agricultural-based nonfood products begin to be implemented by governments.

Epidemics and epizooties increase in frequency and severity, as a result of the accumulated effects of the mismanagement of ecosystems, the introduction of new pests, the lack of action to adapt and mitigate the phenomena associated with climate change, and drastic changes in the pattern of land use and technology. The quality of export products is strictly monitored, as is that of foods sold in internal markets.

Climate change remains a concern, but shows signs of increasing, in temperature and in the frequency of extreme events. In LAC there is already a greater capacity to implement adaptation and mitigation measures, and this capacity grows throughout the period.

Transnational companies have increased power over technological development. Traditional innovation polices become inadequate, since the state is no longer the main promoter of R&D activities. Moreover, problems emerge related to social development (such as job losses as a result of constant technological modernization), the environment, and excessive control over the life of the common citizen by these companies, which require governments to make institutional innovations. The situation of global climate change also requires new and vigorous policies designed to protect the environment and modify agricultural production systems.

The governments of the most developed countries in the region allocate a large part of their fiscal resources to implementing an unemployment insurance system. These governments also offer incentives to corporations not to lay off their employees as a result of technological changes, but to retrain them instead to operate the new technologies. In 2025, governments establish a goal for gradual reduction of the work week within the next ten years.

Most of the countries in the region are in an acceptable situation from the standpoint of the their food quality standards and regulations and their enforcement. This is reflected in the reasonable efficiency of production systems, products, and services to meet the needs of their users. However, the systems are not necessarily environmentally sustainable, nor are the products, subproducts, and wastes in general, and this has a negative impact on the environment.

In general, stability and consistency among social, environmental, and foreign trade policies progress considerably for most LAC countries during this period, and various policies initiated in the course of the previous period also improve.

Education of stakeholders in production systems under the responsibility of the public education system ensures a critical mass of educated persons capable of meeting the objectives of international competition. Strategic alliances between both national and international companies and academic centers of excellence help improve the quality of public education at all levels.

The more developed countries make major advances in bio- and nanotechnology. In biotechnology, there is a much better understanding of the systemic impacts of the manipulation of genes in the second half of the period. This allows for greater efficiency in the use of these techniques and for a reduction in the negative effects on the environment. Biotechnology goes back to the technological base of genetic improvement processes, integrated into conventional processes. Nanotechnology for its part realizes its first successes with intelligent systems for monitoring crops and livestock, by using nano-electronic sensors based on DNA and other molecules. There is also integration of the two disciplines for development of environmental remediation systems, although these technologies do not develop fully. Biotechnology is also used successfully to develop plant biomass adapted to the needs of agroindustry, producers, and consumers. Moreover, other alternative forms of energy (wind, photovoltaic, hydrogen, etc.) begin to arrive on the markets. Some of them, which are more economical than biofuels, threaten to displace them from the market. These advances are made most often by the large transnational companies that export know-how to the less developed countries.

3.4.1.2.2 AKST systems

The division of labor between the public and private R&D sectors is expanded in the few countries that still have public research institutes. Public institutes focus primarily on a research agenda for the poor segments of rural producers and consumers.

For private transnationals that dominate R&D, research is centered primarily on all those technologies most directly geared to immediate application. These companies also maintain a portfolio of basic science projects oriented to new applications of biotechnology, nanotechnology, and their integration. Profitable applications based on knowledge generated by these initiatives are obtained with increasing speed, or in other words, the time between generating basic knowledge and its technological application is shortened.

Public AKST organizations still active in LAC rely increasingly on more basic knowledge generated by transnationals. In LAC, transnational companies also play the most important role in AKST. For this reason, there is no problem in incorporating advances in formal knowledge; actually, the process of obtaining advances in knowledge already has the incorporation of those advances built in, because these companies use the scientific skills of persons throughout the world.

The large companies do not save resources for AKST activities, because they need to continuously renew all the available technologies for the agricultural sector so that they will be in a better position to displace their competitors on the technology market.

Governments continue to perform the function of suppliers of financial resources for development of technologies for the poor. Transnationals also provide financing for this purpose, to enhance their corporate mage in public opinion.

There are practically no more spaces—except for marginal ones—for technological development by public organizations, which concentrate on basic and applied research. The public research that is done is directed to vulnerable social groups and "social" farm products such as rice, yucca, and beans.

R&D is highly successful in developing products that consumers throughout the world are eager to buy. These products are extremely varied, to satisfy all tastes. Consequently they form a large mass of constantly changing products, virtually on a daily basis.

The companies also develop technology for all the components of production chains, from producers of inputs up to distributors of processed products. Although these products are developed and produced efficiently, their effectiveness is more problematical, because markets and consumers constantly want consumer products to have new attributes. In other words, the effectiveness of a product is ephemeral.

The technologies developed are adapted to large companies that compete on markets for agriculture-based products (but not necessarily agricultural products in the traditional sense of the term). For traditional agricultural production systems, some low-intensity technologies are also developed; these technologies take into account their possible environmental impact and also serve to mitigate climate change or to adapt to it, or to do both.

3.4.1.2.3 Agricultural production systems

The process of incorporating knowledge into agriculture, initiated during the previous period, thus continues. This process occurs by incorporating new inputs into production systems or because of the need to comply with regulations or meet demands for quality. Its development is promoted by more favorable conditions for investment in education, greater availability of resources for agriculture, and more open markets and borders.

In many LAC countries, farm production is directed to external markets, especially those made up of countries with greater purchasing power and vigorous domestic markets.

A reasonable proportion of small agricultural producers manages to gain entry to markets, with the result that their improved education is reflected in improved production systems and competitive capacity. Many others, however, that do not achieve this comparative advantage of improved education are displaced from their rural work to the cities.

The countries in the region generally have adequate resources consistent with their size, economy, and intellectual and technological capacity. Transnationals are monopolies that govern the use of natural resources, such as water and fertile soils, for agricultural activity.

The large agriculture-based corporations experience trade competition similarly to transnationals that dominate the creation of agricultural technology, because they constantly need to produce new innovative products to satisfy their markets. The products are developed on an agricultural basis, but they have strong components of bio- and nanotechnology. They include, by way of example, fiber crops with thermodynamic properties, monitored by nanosystems, plants that synthesize HIV innoculations and microorganisms that remedy environmental contamination. These corporations use as inputs commodities produced on huge tracts of land with highly mechanized and automated techniques.

The large corporations frequently integrate all the processes for agricultural production and production of inputs, and other times they outsource them. They build highly competitive, more regionalized production chains that are dedicated to the integral production of specialized, differentiated products, to meet social demands for more cultural diversity and preservation of the identity of peoples. The performance of these corporations, in terms of efficiency and effectiveness, is very high, because increased trade competition requires them to make large investments to mitigate the risk of losing markets.

3.4.1.2.4 Results of interaction among the systems

The openness of markets and borders creates a climate for investment in agriculture. National and transnational companies consolidate their control over production chains and the markets they serve. More production units manage to operate in this sphere, thereby improving their income. Nonfood imports, the monopoly over natural resources, and an intensification of the effects of climate change drive small farmers out of the circuit. These factors all exacerbate income inequality. More resources are invested in education, however, which are used to a great extent to retrain a large part of the rural population of displaced producers as skilled workers for industry. Partly as a result of these policies, the proportion of poor in the Latin American population is considerably reduced.

In this situation of growth, various social groups have greater access to education, health, and food security, although major differences persist among LAC countries in terms of social and economic development. Access to jobs is still difficult for less skilled workers. Government intervenes to provide food, housing, and transportation for the unemployed. In societies in general, the value assigned to work changes, due to the development of a market geared to recreation and leisure activities.

There is a sharp drop in urban food security problems in LAC, even in countries with a lower per capita income that rely on food imports. There is virtually no urban food security problem in LAC, or in other words food is regular, accessible, and available in the cities. As for food safety, the main sources of contamination are controlled by sophisticated health surveillance mechanisms.

At the start of this period, the environmental sustainability of production systems becomes a priority for societies, and especially in the countries most vulnerable to environmental disasters related to climate change. In addition to threats to sustainability related to poor management of agricultural systems, they are also threatened now by the consequences of climate change. During this period, the environmental sustainability of agriculture is also affected by highly intensive competition among markets that demand more and more new products derived from exploitation of natural resources. The intensive agricultural practices reduce the elasticity of the response of many ecosystems, and lead to various problems in maintaining the efficiency of agricultural production systems over the long run.

3.4.2 Order from Strength

3.4.2.1 2007-2015

3.4.2.1.1 Context of AKST systems and agricultural production

International trade in agricultural products in the region is regulated by tariff and nontariff barriers. The latter ones are designed to reduce the risk of bioterrorism. The possibility of evolving towards a free trade system is remote.

The less developed countries have increasingly less capacity to invest in agricultural innovation. As a result, they are unable to compete on markets for differentiated agricultural products, and the best that they can do is to continue exporting commodities, in more and more difficult circumstances because of the barriers imposed.

Consumers in the more developed countries both within and outside the region are increasingly demanding in terms of quality, safety, functional properties, and environmentally friendly production methods for food and nonfood products. It is more and more difficult for less developed countries to satisfy these demands, but some of them serve special, high-value markets, such as markets for products from the jungles of the Amazon, Chaco Paraguayo, or the Bolivian salt desert, or from Patagonia—albeit on a limited scale. The domestic LAC markets are primarily made up of low-income consumers, who demand low-priced food.

Despite the massive use of pesticides throughout the region, the frequency, severity, and presence of new pests and diseases continue, and the situation in some countries is worsened by changes in land use patterns, climate changes, and the lack of remedial action.

The temperature and frequency of extreme climate events persist. Most countries in the region do not perceive the threat of climate change, and hence the need to direct agricultural R&D to that end. Social organizations that warn the public of the coming danger are not echoed by government authorities. There is also generally a low capacity to mitigate or adapt to climate change in most countries, because most of these countries give no priority to action in this area.

At the outset of the period, some LAC countries adopt highly diverse measures for technological innovation, social development, environmental protection and biosafety. By the end of the period, as a result of the relationship with developed countries outside the region and their dependence on external resources, most of the LAC countries adopt more coherent biosecurity policies based on protocols imported from more developed countries, the implementation of which is completely subsidized by these developed countries. However, as a rule, these policies are not seen as stable in most countries, and in highly import-dependent countries, this stability is very much weakened.

Management of these policies is also precarious, but due to the developed countries' concern over bioterrorism, from midway through the period onwards, a slow transition towards establishing regulations and quality standards and enforcing compliance with them begins, to reduce the risk of terrorist acts related to the food supply or agricultural products. Early in the period governments, and by the end of the period transnationals, take responsibility for managing health standards and antiterrorist measures. Transnational corporations are only capable of exercising this control in the major cities.

Public education does not lead to good results, especially in the less developed countries. Similarly, private education often offers defective, poor-quality courses and programs.

In more developed countries, there is a major social concern that science should provide ways to avoid any biological, physical, or chemical threat from less developed countries. The developed countries invest large amounts of private and public funds to develop new technologies (nanotechnology and biotechnology) to reduce this threat. In most of the less developed countries of LAC, due to the fact that many basic needs are not met and education levels are generally low, development of science is limited. These countries as a rule do not place value in traditional knowledge as a source of agricultural innovation.

3.4.2.1.2 AKST systems

In the few LAC countries that have the capacity for technological innovation, efforts and resources are channeled to biosafety. The larger countries, many of which are members of economic blocs, establish health barriers to food imports, but without repercussions on the focus of AKST. In view of the scarcity of economic resources in the region, R&D is mostly directed to ensuring food supplies and economic efficiency. The sustainability of products and processes and their environmental impact are not given priority by the public or private sector.

The capacity of the different LAC countries to incorporate advances in formal knowledge into agriculture varies. Some, such as Argentina, Brazil, and Mexico, even apply their advances in biotechnology and nanotechnology to agribusiness, while others are limited to adapting or importing technology. The few countries with the capacity to generate technologies do not incorporate traditional knowledge.

There is also a loss of personnel and management capacity in public R&D. Personnel migrates to other jobs either abroad or with transnationals. Public R&D institutions have difficulties in establishing policy lines, defining priorities, and especially coordinating the whole research effort. By the end of this period, there is a wide gap between the scientific and technological capacity of the LAC countries and that of the developed countries such as Japan, Germany, and the United States. Some countries in the region begin to import technology from the developed countries, to meet needs in some areas regarded as strategic. Because of a shortage of financial resources, most governments in the region reduce public investment in education, and in science and technology. There are financial resources to use for international support in solving problems, mainly related to biological security. The protocols, patents, and genes generated in these projects are the property of the donor organizations.

Throughout LAC, public R&D institutions give way to transnational companies. In some countries, they still perform the function of generating knowledge and technology in areas relevant to production, that private research institutions are not interested in. For instance, in the genetic improvement of corn, they develop pre-technological products, i.e., intermediate products in the crop development process, as an input for processing the final technological products (Castro et al, 2006). Public AKST organizations also take on the basic research that the private sector is not interested in doing.

Due to these many limitations, public R&D institutions are unable to develop technological products adapted to the demands of their customers and users, whether private transnational or national organizations. The most vulnerable social groups are not given any consideration at all in generating technologies.

3.4.2.1.3 Agricultural production systems

The lack of investment in education, the reduction in resources for agriculture, and the lack of openness of borders and markets lead to a situation that discourages incorporation of know-how into agriculture. Fragmented knowledge on use of inputs and machinery is incorporated on a limited scale, and only among the partners of enterprises, for the purpose of improving the productivity of production systems. Exporting firms and quality standard certification companies also require implementation and verification of a series of practices to meet market requirements, and the partners of the enterprises (medium-sized and small-scale producers) find themselves forced to incorporate complex know-how associated with these product and process standards.

Trade barriers limit agricultural markets for LAC countries. Few countries export commodities to countries with greater purchasing power, because the costs of product certification, as a prevention against any biological threat, are high. A small number of countries and organizations has an opportunity to participate in "Latin-American" or "Amazon" markets, which also demand safety guarantees for the products offered. By the end of this period, a very small specialized market begins to open up for products of traditional production systems.

The internal LAC market has two segments: (1) the segment of high-income consumers, which is constantly shrinking in size, due to the poor economic performance of countries, but which requires goods similar to those of consumers in more developed countries; and, (2) the segment of poor consumers, which is an expanding segment for which the most important factor is price. A considerable number of countries have only the segment of poorer consumers for its goods, and relies increasingly on imports (agricultural imports in general, but especially foodstuffs), to feed its people.

External markets, the high-income market, and part of the poor domestic market are supplied with products from large, technified production systems. The niche markets are supplied by small production systems that nonetheless have a high degree of biosecurity technology incorporated into them.

The poorest domestic markets are supplied by production systems with little technology incorporated, with no links to production chains, and with little concern for biosecurity. This means that a large part of the people in these countries consumes food of poor bromatological quality.

The stakeholders in the production systems are not generally organized into stable associations, and this leads to a diminished resource management capacity, a weak position on agricultural markets, and poor performance by production units.

In the more developed countries of the region, the economic losses of the more vulnerable production systems are offset by aid policies or by an insurance mechanism. As a rule, however, the most vulnerable systems—which the large agricultural corporations of some countries are not part of—do not have financial resources to protect themselves from risks related to epidemics or the impact of climate change, for instance. The large transnational companies that do their own R&D operate at high levels of efficiency and effectiveness. In other words, they produce with a high cost-benefit ratio, as demanded by their consumer markets, while family production systems are pushed towards increasingly less profitable agriculture.

3.4.2.1.4 Results of interaction among the systems

Income inequality rises, as a result of domination of agribusiness investment in LAC by the large transnational companies, and also because of the reduction of public investment in education, science, technology, and rural development. Only a small group of producers with better ecological and economic conditions materializes partnerships with these companies, while the vast majority of small production units are left out of the playing field.

There is a general deterioration in the capacity of countries to guarantee the sustainability of their agricultural production systems—especially the most vulnerable ones—and this is dramatically reflected in a reduction in access to jobs, housing, health, and education, and a decline in food security. Many unemployed rural workers and small bankrupt landowners move to the cities, where the generalized reduction in productive activities is also felt. Governments as a rule are not able to provide social protection to a large and growing poor population in urban settlements. In many cities, there is an atmosphere conducive to social protest and vandalism. Insecurity increases both in cities and in the rural areas. Along borders with more developed countries like the United States, and also with some Latin American countries with higher living standards, there is an increase in fatalities resulting from thwarted attempts to enter a world in which there are “greater opportunities”.

As for urban food security, the supply of food is inadequate and a fraction of it has a high contamination risk.

The status of climate change tends to be critical, because temperatures are rising, as is the frequency of extreme climatic phenomena. The primary cause of this situation has to do with the specific energy matrix of the more developed countries and also with massive imports of raw materials from poor countries, reflected in the growing exploration of their natural systems and in the exposure of their native forests. Environmental sustainability and adaptation to climate change are not concerns of governments, except in the more vulnerable countries, which are usually the less developed ones.

3.4.2.2 2016-2030

3.4.2.2.1 Context of AKST systems and agricultural production

Both the countries of the region and those outside LAC continue to use all types of barriers to agricultural trade, encounter difficulties in making their national production systems competitive, and face ongoing threats of bioterrorism. The LAC countries with a greater presence on agricultural markets have compulsory certification systems, exert strict control over the production process, and impose patterns of technology to manage epidemics and to guarantee the quality and safety of foods.

The markets are increasingly more sharply divided, with developed countries outside LAC dominant in trade competition and on world markets. Very few LAC countries are able to supply commodities to external markets. The less developed countries and the poorest ones have little access to these markets, so most of them turn to their domestic markets. These markets have a high percentage of low-income consumers, who are more interested in low prices than in food quality.

Management of agricultural pests and diseases relies mainly on the use of expensive, specialized external services and inputs. There is a reduced capacity in most LAC countries to implement preventive health measures or measures to contain diseases, and to adapt to and mitigate climate change. For these reasons, epidemics in the region increase.

In LAC there are even greater rise in temperatures than in the preceding period, and also more frequent and intensive extreme climate events. Their strong impact on the region is largely responsible for the highly reduced capacity to adapt to and mitigate climate change.

The situation of governance is highly varied in the region. In many countries, the general situation of survival is aggravated by corrupt politicians who have joined forces with groups that engage in illegal activities, and that frequently often offer one of the few opportunities for survival for many urban dwellers. In a few countries, there are governments that endeavor to follow consistent, sustainable policies, but these efforts are hampered by the shortage of economic resources. This is seen in the inability to intervene proactively to cope with various types of problems, such as social disaggregation, epidemics, natural disasters, and the like.

With resources becoming increasingly more scarce, most countries in the region experience enormous and mounting difficulties in ensuring social order and productive capacity and in guaranteeing the supply of essential services, such as health, unemployment insurance, education, housing credits, and the like. Laws on environmental protection, trade security, the protection of knowledge, and biosecurity, among others, remain unenforceable. The more developed countries feel threatened by this state of affairs, and create funds to alleviate the situation in the countries suffering most, by sending professionals, products (such as pharmaceuticals) and equipment to help these countries. This aid, which begins around 2022, ends when the period comes to a close.

Due to the deterioration of both economic resources and governance in LAC countries, their capacity to impose regulations and quality standards on food, which they had during the previous period, is diminished. Some of them make an effort to reverse this situation, but achieve meager results. Assistance by developed countries to recover that capacity is inadequate and limited in time.

The education of stakeholders of production systems in the public system of education does not generally produce good results. Private education is usually expensive and of mixed quality, because there are relatively few schools that offer a quality education.

Developed countries make enormous scientific progress. In the sphere of biotechnology, there is a sharp increase in the understanding of the systemic repercussions of gene manipulation. This leads to greater efficiency in the use of these techniques, which in developed countries is reflected in the decrease in negative effects on the environment. Biotechnology becomes the basis for genetic improvement projects; the use of conventional improvement systems moves to second place. Nanotechnology in turn is used successfully for the first time in intelligent systems for monitoring crops and livestock and food processing systems. These systems rely on the use of electronic nano-sensors based on the characterization of DNA, which are especially designed to detect threats to biosafety or biosecurity in raw materials or processed foods.

Nanotechnology is also used to develop systems for tracing origin and preservation of identity. These systems are sold to poor countries that want to export their raw materials to rich countries and so must comply with the identity preservation requirements for exports. This technology is also used to generate strict control protocols for biosecurity and biosafety in international transactions.

Biotechnology is also used to produce plant biomass adapted to the needs of agroindustry, producers, and consumers in LAC countries in a better economic situation. Moreover, other sources of energy cheaper than biofuels begin to be developed and threaten to take over their market share. These advances are realized in most cases by large transnationals that export their know-how to less developed countries.

3.4.2.2.2 AKST systems

Scientific activity, virtually abandoned in LAC countries, is left on its own. In many countries the scarce resources of the people encourage the formation of markets for traditional products. For instance, expensive medicines manufactured by international laboratories are replaced by active principals obtained directly from plant biodiversity. However, since there is no interaction between formal and traditional knowledge, the systematization of the latter and its incorporation in formal systems are reduced. The activity of generating know-how and technology is left to the developed countries outside LAC.

The capacity to incorporate advances in formal knowledge is in the hands of large transnational corporations, because there are actually no public or private research institutions or universities that perform this work effectively. At the outset of the period there is a fleeting attempt to incorporate traditional know-how into efforts to generate agricultural products.

R&D resources come from major transnational corporations, which tend to focus on their short-term interests and the needs of markets outside the region. There are virtually no other sources of funds to sustain sizeable investments in R&D. The focus of the large corporations is on the competitiveness of commodities and biosecurity protocols. These are produced with technologies generated in other countries, which are directly applied or adapted to the conditions of LAC and exported to wealthier countries outside the region.

Almost all the R&D produced by large corporations is directed to improving successful products, such as transgenic varieties, or to testing new products, to serve external and internal markets. For the R&D activities of these corporations, the countries in the region have a comparative advantage in that they can explore the environment without facing protests from environmentalist organizations, taxes are low, and there are generally few restrictions to such exploration. Locally important food crops, such as beans and yucca, are not the subjects of the R&D done by these corporations. However, the technologies generated by the corporations are not the best suited to the diverse needs of the countries of the region, either in terms of sustainable development, or their culture or production conditions.

3.4.2.2.3 Agricultural production systems

The slow economic growth of the region makes it much more difficult to incorporate know-how into agriculture, and especially as required for the most vulnerable production systems. Moreover, the large corporations no longer operate as organizations dedicated to a broad sector of activity, such as production of inputs, for example, but instead they operate as large, well-coordinated production chains, ranging from production to sale of these same inputs, including technology, and including the production and sale of agricultural products. Know-how is automatically incorporated into these chains as part of the whole process.

Production systems that do not participate in these chains do not have an adequate supply of technology to solve the problems of agricultural pests and diseases or to adapt to higher temperatures, nor do they have the resources to incorporate innovations when there are a few available.

The vast majority of LAC countries lose a great deal of their competitive capacity on external markets, due to the following factors:

- The rich countries become increasingly closed to guarantee the best markets to their own agricultural producers;
- The rapid change in the technological base of economic development, increasingly more dependent on expensive technologies, such as biotechnology and nanotechnology, information sciences, geomantics, and on their incorporation, which are not affordable for all countries of the region;
- The creation of new products with these technologies incorporated into them, that are not dependent on the use of commodities—the principal exports of LAC, which have experienced a sharp drop in international prices;
- The limited capacity of the region to maintain agriculture free of pollutants, diseases and pests.

Few LAC countries, especially the largest ones, sell their agricultural production on external markets. In all the LAC countries, the domestic market is an important target for agriculture. For most of the countries, that market is virtually the only market on which the large corporations participate as chains. Small-scale vulnerable producers supply the poor on local markets, or sustain themselves. It is increasingly more difficult for these producers to become part of production chains, due to their reduced capacity to satisfy certification and biosecurity and biosafety requirements.

In view of the ongoing poverty crisis and social and productive vulnerability, the stakeholders of vulnerable production systems are reliant on assistance to mitigate social and natural emergencies.

The financial resources available for agricultural production are channeled to economic and political power groups, and not to small-scale producers, which are generally family or traditional and indigenous farmers. The allocation of resources to agriculture tends to diminish during this period throughout the region, and especially in the poorest countries, as a result of poor governance.

Medium-sized production systems, which are dependent on government support, are efficient, but unable to meet market demand. Consequently, they frequently lose market shares to multinational production chains, which export their products to the region. The performance of these medium-sized systems deteriorates, as they need to reduce their production costs more and more to keep their market share.

3.4.2.2.4 Results of interaction among the systems

Because of a lack of proactive measures to mitigate the effects of climate change, extensive tracts of land are increasingly vulnerable to those effects, making investments more risky. Agribusiness stakeholders wage an aggressive competition to gain access to natural resources. Investments in agriculture are dominated by transnational companies, which in many cases receive support from governments. The result is a volatile land and water market and the consolidation of natural resources in a few hands. All of this leads to an increase in income inequality.

Public resources for education decrease, which increases the number of people without access to information and to collective organizations to defend their interests. This creates conditions that exacerbate income inequality and deepen social inequality. The income gap expands in some countries and remains stable in others, with an eventual improvement resulting from the delivery of resources, in the form of land titles for small farmers, for instance. In this way, an attempt is made to attenuate the heavy migration from rural areas to cities and other countries, which grew in the course of the previous period.

As a rule, for a growing number of persons, access to health, employment, education and food security becomes more difficult. A segment of persons employed by the major corporations is created, as compared with persons who work for national organizations, the government, or independently. The middle class loses its status, since it becomes more impoverished. The situation of social disaggregation, violence, and insecurity worsens considerably.

Although the bromatological quality of foods accessible to the urban poor is maintained by the standards of the previous period, the quantity of food for the poor in large urban centers decreases, mainly for the following reasons: (1) the number of urban poor is on the rise, as a result of the lack of opportunities and jobs; and (2) there is a strong internal migration from rural areas to cities. The wealthiest countries, even in LAC, institute drastic measures to contain this migratory movement.

The resilience[22] of ecosystems diminishes considerably, especially in poor countries. In these countries, natural resources are exploited virtually without restrictions. There is no capacity to adopt measures to recover degraded land or to mitigate and adapt to climate change, which is not a priority for the governments.

3.4.3 Life as it is

3.4.3.1 2007-2015

3.4.3.1.1 Context of the AKST systems and agricultural production

Trade barriers are used by developed countries as a mechanism to defend the competitiveness of their agricultural products. Minor victories in reducing barriers by agricultural commodity-producing countries are offset by new social or environmental barriers.

The LAC countries already established on commodity markets (Argentina, Brazil, Chile, Colombia, Ecuador, Mexico, etc.) try to gain access to more dynamic markets (United States, China, India) and the market for differentiated products. However, these countries still cannot compete on markets for differentiated agricultural products, because of their increasingly limited capacity to invest in technological innovation for agricultural production systems. These countries continue to export commodities and diversify the portfolio of products by including biofuels, such as alcohol and biodiesel.

Consumers in the richer countries both within and outside the region demand more and more quality, safety, functional properties, and environmentally friendly production methods for food and nonfood products, but they are not yet prepared to pay the cost associated with these demands. There are market openings for some differentiated products, such as products of the Amazon jungles, or the Chaco Paraguayo, or the salt desert of Bolivia, or from Patagonia. The internal LAC markets primarily consist of consumers with few resources, who demand low-priced foods and of niches for high-income consumers, with their demand for differentiated products.

In most of the region, there is an increase in either the frequency or the severity of agricultural diseases and pests, as a result of the lack of incentives to use good management practices in production systems and the lack of a national governmental structure with the capacity to implement regional cooperation to prevent and mitigate the impacts of new epidemics and losses in biodiversity.

In some parts of the region, there are huge changes in the pattern of land use, such as large tracts of monocultures of oleaginous crops and sugar cane for production of biofuels that lend themselves to the manifestation of new epidemics.

The temperature is rising at the rate of 0.22°C-0.24°C per decade, and the frequency of extreme phenomena is growing. There are relevant but highly variable effects on agriculture and the systems in the region, especially as a result of the frequency with which these phenomena affect

[22] Resilience is the capacity of a socio-environmental system to absorb disruptions, deal with changes, and still essentially maintain the same function and structure. Resilience depends on the variability and flexibility of the system (Carpenter et al., 2005).

each country, their economic consequences, and the equally variable capacity to mitigate and adapt to them.

Countries in the region with a more developed scientific research structure perceive the threat of climate change and thus the need for R&D in this area. But there are still financial and management limitations in obtaining results applicable to adaptation to and mitigation of the climate problem.

Some LAC countries adopt measures of technological innovation, social development, environmental protection and biosecurity, but due to political and budget limitations, achievements fall short of expectations. Changes in government generally lead to changes in management of public institutions, which frequently interrupt the continuity needed to obtain results. Either because of their own internal conviction (the case of countries more dependent on agribusiness) or because of their dependence on external resources, the countries of the region adopt more coherent biosecurity policies based on protocols imported from more developed countries, which fully subsidize implementation of such policies.

There is a slow transition towards implementation of food quality standards and regulations, and enforcement of them. Governments are initially responsible for management of health standards and anti-terrorism measures, but towards the end of the period, transnational companies are as well. During this period, transnational corporations are only interested in the most economical production chains, and this can lead to problems in the consumption of some types of foods produced by family farmers.

The education offered by the public school system, especially in the poorest countries, does not produce good results, even when governments give it high priority. Private education frequently offers defective and poor quality courses and teachers. There is strong social pressure to improve the educational structure of the region.

While rich countries make major investments in basic science to develop new technologies, such as biotechnology, nanotechnology and information science, Argentina, Brazil, Chile, Colombia, and Mexico make limited investments, and the other countries very limited ones. Consequently, the region moves further away from pioneering scientific development, capable of sustaining important advances in production technologies for agricultural systems and agriculture, and from the development efforts needed for product differentiation and an improvement in the competitive capacity of countries.

Few people recognize the value of traditional knowledge in LAC. It is appreciated by NGOs that advocate environmental sustainability and social inclusion, and also by a few large private companies that are interested in this knowledge to create new products, such as pharmaceuticals or plant-based insecticides, intensively used by agricultural production systems.

3.4.3.1.2 AKST systems

As a result of scarce economic resources in LAC and the social problems of its population, R&D for the most part goes to ensure the food supply and economic efficiency, with priority given to increasing productivity in agriculture. Environmental sustainability, differentiation, and product quality are not priority items in the public or private sector, but instead are issues addressed by personal initiatives in R&D organizations.

The capacity to incorporate advances in formal knowledge into agriculture varies widely among the different LAC countries. Some, such as Argentina, Brazil, and Mexico, even apply their limited advances in biotechnology and nanotechnology to more dynamic production chains in agribusiness. The poorer countries, with limited R&D resources and infrastructure, are confined to adapting or importing technology. The few countries with the capacity to generate technologies incorporate little traditional knowledge during this period.

Public R&D organizations have problems establishing lines of action, defining priorities, and especially coordinating the entire research effort. There is also a loss of personnel and technical and management capacity in the public R&D system, in some cases because of the retirement of professionals, and in others due to a shift to other more remunerative jobs.

As a result of limited public and private investment in research and the priorities set by R&D institutions, at the end of this period there is a wide gap between the scientific and technological capacity of LAC countries and that of developed countries such as Japan, Germany, and the United States, and also among the countries in the region themselves. For some areas of application regarded as strategic, a few countries in the region begin to import technology from rich countries, which leads to renewed interest in LAC in renovating existing public R&D structures or creating new ones.

The situation in the different countries in the region continues to be widely disparate. Brazil, Mexico, and Argentina, traditional exporters of agricultural commodities, invest more public and private monies in R&D than the other countries. However, these regional investments continue to be proportionally lower than those of other regions of the world, except for Africa. In certain export production chains and in countries where they exist and where laws to protect innovation are in force, an increase in private investment in research is observed.

Due to the scarcity of financial resources and the competition for them with other areas such as health and security, most governments of the region reduce public investment in science, technology, and education. There are financial resources to use for international support in solving problems related primarily to environmental sustainability, social inclusion, and biosecurity.

In LAC countries without relatively institutionalized public AKST structures, there are technology transfer and adaptive research programs in operation. In countries with more institutionalized public AKST structures, competition over work spaces is triggered between the public and private sectors, principally in relation to generation of technology to make production chains more dynamic. This competition between public and private institutions is driven by the economic return on AKST investment, as a result of knowledge protection laws.

In commodity-exporting countries in the region, the technologies generated by public and private AKST systems are oriented more toward intensive agriculture for export, large- and medium-sized agricultural producers, agroindustry, and input suppliers.

As a result of strong pressure by international and national public opinion, in countries with fragile, threatened ecosystems, such as the Amazon, or with semiarid or arid zones, as found in Brazil, Peru, Ecuador, and Mexico, research programs include aspects related to protection and conservation of the environment. The technologies generated are therefore adapted to these conditions, but few take into consideration the most vulnerable social groups, including peasants, subsistence farmers, or indigenous communities.

3.4.3.1.3 Agricultural production systems

The limited openness of borders and markets associated with social control of certain technologies, such as transgenic technology, creates a situation that works against incorporating knowledge into certain agricultural activities. Agricultural enterprises increasingly incorporate fragmented knowledge on use of inputs and machinery to improve the efficiency of production systems, generally by reducing costs.

Export and product origin and quality certification companies also require the application and verification of a series of quality attributes to meet market requirements. Producers are required to include complex know-how associated with product and process standards.

On the internal LAC market, there are two segments: (1) high-income consumers, a small segment but one that demands quality goods similar to those of consumers in richer countries; and (2) a large segment of poor consumers, for whom the most important factor is price. A considerable number of countries only have the poor consumer segment for their goods, and increasingly need more agricultural imports in general, but especially food, because they are unable to meet the growing demand of their population.

Commodity production systems consist primarily of large capitalist corporations that produce for the external market and for mass domestic consumption. A considerable proportion of small-scale producers are linked to large production chains, such as the ones that participate in the poultry chain which is highly fragmented but efficiently coordinated. Others manage to find market niches for products with a high value added, either on domestic markets or on markets in wealthier countries.

The problems of inclusion of farmers displaced by production chains, and without access to factor markets (land, water, and other inputs) and product markets, persist. Conflicts over development models and among organized social groups, the absence of public policies, and the shortage of resources constrain efforts to plan and implement programs geared to these social segments.

Limited openings in markets and borders and a short supply of public resources work against a healthy climate for investment in agriculture, although this is the economic sector that contributes the most to the economies of the countries of the region.

Investment of resources in agribusiness fluctuates on the basis of the prices of export commodities, which go from boom to crisis situations based on price variations. Agribusiness is still the main source of income for many LAC countries, however.

Since the main economic activity in the region is the competitive production of commodities for the international market, production systems focus on increasing their productive efficiency on the basis of comparatively lower production costs. To achieve this objective, the major corporations frequently take over and integrate all agricultural production, agroindustrial, and input production processes. Highly competitive national and multinational production chains are strengthened, for products such as soybeans and sugarcane, driven by the demand for biofuels.

Efforts to develop systems to produce specialized and differentiated products, to meet social demands for higher quality products, are timid. There is a moderate increase in organic production systems, although it is limited by the lack of an efficient certification structure. Product differentiation is restricted by the lack of a structure and R&D capacity in technologies for processing agricultural products.

3.4.3.1.4 Results of interaction among the systems

Continued production of commodities for the external and internal market prolongs income inequality, caused by competition to reduce production costs. Thus, small-scale producers are prevented from participating in the most dynamic sector of agribusiness. Inequality persists because of a reduction in public investment in education, science and technology, and rural development.

Social inclusion and agrarian reform programs are not successful in raising the income of most peasants and small farmers, due to widespread social conflicts and management and continuity problems. Only a small group of producers in the best ecological and economic conditions improve their income profiles, because they form partnerships with companies that are in production chains or manage to produce for market niches for differentiated products with a high value added.

There is still a considerable degree of social inequality at the end of this period, which is expressed in differences in the access to employment, food security, education, and health, by various social groups, including large producers, small family producers, agricultural wage-earners, and subsistence farmers.

The effects of climate change, the intensification of pests and diseases associated with them, and the shortfall in financial resources contribute to a slight increase in social inequality that prevails to the end of the period. This is the general situation in LAC, but in a few countries improvements are beginning to be seen, as a result of changes in and more stable development policies.

Food security problems in the region are much more the result of demand problems caused by consumers whose economic resources do not allow them access to the market, than due to the food supply. The region has the capacity to produce sufficient quantities to supply its domestic markets and also to create an exportable surplus, especially in agricultural commodity-exporting countries, such as Brazil, Argentina, Mexico, and Colombia. For the low per capita income countries that are highly dependent on food imports, the prices of these products increase, causing urban food security problems.

Production of export commodities is generally based on the use of environmental factors, such as water and soil, and on biodiversity. There is strong pressure by organized social groups to protect the environment, but resources available to implement effect protective measures are inadequate. Private enterprises, and mainly producers of export commodities, refuse to include environmental preservation costs in their production costs. In both the poorest countries in the region and in peasant production, where economic efficiency is low, environmental sustainability is generally font a concern for production systems, except in some traditional or indigenous cultures. Deforestation continues, as does the intensive use of fertilizers and herbicides and the expansion of arable land, as a result of incentives to produce biofuels.

3.4.3.2 2016-2030

3.4.3.2.1 Context of AKST systems and agricultural production

After a long period of negotiations in the World Trade Organization, developed countries begin to reduce trade barriers previously used as a defense mechanism against the competition of agricultural products. Agricultural commodity-producing countries have to neutralize environmental barriers imposed out of fear of harmful environmental and climatic effects resulting from the expansion of land planted to grain crops and energy products.

The LAC countries already established on commodity markets, i.e., Argentina, Brazil, Chile, Colombia, Ecuador, and Mexico, among others, manage to gain access to the most dynamic markets (United States, China, India), and, on a smaller scale, to the market for differentiated products. The economic results obtained allow these countries to increase their capacity to invest in technological innovation for agricultural production systems and thus to compete on some differentiated agricultural product markets. These countries continue to export commodities in addition to a portfolio of bioenergy products such as alcohol and biodiesel.

Consumers in wealthier countries both within and outside the region gradually demand safer and higher quality food and nonfood products that are also have functional properties and are produced according to environmentally friendly production methods, and they are willing to pay the cost associated with this demand. Internal LAC markets are composed mostly of low-income consumers, who want low-priced food, and of a middle class capable of demanding differentiated and healthy products at higher prices. Niches for high-income consumers with differentiated demands increase.

In most of the region, an increase in the frequency or severity of pests and diseases, seen in the previous period and aggravated by rising temperatures, leads to improvements in the development and use of best practices for management of production systems, and to improvements in the national governmental structure for preventing and mitigating the impact of new pests or diseases, or even epidemics, both on a domestic level and through regional cooperation.

Major changes in the pattern of land use—for example, large tracts of land planted to a single oleaginous crop or sugarcane for production of biofuels—lead to the appearance of new pests and diseases, which in turn result in the creation of public policies and research plans to mitigate the effects of these pests and diseases. Similarly, government have planned adaptation policies in regions already highly affected by early manifestations of climate change, such as floods, droughts, heat waves, and the like, and these policies create an environment that is conducive to the proliferation of epidemics. Thus progress is made in dealing with the coexistence of agricultural production and epidemics in the region.

The temperature rises at the rate of 0.22°C-0.24°C every ten years and the frequency of extreme events increases. This has important but disparate effects on agriculture and production systems in the region, mainly due to the equally disparate capacity of countries to adapt to or mitigate these effects. At the same time, many countries expand their capacity to live with these phenomena.

The countries of the region that have a more developed research structure apply the results obtained from public policies designed to mitigate the impact of climate change, to guide agricultural development. Financial and management limitations still affect the ability to obtain results that can be used for adaptation to or mitigation of the climate problem, mainly in the poorest countries in the region.

Many LAC countries adopt measures of technological innovation, social development, environmental protection, and biosecurity, but in some countries political and budget restrictions cause the results to fall short of expectations. Democratic changes in government usually lead to management changes in public institutions, which in turn disrupt the continuity needed to obtain valid results. As a result of the creation of an environmental conscience, the countries of the region implement more coherent biosecurity and environmental protection policies based on both domestic protocols and protocols imported from rich countries, which subsidize all of part of the relevant implementation costs.

The transition to establishing regulations and quality standards for food or agricultural products and their enforcement, initiated in the previous period, continues. Governments, working in partnership with transnationals producing agricultural inputs and major stakeholders in the wholesale and retail trades, are responsible for management of health and biosecurity standards. Governments take on the task of supervising and assisting family-based agricultural units, with encouraging results.

Strong social pressure to improve the structure of education in the region has a positive impact on the quality of public education, especially in the poorest countries, which obtain good results. Private education improves as well.

While developed countries far from the region make major investments in basic science to develop new technologies, such as biotechnology, nanotechnology and information science, the LAC countries also boost both investment in basic science and transfers of know-how from developed countries. Consequently, in some countries of the region and in certain fields of research, there is pioneering scientific development, that enables them to acquire the capacity to make important progress in production technologies for agricultural systems, agriculture, and product differentiation, and in improving their competitiveness.

In LAC, NGOs that defend environmental sustainability and social inclusion, large private companies, and public R&D institutions recognize to varying degrees the value of traditional knowledge, which they seek for use in creating new products (such as pharmaceuticals or plant-based insecticides), cosmetics, and nutraceuticals.

3.4.3.2.2 AKST systems

R&D resources remain scarce in LAC. As a rule, R&D is largely channeled to ensuring food supplies and economic efficiency. Priority is given to increasing productivity in agriculture or reducing production costs or both, in order to ensure that the commodities produced are competitive. In the larger countries of the region, environmental sustainability, differentiation, and the quality of products are on the public sector's research agenda.

The different LAC countries still have varying capacities to incorporate the advances in formal knowledge into agriculture. Some, like Argentina, Brazil, and Mexico, even apply their advances in biotechnology and nanotechnology to the most dynamic agribusiness production chains. Poorer countries, with limited resources and R&D infrastructure, are confined to adapting or importing technology. Countries with the capacity to generate technologies incorporate traditional knowledge in this creative process.

Public R&D organizations in the countries of the region with a long tradition in scientific research are better able to manage strategic R&D tools, because they coordinate the research effort. In these countries, a new generation of researchers replaced the former one and there was an increase in technical and management capacity in the public R&D system. By the end of this period, the gap in scientific and technological capacity existing among the LAC countries and between them and developed countries, such as Japan, Germany and the United States, is narrowed.

Brazil, Mexico, and Argentina invest more public and private resources in R&D than the other countries, but investments in the region are proportionally lower than in the other regions of the world. In specific export production chains and in countries with legislation to protect innovation, private investment in research is comparable to public investment.

Despite the persistence of scarce financial resources and competition with other areas of government, such as health and security, governments of the region gradually increase public investment in science, technology, and education. There are financial resources for international assistance to help solve problems related primarily to environmental sustainability, social inclusion, and biosecurity.

In the few LAC countries that do not have more institutionalized public AKST structures, there are technology transfer and adaptive research programs. In countries with more institutionalized AKST public structures, there is competition for work space between the public and private sectors. This is focused mainly on generating the technology for more dynamic production chains. This competition between public and private organizations is driven by the economic return on investment in AKST, derived from knowledge protection legislation.

In the region's commodity exporting countries, the technologies generated by public and private AKST systems are oriented more to intensive export agriculture, large and medium-sized agricultural producers, agroindustry, and input suppliers. There are programs directed to adding value to family agricultural production and developing differentiated products.

Due to continued pressure by world public opinion, in all countries, and especially in those with fragile, threatened ecosystems like the Amazon, or with water-stressed areas, such as the semiarid or arid regions found in Brazil, Peru, Ecuador and Mexico, research programs on environmental protection and conservation and on recovery of formerly degraded areas are developed. The technologies generated are therefore adapted to these conditions and take into consideration the most vulnerable social groups, such as peasants, subsistence farmers or indigenous communities.

3.4.3.2.3 Agricultural production systems

Fluctuating economic growth affects the region's production chains differently. Large corporations form extensive, well-coordinated production chains, which incorporate everything from the production and sale of inputs, including technology, to the production and sale of the end products. Know-how is automatically incorporated into them, as part of the process. Competition on the international market is the determining factor for including innovation in these chains.

The most vulnerable production systems that do not participate in these chains seek diverse sources of technology to solve efficiency and quality problems, which is critical to gain market access. There are public credit resources for incorporating any innovations that are available. Throughout the region, commodity-producing systems made up of large capitalist companies are established to produce for the external market and for mass domestic consumption.

A considerable proportion of small commercial producers is linked to large production chains, such as the ones that participate in the highly fragmented but efficiently coordinated poultry chain. Others are able to participate in market niches, producing products with a high value added in their own country or in wealthier countries.

Many of the problems related to inclusion of farmers displaced by production chains, without access to factor markets (water, land, and other inputs) and product markets, are solved by persistent efforts and an improvement in public policy results. More open markets and borders and greater availability of public resources lead to an increase in investment in agriculture, the economic sector that contributes the most to the economies of the region.

Investment in agribusiness still fluctuates on the basis of export commodity prices, but the fluctuations smooth out due to better coordination between stocks, production management and commodity prices. Agribusiness gains strength as the primary source of income for most LAC countries.

In the performance of productive systems, the focus is on increasing productive efficiency, based on increases in productivity and lower production costs. The large corporations integrate all the agricultural productive processes, agroindustrial processes, production of inputs and the wholesale trade, leaving to third parties only the retail trade. Highly competitive and increasingly national and multinational production chains are strengthened, driven by the demand for biofuels, such as biodiesel from soybeans and African palm and ethanol from sugarcane. Productive chains for meat and fruit become part of the economic portfolio of the region.

Efforts to develop production systems for specialized and differentiated products are stepped up to meet the demand for high quality products. There is a sharp increase in organic production systems, stimulated by implementation of a certification structure. Product differentiation begins to produce results based on the growth of an R&D structure with the capacity to develop technologies for processing agriculture products.

3.4.3.2.4 Results of interaction among the systems

Despite the fact that the consolidation of commodity production for external and internal markets exacerbates income inequality, by hampering the participation of small-scale producers in the most dynamic sector of agribusiness, social inclusion programs and research on family agriculture and agrarian reform lead to an increase in the income of many segments of peasant farmers. In addition, an expanding group of producers forms partnerships with companies inserted in productive chains or produces differentiated products with a high value added for market niches, and so it manages to improve its income profile.

During this period, there is a considerable improvement in access by the people to health, employment, education, and food security in most of the countries. However, social exclusion and lack of access to basic services are still prevalent in many countries.

When problems of food security do occur in the region, they are caused by pests, diseases, epidemics, and climate and environmental disasters. However, the region generally, and especially commodity-exporting countries like Brazil, Argentina, Mexico and Colombia, have the capacity to produce for both their national markets and to generate exportable surpluses. All of the countries still experience urban food security problems because of a lack of access to the food market. For countries with a low per capita income that still rely on food imports, the prices of these imports increase, causing food security problems.

Organized social groups continue to exert strong pressure for measures to protect the environment, and they receive international funds to implement effective measures to this end.

Private enterprises, and mainly export commodity producers, partially incorporate environmental conservation costs in their production cost, because they share this environmental cost with the national government.

In the poorer countries of the region, and in peasant production, an improvement in economic efficiency, outside resources, and technical and management assistance include environmental sustainability as an objective of production systems. As a result, deforestation diminishes, the use of fertilizers and pesticides improves, and use of arable land for large-scale production of biofuels stabilizes.

3.4.4 Adapting mosaic

3.4.4.1 2007-2015

3.4.4.1.1 Context of AKST systems and agricultural production

The concern over climate change and environmental sustainability is reflected in changes in various policies and regulations in some LAC countries in the early part of the second decade of the millennium, and in countries with better governance capacity.

Initially, changes in regulations affect trade among countries, including LAC, through a curious combination of barriers: on the one hand, nontariff barriers hinder agricultural imports of doubtful environmental and social sustainability; and, on the other, subsidies are granted for agricultural products with environmentally friendly characteristics.

Barriers hamper trade among countries. Moreover, as regards external markets, the LAC countries see their competitiveness in agribusiness weaken on some markets, and especially the European ones, that require guarantees on the environmental sustainability of the production process. New and differentiated products are not demanded by the "new consumers."

Agricultural production declines in many countries, due to climate effects. Social movements in LAC in favor of greater environmental sustainability also favor consideration of ecosystems and strict development rules in each country. All of these factors further reduce the productive capacity of agriculture, and leads it to focus more on the domestic market, and especially local markets. Thus, external markets are no longer the target of agricultural products for many countries.

Climate change contributes to the sharp rise in epidemics and the emergence of new pests, leading to considerable losses of human and animal lives and a substantial decline in crops. These losses are scattered unevenly across LAC, and also affect countries that contribute only slightly (in terms of CO_2 emissions, for instance) to global warming and the severity of extreme events.

This scenario begins to take shape following major temperature increases in various regions of the world, and extreme weather events of an unprecedented intensity are observed by the end of 2010. Countries prove incapable of dealing with the crises triggered by these changes.

Governance ranges from mediocre to acceptable in the countries of the region. The profound institutional innovation required takes place under the pressure of strong mobilization of different social groups, which force governments to share all of their decisions and action with these groups.

Following the global trend, some LAC countries begin to modify their policies to create more sustainable systems, based on lessons learned from the relationship between socioeconomic and environmental systems. Some of the larger countries of the region, such as Brazil, Mexico, Argentina, Peru, and Colombia, are very much affected, however, since some of their ecosystems and people have been subjected to extreme conditions for a long time, a situation aggravated by climate change. For the first three countries, it is difficult to make the transition to a new paradigm, since they have commodity export-oriented economies and agriculture. For poorer or smaller countries, where agriculture already concentrates on products for local markets or niches, such as Costa Rica for ornamentals, and Bolivia for quinoa, this transition is easier.

Agricultural development policies are designed to facilitate a change in the productive paradigm through specific R&D activities and the transfer and dissemination of the necessary traditional and conventional know-how and technologies.

Policies for the integral development of biofuels and other renewable sources of energy are established within a framework of environmental sustainability. Similarly, laws are adopted to encourage agroecological-based agricultural production systems, and fees or taxes are imposed to limit agricultural operations that use large tracts of land or single crop techniques. In the middle of the period, policies to facilitate access to land for small landless producers are defined, as a way of minimizing the impact of the climate on this vulnerable social group. Various conditions are facilitated, and credit, technical assistance, training, and the like are provided, so that they can produce at least the basic food they need to survive in such adverse circumstances.

Education is a key element for making the institutional changes needed by this new society. By about 2010, most LAC countries invest on average 13% of their GDP in education.

At the outset of this period, many countries see the emergence of groups of scientists who advocate more a systemic approach to agriculture. In their view, for instance, research on the biosecurity of transgenics should take into account the possible systemic repercussions of genetic manipulation on the cell and the environment. These groups argue that agriculture needs to use more environmentally friendly practices.

Advances in scientific knowledge, including biology and nanotechnology, continue. Major investments are also made in R&D on the environment and its effects on agriculture. Research in this field provides the technological basis for certification of environmental protection for agricultural products.

To reduce the risk of new environmental disasters, various international organizations, including the World Bank, UN, UNESCO and WHO, step up efforts to organize and empower traditional communities around the world.

The knowledge of these communities begins to be more highly valued. Numerous initiatives for environmental protection and for certification of the environmental safety of products and production processes are proposed. In many Latin American countries, there are numerous initiatives to systematize traditional knowledge and elucidate its principles.

3.4.4.1.2 AKST systems

One of the demands for R&D is development or improvement of agricultural processes such as the following ones: (1) biological control of pests and diseases; (2) control of the application of nutrients and residues to soils in the productive system; (3) elimination or reduction of agricultural and agroindustrial residues or waste; (4) identification and use of natural sources of soil fertilizer; (5) supervision of safety and quality in processing foods; and (6) generation of productive processes with a lower environmental impact. Processes for increasing productivity continue to be given importance, but environmental aspects are also prioritized now. The following topics linked to the environment and ecosystems are now considered as priorities: (1) on-site prospecting and conservation of germplasm; (2) economic valuation of biodiversity and natural resources; (3) sustainable economic exploitation of biodiversity; (4) traditional knowledge of biodiversity; (5) management of fishing resources; (6) conservation-oriented agriculture, management, and zoning; (7) management of water quality and use; and (8) management of forest resources.

The existence of barriers promotes research on certification of origin systems and ecological labeling of foods. A large part of R&D is channeled to research on adaptation to climate change. In defining R&D priorities, consideration must be given first and foremost to the need to adapt to and mitigate climate change and environmental sustainability.

Added to the social groups that have been the traditional users of agriculture research are now small-scale producers, subsistence farmers, and indigenous communities, as high priority groups for R&D.

In LAC, all countries are interested in and share efforts to ensure that R&D is used to offer responses to the demands of these social groups. However, only a few countries have the infrastructure, trained scientists, and financial resources to achieve advances in this area. Incorporation of know-how is partly limited by these resources. It occurs only after an evaluation of its potential repercussions on socioeconomic and environmental systems. Everyone involved in scientific work makes major efforts to incorporate traditional knowledge into formal AKST systems, while guaranteeing the rights of traditional/indigenous communities.

In some LAC countries there are sufficient but not optimum R&D resources. In allocating these resources, priority is given to major environmental protection objectives, sustainable agricultural practices, and the safety of the consumer. These resources are for the most part national government monies or social funds, but a small portion comes from regional sources.

Strict biosecurity protocols are defined for research in biotechnology and nanotechnology. Research in these sciences is uninterrupted, but progresses slowly.

R&D management is important, so that it is channeled correctly to meet environmental protection objectives. Various social groups gain full participation in the integral process of agricultural R&D.

R&D is concentrated in research institutes and public universities, which work in a highly participatory way with users and other organizations interested in R&D and its social repercussions. Private firms cooperate to some extent with these organizations, but their sphere of action is more restricted by laws limiting their concentration (i.e. to prevent a few firms from controlling the entire market). They are oriented more towards solving problems related to productivity and reduction of production costs in productive systems and their environmental externalities. Towards the end of 2015, the vast majority of private R&D firms become aware of the existence of important environment-related markets that are worth exploiting.

In a situation of scarce resources, R&D endeavor to achieve efficiency in their use. Yet effectiveness is more important than efficiency. In other words, the emphasis is on R&D products and how well they adapt to the need for a reduced environmental impact, and only secondarily on optimization of the use of financial resources to obtain them. In the beginning of the period, few technologies are available for the wide range of R&D users. By the end of the period, capacity increases, as does the understanding of the needs of these users. There is also an increase in the stock of different technologies available and adapted to different users.

By the end of the period, after several years of effort, agricultural technologies that are better suited to different production systems, crops, and social, economic, and ecological conditions are developed.

3.4.4.1.3 Productive agricultural systems

Policies that emphasize local sustainable development require a considerable input of agroecological knowledge, as well as the parallel development of diverse theories on the valuation of natural resources and environmental services, as an integral part of the methodology needed to estimate the economic efficiency of the new productive systems. These policies also require a high degree of social mobilization in order to be accepted. This makes the relevant technological innovation processes highly dynamic.

Networks of advisory services are established, including public or private NGOs, for multifunctional and sustainable management of production systems, dissemination of technology, and facilitation of access to resources on the part of agricultural production systems and especially the most vulnerable ones.

The local markets served are very limited in volume and global scope. In reality, the countries of the region have imposed reciprocal trade obstacles. Agricultural production chains are encouraged to incorporate the more vulnerable productive systems and to support them in this effort. These chains also become more limited in their geographical scope, and this facilitates the insertion of small-scale producers. The participants in these chains work to improve the productive processes and products, always with the environment as the reference point.

The pursuit of environmental sustainability as a priority objective has a strong effect on access to productive resources, for the following reasons: (1) it notably restricts the use of natural resources, such as fresh water sources, for instance; (2) it makes it easier to obtain development credits, in order to facilitate in turn the purchase of land by farmers; and (3) it demands an enormous effort to provide basic training in cultural, scientific, and technological aspects, in order to successfully rationalize and modernize production systems.

In general, the productive systems supply relatively small nearby urban groups, because they do not have the capacity to guarantee the supply of food in the amounts and with the regularity required by populous urban centers. The largest LAC cities, including Mexico City, São Paulo, Rio de Janeiro, Buenos Aires, Caracas, Santiago, and Bogota, are abandoned by thousands of citizens without employment options or food. Many people loot supermarkets or urban stores. Others go to the countryside, and try to sustain themselves directly with certain crops that are resistant to natural disasters, and especially food crops such as rice, beans, corn, and yucca. This is another source of agricultural losses.

3.4.4.1.4 Results of interaction among the systems

Following the serious effects of climate change, a drastic change in agricultural production systems occurs. Many of the major single-crop commodity systems do not survive these changes. However, smaller integrated production systems manage to remain in operation and become stronger in this scenario. Thus, in rural milieus, the rich and the poor—at least as regards the owners of the land—trade their positions of relative wealth. In many cases, the percentage of rich and poor also changes. The most vulnerable groups, i.e., subsistence farmers, rural wage-earners, or communities that produce for their own consumption, especially in areas that were subjected to climatic stress such as frequent floods or droughts, are the most affected by climate change. Many leave their homes and seek refuge in the cities, where there is generally not enough food and jobs for everyone.

Moreover, the effects of climate change and the failure of many large-scale enterprises also displace unskilled workers, who previously worked in sugarcane production in Brazil, for example, or in oil palm production in Ecuador or Colombia.

With regard to income inequality, results are mixed. However, when we look at small, medium, and large landholders in agricultural production systems, we see that ownership of the land changes hands. Many rich owners leave the business and become poor, while small owners growing crops and crops systems with a lower environmental impact become stronger and grow. Rural workers, however, are frequently left without employment and need assistance to meet their basic needs. Their situation improves with policies that facilitate their access to land, water, credit, and know-how. But the employment issue is not totally resolved, because economic fragmentation causes a sharp drop in agricultural production and job creation.

Access to basic education, health, employment, housing, and food security are objectives pursued in a heterogeneous way by the countries of the region. In the fields of education, health, and housing, the countries pioneering in social and political change begin to reap their first successes towards the end of the period.

Access to food in the quantities and with the regularity needed in the cities becomes a major problem, because the number of persons without regular access to sufficient quantities of food to meet their basic needs increases. This access is even more difficult for the poor, because the reduced supply of food leads to increased prices.

During this period, agriculture undergoes a major change of objectives: it shifts from a strongly productivist approach to a profound environmentalist conviction. The quantity of chemical products applied to agriculture, such as fertilizers and pesticides, is reduced. Environmentally friendly practices and biodiversity gain ground, and although they do not always lead to greater productivity or a higher yield in the short run, they guarantee a continued supply of agricultural products in the exploited ecosystems. There is also more control over health standards, and products are required to be free of contaminants related to production technologies. These changes in agriculture mean that environmental sustainability begins to show signs of improving towards the end of this period, after a profound crisis during a good part of the previous years.

3.4.4.2 2016-2030

3.4.4.2.1 Context of AKST systems and agricultural production

International trade barriers, and especially nontariff barriers, continue in place, but countries agree not to impose restrictions on the exchange of information. The methods and procedures developed in the previous period for ecological labeling of foods are perfected and extended.

Continuing the trend of the previous period, competition among countries virtually comes to a halt. Countries produce primarily for their domestic markets, without large surpluses. In a few cases, especially when a country afflicted by natural disasters or social crises needs assistance, food is exported and imported. In a few cases, there is also specialization of agricultural production by country, based on its tradition, culture, and agroecological capacity. Consumers, both within and outside LAC, increasingly value products with certification of origin and environmental protection. There is also a growing demand by consumers for nutritional and safe foods.

In certain countries or regions, pests and diseases, as well as epidemics, are almost permanently reduced by improved socio-environmental management, use of appropriate technologies, mitigation of the loss of biodiversity, and improvement of soils. The results are: (1) an increase in production and marketing of healthy, higher-quality products; and (2) a greater added value in these products.

The status of climate change is still worrisome throughout the period. Many countries encourage agricultural R&D on adaptation to climate change and implement production systems specifically designed for that purpose. There is a more robust capacity to adapt to and mitigate climate change.

Optimum governance conditions are consolidated in most of the region towards the end of the period. Agricultural development policies are pursued. Laws are adopted to limit the size of large corporations, applicable to both existing ones and new corporations that may be established, by restricting their acquisitions of and mergers with other companies. The purpose of this legislation is to guarantee a better balance of power among the different social stakeholders. A considerable portion of the fiscal resources obtained are used to implement initiatives for designing and establishing a new society. Many countries adopt regulations pertaining to the "Local Commerce Regionalization Initiative" (Carpenter et al., 2005), permitting cooperation among transnational companies if they use local products and if the value added is appropriate for all the partners.

Strict standards and regulations on the composition, origin, and environmental safety of foods are applied both domestically and to erect trade barriers.

A concern over the environment leads to restrictions on the participation of biofuels in the energy matrix of countries, to prevent the expansion of agricultural land. Alternatives, such as nuclear energy and solar energy captured and powered by nanotubes, emerge in the middle of the period, as clean, mastered alternatives to meet the energy requirements of a growing world population. An extensive debate begins on meeting energy needs by using these alternative sources of energy instead of biofuels and the consequent expansion of agricultural land.

The processes and activities initiated in the previous period to improve education are pursued. Local educational systems achieve good results, after overcoming problems related to financing and teacher training.

In LAC many cooperative work arrangements are consolidated, in view of the realization that R&D is increasingly more expensive but essential for the development of the countries of the region. These arrangements even include the foundation of regional R&D institutions to achieve a critical mass of researchers and increase the probability of important progress in the new technologies (biotechnology and nanotechnology). They are also a way to considerably reduce operating costs.

There are many projects shared among countries that were designed to obtain the scientific support of this guarantee of the production and supply of healthy, quality food. Biotechnology and nanotechnology are used to generate knowledge on the reaction and resilience of ecosystems (Carpenter et al., 2005), but the interaction between them is not yet fully understood. This is reflected in the scant attention given to the impact of this interaction that results in episodes of contamination of many natural resources found in different countries. In other words, there is generally no awareness that waste products thrown into a river that runs through many countries is going to cause the contamination of drinking water in other communities, for instance.

By the end of the past decade, indigenous and local communities begin to reap substantial benefits from the appropriation of formal knowledge in the most widely varied areas. As a result of this and the fact that they are highly organized, they receive monetary income from various products derived from agriculture or biodiversity obtained on the basis of this knowledge.

The failure to care for common resources, such as oceans, cross-border rivers, the atmosphere, wildlife, etc., enhances the value attached to traditional knowledge. It is increasingly more systematized and its principles are elucidated by scientists from the communities themselves, who use formal knowledge in this effort. These situations that are so favorable to traditional knowledge are not found uniformly throughout the world or even throughout LAC.

3.4.4.2.2 AKST systems

The existence of barriers promotes R&D on origin certification systems and ecological labeling of foods, and the relationship between environmental services and climate change, and its reciprocal effect on agriculture and ecosystems. There is also a greater interest in (1) conservation and management of pollinating insects; (2) prospecting for and the sustainable management of plants; (3) identification and study of current and potential exotic invasive species; (4) the use of genetically modified organisms and their impact on agrobiodiversity; and (5) the impact of agricultural nanotechnology on human health and the environment. An important concern for R&D during this period is the development of sustainable productive systems capable of large-scale food production. R&D systems are directed to all social groups, but focus especially on the most vulnerable groups. The free exchange of information and scientists among countries, and the growing value attached to science guarantee the technical capacity of the R&D system in many of the LAC countries. Biotechnology and nanotechnology are disciplines that play an important role in R&D projects. The incorporation of traditional knowledge increases.

Society's confidence in science mounts. The control of social stakeholders over R&D activities implemented in the previous period slackens in this period, so that advances in basic disciplines may be incorporated, thereby contributing to the understanding of the environment and its friendly use. Resources available for R&D continue to be adequate but not abundant. There are some additional resources derived from accreditation services and certification of products by some R&D institutions. There are difficulties in obtaining outside resources for R&D. Social participation in generating know-how and technology for productive systems expands. The coordination of efforts among the various stakeholders with different interests and the need for a focal point for similar programs and projects are sources of considerable inefficiency in the use of financial resources, infrastructure, and capacity. At the outset of this period, private R&D organizations, greatly reduced in size and power, begin to participate more actively in R&D, in cooperation with public organizations.

R&D achieves important progress in understanding and managing ecosystems. Environmental services improve as a result of the better understanding of their repercussions on the environment. The efficiency and effectiveness of scientific activity have gained considerable ground in comparison with the previous period: efficiency, because it is necessary to rationalize the use of scarce resources; and effectiveness, because the competition of many stakeholders, including users, in defining and obtaining a technological solution makes it possible to build transdisciplinary structures that are better adapted to the needs of these users. The time between creation and implementation of a project, however, becomes longer, due to the application of rules of collective participation in this implementation. There are cases where the result is delayed so much that it is no longer relevant for users. There are also many cases of duplication of efforts, caused by the fact that the local and decentralized systems do not have adequate communication and integration mechanisms.

The participation of so many stakeholders in developing know-how and technologies is also a factor that has a positive influence on obtaining appropriate technologies, but at times they are not applied to the interested systems, either because of delays in obtaining them, or because the information on their existence is not adequately communicated.

3.4.4.2.3 Agricultural production systems

The incorporation of knowledge into agriculture is actively pursued by all stakeholders that can benefit from it. Decision-makers are also moving in this direction to reduce the negative impact of the transition that occurred in the previous period and to stimulate grater agricultural production. Policies emphasizing local sustainable development allow for more agroecological knowledge to be included.

The markets served are essentially domestic. A few specialized markets are established as a result of the gradual specialization of countries in a few agricultural products, which have comparative advantages in terms of culture, tradition, agroecological conditions, and the like.

Most stakeholders in vulnerable production systems are highly organized, as a result of decentralization of rural development planning and the greater weight given local proposals. The development of community organizations incorporates social organizations promoted by production chains or cooperative movements in the communities.

There are resources to support agriculture, with a view to protecting it from natural disasters. But these resources are not abundant, since there are many social demands and economic resources for this purpose are limited. During the last decade of the period, both agricultural production systems and cities suffer from limited access to water, especially in the semiarid zones of Latin America, in Brazil, Mexico, Argentina, Peru, and Colombia. This reduced access displaces subsistence farmers and reduces agricultural production in many countries.

The products and processes of practically all agricultural systems are healthier and more environmentally friendly. As in the previous period, there are problems in obtaining food in the quantities and with the regularity needed to feed the entire population.

3.4.4.2.4 Results of interaction among the systems

Agricultural income does not increase very much, as a result of the dynamics of the local markets themselves. The policies designed by countries to reduce the gap in agrarian income in the previous period are improved and show promising results. The narrowing of the income gap indirectly induces many who had migrated to urban centers to return to the rural milieu, thereby partially alleviating the problem of food supply to the urban poor.

With regard to education, health, and housing, countries improve access to these sectors towards the end of the period. Access to employment is somewhat better than in the previous period, because agricultural systems acquire greater capacity and experience, and thus are more efficient that in the previous period. Many of these systems also achieve economic sustainability by the end of the period.

Healthy food is guaranteed for the urban poor, who have the means to acquire it in the cities, but the total food supply is not guaranteed, in the quantity and with the regularity needed during this period. The increased population and demand for food causes major social conflicts, causing many countries to include in their constitutions the guarantee of available food. This only partially solves the problem of a shortage of food, which is democratically distributed among the poor.

The result in terms of environmental sustainability is an improvement in the protection of ecosystems locally. However, common natural resources shared by various countries frequently suffer from the impact of different management systems, and also at times from neglect, which has a repercussion on other societies.

3.4.5 TechnoGarden

3.4.5.1 2007-2015

3.4.5.1.1 Context of AKST systems and agricultural production

The governments of various European countries begin to eliminate agricultural subsidies and tariff barriers, as a result of pressure exerted on the WTO and other international organizations by poorer agricultural countries. This liberalization produces a strong flow of imported foods and the consequent expansion of supermarkets in some LAC countries.

Throughout this period, the nontariff barriers of biosecurity and environmental protection are implemented and strengthened. These include certification of sustainable production processes in the country of origin of agricultural products and of low environmental impact, as a result of their use.

Although the diversification of agriculture, which occurs initially in the rich countries, leads to greater environmental sustainability, it also discourages them from food production, which becomes even more concentrated in the poorer countries. The poorer countries in turn, which were already dedicated to agriculture, but as commodity producers, now shift to producing differentiated products with a greater value added, and also begin to diversify their agriculture. This latter movement is seen especially in the countries with greater biodiversity, as in the case of the countries that share the Amazon biome in the region.

The free circulation of information and persons in the world enhances the diversity of consumer demand for differentiation of foods by taste, appearance, nutritional value, nutraceutical properties, healthfulness, etc. In many countries consumers require certification pertaining to the food processing method (without agrotoxins, child labor, GMOs, animal suffering, etc.). The food traditions of other cultures is now familiar to many consumers. This means that there is an increasing demand for the inputs needed to prepare this type of ethnic meal in specialized restaurants. Traceability requirements also grow. In LAC, the increased education of the people and availability of information on food also serve to augment consumers' requirements.

Despite the implementation of more controlled production systems, agricultural epidemics increase in frequency and severity, and new pests emerge, mainly due to the effects of climate change. At the outset of the period, there are few LAC countries with the capacity to prevent and adapt to epidemics and pests. This capacity increases, however, throughout the period, as a result of abundant resources, the efficiency of international biosecurity barriers, and better governance in the countries.

The status of climate change is a source of concern throughout the period. Societies are aware of the possible repercussions of climate change on production systems. A decade of droughts and floods reinforces the concern over the effects of human action on the climate and environment, enhancing the value of environmental services in those countries. A visible consequence of this growing appreciation is that agricultural production processes begin to be monitored by consumers in the richer countries, who organize to ensure that these processes comply with low environmental impact standards and procedures, and to demand compensation—for preservation of forests, for instance—for agricultural operations. This leads to strict global regulations for the preparation and import of agriculture-based products.

Many LAC countries make great strides in their institutionality throughout this period. Despite changes in government with different parties in power, in many of these countries there are more stable and coherent policies, especially in the field of development, which is now seen as a multidimensional economic, social, and political phenomenon.

Many Latin American countries implement compensatory policies for the poor at the outset of the period. In a few countries, these policies are not accompanied by employment policies, and so the improvement in the social and economic condition of these groups is ephemeral. For the majority of social groups, more consistent, successful, and lasting policies for employment, education, and health are implemented. Many countries have laws protecting investment in science, creating an incentive for that activity.

With regard to the environment, many countries move in the direction of an institutionality that allows for the managed exploitation of natural resources. This institutionality applies rules on ecosystems and segments of ecosystems that may or may not be exploited, and regulates the type of exploration possible, conditions for exploitation, and so forth. Participation in the global market leads to rapid improvement in regulations and standards and the rigorous enforcement of them, to comply with food quality standards.

In some LAC countries, little progress has been made in the field of education. But even in those cases, there is a slight improvement, a continuation of the trend observed in the previous decade. In a large part of the countries, there is fortunately a notable gains in education, and even stakeholders in the most vulnerable agricultural production systems show a considerable improvement in their level of education by the end of this period.

At the start of the period, there is still a distrust of the true intentions and uses of science, However, certain successes towards the end of the period lead to renewed enthusiasm over the benefits of scientific activity. There is progress in the world and in LAC in establishing conditions for scientific activity, especially considering the major ethical dilemmas besetting this sector in the new day and age.

R&D applied to agriculture in the global sphere develops along two lines: one is a deeper understanding of the effects of anthropogenic action on ecosystems, with a view to reducing such action; and, the other is putting a specific value on environmental services, as a way of creating policies to promote the diversified use of the land (agricultural production and environmental services). Major efforts are made to advance knowledge of biology, nanotechnology, and the information sciences, and the integration or interrelationship among them.

The rich countries, especially European Community members and the United States, pursue their course of intensive scientific and technological development oriented to technologies such as biotechnology and nanotechnology and information technologies. The development of new products is a critical factor in international trade competition. On many occasions, and even to guarantee genetic variability, research organizations use biodiversity resources in the hands of less developed countries, especially in Latin America and the Caribbean.

Laws on biodiversity in most countries are relatively inefficient, even in those countries that have ratified relevant international conventions like the CBD. Thus, traditional knowledge is little valued, and remains isolated from formal knowledge in the vast majority of cases. The enhanced value of environmental services gradually changes this picture.

3.4.5.1.2 AKST systems

The concern over the environment and environmental sustainability in agriculture grows throughout the period, as a result of increased temperatures and more frequent extreme climate events in the region. Consequently, R&D in LAC gives high priority to knowledge about the environment and its relationship with agriculture. This concern materializes in a heavy investment of resources for research on this. Various R&D programs initiated also specifically focus on adaptation to or reduction of the impact and mitigation or reduction of the causes of climate change. By midperiod, investment in research designed to measure and assesses the value of environmental services and biodiversity also increases.

R&D priorities include development of processes for: (1) control of residues and nutrients added to soils of productive systems; (2) treatment and recycling of agricultural and agroindustrial waste; (3) precise evaluation of the need for inputs, water, etc. for plant growth (precision agriculture); (4) safety and quality guarantees in food processing; and (5) creation of varieties and strains adapted to hostile environmental conditions. All of these processes are complementary and designed to increase productivity. The following topics linked to the environment and ecosystems are priorities: (1) the economic valuation of biodiversity and natural resources; (2) sustainable economic exploitation of biodiversity; (3) management of fishing resources; (4) management of the quality and use of water; and (5) management of forest resources.

In terms of the social groups targeted by R&D, by the end of this period an important change occurs: R&D is no longer directed preferentially to large and medium-sized traditional producers, but instead it is geared to end consumers, agroindustry, and policymakers on a priority basis, and only secondarily to merchants and subsistence farmers (Castro et al., 2005; Lima et al., 2005). Indigenous communities and small-scale producers are not important to R&D organizations at the outset of the period, but this situation changes over time due to the growing interaction between research institutions and these communities.

A growing awareness of the importance of science and R&D also means that LAC scientists receive greater financial and token compensation for their work. They work in close cooperation, forming multi-institutional research networks with scientists in many LAC countries and in countries outside the region as well. In this way, advances in knowledge within LAC and the incorporation of knowledge generated in other regions of the world are facilitated.

Throughout virtually the entire period, traditional knowledge is not given serious consideration as a source of technologies for formal systems in LAC. In 2013, with the impact of climate change in LAC, many countries begin to debate the advisability of using traditional knowledge to define practices to adapt to extreme weather phenomena. Little by little traditional communities begin to be seen as sources of knowledge on the different biomes and the environmental services provided by them. This realization is confined to a few countries.

Thanks to sustained economic growth, during this period most LAC countries have financial resources for long-term investment, for instance in R&D. They also have a critical mass of internationally reputed scientists in specific fields. The R&D project management and implementation process is increasingly professionalized. It is based on detailed studies of the future and on long-term planning. This process also increasingly includes other stakeholders interested in the results of R&D activities.

Research and development activities form an arena where public and private R&D organizations compete and cooperate. These two sectors have the financial and human resources needed to perform well. They establish a division of labor according to which some of the more profitable commodities, such as corn, tobacco, melons, papaya, wood species, and cotton, in addition to most of the products with a high value added, are the purview of the private sector, while species such as rice, beans, coffee, citrus fruits, wine, yucca, mango, bananas, and cashews are of strategic importance to the public sector. The two sectors cooperate in some areas of research, such as soybeans (Castro el al., 2006).

Research in LAC produces important results for agriculture. In food chains, there are advances in certification, traceability, and food safety in general. There are also important developments that have to do with biofuels. The successful experience of Brazil with alcohol as a replacement for gasoline is used as an example for the development of other plant-based energy sources, such as oil from oil palm, which is used as a substitute for diesel in Brazil and other LAC countries. As a result of heavy investment in the environment, by around 2015 difficult issues having to do with the economic valuation of biodiversity and natural resources in the provision of environmental services and for sustainable agricultural production begin to be resolved. Important efforts are also made in the area of management of forest resources and the quality and use of water, which becomes a source of concern on the heels of climate change effects observed in the course of the period.

The technologies generated by public and private R&D and by broad social participation in the research process are usually adapted to the systems served by them. These technologies also come close to an ideal of what the most appropriate technologies for sustainable development would be. This is true even in the case of more vulnerable social groups that were not given priority at the beginning of the period.

3.4.5.1.3 Agricultural production systems

The situation created by extemporaneous changes in the climate encourages the intensive incorporation of relevant knowledge into agricultural production systems. The countries of the region approach the incorporation of knowledge and nature itself with widely varying degrees of intensity.

In this scenario, the incorporation of knowledge into agriculture is a business matter, and producing enterprises do it by training their workers in the use of new techniques and inputs to improve the productivity and sustainability of the systems. The enterprises also require the implementation and verification of a series of practices to comply with market requirements. Similarly, the stakeholders of smaller production systems are organized in associations, so that they can comply with rules of efficiency and standards and certification requirements.

Genetically modified organisms are used more frequently by a growing number of producers throughout LAC. The costs of using these technologies are reduced and thus their use becomes more widespread throughout the region. At the outset of the period, the use of transgenic organisms that leads to an increase in use of environmentally harmful inputs, such as herbicides, for instance, causes conflicts with all those who defend environmental protection within and outside the region. Towards the middle of the period, some cases of contamination in units producing biopharmaceuticals cause a wave of social rejection of this type of biotechnology. However, the introduction of new agricultural varieties adapted to hostile environments and of transgenic organisms capable of acting as bio-remedials (for instance, in cases of contamination of the soil by toxic substances) or of preventing soil erosion lead to the dissemination of transgenic organisms and their acceptance by LAC and its markets.

The major production systems, which are highly technified, serve external and internal markets. These systems are an integral part of large production chains; they are highly coordinated and have an in-depth knowledge of the markets served and consumer demands. Most of the small farmers, and also some groups that in the beginning of the period practiced subsistence agriculture, manage to insert themselves in some of these chains or to participate in certain market niches, with the production of goods, such as frog legs, for a very limited public. The number of subsistence producers is sharply reduced.

Since the very beginning of the scenario, plentiful resources are allocated to promote and disseminate use of know-how in agricultural production systems. The production systems receive considerable resources to improve their economic efficiency and product quality, especially in the form of credits and know-how, rather than land. The goal is to increase the productivity of agricultural production systems. Moreover, some of these systems also provide one or several environmental services, which are encouraged in many LAC countries by the end of the period.

Due to the influence of climate change, some regions begin to experience problems in purchasing water in the quantity and with the regularity needed to ensure the effective performance of their production systems.

Large productive systems that use modern production and management methods succeed in operating with great efficiency and use advanced processes to produce high-quality products. Thus, they also become more competitive. A large component of know-how and technology is incorporated into these products and processes. Although the external market still prefers commodities to differentiated products, the latter go to the broad LAC internal market. This situation does not change until the end of the period, when a few important developed markets begin importing a greater percentage of differentiated products from LAC.

The production systems of small farmers are inserted in the major chains by private national or transnational corporations as suppliers of inputs. They are also inserted as producers of raw materials in other chains (or in other words, as independent components that are not coordinated by another component, as is the case in the first situation described). These small systems are dedicated to producing commodities or a few differentiated products.

The vast majority of these independent production systems inserted in production chains are also successful overall. However, this is not the case in situations in which unforeseen factors, such as rising temperature, natural disasters, or epidemics, threaten the performance of these systems.

3.4.5.1.4 Results of interaction among the systems

The improved performance of productive activities, especially in terms of economic efficiency, begins to have a positive effect on income inequality. The need to substantially improve the quality of products and services and to pay more attention to their environmental consequences generally has a good effect on market prices.

During this period, there is generally a considerable increase in the indicators of greater social equality. Access to education, employment, health, and food security improve. In a few LAC countries, this progress is more limited.

Positive changes are recorded in urban food security and safety indicators, because there is a better understanding and monitoring of the handling, packaging, and processing of foods. The incorporation of environmental adaptability in many varieties and strains leads to a widespread increase in the availability of food, and thus to a decrease in prices for urban consumers.

In the beginning of the period, agriculture in both rich and poor countries is heavily based on exploitation of ecosystems to produce processed foods or raw materials. In other words, the products generated are commodities or differentiated products and always derived from human action on nature. Little by little, starting in Europe and then in the United States, global agriculture is diversified and begins to include environmental services as one of its functions. These services range from protection of water sources, carbon sequestration, and protection of habitats for pollinators, such as birds and bees, to the reduction of pollution generated in agriculture and simple conservation of plant and animal species. As a result, there is an improvement in indicators for environment sustainability in agriculture.

3.4.5.2 2016-2030

3.4.5.2.1 Context of AKST systems and agricultural production

Free global markets are consolidated. Biosecurity and environmental protection barriers are further strengthened.

Competition for markets gives priority to product differentiation obtained by incorporating environmentally friendly technologies. LAC increases its participation in these markets. Consumers throughout the world are willing to pay higher prices for products linked in some way to environmental protection initiatives. Thus, certification that products are developed by organizations that provide an environmental service of some kind is a factor adding to the value of the product.

LAC still participates in commodity markets, especially food commodity markets, where rich countries are major importers, since in some of those countries agriculture has disappeared. This group of countries continues to use, when necessary, the raw materials produced by less developed countries, to produce new products by chemical and/or molecular manipulation.

Consumers worldwide, including LAC consumers, are on the alert to prevent any environmental threats, because there are a few severe natural disasters that occur about midway through the period that cause devastation in various parts of the globe. Thus consumers value any products made with a concern for environmental and ecosystem conservation, whether it has to do with the production processes used or the fact that the systems producing such goods offer environmental services. But consumers also demand new and original types of foods, while at the same time they are attentive to issues related to health and contamination, and issues involving new genetic or molecular manipulation.

Thanks to the implementation of prevention and monitoring technologies and more sustainable practices, epidemics caused by known agents are more controlled and the time between successive outbreaks is longer. However, epidemics caused by unknown vectors are particularly intensive and difficult to control, although technological development as a rule allows for a prompt solution to these pests as well.

The status of climate change is worrisome until the end of the period, when the rate of increase in temperatures begins to decline. This reverse is the result of a major effort to develop sustainable technologies that are intensively used by production sectors in countries.

In most countries, governance is nearly optimal, with stability and consistence in policies, regardless of the government in power.

The concern over environmental services and the environment and its protection leads many countries to issue laws to guarantee an economic return for entities that can prove that they provide a specific environmental service to the country and the world. In addition to environmental protection, these laws provide work for many unemployed workers, who would otherwise move to the cities.

When LAC governments observe this unforeseen consequence of their environmental protection policies, they pass laws to allocate land for the sole purpose of environmental preservation and ecosystems. These lands, owned by the government, are managed by persons selected from the poor, based on proposals that these managers make for the sustainable use of these properties.

In LAC, there are policies to encourage tourism that promise a return to nature, with farms that function in the same way as they did in the mid 1900s and that resemble large entertainment parks, where tourists interact with persons and not machines. Activities involving visual arts or the culture of body aesthetics are also strongly promoted, as an ideal way to prevent the deterioration of health or to reduce the mortality rate.

The economic return on investment in R&D is guaranteed by sustainable policies for protection of knowledge and by good management of these policies. Education is increasingly guaranteed and valued. It is offered partly by the state and partly by corporations that employ highly qualified professionals. They must have increasingly complex advanced degrees to meet the performance standards required by systems that apply knowledge at increasing rates of intensity.

Improvements in regulations and standards and their enforcement are completed.

Unemployment grows as a result of the intensive incorporation of technology in all activities. However, this growth is offset to some extent by policies providing incentives for new economic pursuits. Large properties are taxed heavily, so that governments will have the resources to establish and maintain unemployment insurance for those out of work in such a technified world. There are also incentives to discourage corporations from laying off employees as a result of the incorporation or modification of technology.

R&D provides the basis for the valuation of environmental services based on research that uses biotechnology and nanotechnology. Public institutions in some LAC countries participate in this research.

There are enormous advances in virtually all areas of application of biology—animal and plant production, processing of quality, healthy foods, biomanufacturers of industrial raw materials, the environment, production and use of the biomass, and new nonfood products—and also of nanotechnology—animal and plant monitoring and therapies, monitoring of food processing, detection of pathogens, virus, GMOs in raw materials and processed goods, identity preservation systems, and environmental treatment and monitoring systems.

Biotechnology, nanotechnology, and soil science are integrated and produce spectacular results in the area of environmental remediation.

Varieties and strains adapted to hostile environmental conditions, such as plants resistant to drought and salinity, are developed for agriculture by genetic manipulation. These are a few examples of the advances that take place in LAC.

Concern over the handling of environmental services increases in all countries, and gradually leads to an enhanced appreciation for traditional and local knowledge. To better guarantee the continuity of these services, many practices of indigenous and traditional communities are appropriated. Many of these communities receive economic benefits from this knowledge, because there are stronger laws that guarantee this. Conservation of biodiversity is also regarded as an environmental service. It includes preservation of river basins and the reduction of environmental contamination, because the importance of living in harmony with different animal and plant species for the preservation of many ecosystems is a matter of common knowledge. In various LAC countries traditional knowledge is also highly relevant, especially in relation to its interaction with formal science, to enhance the understanding of biodiversity and its uses.

Enormous advances in science once again bring out global fears regarding the ethical limits of scientific activity and technological innovation. Innovation of products and processes generates a debate among various social groups regarding the use of nature, as known and appreciated. Advances in science and its applications also give rise to more practical problems, because the latest technology is almost completely autonomous and no longer requires as much labor as before, especially relatively unskilled labor. During this period, however, the average skill level is high at the level of secondary education. Thus there is social pressure to reduce the pace of scientific development, and LAC is not exempt.

3.4.5.2.2 AKST systems

R&D priorities for LAC are as follows: (1) application of recent advances in valuation of environmental services, to define protocols that make environmental protection an activity that supplements agriculture; (2) application of advanced biology and nanotechnology to production of food and new materials, that can be used in many productive areas, such as health, pharmaceutics, agriculture, industry, etc.; (3) use of microorganisms for environmental remediation; and (4) improvement of nanosystems for monitoring diseases and for application of therapies to animal or plant groups, identity preservations systems, and tracing and monitoring and environmental recovery systems. Priority is also given to developing alternative technologies that allow for the continuity of agriculture even under the impact of climate change and that prevent increases in the frequency and intensity of these effects by reducing the factors that contribute to climate change today.

All social groups are focused on R&D in LAC. The capacity of professionals in science and technology in LAC is growing day by day, as a result of their daily participation in the global development of science and technology, through publications, attending congresses, and joint projects. The time lag between an advance in one area of knowledge and its application to productive activities is considerably shortened.

There is a keen interest in systematizing traditional knowledge, which is massively explored by formal science under the protection of national, regional, and international laws or agreements that guarantee the rights of traditional/indigenous peoples and the harmonious interaction between these two types of knowledge. This interaction is strongly driven by a common concern for the environment.

All productive and economic activities depend on the continued progress of R&D. Governments and corporations give priority to investment in know-how and technology. There are abundant resources available for this purpose. Management of R&D is regarded as a strategic factor in the competitiveness of companies that develop agricultural technology. This has to do with the fact that the time span between the design of a new product and its entry on the market becomes shorter and shorter.

Society participates more in research, since private R&D organizations feel the growing pressure of public opinion that is concerned over their power. This participation is mainly in management processes, but it is limited in the case of technological development projects, due to the specialized knowledge required.

Public and private organizations still work in cooperation, but the role of the private sector in R&D becomes more pronounced. In terms of products and services developed, this means that now there are few species of plants and animals that the private sector is not interested in, and that are left for public research. Interest of the private sector in basic science also increases, because of its capacity to generate knowledge that serves as a basis for future practical applications. There is a huge number of plant and animal species with sequenced genomes. Functional and structural genetics also make great strides in understanding gene functions. These advances are achieved to a great extent as a result of the cooperation between public and private science.

Research is increasingly more effective, i.e., capable of generating the innovative products or services demanded to address equally novel problems in production systems, ecosystems, and their interface in brief time periods. But the plentiful resources lead to a lack of concern with the efficiency of R&D, which becomes increasingly more expensive, even in situations that lend themselves to a more rational use of resources to obtain a certain outcome.

As for products and services obtained from R&D, they are now virtually problem-specific or demand-specific, because they are designed to solve a specific problem or to meet a specific demand of a social group. This extensive portfolio of products and services is also one of the reasons for the low efficiency of R&D activities in certain circumstances.

There is a sharp improvement in the understanding of social, economic, biological, and ecological systems. Technologies are increasingly better adapted to the systems where they are to be applied, although this adaptation is not yet perfect. New problems arise periodically in these systems, as a result of the unforeseen interaction of new technologies and their repercussions on the emerging properties of these systems.

3.4.5.2.3 Agricultural production systems

Throughout this period, new knowledge was intensively incorporated into production systems. Various other human activities considerably mitigate climate change. Relevant technological changes introduced in production systems contribute to this mitigation. There are also important advances in adaptation to climate change effects.

In this scenario, companies manage the incorporation of know-how into agricultural by training their workers in the use of inputs and new techniques, to improve the productivity and sustainability of the systems. Companies also require their employees to use and check a series of practices to comply with market requirements. Company employees or partners are required to incorporate a pool of complex knowledge associated with the standards applicable to products and production processes.

The large, highly tecnified production systems serve the external and internal markets. These systems are part of major production chains, which are highly coordinated and have an in-depth knowledge of the markets served and the consumer demand that influences those markets. Processors of basic agricultural products participate as suppliers of pre-treated raw materials (in other words, products that are subjected to some processing following primary production) for these major production chains. Virtually all the systems include new activities not in the agricultural sector, such as environmental services, tourist operations, or operation of rest homes, to give a few examples. These activities are integrated into the agriculture-based activities and serve both internal and external markets.

The major production systems and independent producers are well organized to defend their interests, with strong professional support. Most of the independent producers manage to insert themselves into the chains and markets, but there is still a displacement of small-scale producers to the cities.

The policies of abundant resources available for incorporating knowledge into production systems remain in effect. The region tends to become standardized in its technological efforts, and plentiful resources prevail throughout most of the region. Problems of access to water are solved by new technologies to reprocess wastewaters and by desalinization of salt or brackish waters. Land as a resource and environmental protection are ensured through the successful use of degraded environments considered as hostile to life in the past.

The major production systems, which use modern production and management methods, operate with great efficiency and produce high-quality products using advanced processes; this enhances their capacity to compete on markets. A large component of knowledge and technology are incorporated into these products and processes, thereby generating countless differentiated products.

Smaller-scale production systems (no longer called "small producers") participate as suppliers of preprocessed raw materials for large production chains. The vast majority of the production systems are successful.

3.4.5.2.4 Results of interaction among the systems

If only agriculture-based productive activities are considered, income inequality is sharply reduced in this period, as a result of the insertion of many producers, considered as small producers in the previous period, into powerful production chains and transnationals. Thus all the social groups participating in this activity obtain higher incomes. However, wage-earners who were working in the fields before the work was completely technified lose their jobs and migrate to the cities, which are now faced with an increased demand for food and basic services.

Access to education, housing, and food security are guaranteed by governments in different ways. Employment, however, is not guaranteed, although the diversification of agriculture has contributed to its increase and governments have implemented powerful mechanisms to create alternative labor markets and provide compensation for the unemployed.

Urban food security is supported by abundant, cheap, diversified food that meets high health standards. The sustainability of agricultural production systems gradually increases throughout the period, as a result of the application of more sustainable technologies, but also because agriculture has another paradigm, since environmental services are almost always provided along with the conventional production systems. Another important reason for this growing, yet incomplete sustainability is the use of regulatory procedures and standards in the technified countries of the region. There are also isolated cases of newly emerging environmental problems, resulting from technological solutions to previously existing problems.

3.5 Implications of the Scenarios for Innovation and Development Policies

The purpose of this chapter is to help answer the following question, with specific reference to Latin America and the Caribbean and alternatives for the future development of the region:

"How can we reduce hunger and poverty, improve rural livelihoods, and facilitate equitable and environmentally, socially, and economically sustainable development through the generation, access to, and use of agricultural knowledge, science, and technology?"

On the basis of these alternatives, it is possible to propose nonprescriptive recommendations as to how science and technology can best contribute to this development.

The five scenarios constructed to answer this question show that knowledge, science, and technology can contribute to the changes suggested in the question, but in different ways, depending on each alternative scenario considered.

The scenarios also make it clear that this contribution will be more likely and facilitated in situations in which other political, economic, and social conditions are also present. In each scenario, the direct influence of these conditions, and the interaction among them, will guide the action of formal AKST systems, and the use of traditional knowledge, and hence determine their contribution to sustainable development, as proposed in the question that generated this critical evaluation (IAASTD).

According to the *Global Orchestration* scenario, society has abundant resources, it is guided by market forces and is highly interconnected, but is concerned only on a reactive basis with the impact of human action on the environment. Formal AKST systems are characterized by uncontrolled generation of new products, which increasingly incorporate more technology to meet ever more sophisticated demand. Little if any use is made of traditional knowledge. As a result of the high degree of technology incorporated into the system, there are unemployment problems. Due to the careless exploitation of natural resources, the impact of human action intensifies, generally leading to highly negative consequences for agriculture and human life.

In the *Order from Strength* scenario, society is fragmented, and there is a pervasive distrust of the rich, and generally developed countries on the part of the poor and generally undeveloped countries; highly restrictive governance conditions and largely inadequate policies prevail in LAC, and there is a strong trend towards aggressive exploitation of the natural resources of the poor countries by the richer countries. The region even loses its capacity to generate technology independently, and becomes increasingly dependent on other regions. The incorporation of traditional knowledge in this scenario is only peripheral and marginal. As a result, LAC becomes a mere supplier of inputs for the rich countries. There is an enormous social and economic crisis, and the environment is subjected to unprecedented impacts.

The *"Life as it is"* scenario presents a world in which countries are integrated, but not to a great extent. The course of action is defined by the market, but not fully, and a division among countries persists, but it is still possible to conceive of change in the long run. There is both a proactive and a reactive approach to interaction between man and nature. In other words, it is a pluralistic world, in which none of the variables considered dominates others in its influence on the scenario. In these circumstances, the AKST system also obtains relatively mediocre results in its efforts to achieve any of the major sustainable development objectives referred to in the initial question that the chapter endeavors to answer, although the results are positive in the area of social development and environmental sustainability. Application of traditional knowledge improves towards the end of this scenario

The *Adapting Mosaic* presents a world in which immense institutional changes occur, including asymmetries of power among social stakeholders, paradigms for exploration of natural resources, generation of socio-political agreements, and distribution of wealth among social segments. All of these key elements for social and economic life are transformed. It is also a fragmented world, as in the *Order from Strength* scenario, but this fragmentation is not oriented towards domination of a fragment—or a region or river basin—over others. Each fragment seeks its own ways and places to deal with the environment, to reduce the impact on it. This entire transformation generates major crises and difficulties, affecting even urban food security in this scenario. There is also duplication of efforts, with a weak capacity to learn from imitation in many fragments, and delays in arriving at solutions. But there are also improvements in some indicators, and especially in the environmental impact. According to this scenario, formal AKST systems are initially viewed with distrust, but they clearly make an important contribution to achieving the objectives pursued by social groups, and so this distrust diminishes towards the end of the period. The empowerment of all of the more vulnerable social groups enhances the value of traditional knowledge, which is used in the *Adapting Mosaic* world.

The *TechnoGarden* scenario depicts a world in which countries are highly interconnected and motivated by a strong concern for the environment—with a pro-active approach, to prevent impacts on the environment. It is a world in which the actual concept of agriculture is transformed to include protection for environmental services. Environmental problems are solved and prevented by incorporating a high degree of technology. However, as in the *Adapting Mosaic*, there is also an interest in improving the quality of life of all segments of society and AKST institutionalizes this concern in its practices. Thus, new technologies are adapted to the different social groups, but also to different environmental conditions. Traditional knowledge is valued, and is used and systematized to a great extent in this scenario. Consequently, many sustainable development indicators improve, although in this world an optimum solution to the environmental problem is never found.

What are the implications of these scenarios for AKST and sustainable development policies, that could prevent the negative situations described in them, and what possibility is there for facilitating such action and ensuring interaction that would foster sustainable development?

In the following section there is a brief presentation of the implications for innovation policies and social development policies in support of vulnerable social groups under each scenario. It is important to point out that although each scenario is described in the present tense, these scenarios should not be regarded as predictions, but rather as possible future situations.

The policy implications were devised on the basis of the different scenarios, but also in consideration of the current situation of vulnerability in each country with respect to the different variables involved in them (this situation was described at the beginning of the scenarios, in Table 3-3). The line of reasoning followed is that even though we cannot accurately say that the most vulnerable countries today will have the same degree of vulnerability in future, this comparison makes it possible to indicate which countries have a greater or lesser probability of overcoming risks or taking advantage of future opportunities.

3.5.1 Global orchestration

3.5.1.1 Implications for innovation policies

The absence of barriers could lead to a reduction in product prices, and so productive efficiency would be very important in this scenario. However, competition is also based on quality differentiation. According to this scenario, there is a great diversification in the demands of end consumers, who, like the major corporations that govern this scenario, are generally relatively unconcerned about the environment.

This is a scenario where there is tremendous competition among countries, based on the constant development of new, differentiated products through the use of technology. On the one hand, this involves risks, even for the countries with the greatest current capacity to generate knowledge, such as Brazil, Argentina, Chile, and Mexico, because the gap between these countries and the developed world widens, especially in terms of investment in new technologies. The demand for product differentiation cannot be met at the level specified in the scenario, with the current capacity of the LAC countries. To maintain this capacity at the required levels, there would have to be a heavy investment in R&D. For those countries that have a very limited capacity to generate know-how today, it is important to make an effort to achieve independence in generating know-how and technology in this scenario.

There is also a greater risk of epidemics, of the effects of climate change and of negative impacts on environmental sustainability, in comparison with the *Life as it is* scenario, for the reasons set forth below.

With regard to epidemics, the countries of Central America and the Caribbean are more vulnerable (in view of their current capacity to prevent known and newly emerging pests). They could damage agriculture and human health, and cause important losses. The research agenda should include development of technologies to prevent and eliminate these epidemics, or to find ways to adapt to or live with them.

Policies that guarantee inclusion of environmental problems on the research agenda for the region—especially for the megadiverse countries, such as Bolivia, Brazil, Colombia, Costa Rica, Ecuador, Mexico, Peru, and Venezuela—should be implemented over time, and mechanisms to inform end consumers and make them aware of the risks to the environment involved in this scenario should be established.

Requirements pertaining to quality, traceability, and safety of foods entail costs that may be too high for small enterprises to bear. It is important to build policies and strategies to guarantee access to low-cost technologies that enable producers to meet these requirements.

3.5.1.2 Implications for sustainable development policies

Global Orchestration describes a world in which knowledge and its constant accumulation is the key factor of development. This involves a risk for more vulnerable segments of the population in the poorest countries of the region that are importers of food and agricultural products and/or that have a reduced capacity to offer quality education.

Policies to reduce the vulnerability of these countries and their people by reducing their dependence, primarily on food, are extremely important. A short-term, but less recommended alternative would be to guarantee food security in the most vulnerable countries, which are the current food importers.

An effort to guarantee quality education for the people in these countries in a consistent and lasting way would be another way of reducing risks. It is important to also bear in mind that this effort would be facilitated in the world described by this scenario, in which education and knowledge are the basis for the development model.

There is a large migration from rural areas, which will increase urban poverty. Policies to offset this phenomenon would have to be implemented, primarily in the poorest countries.

3.5.2 Order from strength

3.5.2.1 Implications for innovation policies

In this scenario, the key element is the existence of barriers and the division between groups of countries. This division causes an increase in all the types of vulnerability found in LAC countries today.

In a scenario of scarce resources like this one, the R&D agenda focuses on efficiency and is governed by a business-like approach, and the safety of commodities. At risk of disappearing in this scenario, generation of technologies must find creative forms in terms of implementation, but also to ensure the conditions, i.e., financial resources and capacity, needed to develop them.

In view of the weakness of R&D in the public sector, policies are needed to ensure that it is adequately maintained/restructured, in order to generate capacity in line with national and international demand. This applies even to countries that currently have a greater capacity to generate technology, such as Argentina, Brazil, Chile, and Mexico.

The poorer countries need policies for innovation and coordination of research with extension services and technology transfer, which make it possible to generate, adapt, and adopt technologies suitable for the most vulnerable sectors.

In the case of technology transfer and extension services, greater financing as well as a restructuring of capacities, infrastructure, procedures and focal points are required. In this scenario, the system is in a very weak condition today in most countries. Even the few countries that invest most in these activities, i.e., Cuba, Brazil, Paraguay, and Peru, need some assistance to remain efficient.

The research agenda for a weakened R&D system in which demand is diversified and problems are acute requires a major effort to establish priorities for the allocation of scarce resources. Epidemics, the impact of climate change, and food security are competing for these resources. This means that R&D must receive and act on strong, precise, leading strategic proposals on the areas of research to pursue.

3.5.2.2 Implications for sustainable development policies

According to this scenario, international trade restrictions are one of the principal factors determining the sharp decline in virtually all conditions in the LAC countries. Consequently, policies to reduce barriers to Latin American agriculture are needed. On a global level, policies promoting multilateral relations would be important, as a way of avoiding such a negative scenario, especially for the most vulnerable countries.

The division among countries and regions in this scenario calls for regional cooperation to overcome intra-regional weaknesses in capacity and infrastructure, among other things; thus governments should give consideration and priority to this.

In view of the greater risk of epidemics, the effects of climate change, and environmental deterioration, special policies are also required to train and assist the most vulnerable groups to overcome the vulnerabilities prevalent in this scenario. The Central American and Caribbean countries are most affected by these negative influences. There are also losses in South America, due to climate change.

As for food security, which is highly compromised in this scenario in virtually every country, food importing countries have the option of planning and implementing policies to overcome their dependence or, if this objective cannot be attained, to establish mechanisms to assist their most vulnerable population segments.

Education policies to facilitate access by the most vulnerable sectors also need to be implemented, as do policies to compensate for the impact of migration and food security, mainly in the poorest countries.

3.5.3 Life as it is

3.5.3.1 Implications for innovation policies

In view of trade restrictions and to make agricultural products more competitive, product differentiation is needed based on innovation, but this is only presented as a strategy towards the end of this scenario.

In view of the heterogeneity of the region, R&D must also focus on increasing efficiency, by reducing production costs and increasing productivity, or both, and on producing low-cost foods for domestic consumers and low-income countries.

Moreover, it is important to meet the technological needs related to improvements in the quality of products produced by the most vulnerable groups in response to the more exacting demands of better educated consumers.

Research is needed to adapt to and mitigate the effects of climate change and to prevent and manage pests and diseases, while preventing environmental deterioration, so that production efficiency and productivity will not decline. Since the scenario is based on the current reality, the South American countries are the ones with the greatest capacity to deal with these impacts in this area and in the future.

Countries that have the greatest capacity to generate knowledge today—in South America: Argentina, Brazil, Chile; and in the Andes: Venezuela; in Central America: Mexico and Panama; and in the Caribbean: Cuba, Trinidad and Tobago—also are most likely to generate the knowledge demanded by this scenario.

In this scenario, many countries in the region are limited to importing technology, in a world where productive processes and trade are integrated. This means that these countries have a limited capacity to cope with unforeseen risks, and are therefore more vulnerable. It is important to plan and implement mechanisms to improve their capacity to produce know-how and technology, through specific programs or well-defined objectives, and to consider alternatives for sharing the scarce available resources.

Environmental and social issues are not adequately taken into account by all countries in their research activities. At the outset of the first period, a few countries include this concern in their portfolio of R&D projects. However, this effort needs to be stepped up to enhance knowledge of ecosystems and of the impact of agriculture on them and on environmental services.

The application of traditional knowledge is only just beginning to be seen towards the end of the period. R&D should therefore be guided by proactive policies to incorporate this knowledge in generating know-how and technology.

Specific funds and project portfolios geared to more vulnerable population groups would be important alternatives to consider, to ensure that R&D pays attention to these groups. Research organizations should also acquire more expertise than they have today regarding the technological demands of the neediest social groups whose livelihood is agriculture.

Transnational companies become a relevant stakeholder in R&D, and the public sector loses ground. Integral management and investment policies in public R&D need to be implemented, to ensure that not only short-term economic demands are considered in this research. Moreover, it is important to implement proactive mechanisms to increase participation of private organizations in generating know-how and technology in strategic economic and social areas, when the countries do not have the capacity to do so. The technology produced must be accompanied by a transfer of the capacity and knowledge needed to continue this process.

3.5.3.2 Implications for sustainable development policies

In view of persistent management instability, as a result of changes in administration without policy continuity, stability mechanisms are needed in government management, to ensure the continuity of long-term policies. This is particularly important to ensure quality education, which requires consistent and sustainable policies, especially in certain countries that are weak in this area.

As a result of climate change and the increase in food prices, some countries must implement policies to ensure access to quality food.

For poor countries and peasant production, specific policies are needed to assist them in incorporating sustainable practices in their production processes.

3.5.4 Adapting Mosaic

3.5.4.1 Implications for innovation policies

This scenario is based on huge climate changes and social crises, which governments are unable to manage without the assistance and empowerment of various social groups. These changes will probably have a greater effect on South America, because of its larger size and environmental restrictions on production of crops without irrigation, than on Central America, due to its smaller production capacity, which will also have repercussions on food security in the countries of this subregion.

The environmental issue and the reduction of the effects of climate change are critical to this scenario, which seeks to achieve these objectives by using all types of available knowledge, including biotechnology and nanotechnology, agroecology, and traditional knowledge. Thus the R&D agenda should already be oriented to these objectives. There is also a need to find ways to promote interaction and synergy among the different types of knowledge that can presumably bring benefits to all, in the form of reduced environmental risks. In LAC, countries that already have the capacity to generate technology today (Brazil, Venezuela, Trinidad and Tobago, Cuba, Chile, Argentina, Mexico, and Panama) and technology transfers (Cuba, Brazil, Paraguay, Peru) will possibly be in a better position to engage in this interaction and achieve this synergy.

R&D should be directed to understanding and solving environmental and climate change problems. This requires an understanding of the interaction among ecosystems, and between them and the new technologies, and of the possible international effects on shared natural resources.

In view of the longer time required to obtain results in a scenario which requires that R&D consider all affected groups, where the environment imposes restrictions on the independent development of science, and where there is a need for more efficient use of resources, it is imperative to focus on improving management of R&D, with the integration of all stakeholders.

3.5.4.2 Implications for sustainable development policies

Adapting Mosaic is a scenario that requires many institutional changes, which is strongly reflected in governance and development policies in countries. While some countries that are generally less vulnerable today, such as Argentina, Brazil, Chile and Mexico, will have major difficulties in adapting their laws, regulations, and practices to the new times, other countries will have similar problems because they do not have political stability and efficient governments. These are the countries with the worst problems of governance and integrated development policies at the present time. All of these countries should consider the possibility of designing stable policies aimed at improving environmental protection, providing greater access to quality education, and increasing the capacity to guarantee food security to their people in future.

Food security and the common environmental issue are the two major sources of concern in this scenario. For the first, it is important to identify alternatives that will not jeopardize environmental protection but will provide the growing, increasingly educated, hence more demanding population to have access to quality foods.

The scenario offers conditions for public support to facilitate initiatives to protect common natural resources, as part of the same environmental protection approach it favors.

3.5.5 TechnoGarden

3.5.5.1 Implications for innovation policies

This scenario is triggered by a strong impact on climate change, together with social movements initiated in European countries in favor of diversification of agriculture, and geared to protecting the environmental services of ecosystems. Societies cope with their problems by anticipating and identifying specific technological solutions.

Agricultural diversification is already beginning in the megadiverse LAC countries (Bolivia, Brazil, Colombia, Costa Rica, Ecuador, Mexico, Peru and Venezuela. But not all of these countries initially have the capacity to conduct the research needed to obtain an adequate economic return from different environmental services. Brazil, Colombia, Mexico and Venezuela are in the best position to do so.

Environmental protection, an understanding of ecosystems and the environmental services they provide, the correction of anthropogenic aggression against nature, interaction among the different socio-economic, cultural, and environmental systems, and the creation of differentiated products by technological innovation (always with a low environmental impact), and new processes for diversification of agriculture constitute the main items on the technological agenda in this scenario.

This is a scenario that gives preference to the growing integration of knowledge of all kinds, whether formal or traditional. Thus, more than in any other scenario, this world is governed by knowledge, which at the same time strongly drives it, leading to the development of a new understanding of the systems and their integration.

It is also a world in which all social groups are covered by R&D, while at the same time the development of new products and processes intensifies, as does the anticipation of problems, especially in relation to the environment. Consequently, a large capacity for management and planning of the development of know-how and technologies is also needed. Here the scenario differs from *Adapting Mosaic,* where the issue of the speed of technological development is not as important.

3.5.5.2 Implications for sustainable development policies

In the world of *TechnoGarden,* agriculture is only one part of the agroindustrial complexes that offer differentiated products based on technology as well as environmental protection processes. There are no more small-scale producers, as they were displaced to the cities.

This means that new institutions and institutional arrangements need to be created to support this new paradigm, but they are also required to monitor its benefits and risks for society. Countries that already have the capacity to generate technology and megadiverse countries, which are encountering environmental protection pressures and already have relevant laws, will find it easier to adapt to the new paradigm.

Unemployment is one of the major problems in this scenario. It will have a greater impact on countries whose current population is characterized by low levels of education, such as Peru, Bolivia, Honduras, and the Dominican Republic. In these countries especially, policies that will lead to the creation of new job opportunities can be implemented, in areas such as diversification of agriculture, enterprises related to the new agriculture-based products, or reductions in workload.

Despite the concern over the environment, new environmental problems emerge, as a result of the technological solutions tried out in this scenario. R&D needs to be oriented to achieving a systemic, in-depth understanding of ecosystems, biological systems, and their interaction, and also to adequate monitoring of these ecosystems and the impact of technologies on them, which is already included in this scenario as a way of solving these problems.

References

Anderson, K. y E. Valenzuela. 2006. Do global trade distortions still harm developing country farmers? Pol. Res. Working Pap. 3901. World Bank, Washington DC.

Bora, B., A. Kuwahara, y S.Laird. 2002. Quantification of non-tariff measures. Policy issues in international trade and commodities. Study Ser. No. 18. UNCTAD, Geneva.

Byerlee, D., y K. Fischer. 2000. Accessing modern science: Policy and institutional options for agricultural biotechnology in developing countries. 24th Conf. Intl. Agr. Econ. Assn. 14-18 Aug 2000. IAAE, Berlin.

Carpenter, S.R.P., P.L. Pingali, E.M. Bennetty and M.B. Zurek (eds) 2005. Ecosystems and human well-being: Scenarios, Vol. 2. Island Press, Washington DC.

Castro, A.M.G. de, S.M.V. Lima, J.S. Silva, A. Maestrey, J. Ramirez Gaston, J.S. Guerra et al. 2005. Proyecto quo vadis: El futuro de la investigación agrícola y la innovación institucional en América Latina y el Caribe. 1ª. ed. IFPRI, Red Nuevo Paradigma, Quito.

Castro, A.M.G. de, S.M.V. Lima, M.A. Lopes, M. dos S. Machado, y M.A.G. Martins. 2006. O futuro do melhoramento genético vegetal no Brasil: Impactos da biotecnologia e das leis de proteção de conhecimento. EMBRAPA, Brasilia.

CEPAL. 2006. Panorama de la inserción internacional de América Latina y el Caribe. CEPAL, Santiago de Chile.

CEPAL. 2005a. Panorama social de América Latina. Cepal, Santiago de Chile.

CEPAL. 2005b. Objetivos de desarrollo del milenio: Una mirada desde América Latina y el Caribe. CEPAL, Santiago.

De Ferranti, D., G. Perry, W. Foster, D. Lederman, y A. Valdés, 2005. Beyond the city: The rural contribution to development. World Bank, Washington DC.

Dutfield, G. 2001. TRIPS-related aspects of traditional knowledge. Case Western R. J. Int. L. 33:233-275.

Filgueira, C., y A. Peri. 2004. América Latina: los rostros de la pobreza y sus causas determinantes. Proy. Reg. Pobl. CELADE. CEPAL, Santiago de Chile.

Fischer, G., M. Shah y H. van Velthuizen. 2005. Climate Change and Agricultural Vulnerability. IIASA, Vienna.

Jaffee, S., K. van der Meer, S. Henson, C. de Haan, M. Sewadeh, L. Ignacio et al. 2005. Food-safety and agricultural health standards: Challenges and opportunities for developing countries export. World Bank, Washington DC.

Jones, J.G.W. 1970. The use of models in agricultural and biological research. The Grassland Res. Inst., Hurley.

Kaufmann, D., A. Kraay, y M. Mastruzzi. 2006. Governance matters V: Aggregate and individual governance indicators for 1996-2005. World Bank, Washington DC.

Kjöllerström, M. 2006. Gasto público en el sector agrícola y las áreas rurales: La experiencia de América Latina en la década de los noventa. *En* B. Fernando Soto et al. (ed) Políticas públicas y desarrollo rural en América Latina y el Caribe: El papel del gasto público. RLC-FAO, Santiago de Chile.

Lima, S.M.V., A.M.G. Castro, M.S. Machado, N.A. Santos, M.A. Lopes, J.R.P. Carvalho et al. 2005. Projeto Quo Vadis: O futuro da pesquisa agropecuária brasileira. 1ª. ed. EMBRAPA Informação Tecnológica, vol. 1. Brasília.

Morón, C., L. Alonso, y M. Crovetto. 2005. Cambios en la estructura del consumo de alimentos y nutrientes de América Latina: 1979-1981 a 1999 a 2001. RLC-FAO, Santiago.

Niosi, J., y S.E. Reid. 2007. Biotechnology and nanotechnology: Science-based enabling technologies as windows of opportunity for LDCs? World Dev 3:426-438.

PANAFTOSA. 2006. Situación de los programas de erradicación de la fiebre aftosa. América del Sur, 2005. PANAFTOSA, Río de Janeiro.

Ramírez-Gastón R., J., J. de S. Silva, A.M.G. de Castro, S.M.V. Lima, V.A. Hart et al. 2007. Proyecto Quo Vadis: El futuro de la innovación tecnológica agraria en el Perú. Ethos Consult SRL y Red Nuevo Paradigma, Lima.

Renard, M.C. 1999. The interstices of globalization: The examples of fair coffee. Sociol. Ruralis 39: 484-500.

RICYT. 2007. Indicadores por país. Available at www.ricyt.org.

RLC-FAO. 2004: Tendencias y desafíos en la agricultura, los montes y la pesca en América Latina y el Caribe. RLC-FAO, Santiago de Chile.

RLC-FAO. 2006. GPR-Base de datos de estadísticas e indicadores de gasto público agrícola y rural. Available at http://www.rlc.fao.org.

Saldaña, A., J.A. Espinosa, G. Moctezuma, A. Ayala, C.A. Tapia, y R.M. Ríos. 2006. Proyecto Quo Vadis: El futuro de la investigación agropecuaria y forestal y la innovación institucional de México. INIFAP, México.

Santamaría G., C. Guerra, J. Macre, V. Guillén, I. Ruiz., J. Souza et al. 2005. Escenarios futuros para la tecnociencia y la innovación agropecuaria y forestal en Panamá. Inst. Investigación Agropecuaria de Panama, Ciudad de Panamá.

Schneider, C., A. Roca, C. Falconi, A. Belotto, y A. Medici. 2007. Avian and human pandemic influenza: Addressing the need for integration between health and agriculture in the preparedness plans in Latin America. No RUR-0702. IDB, Washinton DC.

Simonis, F., y S. Schilthuizen. 2006. Nanotechnology: innovation opportunities for tomorrow's defence. TNO science & industry. Available at http://www.futuretechnologycenter/downloads/nanobook.pdf.

Varela, L., y R. Bisang. 2006. Biotechnology in Argentine agriculture faces world-wide concentration. Electron. J. Biotech. 9:3. Available at http://www.ejbiotechnology.info/content/vol9/issue3/full/27/.

WIPO. 2001. Intellectual property needs and expectations of traditional knowledge holders. WIPO Report on fact-finding missions on intellectual property and traditional knowledge (1998-1999). WIPO, Geneva.

World Bank. 2003. World Development Indicators. Wolrd Bank, Washington DC.

4

AKST in Latin America and the Caribbean: Options for the Future

Coordinating Lead Authors:
Inge Armbrecht (Colombia), Flavio Avila (Brazil)

Lead Authors:
Jorge Blajos (Bolivia), Patrick Lavelle (France), Dalva M. da Mota (Brazil), Lucia Pitalluga (Uruguay)

Contributing Authors:
Sergio Salles Filho (Brazil), Jorge Irán Vásquez (Nicaragua)

Review Editors:
Edelmira Pérez (Colombia), David E. Williams (USA)

Key Messages

1. Interaction of systems. While the great diversity of AKST systems is a major strength in Latin America and the Caribbean, these systems need to interact with each other. The interaction can be achieved through the generation, sharing and dissemination of experiences and by combining and developing the knowledge inherent in the three production systems (conventional, agroecological and traditional) in order to overcome weaknesses and share strengths.

2. Systemic vision. Develop a systemic vision of production systems by combining the strengths of the agroecological, conventional and traditional approaches in assessing (in the short, medium and long term) the results of all three in terms of cost-benefits and in light of the goals of IAASTD, in other words, to ensure environmental, social and economic sustainability. This approach considers the social, economic and environmental impact of the application of technologies to agricultural, forestry and livestock production, regulatory and support environmental services and farming extension services.

3. Prevention and environmental restoration. Poor management of water, soils and forests is a serious problem throughout Latin America and the Caribbean. The AKST systems should therefore focus urgently on prevention, conservation and environmental restoration and seek to reverse the degradation caused by intensive agriculture. One option may be to adopt technologies that help to restore the natural capital (soil, water, etc.) while at the same time achieving the goal of production of environmental goods and services.

4. Recovery and conservation of traditional knowledge. Promote and intensify efforts to rescue, develop and preserve ancestral knowledge by and for local and indigenous communities by empowering local communities and combining their know-how with agricultural knowledge. Promote training and the generation and ownership of fresh knowledge by local producers and consumers. Develop mechanisms for more effective organization (developed by AKST) of small- and medium-sized producers.

5. Biodiversity. AKST systems should focus their strategies on the conservation and proper management of biodiversity. Biodiversity, at different levels (genes, species, ecosystems and landscapes) performs ecological functions, which are the functions that produce environmental goods and services. Biodiversity is an important source of opportunities for the development of new products and new economic activities. Through proper management, it is able to respond to the growing demand for food and other products in a context of economic and climatic changes. Environmental services (such as carbon sequestration, ecotourism, landscaping, or the storage and purification of water) need incentives for conserving biodiversity, beginning with the preservation of natural habitats and the diversity of landscape ecosystems.

6. Participatory approach. Promote a participatory approach in processes for the generation and socialization of knowledge and in the various development strategies. This approach will help to reconcile the different expectations of the various actors, producers, researchers, officials and others.

7. Research for small-scale producers. AKST systems should prioritize research (basic, applied, adaptive and strategic) to meet the demands of small producers: to improve the quality of life of local populations and to promote social and gender equality and a healthy environment and not only to improve productivity. Without an understanding of the underlying environmental and social mechanisms that result in inequality, hunger and environmental degradation, it is difficult to address the root causes. Basic research is therefore also necessary. One option is to promote interdisciplinary research to identify the relationships that were never established between the socioeconomic environment, productive landscapes, the biodiversity that is present in these landscapes and the ecological functions that it performs.

8. New institutional mechanisms for knowledge sharing. Create institutional mechanisms to promote knowledge sharing between AKST stakeholders. The synthesis of knowledge and its socialization/dissemination within the three production systems (conventional, traditional and agroecological) requires the use of new institutional tools tailored to each situation. It would be particularly useful to institutionalize the knowledge sharing systems used by private organizations (NGOs, foundations, etc.) and various research and development programs. This would allow for the continuous training of all sectors of society geared to the needs and technologies for the sustainable management of resources.

9. Strengthening of networks. Promote cooperation among AKST systems in the countries of Latin America and the Caribbean in the sharing of knowledge and skills. Given the limited human, financial and infrastructural resources available at the national level, research programs (platforms) among AKST stakeholders must be integrated by strengthening existing regional networks and cooperation programs. Regional networks and programs should also extend their activities to other AKST stakeholders, since they are currently mainly restricted to public actors, particularly Nacional Agricultural Research Institutes (NARIs).

10. International cooperation. Expand and strengthen cooperation activities and promote joint research by AKST systems in Latin America and international centers (for example, the Consultative Group on International Agricultural Research (CGIAR) and the Tropical Agricultural Research and Higher Education Center (CATIE)), national research institutes and universities,

among others, in developing countries. The AKST systems in Latin America and the Caribbean must place emphasis on the development of research projects and on the training of personnel in centers of excellence so that the region could reduce the current technological gap and not remain on the margins of the major technological advances being made in other parts of the world. These efforts to promote cooperation must be aimed at strengthening the technical and scientific capacities of AKST stakeholders and thereby improve their impact on reducing poverty and hunger in the region.

11. Emerging technlogies. Channel research in new fields of knowledge (biotechnologies—molecular or other—and nanotechnogies, among others) towards the achievement of the goals of reducing poverty, hunger, malnutrition, human health and environmental conservation. Give priority in this process to the development of products based on these new technologies that benefit small producers (family businesses) by seeking to maximize their social, economic and environmental impact while observing the precautionary principle. AKST systems must exploit the advantages of these emerging technologies while ensuring their biosecurity.

12. Biosecurity. Contribute to the strengthening of national biosecurity committees. AKST systems must act effectively in the development of impact analysis and assessment of the potential risk of the products they research to ensure that their adoption would not cause problems for the environment or for consumers. AKST systems must ensure the biosecurity of the results of their research programs, based on the principle of precaution.

13. Organizational models. Create and/or strengthen AKST organizational models. Given the limitations of the organizational (structural) models of the various actors that comprise the AKST in many countries of Latin America and the Caribbean, one option would be to create new models or to strenghten the existing ones. Emphasis should be placed on sharing experiences in the adoption of different organizational models by the governments of the region and AKST stakeholders that enhance the efficiency and effectiveness (while preserving democracy and equity) of their respective systems.

14. Models of governance. Strengthen and modernize management models. One of the main problems of AKST stakeholders is the absence of models for the management of their organizations. Since there are many successful experiences in AKST management in the region, these experiences should be shared among AKST stakeholders.

15. Interaction between AKST systems and the society. Promote interaction between AKST systems and consumers, movements and social organizations. Promote structures that facilitate dialogue between them and other social actors and AKST systems. There must be greater participation by social actors in the identification and especially in the selection of research priorities and in the evaluation of results.

16. Intellectual property. Enactment of national legislation that recognizes collective intellectual property rights. National TRIPS legislation provides for the possibility of formulating strategies within the established multilateral framework. However, legal recourses may not be the most appropriate ones for protecting the knowledge that traditional (ethnic) communities have accumulated over centuries and must be adapted to their new purposes.

17. Teaching and training programs. Teaching curricula may be revised to include elements of AKST. For its part, AKST must also evolve to adapt to the transitions in formal education (from primary school to university) and in continuing education and training programs for all members of society. Advantage must also be taken of other types of training and information such as the communications media, which can be associated with this effort through special programs. In this way, it may be possible to preserve crops and agricultural knowledge and research by and for local and indigenous communities.

18. Inclusion of women in AKST systems. Promote increased participation by women in the management of organizational models, the generation and socialization of knowledge, and in the various development strategies. The participation of women in each aspect of AKST has increased over the past 15 years but is still very limited, despite the increasing numbers of women in leadership positions in both productive activities and producer associations and organizations.

19. Accountability to society. Impact studies must be carried out by AKSTs to account for investments and to demonstrate to society the importance and impact of the products they generate. These studies should include an analysis of all the impacts of their products (economic, social, environmental and others) together with a program to communicate their results that is geared to the various stakeholders (actors—individual and collective—have very different levels of training and access to information).

20. Formulation of public policies. AKSTs should participate proactively in the formulation of public policies related both to the system itself and to the policies supporting the system. AKST stakeholders, including publicly funded national agricultural research institutes, have traditionally had little real participation in the preparation of the legal framework in which they operate (biosecurity and intellectual property legislation, financing of research and development, credit policies, etc.). Generally speaking, their impact on public policies is limited to submitting reports with the results of their research.

4.1 Conceptual Framework

For purposes of the IAASTD, the agricultural production systems in Latin America have been classified as follows (Chapter 1):

1. Traditional/indigenous production systems;
2. Conventional/agroindustrial production systems;
3. Agroecological production systems.

Each system of production is associated with a body of knowledge, science and technology that sustains and promotes it. Together these bodies of knowledge comprise the system of agricultural knowledge, science and technology. However, while this store of knowledge belongs to very different institutional and social systems, they are—or will have to be—permeable and must interact with each other, and it makes no sense therefore to establish vertical limits between them.

Chapter 4 identifies the principal options for making AKST work more effectively to achieve the goals of reducing hunger and poverty, improving rural living systems, improving nutrition and human health; and promoting equitable and sustainable environmental, economic and social development in Latin America and the Caribbean.

In terms of structure, the chapter has been divided into two main sections:

1. Options for enhancing the impact of AKST systems.
2. Options for strengthening the capacities of the System to generate, socialize, access and adopt AKST.

The system of knowledge and the institutions and organizations concernid with its generation and socialization are very separate and in most cases do not interact with each other. None of the systems of production, in their current state, whether conventional, traditional or agroecological, contributes at the same time to meeting the requirements of environmental sustainability and social and economic development. Thus, for example, it is evident that the conventional system has negative impacts on the environment, that the traditional system is failing to bring populations out of poverty and that agroecological systems still have not acquired the technological maturity that would make them acceptable and applicable under any conditions. However, Badgley and colleagues (2007) found, in a quantitative meta-analysis, that organic agriculture could today succeed in feeding the human population of the world (Badgley et al. 2007).

The different social groups in Latin America and the Caribbean exhibit a sometimos marked separation between land-use methods and the AKST storehouses on which they rely. The knowledge is generated and acquired in five main types of institutions that are generally separate and which can be completely unaware of the knowledge possessed by other types. This is the case in the institutions identified in the diagram in Figure 4-1 in which local knowledge (disseminated locally within the family and social groups) has very few or no links at all to the conventional/agroindustry model (see Chapter 1) taught in universities and centers of advanced learning.

The future development of agriculture in Latin America will depend on improvements in each one of the three bodies of knowledge mentioned above and, more than anything else, on the incorporation into each one of them of the elements needed to mitigate the negative effects of each one: the negative environmental impacts of some, and the low productivity or incapacity of others to reduce poverty and inequality. The relations expressed in the triangle in Figure 4-2 are explained by the following examples:

Example 1. Pole 1 represents a system of traditional agriculture in tropical forests of Latin America and the Caribbean (clear, slash and burn), where traditional local AKST is used. The introduction of the practice of leaving land fallow and improved with the planting of vegetables shifts this category towards number 1; a situation in which the availability of good quality forage reduces the pressure on pastureland and therefore allows degraded areas to recover and/or the need to transform more forest into pastureland. The use of improved varieties and the inoculation of beneficial organisms (e.g., Rhizobium or Bacillus thurigiensis) would move them towards pole 3.

Example 2: pole 2 is an agroforestry system based on an agroeological AKST, using greenery of multi-use leguminous plants and annual crops of maize. The addition of chemical fertilizers (e.g., P, K) to organic fertilizers in order to improve the balance between the supply of available nutrients and the needs of plants, use of better selected plants and crops that trap certain pests (e.g., rows of okras between maize) would take it towards pole number 2.

Example 3: lastly, pole 3 is a soy monoculture based on a conventional AKST with annual plowing, fertilizing and pest control with chemical products. The abandonment of arable land and the movement towards a system of reduced plowing and the application of organic fertilizers and plant cover move it towards pole number 2.

The methodology used to identify options for improving the impact of the system of scientific and technological knowledge in agriculture was based on a double entry matrix in which each option proposed was analyzed in the context of the sub-regions and the goals of IAASTD. The options for the future were analyzed schematically based on the three extreme systems of agricultural production (and the bodies of knowledge that sustain them) (See Figure 4-1)

This chapter seeks to identify the principal options for making AKST work more effectively to achieve the goals of sustainability in Latin America and the Caribbean. It is therefore necessary to seek options for: (1) improving the impact of the AKST. This section contains four parts: diversity of AKST in Latin America and the Caribbean; sustainable environmental and socioeconomic development; climate change and bioenergy; and biodiversity. (2) Strengthen capacities to generate, socialize, access and adopt AKST. The options in each one of these two parts are presented below.

4.2 Options for Strengthening the Impact of AKST Systems

4.2.1 Diversity of AKST bodies in Latin America and the Caribbean

AKST systems must interact more and differently. This goal could be achieved by exchanging experiences and comparing the different types of know-how and skills in order to address weaknesses and share strengths. The great diversity of AKST systems in Latin America and the Caribbean is its main strength. One type of knowledge does not exclude the other.

4.2.1.1 Integration of AKST systems

The management options being pursued in Latin America and the Caribbean combine in different proportions the

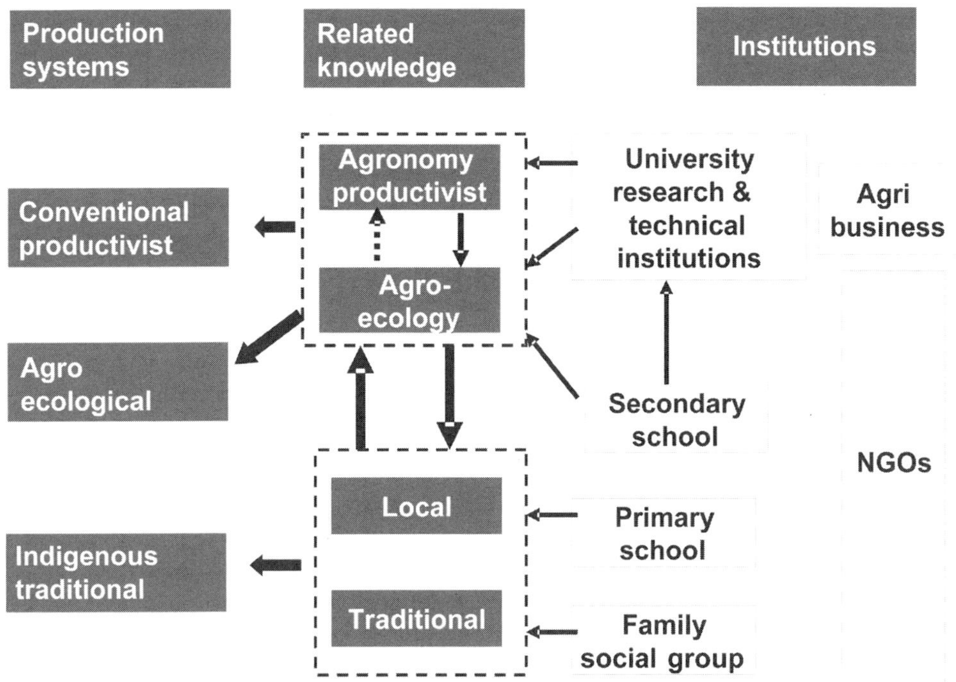

Figure 4-1. *Relationship between production systems, types of knowledge used and institutes involved in its generation and dissemination.* Source: Authors' elaboration

various types of knowledges and technologies (see Chapter 1 and Figure 1-1). The conventional approach taught by universities and advanced technical institutes strongly advocates agroindustry; the agroecological approach espoused by universities and some NGOs serve to create more diverse systems in terms of the production of environmental goods and services; and the local/traditional know-how imparted by families and local social groups is based on extraction and low-input family-based agriculture. It is clear that in the particular case of each management typology, the AKST system used employs different proportions of each type of AKST. While projects undertaken have shown how systems interact with each other (FAO, http://www.fao.org/ag/agl/agll/farmspi/; Settle et al., 1996), greater effort is needed to expand and institutionalize these initiatives.

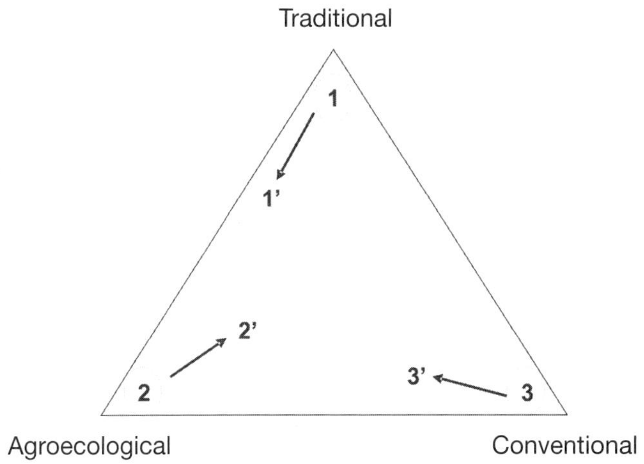

Figure 4-2. *Types of production systems in space defined by types of AKST (triangle) bodies used and possible trajectories depending on implementation of alternative AKST systems.* Source: Authors' elaboration

It is argued that the different systems replicate their drawbacks and reinforce their potential through the integration of elements of other typologies directed always towards the achievement of development and sustainability goals. The option proposed therefore argues that it is necessary to bring about changes in the respective systems that bring them closer to other systems in order to take advantage of their strengths and to optimize the practices of each of the three groups identified. In an ideal world, the differences between locally observed practices should not depend on access to resources and economic assets and possible access to formal education, but rather on how producers can adapt to the restrictions imposed by the environment and to market conditions.

There is need for partnerships between researchers, extension workers, producers and producer associations for the pursuit and sharing of research. This would be only one step in a very complex process that requires more than partnerships. The conventional model that separates those who conduct research from those who disseminate the results and, above all, from those who use the resulting product showed that many of the alternative technologies generated are of little use, especially to traditional producers (families, indigenous groups) (Salles Filiho and Souza, 2002; Embrapa, 2006).

It is also necessary to encourage a debate on what should be the role of each of the agents of AKST and which scientific, technological and innovation policies should be applied in rural areas of Latin America. Parallel initiatives result in wasted efforts, resources and time and do not allow for the development of a common approach to dealing with challenges that have already transcended national frontiers (FAO, 2003). An ongoing training program is recommended for the design and implementation of scientific, technological and innovation policies. This is necessary in order to meet the challenges of social inclusion and the new market demands in which protecting the environment and the health of consumers are prerequisites (Embrapa, 2004; FAO, 2003).

4.2.1.2 Priority to research that supports development and sustainability goals

AKST systems should give priority to research whose practical results demonstrably contribute to the goals of IAASTD, that is to say, that can point to improvements in the quality of life of local populations, promote social and gender equality (in other words, respect for the differences between men and women), a healthy environment and increased productivity. Basic research is essential to understanding the underlying ecological and social mechanisms that create inequality, hunger and environmental degradation.

It is clearly not possible to achieve development and sustainability goals one at a time, since the goals are all interconnected. In Amazonia, for example, the promotion of cattle raising as the only land exploitation system can lead to satisfactory living standards for certain groups, but the disadvantages in terms of gender equality and equality between social groups and the degradation of environmental functions make it unsustainable. Only a holistic vision in which the different goals are all considered together and the best trade-off sought between them in the socio-economic and biophysical conditions of the land can achieve this goal.

The need for a holistic approach to research is related to the analysis of the relationships between the different parts of the system on the property but also between the socio-economic environment and the landscape created by human activities that tranform the natural environment. In the mountainous regions of Central America, for example, primary forests are being transformed into a mosaic of secondary forests, coffee plantations, pasture land and fields of maize. The proportion of this type of use depends on market conditions, means of access to land ownership and on many other socio-economic variables. Depending on the diversity of the landscape, the degree of transformation and the intensity of use of the land, this landscape may include different levels of biodiversity that in turn will participate in different ways in the provision of environmental goods and services (for example, soil conservation, storage and purification of water, carbon sequestration or biocontrol of pests) (Mattison and Noris, 2005). Implicit in this representation is the improvement of human well-being with sustainable production at a high level of environmental goods and services.

It is essential to understand the relationships between these different entities, identify the drivers and the threshold impact in relationships in order to model this system of interactions and improve the management of all resources, whether human, economic or ecological. For example, it is known that transformed landscapes are capable of resisting the invasion of (invasive) species if the proportion of the natural ecosystem does not fall below a certain threshold and if the trend toward more intensive land use is slowed. However, the inner workings of the mechanisms affecting these qualities of the various agroecosystems (and various landscapes) are not known, although the positive role of biodiversity is well established (Kennedy et al., 2002).

It is also necessary to prioritize the research on options for increasing incomes (returns) and conserving biodiversity with a gender perspective. Initiatives targeted to women improve family incomes and help realize the potential of the know-how accumulated by them and which is only now beginning to be recognized (Cavalcanti and Mota, 2002).

4.2.1.3 Development and strengthening of agricultural programs to generate and increase the value of knowledge for local and indigenous communities

Of the three types of knowledge of AKST that have been identified, the traditional/indigenous is the least formalized and thus the most threatened. Preservation of the cultural services and biodiversity that this system sustains can be done by the development and strengthening of educational programs, crop preservation and knowledge retention, and agricultural research by and for local and indigenous communities. This, in turn, can be achieved by empowering local communities and combining their know-how with agroecological expertise, taking into account the fact that local and indigenous know-how is generated and disseminated within small social groups (family, town, association).

Rarely is this local and indigenous know-how broadly formulated and recognized outside the local environment. This situation makes it difficult to use and develop the capacity to observe and understand the functioning of the ecosystems developed by these populations (Veiga and Albaladejo, 2002). Knowledge of the functions of local biodiversity and other natural resources would be very useful in developing agroecology that depends to a great extent on intimate knowledge of the natural conditions that are peculiar to each region/crop. This knowledge should also enrich conventional knowledge to help correct any negative environmental impacts of these practices without reducing their productivity and economic value. To this end, one promising option would be to put in place instruments to regulate access to traditional knowledge.

Properly focused, traditional knowledge, science and technology can lead to development and social well-being. To achieve better coordination between higher education programs and programs in science and technology, both in research and in the transfer of knowledge, requires a reorganization of academic and scientific research institutions in all areas and ending the isolation and dispersion that currently exists. It is necessary to strengthen educational and occupational training programs that promote and respect diversity and differences and permit advantage to be taken and use made of the positive elements of the agricultural revolution that is under way, while also combating and managing the crushing force of this very revolution as we enter into a new paradigm of agricultural science and technology (Sanchez, 1994).

4.2.1.4 Promotion of advances in agroecology as cutting-edge technology

Agroecology needs incentives for it to become cutting-edge technology, while evaluations (short, medium and long-term) are made of its results in terms of cost benefits. Agroecology has the potential to efficiently reverse the trend towards degradation of the environment and to contribute to food security, as it has demonstrated, despite the limited investment and attention that it has received (CIFAA, 2006).

Agroecology seeks a balance between, on the one hand, agroindustry that holds the promise of achieving the goals of poverty reduction through the production of food and other agricultural, forestry and livestock goods and, on the other, the maintenance of biodiversity and other environmental services while also pursuing the goal of social well-being. The basic paradigm is that the more similar the agricultural, forestry and cattle-farming ecosystems are to the natural ecosystem the more sustainable are medium- and long-term production and other environmental services, such as the recycling of nutrients, carbon sequestration in soils, and water percolation, detoxification, regulation and storage (Altieri, 1995). Agroecology cannot be developed without the active participation of an entire social network. One possibility is therefore to introduce agroecology into agricultural research and extension programs and in rural schools. It is also necessary to promote internal changes in the institutions seeking to adopt this new paradigm (Sanchez, 1994; Caporal and Costabeber, 2004; Guzman, 2004; EMBRAPA, 2006). Local and regional agroecological experiences with change must be systematized by promoting exchanges between different regions and between the different social actors. The crisis of the agroindustry model requires new alternatives that are less harmful to the environment and to producers and consumers (FAO, 2003; Caporal and Costabeber, 2004; Guzman, 2004). However, extension activities and the sharing of experiences are not sufficient by themselves. Knowledge and innovation must also be further developed with a view to mainstreaming the agroecological approach into production systems (Vandermeer, 1995).

In order to develop ecological production systems compatible with the sustainable management of natural and human resources, a development style consistent with these goals must be sought. For this reason, basic research should also be complemented by market research and economic studies to both demonstrate and increase the profitability of agroecology (Vandermeer, 1995; Swift et al., 1996). Comprehensive studies are also needed to analyze and objectively test the assumptions about the low productivity of unconventional systems (for example, organic productions (Badgley et al., 2007). Agroecology should be viewed as a strategic factor for development in Latin American countries, not only as a factor for economic development but also as a key input in social and environmental policy. This requires a series of standard-setting initiatives, institutional reorganization and the allocation of significant economic resources as countries progress towards their broad objective of sustainable development.

Most Latin American countries lack regulatory frameworks for the implementation of an incentive system for agroecological production. It would be helpful if such frameworks included a review of the external factors affecting agroecological production compared with conventional production. This means that, among other things, governments should offer institutional support in the form of technical assistance to producers for production, processing and marketing; a reduction in the costs of certification; and the launching of demonstration projects that facilitate the supply of inputs and services, manage financing, promote arrangements for marketing and finaning of production, and help provide the equipment necessary for production (Ortiz, 2004).

4.2.1.5 Redirecting new areas of research towards development and sustainability goals

AKST systems in Latin America and the Caribbean can seek to increase their technical training in the new areas of knowledge and those generated under the conventional system, although with a critical and cautious attitude (applied to biotechnology, niche agriculture, nanotechnology, organic farming). In order for advances in new technologies to be useful in tropical conditions, they must be adapted and improved for the particular conditions of the agroecosystems. The research priorities for these new areas should first consider the environmental and social development goals and not the profit potential. Consequently, a critical evaluation is needed in order to determine whether or not reputed leading-edge technologies satisfy the goals of sustainable development and which sectors benefit. Conventional knowledge has made it possible, through the Green Revolution, to fulfill urgent needs for an increase in the production of foodstuff (Wood et al., 2005), but this has not necessarily meant greater access by the poor to food (see chapters 1 and 2). This type of know-how must continue to be developed with special emphasis on those types of research that take particular account of the need for the development of a multi-functional agriculture (one that provides ecological services) that meets the socioeconomic and environmental Millennium Goals. Synthesis with the agroecological approaches mentioned in the above paragraph is an essential phase of this process.

Educational systems should also promote solid ethical principles. It is becoming increasingly necessary to incorporate into educational programs a philosophical discussion of the ethical principles of justice, equality, reciprocity, autonomy and responsibility, applying them to the topic of the management of resources, both private and public (Hardin, 1998).

4.2.2 Sustainable environmental and socioeconomic development.

There is currently no state policy that does not set as a priority goal sustainable environmental and economic development. Achieving this goal requires greater efforts in the search for technical solutions, a more practical knowledge of the dynamic of soil and water resources, and urgent reform of management systems to mitigate their negative impact on the environment.

4.2.2.1 Emphasis on the search for more effective solutions to prevent environmental degradation

The degradation caused by improper management of water, soils and forests is a serious regional problem throughout Latin America and the Caribbean. Many unsound management practices have had a severe impact on natural resources (Natural Capital) and environmental services (Adis, 1989; Brown, 1993; Cairns, 1994; Polcher, 1994; Brosset, 1996; Neill, 1997; Rasmussen, 1998; Fearnside, 1999; Ellingson, 2000; Tian et al., 2000; Bierregaard et al. 2001; Portela, 2001). The firm denunciation of these events has led countries like Costa Rica, Mexico and Brazil to develop programs for payment for environmental services (Calle et al., 2002), but the long-term effect of these programs are still not known. Studies are needed on the value and impact of the environmental services provided by ecosystems and on identifying the type of human communities that sustain them (Mattison and Norris, 2005). Such understanding would he helpful in finding strategies for continuity (and respect) of the lifestyles of the local populations directly associated with the management of native ecosystems, thereby balancing the need for rural production with environmental conservation (Daily, 1997; Mattos et al., 2001; Bensusan, 2002; MMA, 2004; Zbiden, 2005).

Another priority need is for research into and dissemination of the use of secondary and other degraded or abandoned habitats. Secondary forests, if properly managed, play an important role as providers of environmental services, protection and maintenance of biodiversity, and protection of water sources and wood products for rural construction, manufacture of domestic utensils, medicinal and ornamental plants, fruits, honey, fiber, oils, resins and seeds, among other things, (Promanejo, 2001; FLOAGRI, 2005). Comparative studies are also needed on agricultural alternatives that do not include burning in Amazonia, since burning has a negative impact on the atmosphere (carbon emissions) and leads to the loss of the nutrients retained by the biomass. New approaches are needed to repair the planting area which are less harmful to the environment and ensure the sustainability of forests, ensuring in particular that the extraction of wood does not negatively impact the conservation of the diversity of plant varieties, or, by extension, of the ecosystem. One option being proposed for the sub-region of Amazonia is the development of a program of study, dissemination and exchange of experiences on the communal management of the natural resources of the Amazon in order to promote public policies that take account of the realities of farmers (co-management and self-management). The experience of local groups in close symbiosis with the forest will contribute knowledge and management styles that would ensure conservation for centuries. Moreover, the combination of this experience is indispensable for the development of agroecological practices. Such experiences should therefore be pursued as an option for promoting conservation and social and environmental sustainability, which are the goals of IAASTD (Barros, 1996; Benatti, 2003; Amaral Neto, 2004).

4.2.2.2 *Study and understanding of the dynamics of basic natural resources*

Water, soils and the biological processes associated with biodiversity are one of the acknowledged bases of environmental and economic sustainability. Generally speaking, there is an urgent need throughout Latin America and the Caribbean to generate technologies for controlling the erosion and degradation of soils and these should go together with the creation of more economic opportunities for small-scale producers, while at the same time recognizing the limitations imposed by the low productivity of labor and the small size of landholdings (Dixon et al., 2001).

Many studies have demonstrated the need to promote practices that closely resemble natural ecological processes for the management of natural resources, control of pests and diseases (Alpizar et al., 1986; Von Maydell, 1991; Kursten, 1993; Jong, 1995; Gallina, 1996; Vohland, 1999; DeClerk, 2000), and the promotion of related biodiversity (Armbrecht et al., 2004). Given that all agroecosystems originated from natural ecosystems, ecological, indigenous and traditional knowledge of agricultural systems (cattle farming, fish farming, growing of crops) must be expanded so as to increase the impact of AKST systems in Latin America and the Caribbean.

Studies must be undertaken and plans developed to manage the impact of agriculture in water basins both in the Amazon and other regions. Expansion of agricultural frontiers is a reality and the use of soils for agriculture leads to chemical modifications of underground and surface waters (Markewitz et al., 2001, 2006). It is necessary to identify which alternatives for agricultural management and environmental conservation minimize these impacts (Markewitz et al., 2001, 2006). In order to achieve sustainability, research and dissemination programs are needed to stabilize the agricultural frontier, to add value to and ensure the sustainability of the resources and environmental services provided by secondary forests, to restore degraded land and encourage the establishment of enterprises through community partnerships for the exploitation of forests and development of non-wood forestry products (Promanejo, 2001; FLOAGRI, 2005).

Moreover, depending on the Latin American landscape in question, the management plan may be supplemented by the introduction, development and dissemination of aquaculture technologies that rely on local ingredients (residues, fruits and seeds) (Mori-Pinedo,1993; Pereira-Filho, 1995) while continuing to further develop local and indigenous knowledge. Aquaculture in Amazonia is based on local particularities (use of local ingredients for fish feed, subsistence and local know-how). The models for enhancing the efficiency of this activity are dispersed and do not include existing production systems. It is necessary to explore fisheries management systems with a view to developing balanced models that would strengthen the capacity of local populations to support themselves.

Traditionally developed strategies are beginning to experience a crisis because of the over-exploitation of resources. The techniques developed from local and scientific know-how (before the former disappear) need to identify ways of restoring balance (Baltazar, 2005). In Andean regions, where intensive and extensive cattle farming is practiced in zones that are extremely vulnerable to erosion, (hillsides, inter-Andean valleys) agricultural, forestry and pastoral technologies need to be developed (Calle et al., 2002). Such technologies would combine the herbaceous, shrub and plant strata to improve production (production of biomass stockfeed for cattle) and to expand the services provided by ecosystems (Murgueitio, 2003). However, it is difficult to achieve the goals of IAASTD as long as the paradigm remains economic growth alone (individual enrichment) without taking into account external factors (environmental and social damage). The intensification of cattle farming should therefore be avoided and emphasis placed on the generation of knowledge for incorporating agrobiodiversity and forestry biodiversity into the production process (Blann, 2006, DeClerk et al., 2006).

4.2.2.3 Improvement of conventional systems in order to reduce and mitigate their negative environmental impacts

Conventional systems may also be gradually transformed into more sustainable systems with the support of the AKST. Given the demonstrably positive impacts of environmentally friendly production, conventional production systems must undergo technical changes to make them less harmful to the environment and to the health of consumers (Fachinello, 1999).

Many regions of Latin America and the Caribbean have large quantities of aquatic and marine resources (e.g., Caribbean, Amazonia, Andes), and what is now required is research into and the dissemination of models of communal management of water resources. There are currently few technical models for an activity that can reduce the pressure on resources at the same time as it generates income for the local population.

For producers working lands on degraded or fragile slopes and who are not prepared to use this land for forest cover, their priority should be to promote sustainable production styles that can be easily adopted with limited resources and which produce relatively quick and attractive returns, either in terms of production or in the use of labor (Dixon et al., 2001). Specific initiatives include: (1) the permanent production of commercial valuable crops; (2) reduced ploughing; (3) greater density of cultivation; (4) contour farming; (5) improved varieties; (6) live hedges; (7) interspersing of crops; (8) dispersed forest cover; (9) mulching (Dixon et al., 2001); and (10) management of invertebrate fauna in a way that is beneficial for the soil.

The benefits of these technologies and the feasibility of their adoption have been amply demonstrated by a series of innovative projects that have been undertaken throughout the system. However, even though such practices may lead both to an increase in yield and to more sustainable management of natural resources, they will have limited impact on increasing family incomes, and unless they are incorporated into diversification and marketing programs, these alternatives must be considered to be only part of the solution (Dixon et al., 2001).

For those systems with high population densities in Latin America and the Caribbean, the development and implementation of effective plans at the community level for the management of natural resources are extremely important. These interventions should include technical assistance and incentives for their adoption, in addition to emphasis on obtaining benefits in the short term for resource management activities, such as the management of water basins and forestry resources, among others (McNeely and Scher, 2003). Other promising interventions should focus on technologies for the conservation of humidity in drier areas and for combating drought and desertification (for example, northeast Brazil and the Central Andean region). There is also need for effective management of water basins (Mesoamerica and northern Andes). Both of these impacts are expected to become more widespread as a result of global climate changes, which is why risk reduction mechanisms must be strengthened (Dixon et al., 2001).

4.2.2.4 Use and control of the application of new technologies

AKST in Latin America and the Caribbean need to prioritize research and the training of native personnel in centers of excellence in order for the region to narrow the current technological gap and not remain on the margins of the great technological breakthroughs taking place in new areas of research (agroecology, biotechnlogy, niche agriculture and biological controllers, for example) in the developed countries. This effort of cooperation must be directed towards strengthening the technical and scientific capacities of AKST actors in the region and should address the needs and particularities of each subregion of Latin America and the Caribbean.

This development and training in the use of new technologies should also be geared towards achieving the goals of IAASTD. This means that AKST systems should direct their efforts to take greater advantage of these technological innovations by placing emphasis on issues of biosecurity, for example in genetic modification programs. The adoption of any new technology (including GMOs) should be preceded by a careful analysis of risks to health, to genetic introgression in localities of evolutionary origin and other impacts on the environment, including considering the possible prohibtion of the release of GMOs in centers of evolutionary origin.

AKST systems should act effectively to carry out impact and potential risk assessments of the products being researched in order to prevent their adoption from causing problems for the environment and for consumers. In other words, they should ensure the biosecurity of the results of their research programs.

One option would be to direct efforts towards the adoption of the precautionary principle (through AKST) to prevent irreversible damage and promoting their observance through national, regional and international agreements.

4.2.2.5 Investment in AKST systems for the development of technological innovations to overcome health problems

This point refers in particular to the introduction of methods for the traceability and safety of foods (possible risks, nanotechnology) and methods for the control and detection of health problems, among others. In order for investment in AKSTs to be efficient, it is proposed to conduct research into the know-how of producer groups and the objective conditions under which they use technologies. Many traditional processing techniques are unsanitary. Meanwhile, health barriers represent obstacles for small-scale producers, who lack large amounts of capital. Alternative approaches must therefore be sought that allow for the strengths of the less strong producer groups in the agricultural economy to be maximized (EMBRAPA, 2006).

4.2.2.6 Development of technologies to strengthen integrated pest management

Promotion of integrated pest management practices (IPM) and technologies that reduce or eliminate agrotoxins. The development of this type of research is today common in many of the AKST systems in Latin America and the Caribbean, but needs further strengthening since it can lead to substantial reductions in the use of agrotoxins in fields. In order to encourage this trend, it is proposed to establish strict rules for the use of agrotoxins, in particular for the protection of men, women and children who work in fields (Nivia, 2003). It is also necessary to combine conventional research with traditional research in order to identify biocontrollers, to develop strategies for the agroecological management of production systems (Buck, et al., 2006) and to improve conventional systems.

4.2.2.7 Land distribution

Access to land is a burning issue throughout the region. Promoting research and providing training in methods of distribution of productive land among social groups and their impact on the sustainable use of land and on poverty would be of great assistance in resolving the numerous problems. More families living sustainably in the countryside leads to greater diversity of decisions and consequently to greater diversity of landscape, biodiversity and crops and facilitates food security and the exploitation of biodiversity (Monro et al., 2002; Dietsch et al., 2004). Comparative research projects are also needed to identify the most sustainable distribution and land-use alternatives and thereby to promote diversity in the modes of distribution and use of land and diversity in modes of access to land and conservation of biodiversity (Almeida, 2006). This type of research could lead to policies that are more conducive to achieving the goals of IAASTD.

For production systems on agricultural frontiers, intervention priorities include the development of a comprehensive database of natural resources and their characteristics within the system as well as the relationship between this information and planning tools together with appropriate resource management policies. This could be strengthened through research partnerships in the development of crop varieties that are adapted to the conditions of agricultural frontier zones (e.g., aluminum tolerance, post-harvest characteristics), and dissemination of the results. Of prime importance, however, are the legalization of land holdings and elaboration of policies that promote appropriate patterns of land use by employing such instruments as land taxes (at the regional and municipal levels); land concessions; easy access to investment loans and operating capital; eligibility for support services; and marketing, extension, and veterinary services, among others (Dixon et al., 2001).

4.2.3 Climate change and bioenergy

Energy efficiency understood as the cost-benefit ratio, i.e., the investment of crop energy in the agroecosystem (or aquatic system) versus the energy benefit obtained from the production (in Kcal) and the diversity of products (National Research Council, 1989) could be considered as a basis for evaluating production systems.

The oil crisis is another factor that reinforces this need for AKST to give priority to the search for more efficient alternative energy sources, in keeping with the characteristics of the various sub-regions of Latin America and the Caribbean. Wind and solar energy are very rarely used in the region but have great potential in the tropics, particularly in rural areas.

Combating the effects of climate change by developing alternative systems of production

Regional studies of the impact of climate change confirm the negative effects that global warming will have on the incomes of producers, particularly small/traditional producers who have less resources to adapt to these changes (Mendelsohn et al., 2006; SEO et al., 2006). Alternative production systems could be used to mitigate the negative impacts of climate change, since agroecological systems have greater capacity to recover from environmental disasters (SOCLA, 2007).

With the help of indigenous/traditional and scientific (agroecological and conventional) knowledge, it is possible to promote research into the use of perennial plants and agroforestry for carbon sequestration (see option 6). Efforts are also being made to promote the development of new plant varieties that adapt better to climate change, in particular to rising temperatures and to variations in the dry and rainy seasons. It is also necessary to identify more efficient methods of water use and management of soils that are vulnerable to erosion (eg. management of plant cover, green fertilizers, wind breaks, drainage) (Murgueitio, 2003).

Latin American and Caribbean AKSTs must strengthen its links of technical and scientific cooperation in the search for joint solutions to mitigate the effects of climate change on producers in the region (Lima et al., 2001). A program of research and development is needed to add value to forestry resources through innovation, commercial agreements, gradual training to take over the productive process, and institutional support. The latter should focus on the reality in which AKST actors operate. The use of forestry products is irreversible and it is therefore essential to develop alternatives for sustainable use based on empirically observed realities (FLOAGRI, 2005). The socio-economic sciences must also be developed to take account of environmental services in the evaluation of production systems (Altieri et al., 2003; Chavarria et al., 2002).

4.2.3.1 Research to evaluate the contribution of agriculture and livestock farming to the production of renewable energy.

Agricultural production for use in fossil fuel alternatives (crops that can be used to generate energy, the conversion of waste from harvests, and biogas) could constitute an opportunity to revitalize primary sectors, provided that it is based on local resources and does not endanger food security. For this, bioenergy (not only biofuels) should be developed for local use and local knowledge and general environmental-friendly principles should be disseminated at the regional and international levels. Existing knowledge should be organized in such programs and new studies undertaken on energy resources based on knowledge of the flora of new tropical forests (Amazonia, rain forests, dry forests, premontane, gallery or riparian forests).

As this know-how becomes better known and more widely disseminated, it will be appreciated more and more by Latin American communities, especially the urban population. Reforestation can be undertaken in degraded zones to produce bioenergy in already cleared areas, given that the extraction of wood to produce energy is one of the causes of deforestation (Homma, 2005). This may offer a good opportunity to develop technologies for the recovery of degraded areas using local elements of biodiversity, plants, earthworms and other elements of soil fauna and microorganisms (Lavelle et al., 2006).

Brazil's new national agroenergy program, for example, offers a series of options for increasing the share of biofuels in the national energy grid with the very active participation of AKST systems (creation of a center and of specific agroenergy research programs) and with supportive public policies. This experience should be shared with the countries of the region (although this has caused some controversy). Responsibility for ensuring equity and sustainability (related to the production of biofuels) has thus far been evaded (e.g., study of cases such as that of Brazil, a world leader in the production of ethanol). The formation of agroindustrial cooperatives or associations of small- and medium-sized producers could help to ensure greater equity in this sector.

Extreme caution is needed and studies undertaken on the large-scale environmental and social consequences of the production of agrofuels in the countries of Latin America and the Caribbean. A program should also be developed to organize existing know-how and experiences in the production of bioenergy (including biofuels) in the various ecoregions. Research is needed too on the ultimate consequences of the addiction of the current dominant development model to gasoline and petroleum with a view to ensuring that crops now used for food are not converted into raw materials for agrofuels (known as biofuels, such as ethanol and diesel) and to prevent more intensive use and further expansion of the agricultural frontier to the detriment of the last places of refuge for the biodiversity of neo-tropical forests.

4.2.4 Biodiversity

Biodiversity is the basis of all current services provided by ecosystems and the key to their sustainable use for the future. Special strategies are therefore needed to prepare inventories, expand knowledge and utilize this resource. Special attention will be paid to its conservation in recognition of the fact that Latin America and the Caribbean has high levels of biodiversity with countries such as Mexico, Colombia and Brazil among the most biodiverse in the world but also with the highest rates of extinction.

4.2.4.1 Development of strategies for the conservation and sustainable use of biodiversity in Latin America and the Caribbean

The development of AKST systems through agroecological and traditional know-how should focus on strategies for protecting the extensive biodiversity (both domesticated and wild) of Latin America and the Caribbean and on the right of the peoples of the region to have knowledge of them, access to them, and to use them sustainably. Biodiversity holds the greatest potential for the development of new products (plant breeding) and can satisfy the emerging demand both for food (quality and quantity) and for other products. There is need for management of a common research system and for the comparing of experiences in order to promote the regulation of and greater access to traditional knowledge.

Traditional knowledge has sometimes been used by the different actors for economic ends while the population at large receives no reward of any sort (Santilli, 2002; Lima and Bensunsan, 2003). It is necessary to work towards the elaboration of precise and up-to-date inventories of biodiversity, to establish reference databases and to train people in Latin America and the Caribbean in these fields. This knowledge will enable the region to assess the impact of public policies in this area and also the impact of different types of land and landscape use on biodiversity. Many biodiversity crops cultivated by small-scale producers have significant potential to capture international niche markets. The evidence suggests that there could be substantial value-added if products such as alpaca and llama wool, quinine, specialized varieties of potato and others are introduced into these markets, especially if there are possibilities of having labels or denominations of origin. The negotiation of equitable commercial agreements can also create opportunities for increasing incomes, although its scope is limited.

Diversification aimed at supplying demand in export markets will require improved organization on the part of producers in order to ensure coordination and the fulfilment of the quotas required for export shipments, as well as efficient mechanisms for the supply of inputs (including financing). Technical assistance to ensure adequate quality control, the development of adequate post-harvest handling and packaging, and the creation of effective marketing chains will also be indispensable (Dixon et al., 2001).

The options also include expanding knowledge of hydro-ecology and water dynamics, especially in the Caribbean, coastal areas and rivers. The countries of South America need a multidisciplinary approach to the management of their resources, based fundamentally on the integrated management of water basins, multiple use of water resources, recovery of waste water, and protection of fragile zones that are important for the conservation of water resources, such as wetlands and slopes. Urgently needed also are adequate legal and programmatic frameworks, such as national water resources policies, national water legislation and plans for the use and conservation of water resources.

Inadequate knowledge of the state of water resources is one of the impediments to effective management of these resources. However, the increase in conflicts arising from the scarcity of this resource, as a result of natural disasters, unregulated use and pollution, has sparked interest in this sector which is so important for national development. Elements of conventional knowledge must therefore be incorporated in order to restore diversity, integrity and productive capacity to water systems (DeClerk et al., 2006). Elements of agroecological knowledge (DeClerk et al., 2006) and traditional know-how must also be incorporated to restore diversity, integrity and the productive capacity of soils. At the same time, an inventory of local know-how is needed as a way of protecting such knowledge (Shiva, 2000).

Marketing channels for biodiversity products need to be identified. The extensive biodiversity of sub-regions such as the Andes, Amazonia, Mexico, and Central America, among others, and access to these resources is a strategic factor that cannot only be translated into value added but also into easy and preferential access to generic technologies. The challenge is to elaborate legal frameworks and to develop appropriate institutions for the commercial exploitation of biodiversity products. There are significant opportunities in this sector, since the international markets for biodiversity products and services are fast expanding. Many developing countries could benefit from the expansion of the market for biodiversity products and services and exploit the potential value of biodiversity. This is only true if biodiversity is protected, since it can easily become a non-renewable resource if there are threats, extinctions and vulnerabilities.

Another option is the development of technologies incorporating local knowledge for the conservation and sustainable use of biodiversity (combining traditional and agroecological know-how). The region has great potential based on its natural riches, especially in its biodiversity. Many plant and animal species are native to the region and can form the basis of poverty reduction strategies. Biotechnology and other niches offer opportunities for improving agricultural productivity without increasing the use of agricultural inputs. A key element of this strategy is to ensure the participation of producers and the identification and pursuit of opportunities throughout the food chain.

In order to succeed in the options described above there must be an inventory and study of local and regional biodiversity (conservation and sustainable use and related know-how). The study on biodiversity must be carried out in close association with the taxonomy, evolution, biogeography and ecology, but on its own terms from which it is hoped that new paradigms will emerge (IAVH, 2006). The most notable lines of work are related to the role of biodiversity in organisms, the structure and functioning of ecosystems, their value to and use by man, and their inventory and monitoring. It is important that the value enhancement, monitoring and inventory of biodiversity, including the way in which inventories are carried out, are all done under common research guidelines that ensure that the compilation of data is standardized. The aim is to arrive at estimates that allow for comparisons of critical sites for protection, identification of key and indicative species, improvement of the procedures used in the exploitation of resources, and evaluation of production systems that have higher yields and less impact on biodiversity.

4.2.4.2 *Sustainability of livestock farming*

It is necessary to identify and disseminate options for sustainable and productive livestock farming in Latin America and the Caribbean, such as forestry and pasture land systems, protein banks that use various plant species as a source of energy and of protein for cattle (in other words, use of diverse landscape elements, such as tree barriers, significant shrub and tree biomass), protection of basins, and recycling of excreta in order to mitigate harmful or catrostrophic effects on soils and water. Depending on the type of biomass or ecosystem (since in Latin America there are savannahs, gallery forests, wetlands, foothills, the Brazilian "cerrado" and even the Andean high plateau which, with global warming, are now being used for cattle farming), the AKST should carry out research into and implement ecological principles to maximize sustainability and production. Know-how (local but also the imported variety) and technologies are needed to stabilize agriculture and preserve natural assets.

Environmental management in areas where livestock farming is practised leads to the unregulated occupation of land that is restored as a result of the creation of conservation units in regions of low agricultural potential, which could make the land resource more expensive and stimulate investment to increase productivity in areas already open or degraded in hopes of improving the environmental management of private land (Arima et al., 2005). In sum, when land used for livestock farming is degraded, the inhabitants leave and the land remains degraded. The aim is therefore to reverse the trend so that the land becomes sustainable and the population remains.

On the other hand, it is also necessary to develop criteria for the allocation of financial resources in accordance with the rate of compliance with environmental regulations (Arima et al., 2005) (policy in support of AKST systems). The creation of options in the livestock sector requires technical solutions to increase efficiency in terms of head of cattle per hectare and improve the quality of livestock without degrading the soils, water or biodiversity. This option may limit the expansion of cattle farming in wooded areas (FLOAGRI, 2005), but the harmful effects of intensive livestock farming models will not be mitigated if decisive and comprehensive acion is not taken to achieve the goals of IAASTD.

4.3 Options for Strengthening AKST Capacities

AKST capacities can be strengthened by creating new institutional mechanisms, promoting participatory research, and strengthening existing institutions, provided that they revise their teaching curricula. Special emphasis should be placed on the issues of property rights and gender equality.

4.3.1 Creation of institutional mechanisms for knowledge sharing

The synthesis of know-how and its sharing/dissemination within the three models identified (Figure 4-1: conventional, traditional and agroecological) require the use of new institutional tools that are appropriate to each context. Projects designed to promote knowledge sharing have proven to be effective in many cases. This sharing has led to significant technological improvements with positive effects on the well-being of the participants and to improvements in the environment. However, in some cases the same initiatives can promote the spread of inappropriate and even dangerous technologies and management practices (for example, use of the pesticide Chlordecone, which is banned in Europe but still used in some Caribbean islands (see http://www.minefi.gouv.fr/dgccrf/03publications/actualitesccrf/chlordecone). Thus far there is no legal or institutional tool to regulate such practices.

A great variety of groups have engaged in knowledge sharing initiatives, but their different and uncoordinated forms of organization and their excessive dependence on a few people acting as leaders are a source of weakness and prevent their spread. There is also the problem of the lack of continuity of these initiatives. In order to redress this situation, one option would be to introduce an institutional model that gives these initiatives greater visibility, connectivity and sustainability, while providing stable funding and monitoring the quality of the proposed activities and the competence of participants. An institution that is standardized at the national or regional level, based on the model of the primary or secondary school, with branches in all municipalities, can accomplish these goals.

4.3.2 Adoption of a participatory approach to research

Mainstreaming the participatory approach to systems (styles) of research requires the participation of different actors in the research activities. This is so in order to incorporate local knowledge in combination with scientific knowledge and to seek alternative and common solutions to the problems of producers. This will ensure that producers resolve their own problems, master new technologies and increase their knowledge and awareness of the problems currently being faced (Schmitz and Mota, 2006). To this end, different participatory methodologies and tools (GTZ, CIAT, ASPTA, among others) that have produced good results in different countries should be used. The use of participatory methodologies in research and extension projects has proven to be of great use both for the potential to recover local know-how and for the acquisition of new knowledge, generated in conjunction with scientists. The adoption of technology can then be done much faster and more efficiently (EMBRAPA, 2006; Pérez, et al. 2001).

This means that, for example, greater use will be made of participatory approaches in such areas as the selection of plant varieties and in field tests for new crops that have potential for diversification. Achieving this goal will require significant restructuring of the national research institutions in many countries of Latin America and the Caribbean to put greater emphasis on disbursement mechanisms for research funding and on the training of personnel in participatory methods. The response capacity of research systems must be expanded to meet the demands of the market and while hybrid genetic material might be acceptable for use in diversification initiatives, material that can be replicated on farms is likely to be required for traditional crops (Dixon et al., 2001).

4.3.3 Strengthening of R&D networks

The creation of specialized networks in certain sectors or cultivation of specific crops may be another option for strengthening interaction between countries of the region and between knowledge systems. The creation of networks requires training and the generation of knowledge by the direct participants, namely local producers and consumers, and the establishment of mechanisms for the organization (generated by AKST systems) of small- and medium-sized producers.

There currently already exists within the international scientific community and among donors a recognition that both organized actors and research centers should develop projects that are more directly geared to the generation of technologies and products that contribute to the reduction of poverty, with priorities being subject to change (CGIAR, 2003). Networks must also be established to protect and disseminate innovations that benefit rural populations in accordance with their local conditions and which help to institutionalize knowledge sharing arrangements (Durston, 2002).

Regional and above all subregional cooperation, where planting, soil and climate conditions are more roughly similar, should be significantly strengthened. Knowledge sharing should also be encouraged between other actors in subregional and national innovation systems, in addition to NARIs such as universities, NGOs, cooperatives and producer and private sector associations (Bisang et al., 2000).

A recent report on agricultural research and development cooperation programs in Latin America and the Caribbean clearly points to a lack of coordination between regional initiatives and the need for the organization of similar networks and governance structures for research and development and innovation activities. Cooperative programs, such as Procis (PROCITROPICO, PROCISUR, PROCIANDINO, PROCICARIBE; see option 15 under Key Issues) are increasingly concerned with the organization of research networks and partnerships (FORAGRO, 2006; Salles-Filho, 2006).

The evaluation of these programs implemented at various levels (Evenson y Cruz, 1989; Cruz y Avila, 1992; Avila et al., 2005; Salles-Filho et al., 2006a,b,c) shows that cooperation has resulted in a great deal of spill-over between countries. In other words, neighbouring countries benefit from research undertaken on the other side of the frontier. Evaluations have also shown that these programs need to be evaluated, restructured and extended to other actors so that they could more effectively fulfil their other objectives (Salles et al., 2006abc).

4.3.4 Organizational models

Currently the main challenge in nearly all Latin American countries is to build and strenghten their institutional capacities in order to promote the development of their AKST. While many countries of the region have made significant efforts to modernize the State, in terms of the first and second generation of reforms,[23] the results were incomplete—particularly those of the second generation—and were not part of a coherent set of policies that could help to develop the capacity to create the minimum conditions necessary for the development of AKST in the region. Actors in the system, especially those in the public sector, on the whole suffer more from the absence or unpredictability of the flow of financial resources, the centralization with limited autonomy of centers/stations, deficiencies, low wages and rotation of qualified personnel and the lack of administrative and financial flexibility (Bisang et al., 2000).

These problems are closely linked to the organizational

[23] Initially in the so-called first generation of reforms the emphasis was placed on the objective of deregulation and waste reduction, the size of the state and its intervention in the economy. These reforms were carried out in the late 1980s and early 1990s (as one of the main pillars of the so-called Washington consensus). Later, in the second generation of reforms, emphasis was placed on building the capacity of the state.

models adopted by AKST stakeholders in the region where the most diverse models of organization coexist. When one analyzes the various stakeholders in this system, the differences are more substantial in the public sector where traditional national research or agricultural technology institutes exist (NARI in Chile, INTA in Argentina, INIAP in Ecuador, INIEA in Peru, and INIFAP in Mexico, among others) and agricultural research departments that are directly connected to Ministries of Agriculture (such as the DIA in Paraguay), alongside institutes or organizations with public funding but governed by private law (EMPRAPA in Brazil, NARI in Uruguay and CORPOICA in Colombia, for example) and private foundations, such as PROINPA in Bolivia which participates in the AKST system without depending on public funding. The latter, in comparison to NARIs and departments, have much more flexibility to manage their human and financial resources.

These public stakeholders in AKST systems use basically two research models: (1) *the diffuse model,* in which research is conducted by research centers or stations that cover the most distinctive products (the majority of NARIs), and (2) *the concentrated model,* in which the centers are concentrated in a few products, ecosystems or priority issues (EMBRAPA in Brazil). According to Alves (1985), the use of the diffuse model, which is very common in Latin America and the Caribbean, generates a great deal of information and is unlikely to be concentrated in new technologies and for this reason is a costly process that is feasible only in a rich society whose producers have high levels of education and which is prepared to invest large amounts in agricultural research. Developing countries, such as those in Latin America and the Caribbean, generally do not have the essential inputs for the functioning of that model, but perhaps may be able to develop it.

Salles Filho et al. (2006) found that a number of countries have introduced institutional innovations into their agricultural research systems, which may serve as models for Latin America and the Caribbean. The study by Janssen (2002), with five industrialized countries, shows the diversity of the initiatives and the area of influence of changes, which have produced significant impacts on the financing and organization of research. One of the author's conclusions was that "the new research systems reflect the new conditions that society is imposing on agriculture, science and public sector management".

In sum, the strengthening of AKST systems in Latin America and the Caribbean, particularly in the public sector, requires a review of its models of organization to improve their efficiency, flexibility and focus and thereby increase their impacts on society. In this process, it is important to review the experiences of the region with differentiated levels of success and to adapt them to the situation of each country. These considerations should not contradict the models of participatory research described in section 4.3.2 and in Key Issues (Option 6).

4.3.5 Governance models: Strengthening and modernization of management models

From Chapter 2 (section 2.5.30—Management of the AKST system) we know that management of the system has become complex, particularly since it has been recognized that innovation comes from processes of interaction among social actors. In other words, there has been progress towards a contextual process of innovation, which implies a significant change in the rules of the game and structures of governance, thereby also increasing the vulnerability of traditional institutions.

The general tendency of national systems of innovation—and in particular AKST systems in Latin America and the Caribbean—to involve many different agents and organizations who exchange knowledge and cooperate in order to generate it, makes knowledge networks the new configurations of socioeconomic activity that address the need for interaction as a key factor in the generation and circulation of knowledge. These networks develop into subsystems of the national system of innovation, in other words, into specialized systems within the main system (Pittaluga et al., 2005).

The interactions between the agents in the network emphasize the relationships between users and producers of knowledge and innovations. These networks are the result of the efforts of agents to selectively internalize the various factors necessary to control the collective process of AKST (such as external factors). The simultaneous development of providers and users of AKST and their ongoing and coordinated interaction therefore further stimulate their activity and create a kind of virtuous circle for technological change (Pittaluga et al., 2005).

There are a number of successful examples in the region where AKST activities have been reorganized guided by the general idea of knowledge sharing or network formation. Research institutions have pursued cooperation to take advantage of knowledge sharing and complementarity of skills and other assets, and to emphasize the approach of demand-driven research. Efforts have also been made to strengthen relations among universities, industries and the public in general (Salles Filho et al., 1998).

These institutional reorganizations require novel forms of governance, in other words decision-making methods and approaches to common problems in which the various actors participate. The idea of the network suggests the way in which a variety of actors situated in a labyrinth of public and private organizations with interest in a particular policy connect with each other. The actors in the network share ideas and resources and work out possible solutions to public problems. Connections are thus made that blur the distinctions between the state and society, and it is the network that merges the public and private.

It will also be necessary to establish a new form of governance in the system of Procis (cooperative technology and innovation research programs). These programs represent important cooperation arrangements that still lack a new direction, more particularly in the sense of giving direction not only to researchers from participating countries, but also to other actors so that progress could extend beyond scientific and technological exchange (Salles-Filho, 2006). In addition, there is an increasing need to coordinate research and development activities and innovation at the regional and subregional levels through the organization of networks and other governance structures.

4.3.6 Interaction of AKST and social movements

It is necessary that the AKST systems research social and peasant movements and development of structures to promote dialogue between them and other social actors and AKSTs. Determine through research why social movements have succeeded in having a recognizably positive impact on IAASTD goals. One way to ensure interaction with social movements is to establish a framework for research on these peasant and social movements and the ways in which they relate to other actors, while always emphasizing their importance and potential for bringing about improvements in quality of life, environmental sustainability and conservation of biodiversity. Studies of this type (also involving the actors themselves based on a bottom-up approach) reveal the impact of the democratization of access to land on the quality of life of producers and consumers.

4.3.7 Intellectual property rights (IPR)

The issue of ownership of knowledge generated in underdeveloped countries is currently at the center of an extremely polarized debate on technology and development. A number of options have been proposed to guarantee such ownership.

The result of the generalization and implementation of the TRIPS [24] Agreement of the World Trade Organization (WTO) is a global system in which IPRs will become increasingly strict. Another school of thought holds that there is room for national strategies within this multilateral framework (PNUD, 2001).

This will require legislation using all the available resources provided for in the Agreement. Many governments have begun to draft their own legislation, while at the same time protecting the rights of farmers and of the patent holder as a means of promoting technological research and development on the one hand and agricultural productivity and biological diversity on the other (FAO, 2000).

Those countries that have the advantages of solid agricultural structures and abundant biological diversity as a support for their national economy, in particular, should protect their farmers and rural communities through specific rights adapted to the particularities of the issue in question. The TRIPS Agreement offers sufficient freedom of action to establish a system for the protection of new plant varieties that encompasses protection of the knowledge and practices of farmers and communities (FAO, 2000).[25]

The Biodiversity Convention, signed in Rio de Janeiro in 1992, recognizes that patents and other intellectual property rights may have an influence on the implementation of the Convention,[26] and therefore provides that the parties "... shall cooperate in this regard subject to national legislation and international law in order to ensure that such rights are supportive of and do not run counter to its objectives" (Article 16-5).

Since then some progress has been made on this issue, although the interests at stake are very important. Within the framework of multilateral negotiations, a group of developing countries with a mandate from the Doha Ministerial Conference of the WTO has pushed for an amendment to the TRIPS Agreement that would provide for three conditions to be attached to requests for patents related to biological resources and traditional knowledge: revelation of the country of origin or source; proof of prior informed consent; and evidence of a fair agreement for the distribution of benefits, in accordance with national laws. The industrialized countries and the major industries have rejected these proposals in the WTO. As a result, numerous objections are raised in the negotiations on access and profit sharing within the framework of the Biodiversity Convention each time that the developing countries insist that the parties fulfil their responsibility to ensure that the protection of property rights does not run counter to the objectives of the Biodiversity Convention (Yoke Ling and Shashikant, 2006).

The adoption by FAO of the International Treaty on Phytogenetic Resources for Food and Agriculture in November 2001 marked an important step forward in this field. The Treaty covers all of the most important phytogenetic resources for food and agriculture and is consistent with the Biodiversity Convention. Under the Treaty, countries agree to establish an effective, efficient and transparent multilateral system to facilitate access to phytogenetic resources for food and agriculture and to share benefits in a fair and equitable manner. The Treaty's monitoring body, comprised of the countries that have ratified it, establishes the conditions for access to resources and distribution of benefits in accordance with the "Agreement on the Transfer of Material".

In their national legislation, more and more countries have been adopting laws to ensure that the protection of intellectual property rights does not run counter to the provisions of the Biodiversity Convention. Costa Rica, for example, has adopted a Biodiversity Law that requires decisions taken to protect biodiversity-related intellectual property rights to be compatible with the objectives of this law. The state also grants protection through, inter alia, patents, trade secrets, recognition of the rights

[24] Trade-Related Intellectual Property Rights.

[25] The norms of the TRIPS Agreement allow countries not to patent higher-level plant or animal organisms or essentially biological processes for the production of plants and animals. Signatories are generally required to protect micro-organisms and non-biological or micro-biological processes through patents. Countries must also protect plant varieties by means of patents, through an effective *sui generis* system or through any combination of both. The provisions of the TRIPS Agreement on patents are not always appropriate for protecting living material or related products. A *sui generis* system can offer greater flexibility when a legal framework for protection is developed.

[26] The objectives of the Biodiversity Convention are: "the conservation of biological diversity, the sustainable use of its components and the fair and equitable sharing of the benefits arising out of the utilization of genetic resources, including by appropriate access to genetic resources and by appropriate transfer of relevant technologies, taking into account all rights over those resources and to technologies, and by appropriate funding." (Art.1)

of a party that improves a plant variety, *sui generis* community intellectual rights, trademarks and farmers' rights. But legal channels may not be the most appropriate way of protecting the knowledge about their habitat that ethnic communities have accumulated over centuries, since these are the result of a social construct. The system of patents invites claims on the indigenous and community innovations of developing countries, thereby making them vulnerable to formal representation and patenting by others. To claim, use and defend patents is easier for private industry than for institutes and innovative communities.

The above creates a situation in which the knowledge of traditional communities is increasingly being used for commercial purposes in such sectors as pharmaceuticals and agriculture. Technological developments based on this knowledge have produced a marked increase in the supply of crops of food products and new products related to health, among other uses. These developments were produced without the generators and owners receiving any benefits from their property (Santilli, 2002; Lima and Bensunsan, 2003). The idea of protecting this knowledge is gaining ground. However, many proposals made to protect traditional knowledge have failed. Indeed, two institutional and cultural systems clash in the exchange of traditional ethnic knowledge. One is the system of commercial exchange of knowledge in which privately owned (tangible and intangible property) institutions are created and maintained within a legal framework. This system has relatively transparent information mechanisms and operates at a global level. The other is the system of local community knowledge in which ownership of the knowledge is undefined or collective.

In the last two decades, transnational corporations, academic institutions and independent research laboratories have patented indigenous knowledge or have reached agreements with ethnic groups. Various normative frameworks have been suggested. In all of these, there are compensations for the group that include the construction of health or education centers, or the preparation of brochures to educate the public about these practices and their origin. Occasionally, these contracts provide for the group to have a right to share in the benefits of products derived from their know-how, but these payments have not actually been disbursed in any known case (Zerda-Sarmiento and Forero Pineda, 2002).

The proposals put forward by authors and NGOs have tended mostly to recognize collective intellectual property rights. Some include the idea of creating an international fund to collect and redistribute the income derived from indigenous know-how. The creation of a regional forum consisting of representatives of indigenous communities from different countries would be necessary for elaborating a consensus agreement to regulate bioprospecting and the use of indigenous know-how. This agreement could provide for alternative models of framework negotiations and enforcement mechanisms to regulate transfers of traditional know-how from these communities to transnational corporations, research laboratories and universities. This framework agreement should establish a balance between preservation and the development of systems of community knowledge and its use by science and the market.

Negotiations may be difficult because, despite the existence of cultural hybrids, ethnic groups are not accustomed to thinking in terms of profits or sharing in benefits. A participatory approach to research achieved through the establishment of local research foundations dedicated to the preservation of knowledge and indigenous culture may be one solution. Indigenous groups should participate in the research and documentation of their knowledge, history and oral culture (idem.).

Another of the priorities would be higher education and scientific research programs based on traditional know-how and which offer training and research opportunities to members of their own communities, which would ensure knowledge sharing. Support is also needed in the international debate in national and international forums on the question of protection of the traditional knowledge of genetic resources (Bayão and Bensunsan, 2003).

4.3.8 Promoting the use of models that guarantee food sovereignty and stem or reverse the rural exodus

The strengthening of organizational know-how through AKST systems is proposed so that small-scale producers, local fishermen and indigenous peoples would have adequate and equitable access to land, water, genetic resources and other resources necessary for sustainable food production. There is also need for the promotion of family and community agroecological models both in practice and through policies, and for research and development to guarantee food security, especially in those sectors that are most vulnerable to hunger and malnutrition, through sustainable management of local agroecosystems for the production of food intended mainly for local markets.

One option proposed for the sustainable exploitation of water systems is research into methods for conserving and adding value to fish and other fresh products in order to facilitate their distribution. Many Amazonian, Andean or plain populations encounter problems in transporting quality fresh products to local markets. Serious studies are needed on the quality of the diet of foods traditionally consumed and which risk being substituted by the adoption of new food habits. Studies should also be carried out on the diversity of diet, especially in rural areas, since ease of access to processed foods is causing changes in habits and increased consumption of fats and sugars (e.g., bottled soft drinks) (Maluf, 2004; Maluf et al., 2004).

4.3.9 Market study for the establishment of a direct link between local producers and consumers of foodstuffs in peri-urban areas

In order for this proposal to succeed, research is needed into marketing systems to identify local and regional particularities and to provide updated information to farmers and their representatives. A large part of the problems of farmers lies in poor marketing of their products (Dürr, 2002a,b). Similarly, the agenda of AKST systems must include research and dissemination of studies on an integral approach to the productive chain in the territory in question. Business opportunities could therefore be identified for the various segments of the rural space (Santana, 2002).

Associated with the above option are proposals for the development of know-how for peri-urban agriculture.

Urban agriculture is characterized by the dynamic use it makes of land (Companioni, 2001; Luc, 2006), and its rapid adaptation to the growth and development of the city. It suffers, however, from a problem of image and is rarely recognized as a valid form of use of urban land. By combining urban agricultural production systems with open urban space, it is possible to identify areas in which urban agriculture is more stable (such as right of ways and "non-buildable" land) as well as areas in which it may be temporary (for example, zones for future building). The restricted areas in the center of cities could benefit from more intensive and generally more profitable activities, such as the production of mushrooms, silkworms or medicinal plants. Sites exposed to pollutants could be used for decorative plants instead of risking health by the cultivation and sale of plants for consumption.

Assigning areas within or on the periphery of cities for the exclusive and permanent use of urban agricultura is quite unrealistic and may be doomed to failure in certain countries. Firstly, it ignores the economic reality of the price of land in growing cities. Secondly, and more importantly, it fails to take into account the interactions which urban agriculture may have (and should have if it is to succeed) with other urban activities. If the municipal authorities involve a broader base of stakeholders, they would have more possibilities for developing policies that cover the needs both of the city and of their voters, particularly in the poor and disadvantaged sectors. Moreover, more equitable decision-making promotes participation and acceptance by citizens at all levels. As part of any political initiative for development, structures and processes must be established to identify problems, prioritize actions and undertake and implement activities for the monitoring of programs.

Workers in the urban agricultural sector and poor producers in particular cannot work as effectively as they might have, unless they are organized and their legitimacy recognized. Municipalities would clearly benefit from a better organization and representation of urban producers in local political decision-making processes.

4.3.10 strengthening the capacities of AKST stakeholders

AKST actors in Latin America and the Caribbean are extremely diverse, which makes it extraordinarily complex to generalize for the region as a whole (see section on Ethnic Groups in Chapter 1). The conceptual principle for seeking options for the future is respect, tolerance and valuing cultural diversity, which are a region's human capital. The diversity of ethnic groups is one of the conditions that make it possible to integrate the various types of knowledge so that they could contribute to achieving the goals of sustainability, quality of life and equity.

4.3.11 Restructuring education curricula

At the same time, while strengthening institutions that promote the sharing and synthesis of knowledge (see 4.3.1), it is necessary at the same time to propose changes in the curricula of all educational institutions at different levels to ensure that they accomplish the goal of teaching skills that are clearly aimed at improving the quality of life and promoting environmental and economic sustainability. The reports clearly show how inadequate current systems are to the changing needs of agricultural, forestry and livestock farming activities. This concern is part of a global movement begun since the early 1990s by the United Nations. The Jomtien Conference organized by the United Nations in 1990 established a series of principles to guide the design and development of systems for lifelong learning, and stressed the importance of replacing the current approach to teaching which is based on passive learning substitution with an approach that is based both on knowledge and on logical and rational analysis (thinking). It also recognized that education is the responsibility of all concerned and not only of states. Five types of institutions are considered here: (1) local information obtained by the family, social groups, and the communications media; (2) primary school where children are taught the basic skills to enable them to perceive and evaluate their natural and social environment; (3) secondary school where the cognitive, scientific and technological foundations are introduced through scientific programs; (4) universities that generate, evaluate and disseminate various types of technological knowledge defined as conventional or agroecological; and (5) the numerous initiatives promoted by NGOs, universities and other actors.

Local information educational programs in the media

This type of informal information is probably the most difficult to change because it includes local (experiences of neighboring producers), regional (discussions with merchants, local authorities, extension workers) and national (information through regional and national media) information. One way to improve it is to propose adequate educational programs directed to all actors for them to present in an accessible and synthetic way the different types of knowledge and to take advantage of information technology.

Primary school

Primary school has to provide the minimum foundations for awakening a sensibility for the fragility of the environment and the need to use natural resources in a sustainable manner. To that end, simple lessons should be taught about soils (how it is formed, its dynamics, life and functions), biodiversity (what it is and what it means for our own survival), agriculture and food (how it is obtained, the problem of producing it for a growing population, types of agriculture).

Secondary school

Adolescents can be taught to better understand and to value the contributions of the three main types of knowledge that sustain agricultural production. It is urgently necessary to strengthen the teaching of ecology at all levels, introducing the subject of ecosystems and the environmental services they provide, with special emphasis on soils and biodiversity, so that students understand the mutual dependence between society and nature. Greater emphasis should also be placed on the subjects of genetics, plant physiology, economics, social and community organization, and other subjects that would help students to understand the strengths and problems of conventional agroindustries and other emerging models of agriculture. The description and history of indigenous lifestyles and technologies should also be taught.

Systems of higher education

Initiatives to develop agroecology curricula are multiplying throughout Latin America. In order to strengthen their impact, the creation of a regional system for the coordination of these curricula is necessary and can play a useful role. The content of the curricula is based on the paradigms of ecology and their application to agroecosystems, consideration of the sustainability of the ecological functions that produce environmental goods and services (production, storage of water, carbon sequestration, conservation of soil biodiversity). Curricula should include a synthetic presentation of other systems of knowledge, explaining their goals, restrictions, strengths, weaknesses and prospects for development in the medium and long-term. It is also necessary to consider the teaching of techniques of communication and pedagogy that would permit efficient exchanges of knowledge when studies, experiments or development activities are carried out in this field.

Agricultural and related sciences

Universities and technical institutes that teach intensive agricultural methods have already begun to include in their curricula certain elements of ecology, agroecology and high-yield but more environmentally friendly models of production. The evolution towards systems with better environmental and social impacts could be achieved by strengthening the presentation of traditional and ecological knowledge in order to integrate them into a systemic way of thinking. The pedagogical tools themselves would permit the communication and transmittal of the basic knowledge.

Cross-cutting issues

Certain common (cross-cutting) issues affect the quality of life, environmental sustainability and equity in any of the scenarios or models of governance. A number of key issues have been proposed for the IAASTD goals in multidisciplinary exercises in Latin American and the Caribbean (Red Nuevo Paradigma, 2005). Some of the issues that could be included in the agenda of AKST in Latin American and the Caribbean are:

- Quality of life issues: as previously mentioned, the concept of human development is more than gross or per capita national income. It refers to the creation of an environment in which people can fully realize their potential and enjoy productive and creative lives in accordance with their needs and interests. Consequently, quality of life, in addition to satisfying basic organic needs, consists in expanding the options that people have to live in accordance with their values.
- Environmental sustainability issues: (1) soil conservation and management; (2) sustainable use of biodiversity; (3) nexus indigenous crops—conservation; (4) germoplasm, prospection and conservation in-situ and ex-situ; (5) adding value to biodiversity and natural resources; (6) traditional knowledge of biodiversity; (7) conservation and management of pollinators; (8) ecology of biological control; (9) organic fertilizers; (10) prospection and sustainable management of plants (particularly native plants); (11) urban agriculture; (12) management of fisheries resources; (13) impact of the agricultural sector on fauna; (14) flora and native micro-organisms; (15) impact of the fragmentation of natural habitats (on hydro-biological cycles, soils, biological interactions); (16) impact of genetically modified organisms (GMOs) on the environment and human and animal health; (17) zoning, management and agroecological agriculture; (18) invasive species (existing and potential, exotic and native); (19) management of forestry resources; and (20) quality control and use of water.
- Equity issues: (1) legislation for protection of the rights of all citizens without distinction on the basis of race, age, sex, origin, traditions, ideology, power, or social or economic status; (2) education without discrimination for all under equal conditions; (3) a sufficient number of educational centers by number of inhabitants and by area of influence both in cities and in rural areas; (4) compulsory teaching of human rights, ethics, philosophy and ecology from a very early age; (5) study of the origin of inequality; (6) study of the origin and consequences of extreme wealth and extreme poverty; (7) decision-making power of communities; and (8) provision of spaces for and promotion of democratic organizations in rural and urban communities (discussion and solution of problems that give rise to inequality).

4.3.12 Evaluation of the impact of AKST systems

Society does not have an accurate perception of the importance and impact of AKST systems, which means there is little support in areas where this weakness is most pronounced (Chapter 2). The experiences of impact assessment vary widely in the region, especially since there are no ongoing programs in this area (Alston et al., 2001; Avila et al., 2007). These studies, moreover, focus on assessment of the economic impact of the technologies generated (profitability of investments), without evaluating their other impacts.

Given the multiplicity of the impacts of AKST products, impact assessments must be multi-dimensional, in other words should include analysis of the economic, social, environmental and other impacts (institutional training and policies). Moreover, analysis undertaken prior to the impact of the research proposals may help to improve the quality and usefulness of projects and strengthen the impact of the products generated.

It is also necessary to develop methods of socioeconomic analysis in order to place AKST stakeholders and their new technologies in a socioeconomic context, which is essential to improve their impact. AKST systems must recognize that it is not sufficient to identify the impact of a particular technology or type of knowledge, but that this technology must be placed in a socioeconomic context. Given the complexity of the relationship between knowledge/science and society and that in order to enhance the impact of AKST, particularly in terms of the IAASTD goals, the research agenda must include the full participation of producers, especially the poorest and most marginalized. Profiles of producers and socioeconomic studies of small-scale producers are also needed in order to enhance the efficiency of these impacts.

4.3.13 Participation of AKST systems in the formulation of public policies

Strengthening the impact and capacities of AKST requires greater participation of such systems in the formulation of public policies. Generally speaking, AKST stakeholders include publicly funded national agricultural research institutes, but these have little effective participation in the preparation of the legal framework in which they operate (laws on biosecurity and intellectual property, financing of research and development, credit policies, etc.). Generally speaking, they are limited to sending or submitting reports with the results of their research, which is often demand-driven (reactively or by express request).

AKST stakeholders must adopt a more proactive attitude in this process. The experiences of EMBRAPA (Brazil) with its closer relationship to policy makers in the Ministries of Agriculture, Science and Technology, the Environment and Agricultural Development, in particular, as well as to the Congress have been very successful and have helped to strengthen the impact of that organization on the various segments of Brazilian society. EMBRAPA is participating more and more actively in the elaboration of laws governing intellectual property, innovation, and the protection of farmers, etc., as well as in rural credit policies, among other types of policies.

Participation in the formulation of public policies is vigorously pursued both by EMBRAPA and by FIOCRUZ, and this approach in the case of these two research institutions is indistinguishable from the process of creation and is a function that has been embraced by them throughout their existence in order to strengthen their legitimacy and institutional sustainability (Salles et al., 2000).

However, the active participation in the formulation of public policies seen today in Brazil, in the case of EMBRAPA, is not common in Latin America and the Caribbean. It is nevertheless clear that the strengthening of AKST systems in the region and enhancing their impact depend on the proactive participation of the system's stakeholders.

References

Adis, J., and M.O. Ribeiro 1989. Impact of the deforestation on soil invertebrates from central amazonian inundation forests and their survival strategies to long-term flooding. Water Qual. Bull. 14:88-98, 104.

Almeida, W.B. de. 2006. Terras de quilombolas, terras indígenas, "babaçuais livres", "castanhais do povo", faxinais e fundos de pasto: terras tradiciónalmente ocupadas. PPGSCA-UFAM, Manaus.

Alpízar, A., H.W. Fassbender, J. Heuveldop, H. Fölster, and G. Enriquez. 1986. Modelling agroforestry systems of cacao (*Theobroma cacao*) with laurel (*Cordia alliodora*) and poro (*Erythrina poeppigiana*) in Costa Rica-I. Inventory of organic matter and nutrients. Agroforest. Syst. 4:174-189.

Alston, M.J., C. Chan-Kang, M.C. Marra, P.G Pardey, y T.J. Wyatt. 2001. A meta-analysis of rates of return to agricultural r&d: Ex pede herculem evaluatión and priority setting. Res. Rep. 113. IFPRI, Washington.

Altieri, M.A. 1995. Agroecology. Westview Press, Boulder.

Altieri, M.A., E.N. Silva, y C.I. Nicholls. 2003. O papel da biodiversidade nomanejo de pragas. Holos, Editora Ltda-ME. Ribeirao Preto SP, Brazil.

Alves, E.R.A. 1985. Modelo Institucional da Embrapa. *In* E.R.A. Alves (ed) A. Pesquisa agropecuária: perspectiva histórica e desenvolvimento institucional. Embrapa, Brasília.

Amaral, N.M. 2004. Manejo florestal comunitário na Amazônia brasileira: análise da participación e valorización de saberes de grupos locais na implementación de três projetos pilotos. Dissertación de mestrado em agriculturas familiares e desenvolvimento sustentável. NEAF/CAP/UFPA; EMBRAPA Amazônia Oriental., Belém.

Armbrecht, I., J. Vandermeer, y I. Perfecto. 2004. Enigmatic biodiversity correlations: Leaf litter ant biodiversity respond to biodiverse resources. Science 304:284-286

Arima, E., P. Barreto, y M. Brito. 2005. Pecuária na Amazônia: tendências e implicaçõews para a conservación ambiental. IPAM, Belém.

Avila, A.F.D., L. Romano, y F.L. Garagory. 2007. Agricultural and livestock productivity in Latin America and Caribbean and Sources of Growth. *In* R.E. Evenson, y P. Prabhu (eds) Handbook of agricultural economics: Agricultural development: Farmers, farm productión and farm markets.

Badgley. C., J. Moghtader, E. Quintero, E. Zakem, M.J. Chappell, K. Aviles-Vazquez et al. 2007. Organic agriculture and the global food supply. Renewable agriculture and food systems.

Baltazar, A. 2005. Peixe e gente no alto rio tiquié. Istituto Socioambiental, São Paulo.

Barros, A.C., y A. Veríssimo. (ed) 1996. A expansão daa atividade madeireira na Amazônia: impactos e perspectivas para o desenvolvimento do setor florestal no Pará. Imazon, Belém.

Bayão, R. di, y N. Bensunsan. 2003. questão da proteción dos conhecimentos tradiciónais associados aos recursos genéticos nos fóruns internaciónais. In: Quem cala consente? Subsídios para a proteción aos conhecimentos tradiciónais. Instituto Sócioambiental, São Paulo.

Benatti, J.H., D.G. McGrath, A. C. Mendes de Oliveira. 2003. Políticas públicas e manejo comunitário de recursos naturais na Amazônia. Ambiente & Sociedade 6(2).

Bensusan, N. 2002. Seria melhor mandar ladrilhar? Nurit Bensusan (org.). Univ. Brasilia, Instituto Sociombiental, Brasilia.

Berendse, F., D. Chamberlain, D. Kleijn, y H. Schekkerman. 2004. Declining biodiversity in agricultural landscapes and the effectiveness of agri-environment schemes. Ambio 33(8):499-502.

Bierregaard, R.O., C. Gascon, T.E. Lovejoy, y R. Mesquita. 2001. Lessons from Amazonia. The ecology and conservation of a fragmented forest. Yale Univ, Press, New Haven.

Bisang, R., G. Gutman, C. Roig, y R. Rabetino. 2000. Los sistemas naciónales de innovación agropecupária y agroindustrial del Cono Sur: Transformaciónes y desafíos. Série Doc. 14. PROCISUR, Montevideo.

Bisang, R., G. Gutman, C. Roig, y R. Rabetino. 2000. Los Institutos Nacionales de Investigación Agropecupária del Cono Sur: Nuevos àmbitos y Cambios Institucionales. Série Doc. 15. PROCISUR, Montevideo.

Blann, K. 2006. Habitat in agricultural landscapes: How much is enough. A state of the science literature review. Available at http://www.biodiversitypartners.org). Defenders of Wildlife. West Linn, Oregon, Washington DC.

Brosset, A., P. Charlesdominique, A. Cockle, J.F. Cosson, y D. Masson. 1996. Bat communities and deforestation in French Guiana. Can. J. Zool. 74:1974-1982.

Brown, S. 1993. Tropical forests and the global carbon cycle — the need for sustainable land-use patterns. Agric. Ecosyst. Environ. 46:31-44.

Buck, L.E., J.C. Milder, T.A. Gavin, and I. Mukherjee. 2006. Integrating biodiversity conservation, agricultural production and livelihood benefits in ecoagriculture landscapes: A framework for measuring outcom. Background report for a workshop: Ecoagriculture outcomes: assessing trade-offs and synergies between agricultural production, rural livelihoods and biodiversity conservation at a landscape scale. World

Bank, Ecoagriculture Partners, Cornell Univ., Ithaca.
Cairns, M.A., y R.A. Meganck. 1994. Carbon sequestration, biological diversity, and sustainable development — integrated forest management. Environ. Manage. 18:13-22.
Caporal, F.R. y J.A. Costabeber. 2004. Agroecologia: alguns conceitos e princípios. Ministério do desenvolvimento Agrário – SAF\DATER. IICA, Brasília.
Cavalcanti, J.S.B., D.M. da Mota. 2002. Olhando para o Norte — Classe, genero e etnicidade em espacos de fruticultura. Revista Areas, Espanha.
Calle, Z., E. Murgueitio, y N. Calle. 2002. Enfoques silvopastoriles integrados para el manejo de ecosistemas. CIPAV, Cali.
CGIAR. 2003. América Latina y el Grupo Consultivo sobre Inverstigaciónes Agrícolas Internaciónales: breve síntesis de una asociación existosa. Secretaria del CGIAR. Washington, 2003. 12p.
Chavarría, H., S. Sepúlveda y P. Rojas (ed). 2002. Competitividad: Cadenas agroalimentarias y territorios rurales. Elementos conceptuales. IICA, San Jose.
Chauvel, A. 1996. Réactions du milieu forestier amazonien aux essais de mise en valeur agrosylvopastorale. C.R. Acad. Agric. Fr. 82:91-106.
CIFAA. 2006. Manifiesto sobre el futuro de los alimentos. Comisión Internacional sobre el Futuro de los Alimentos y la Agricultura (CIFAA), Toscana.
Companioni, N., Y. Ojeda, E. Páe, y C. Murphy. 2001. La agricultura urbana en Cuba. p. 93-110. *In* F. Funes et al. (ed) 2001. Transformando el campo cubano. Avances de la agricultura sostenible. ACTAC Asociación Cubana de Técnicos Agrícolas y Forestales. Centro de Estudios de Agricultura Sostenible, CEAS, Univ. Agraria de la Habana, Cuba.
Cruz, E.R., y A.F.D. Avila. 1992. Impactos económicos de la cooperación tecnológica entre los países andinos. Investigación Agraria: Economía. 7(2):283-289.
Decaens, T., J.J. Jimenez, E. Barros, A. Chauvel, E. Blanchart, C. Fragoso, y P. Lavelle. 2004. Soil macrofaunal communities in permanent pastures derived from tropical forest or savanna. Agric. Ecosyst. Environ. 103:301-312.
DeClerk, F., J.C.Ingram & C.M. Rumbaitis del Rio. 2006. The role of ecological theory and practice in poverty alleviation and environmental conservation. Frontiers Ecol. Environ. 4(10):533-540.
DeClerk, F., y P. Negreros Castillo. 2000. Plant species of traditional Mayan homegardens of Mexico as analogs for multistrata agroforests. Agroforest. Syst. 48:303-317.
Dixon, J., A. Gulliver, y D. Gibbon 2001. Sistemas de producción agropecuaria y pobreza. Como mejorar los medios de subsistencia de los pequeños agricultores en un mundo cambiante. Capitulo 7. M. Hall. Available at http://www.fao.org/ DOCREP/003/Y1860S/ y1860s00htm. FAO, Rome.
Daily, G.C. 1997. Nature's services: Societal dependence on natural ecosystems. Island Press, Washington DC.
Dietsch, T.V., S.M. Philpott, R.A. Rice, R. Greenberg, and P. Bichier. 2004. Conservation policy in coffee landscape. Science 303:625.
Durston, J. 2002. El capital social campesino en la gestión del desarrollo rural. Díadas, equipos, puentes y escaleras. Comisión Económica para la América Latina (Cepal) Naciones Unidas, Santiago.
Dürr, J. 2002a. A comercialización de produtos da produción familiar rural: o caso de Cametá. NAEA/UFPA, DED, Belém.
Dürr, J. 2002b. Cadeias produtivas no "Pólo Altamira": um estudo de caso. Versão preliminar. NAEA/UFPA, DED, Belém.
Ellingson, L.J., J.B. Kauffman, D.L. Cummings, R.L. Sanford, y V.J. Jaramillo. 2000. Soil N dynamics associated with deforestation, biomass burning, and pasture conversion in a Mexican tropical dry forest. Forest Ecol. Manage. 137:41-51.
EMBRAPA. 2004. IV Plano Diretor da EMBRAPA: 2004-2007. EMBRAPA, Secretaria de Administración e Estratégia Brasília.
EMBRAPA. 2006. Marco referencial em agroecologia. EMBRAPA Información Tecnológica, Brasilia.
Evenson, R.E., y E.R. da Cruz. 1989. The Impacts of technology PROCISUR Program: An internatiónal study. IICA/BID/ PROCISUR, New Haven.
Fachinello, J.C. 1999. Produción integrada de frutas (PIF). Anais do Seminário Estadual de Fruticultura 1:1-11.
FAO. 2000. Las negociaciones comerciales multilaterlaes sobre la agricultura. Manual de referencia, Capítulo IV, Acuerdo sobre los derechos de propiedad intelectual relacionados con el comercio. FAO, Roma.
FAO. 2001. The international treaty on plants genetic resources for food y agriculture, adopted by the 31th session of the conference of the FAO on nov. 3 2001. FAO, Rome.
FAO. 2002. Agricultura mundial: hacia los años 2015/2030. FAO, Roma.
FAO. 2003. International code of conduct ond the distribution and use of pesticides. FAO, Rome.
Fearnside, P.M. 1999. Biodiversity as an environmental service in Brazil's Amazonian forests: risks, value and conservation. Environ. Conserv. 26:305-321.
FLOAGRI. 2005. Sistemas integrados de gestão participativa dos recursos florestais e agrícolas pelas populações rurais na Amazônia Brasileira.
FORAGRO. 2006. Elementos para el Plan de Acción de Foragro: 2006-2008. FORAGRO, San José.
Gallina, S., S. Mandujano, y A. Gonzalezromero. 1996. Conservation of mammalian biodiversity in coffee plantations of Central Veracruz, Mexico. Agroforest. Syst. 33: 13-27.
Guzmán, E.S. 2004. Agroecología y agricultura ecológica: hacia uma "re"-construcción de la soberanía alimentaria. VI Congreso de la SEAE, Almeria.
Hardin, G. 1998. Extensions of the tragedy of the commons. Science 280:682-683.
Homma, A.K.O. 2005. Amazônia: como aproveitar os benefícios da destruición? Estudos Avançados. 19(54).
Hildebrand, M. E. y S. Marilee Grindle. 1994. *Building Sustainable Capacity: Challenges for the Public Sector.* Harvard Univ. Press, Cambridge.
IAvH. 2006. Informe sobre el avance en el conocimiento y la información de la biodiversidad 1998-2004. En M.E. Chavesy, M. Santamaría (ed) Instituto de Investigación de Recursos Biológicos Alexander von Humboldt (IAvH), Bogotá.
Janssen, W. 2002. Institutional innovations in public agricultural research in five developed countries. ISNAR, The Hague.
Jong, B.H.J., G. Montoya-Gomez, K. Nelson, y L. Soto-Pinto. 1995. Community forest management and carbon sequestration: A feasibility study from Chiapas, Mexico. Interciencia 20:409-416.
Kennedy, T.A., S. Naeem, K.M. Howe, J.M.H. Knops, D. Tilman, y P. Reich. 2002. Biodiversity as a barrier to ecological invasion. Nature 417:636-638.
Kursten, E., y P. Burschel. 1993. CO2-Mitigation by agroforestry. Water Air Soil Pollut. 70:533-544.
Lavelle, P., L. Brussaard, y P. Hendrix. 1999. Earthworm management in tropical agroecosystems CAB Int., UK.
Lavelle, P. et al. 2006. Soil invertebrates and ecosystem services. Eur. J. Soil Biol. 42:3-15.
Lima, A. y N. Bensunsan. 2003. Quem cala consente? Subsídios para a proteción aos conhecimentos tradiciónais. Instituto Sócioambiental, São Paulo.
Lima, M.A., O.M.R. Cabral y J.D.G. Miguez. 2001. Mudanças Climáticas Globais e a Agropecuária Brasileira. EMBRAPA. EMBRAPA Meio Ambiente, Jaguariúna.
Luc, J.A.M. 2006. Cultivando mejores ciudades, agricultura urbana para el desarrollo sostenible. IDRC, Ottawa.
Maluf, R.S. 2004. Consumo de alimentos no Brasil: traços gerais e ações públicas de segurança alimentar. Disponível em http:// polis.org.br/publicações/papers/20006.html.
Maluf, R.S., F. Menezes, y S.B. Marques. 2004. Caderno segurança alimentar. Disponível em: http://www.dhnet.org.br/direitos/sos/ alimentación/Conferências.html.
Markewitz, D., E. Davidson, E. Atlas, R. de

O. Figueiredo, R.L. Victoria, A.V. Krusche. 2001. Control of cation concentrations in stream waters by surface soil processes in an Amazonian watershed. Nature (UK) 410:802-805.

Markewitz, D., R. de O. Figueiredo, E.A. Davidson, 2006. CO_2-driven cation leaching after tropical forest clearing. J. Geochemical Exploratión 88:214-219.

Mattison, H. y K. Norris. 2005. Bridging the gaps between agricultural policy, land-use and biodiversity. Trends Ecol. Evol. 20:610-616.

Mattos, L., A. Baleiro, y C. Pereira. 2001. Uma proposta alternativa para o desenvolvimento da produción familiar rural da Amazônia: o caso do Proambiente. Encontro Naciónal da Sociedade Brasileira de Economia Ecológica (ECOECO), Belém.

McNeely, J.A. y S. J. Scherr. 2003. Ecoagriculture, strategies to feed the world and save biodiversity. Island Press, Washington DC.

Mendelsohn, R., y N. Seo. 2006. An integrated farm model of crops and livestock: Modeling Latin American agricultural impacts and adaptatión to climate change. Yale Univ., New Haven.

Mendelsohn, R., A.F.D. Avila, y N. Seo. 2006. Synthesis of Latin American Project. Yale Univ., New Haven.

MMA. 2004. Relatório final do grupo de trabalho do MMA. Proambiente: Um novo modelo de desenvolvimento rural para a Amazônia. Ministério do Meio Ambiente, Brazil.

Monro, A., D. Alexander, J. Reyes, M. Renderos, y N. Ventura. 2002. Árboles de los cafetales de El Salvador. The Natural History Museum, London.

Mori-Pinedo, L.A. 1993. Estudo da possibilidade de substituición do fubá de milho (Zea mays L.) por farinha de pupunha (*Bactris gasipaes* H.B.K.) em racões para alevinos de tambaqui (Colossoma macopomum CUVIER, 1818). Instituto Naciónal de Pesquisas do Amazonas/Universidade do Amazonas, Manaus.

Murgueitio, E. 2003. Impacto ambiental de la ganadería de leche en Colombia y alternativas de solución. Livestock Res. Rural Dev. 13(10).

Nair, P.K.R. 1997. Directions in tropical agroforestry research: Past, present, and future. Agroforest. Syst. 38:223-245.

National Research Council. 1989. Alternative agriculture. Nat. Acad. Press, Washington DC.

Neill, C., J.M. Melillo, P.A. Steudler, C.C. Cerri, J.F.L. deMoraes, M.C. Piccolo y M. Brito. 1997. Soil carbon and nitrogen stocks following forest clearing for pasture in the southwestern Brazilian Amazon. Ecol. Appl. 7:1216-1225.

Nivia, E. 2003. Mujeres y plaguicidas, estudio de caso en Palmira, Colombia. p. 28-61. *En* Asociación Colectivo de Agroecología del Suroccidente Colombiano-ACASOC. Aportes a la agroecología Colombiana. Litocencoa, Cali.

Ortiz, R. 2004. Análisis comparativo de las modalidades de asistencia técnica del INTA. Enfoques y modelos de extensión, estructuras de costos y beneficios generados. Available at www.pesacentroamerica.org/doc_hssh/investigacion/analisis_%20comparativo_Ramon_Ramiro.pdf. FAO Nicaragua.

Pereira-Filho, M. 1995. Alternativas para a alimentación de peixes em cativeiro. *In* A.L. Val y A. Honczaryk (ed) Criando peixes na Amazônia. Manaus.

Pérez Correa, E., M. Adelaida Farah, y D.L. Maya Vélez. 2001. Metodologías participativas en la formulación y planificación de proyectos de desarrollo rural. Fase de diagnóstico en siete municipios del sur del Huila. Revista Cuadernos de Desarrollo Rural 49:99-113.

Pittaluga, L., B. Lanzilotta, y C. Llambí (ed) 2005. Uruguay hacia una estrategia de desarrollo basada en el conocimiento. Informe Nacional de Desarrollo Humano-2005. Available at http://hdr.undp.org/reports/viewreports.cfm. PNUD, Montevideo.

PNUD. 2001. Informe mundial sobre desarrollo humano. Poner los adelantos tecnológicos al servicio del desarrollo humano. UN, Nueva York.

Polcher, J. 1994. The impact of African Amazonian deforestation on tropical climate. J. Hydrology 155:389-405.

Portela, R., y I. Rademacher. 2001. A dynamic model of patterns of deforestation and their effect on the ability of the Brazilian Amazonia to provide ecosystem services. Ecol. Modelling 143:115-146.

Promanejo. 2001. Manejo de florestas secundárias por agricultores familiares do nordeste paraense. Promanejo, Belém.

Rasmussen, L. 1998. Effects of afforestation and deforestation on the deposition, cycling and leaching of elements. Agric. Ecosyst. Environ. 67:153-159.

Red Nuevo Paradigma. 2005. Proyecto quo vadis: El futuro de la investigación agrícola y la innovación institucional en América Latina y el Caribe. Artes Gráficas Silva. Quito, octubre 2005.

Sánchez, P.A. 1994. Tropical soil fertility research: Towards the second paradigm. XV International Soil Science Congress, Acapulco, Mex. 1:1-24.

Salles Filho, S. 2006. Trajectorias y Perspectivas del Procisur: Informe Preliminar de Evaluación. Montevideo, Octubre 2006.

Salles Filho, S., y K. Kageyama. 1998. A reforma do IAC: Um estudo de reoriganización institucional Cadernos de Ciência & Tecnologia 15:35-58.

Salles Filho, S., R. Albuquerque, T. Szmrecsányi, M.B. Bonacelli, S. Paulino, B. Bruno et al. 2000. Ciencia, Tecnologia e Inovação: a reorganização da pesquisa pública no Brasil. Editora Komedi, Campinas.

Salles Filho, S.y A.C. de Souza. 2002. Agricultura familiar e investimento em desenvolvimento tecnológico. In: Dalmo Lima, Jonh Wilkinson et alii (orgs.). Brasilia, CNPq/Paralelo 15,. 400p.

Salles Filho, S., M.B. Bonacelli, y A. Bin. 2006a. Documento de Evaluación de Procitrópicos. Brasilia.

Salles Filho, S., E. Pedro y PJ. Mendes. 2006b Conceitos, Elementos de Políticas e Estratégias Regiónais para o Desenvolvimento de Inovações Instituciónais. FORAGRO/IICA. Unicamp. Campinas, Brazil.

Salles-Filho, S.L.M., E. Pedro, y P.J. Mendes. 2006c Conceptos, políticas y directrices para el desarrollo de innovaciónes instituciónales en la investigación agropecuaria. IICA, San José, 2006.

Santana, A.C. de, y M.M. Amin. 2002. Cadeias produtivas e oportunidades de negócio na Amazônia. UNAMA, Belém.

Santilli, J.A. 2002. A biodiversidade e as comunidades tradiciónais. Editora Universidade de Brasilia, Instituto Sociombiental.

Schmitz, H., y D.M. Mota. 2006. Métodos participativos para a agricultura familiar. p.75-102. *In* D.M. Carvaló Monteiro, y M. de Abreu Monteiro (ed) Desafios na Amazônia: uma nova assistência técnica e extensão rural. NAEA/UFPA, Belém.

Settle, W.H., H.A. Ariawan, E.F. Astuti, W. Cahyana, A.L. Hakim, D. Hindayana, y A.S. Lestari. 1996. Managing tropical rice pests trough conservation of generalist natural enemies and alternative preys. Ecology 77:1975-1988.

Seo, N., y R. Mendelsohn. 2006. A Ricardian analysis of the impact of climate change impacts on Latin America Farms" Draft Report of the Latin American Project. Yale Univ. New Haven.

Shiva, V. 2000. Stolen harvest. South End Press, Cambridge.

SOCLA. 2007. Memorias, Primer Congreso. Carmen del Viboral, Antioquia, Colombia Agosto 13-15. http://www.agroeco.org/socla/. Sociedad Científica Latinoamericana de Agroecología.

Swift, M.J., J. Vandermeer, P.S. Ramakrishnan, J.M. Anderson, C.K. Ong, y B.A. Hawkins. 1996. Biodiversity and agroecosystem function. p. 261-298 *In* H.A. Mooney et al. (eds) Functional roles of biodiversity: A global perspective. John Wiley, UK.

Tian, H., J.M. Melillo, D.W. Kicklighter, A.D. McGuire, J. Helfrich, B. Moore y C.J. Vorosmarty. 2000. Climatic and biotic controls on annual carbon storage

in Amazonian ecosystems. Global Ecol. Biogeography 9:315-335.
Vandermeer, J. 1995. The ecological basis of alternative agriculture. Ann. Rev. Ecol. Systemat. 26:201-224.
Veiga, I. 2002. Saber e participación na transformación dos sistemas de produción da agricultura familiar amazônica. Anais do V Simpósio Latino Americano sobre Investigación e Extensão em Sistemas Agropecuários – IESA e V Encontro da Sociedade Brasileira de Sistemas de Produción – SBSP. Florianópolis, 20 a 23 de maio de 2002.
Veiga, I., y C. Albaladejo. 2002. Gestão da fertilidade dos solos de uma localidade na Amazônia Oriental. A formalização dos pontos de vista dos agricultores visando um diàlogo entre agricultores e agrônomos. Agricultura familiar: Pesquisa, Formação e Desenvolvimento 1:109-137.
Vohland, K., y G. Schroth. 1999. Distribution patterns of the litter macrofauna in agroforestry and monoculture plantations in central Amazonia as affected by plant species and management. Appl. Soil Ecol. 13:57-68.
Von Maydell, H.J. 1991. Agroforestry for tropical forests. Agroforest. Syst. 13: 259-267.
Zbiden, S. y D.R. Lee. 2005. Payment for environmental services: An analysis of participatión in Costa Rica's PSA program. World Dev. 33:255-272.
Wood, S., y S. Ehui. 2005. Food. Chapter 8, p. 211-238. *In* Millennium Ecosystem Assessment: Conditions and trends. Island Press, Washington DC.
Yoke Ling, C., y Shashikant S. 2006. La batalla por los derechos de propiedad intelectual. Red del Tercer Mundo, 13 de marzo.
Zerda–Sarmiento, A. y C. Forero-Pineda. 2002. Los derechos de propiedad intelectual sobre los conocimientos de las comunidades étnicas. Revista internacional de ciencias sociales, Marzo Número 171. La sociedad del conocimiento.

5

Public Policies in Support of AKST

Coordinating Lead Authors:
Michelle Chauvet (Mexico), Celso Garrido (Mexico), and Tirso Gonzales (Peru)

Lead Authors:
Ruth Pamela Cartagena (Bolivia), Clara Cruzalegui (Peru), Luís A. Gomero (Peru), Dominique Hervé (France), Ericka Prentice-Pierre (Trinidad & Tobago), Ana Cristina Rostrán (Nicaragua)

Review Editors:
Claudio Bragantini (Brazil), Juan Cardenas (Colombia)

Key Messages

Key Messages

1. The objectives of AKST policies are (1) to reduce hunger and poverty, (2) to improve living standards and health for rural people, and (3) to promote development that is economically, socially and environmentally sustainable. To achieve these objectives, policies must move beyond previous models, in particular the one that made the market the central mechanism for allocating and regulating resources and that has had the effect of exacerbating economic and cultural poverty, hunger and inequality.

2. This situation creates the challenge of formulating alternative policies that take account of the economic, social, cultural and ecological heterogeneity that prevails in the various countries of LAC, without ignoring the new situation generated by trade liberalization and economic deregulation. The prerequisites for implementing these policies are: (1) to ensure a stable macroeconomic framework; (2) to establish strategic guidelines that will give priority to expanding and allocating public resources for the AKST system, strengthening the sustainable output capacity of small productive units, with a gender focus, so as to guarantee countries' food security and sovereignty; (3) institutional designs that will decentralize implementation of the strategy, with close involvement of local stakeholders, recognizing and strengthening their culture; (4) permanent mechanisms for monitoring and evaluating the impact of policies so that the instruments used can be reformulated; and (5) designing suitable mechanisms for financing the various policies.

3. The AKST policies proposed here are targeted essentially at alleviating poverty and hunger, reducing inequality, and promoting sustainable development with an emphasis on small-scale peasant/indigenous agriculture and agroecology (treated in its broad sense). To this end, policies must move beyond the models used since the 1990s, which were based on liberal approaches in which markets were the central mechanisms for allocating and regulating resources, and which have merely served to increase rural poverty and hunger.

4. A policy of food security and sovereignty that embraces production, the availability of food, and the development of capacities. The idea is to take policy measures that go beyond mere subsistence and will bring improve the lives of the poor, by giving them sustainable access to productive resources (land, water, biodiversity, credit) with a focus on gender and equality. In this context, we propose a policy that will help restore and strengthen local culture and knowledge in the management of productive and natural resources. This calls for intercultural policy instruments that will support the efforts of small farmers to achieve integral development and will strengthen their productive capacities, taking into account the worldview and the heterogeneity of these people.

5. A policy for sustainable management of natural resources. Such a policy must have instruments for territorial planning and the identification of ecological and economic zones as the basis for establishing rules for the use of land, ranging from conservation to intensive farming, with a view to creating a mosaic of agroecosystems.

6. Policies to encourage and support the transition from conventional and peasant/indigenous farming systems to models of sustainable agriculture. Policy instruments should be designed for each stage of this transition: reducing industrial inputs, making efficient use of energy, enhancing diversification, and promoting agroecological management. Incentives and support measures should seek to maintain the productive efficiency and competitiveness of agricultural systems, and to establish the objectives of each stage and the means for verifying progress.

7. A policy of participation and democratization that will include now-excluded sectors in defining and implementing the AKST agenda. We propose policy instruments that will increase these stakeholders' access to information, help them build or strengthen their capacities to take part in decision making, and provide institutional forums for deliberation and decision. Under these conditions, cooperative networks could be a prime instrument for coordinating the efforts of public and private stakeholders at the local, regional, national and international levels, so as to produce collective benefits that will take account of specific interests.

8. A policy for access to genetic resources and the equitable distribution of the benefits they generate. We propose as a policy instrument the formulation of legal frameworks that will guarantee local communities' access to genetic resources and regulate access for other players. Sui generis legal frameworks will also be defined to promote the recognition of traditional knowledge associated with these genetic resources, and the equitable distribution of their benefits among the communities that are the custodians of these resources.

9. A policy that prevents the use of food crops for purposes other than food in countries that are centers of origin of phytogenetic diversity. In other regions, the instruments will be of a regulatory nature. The instruments for this policy will include a precautionary legal framework where the granting of licenses will be preceded by a case-by-case evaluation of the social, environmental, cultural and food safety risk.

10. Intercultural education policies to promote the building and development of local capacities and skills. The idea is to facilitate rural people's access to labor markets through policy instruments such as community-oriented educational reform that provides for intercultural and multilingual instruction, the training of specialized teachers, the development of adequate physical and IT infrastructure, scholarships for low-income students, training programs and skills development.

11. The availability of financial services is an essential factor of support for activating the AKST system to meet development and sustainability goals. In LAC as a whole and in individual countries of the region, investment in AKST systems is low, and this trend needs to be reversed by strengthening investment in various components of the system, in order to sustain its dynamics and to reduce AKST dependency on technological innovations from outside the region. This increased investment should take place not only at the national level but also at the subregional and regional levels in order to capitalize on experience and minimize duplication in R&D.

12. Differentiated financing policies for the extremely poor and the creditworthy poor. In order to create comprehensive financial systems, we must differentiate between people who are extremely poor and people who are poor but creditworthy. The first group are unable to borrow, and they require specific solutions along the lines of the Grameen Bank in Bangladesh. The second group, on the other hand, can access financial services under certain prerequisites, primarily the resolution of property rights, education, the acquisition of management capacities, etc.

13. Financial support programs for communities to make the transition to a sustainable production system. One very important aspect to consider in financing policies for supporting AKST systems is the fact that communities in many parts of LAC are starting from very backward conditions marked by the immediate demands of subsistence, and they have few resources of their own. Consequently, it is virtually impossible for these communities, by themselves, to meet the challenge of moving from their current condition towards a productive system that is sustainable in both economic and environmental terms. This challenge must be addressed through financial support so that the transition can be made in an orderly and progressive manner.

14. Social spending geared to growth in GDP. Social policies targeted at the rural population should be based on the assumption that social spending in general (and in particular that for promoting AKST) will grow in real terms by at least the same proportion as the increase in GDP, although it would be desirable for it to grow more than proportionally, since LAC faces the challenge of overcoming the severe shortages and needs of rural people and vulnerable groups.

5.1 Objectives and Conceptual Framework

This chapter recommends policy options for supporting AKST in relation to food sovereignty, development and culture; strengthening institutions and developing the legal framework; sustainable management of productive systems; promotion of markets and financing, which will contribute to reducing hunger and poverty in Latin America and the Caribbean (LAC), in light of the goals set by the IAASTD.[27]

The AKST assessment of LAC in Chapters 1 and 2 has identified a number of economic, social and environmental limitations in the management of agricultural production systems, and a series of support policy measures must be designed and implemented to promote the conversion of current agricultural systems to ones that will guarantee sustainability.

To achieve this purpose, we must address the critical points that are hampering system change, relating to capacity development, research and innovation, and the supply of technology, recognizing the opportunities that arise in the productive chains and the need to make trade-offs between domestic market development and export subsidies. Currently one of the barriers to achieving competitiveness in Latin America is the limited capacity of those who manage productive units, and policies are needed to address this through rural schools, technology institutes and advanced training centers, with a new curriculum focused on development and sustainability, consistent with the objectives of each country and recognizing multicultural diversity.

Similarly, we must note that this conversion process will only be possible if research policies are at the same time oriented toward technological innovation based on the sustainable exploitation of biodiversity and natural resources. The challenge is to find new technologies and alternative inputs, and ways of combining them without sacrificing competitiveness, to identify incentives and protection policies for the various stakeholders, and to select readily observable indicators of progress achieved.

Various experiments with productive development projects show that training and research alone are not enough to achieve sustainable development. Those efforts must serve to link producers to the domestic market. To date, technology has been the preserve of export-oriented productive chains. The market for technologies and inputs has historically been controlled by the multinational agrochemical and seed companies, and there have been no alternative companies present on the market offering clean agricultural technologies. This means that policies are needed to encourage small- and medium-sized firms to enter the market under better conditions of competitiveness (Lizarraga, 2002).

The agroecological conversion of farming systems can also be accelerated if there is an increasingly demanding market for safe, high-quality products. This trend is already apparent and is forcing market-oriented producers to initiate or speed up the conversion of their systems. Similarly, public policies can facilitate the process through incentives scaled to performance. What is essential here is a clear governmental willingness expressed in a policy for financing and strengthening the institutions involved in facilitating productive development in the countryside.

Figure 5-1 summarizes and illustrates the interactions between the set of AKST support policies proposed in this chapter for moving forward in the transition to sustainable production.

It is useful to note here the importance of the models that multilateral agencies and international treaties impose on overall policy guidelines. Examples are the problems flowing from the stalemate in negotiations on the agriculture chapter in the World Trade Organization, and the outcome of free trade treaties, which have exposed broad

[27] The goals are to improve rural livelihoods and to promote equitable development, with environmental, social and economic sustainability that will reduce poverty and hunger through the generation, availability and use of agricultural knowledge, science and technology.

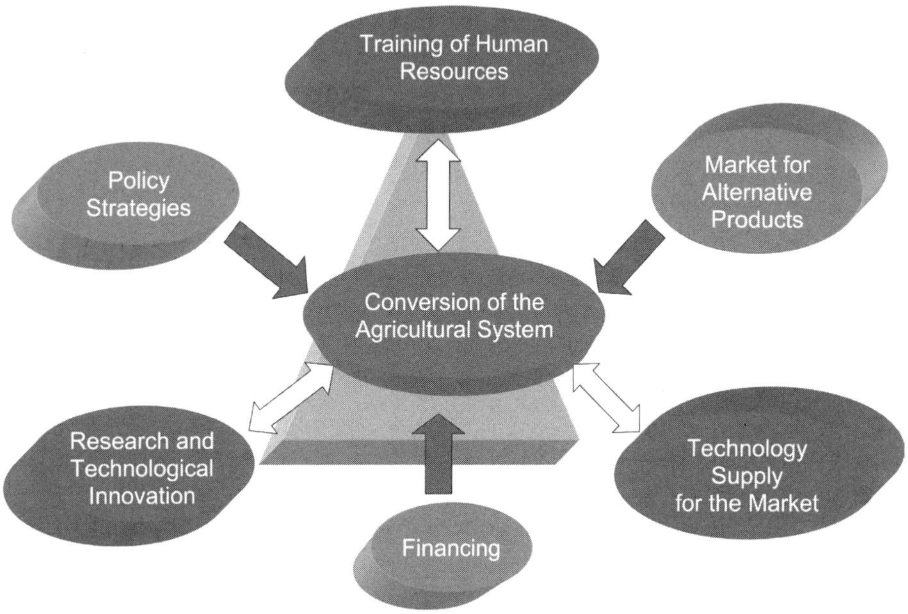

Figure 5-1. *Policies for moving toward sustainable agriculture.* Source: Authors' elaboration.

segments of agricultural producers to unfair competition without any compensation programs. As a result of these policy measures, spending on agriculture has declined as a proportion of public expenditure. These models betray two serious conceptual errors: first, the reduced role of the state, and second, downplaying the role of agriculture and trying to create jobs in other sectors without understanding that rural people have few options apart from agriculture, while ignoring international market distortions. Budget cuts flowing from these approaches are reflected in three indicators: (1) reduced investment in research, extension services and education; (2) few resources for institutional modernization; and (3) scanty investment in human resources (Trejos et al., 2004).

The private sector has not been able to make up for these cutbacks in public spending on the productive sector, which is oriented toward producing food for the internal market, and this has left a significant investment gap.

Finally, in this introductory overview, we must stress that this set of policies presupposes that social spending in general (and in particular that for promoting AKST) will grow in real terms at least by the same proportion as GDP (Gonzalez and Avila, 2005), although it would be desirable for it to grow more than proportionally, since LAC faces the challenge of overcoming the severe shortages and needs of rural people and vulnerable groups.

5.2 Public Policies for Food Sovereignty: Development and Culture

In a setting of nutritional vulnerability, food sovereignty is proposed as a medium- to long-term goal for combating hunger and poverty, but one that also has to do with other aspects such as access to land ownership, basic natural resources, credit, markets, education, health services, women's participation, etc.: in other words, the capacity to decide what, how and when to produce in a sustainable way.

Developing policies to achieve this goal will require a dynamic vision that, starting from the current situation, will involve intermediate phases and instruments to subsidize access to food in extreme cases. Food security concerns itself only with the immediate supply of food, providing or guaranteeing access to food by means such as keeping prices low and providing food stamps. A number of government programs have been confined to this goal, but they have not been effective in resolving the problem of hunger and poverty.

In this first section of the chapter we put forward some policies and instruments relating to both food security and food sovereignty programs, the importance of women's participation, and the role of development and culture in achieving development and sustainability goals.

5.2.1 Food security

An initial issue for AKST support policies is that the rural people should have a reasonable level of security in their access to basic needs, particularly food. In LAC this issue is generally addressed through social policies, particularly those relating to food security. There has been much debate on this issue (see Chapter 1).

These social policies in Latin America have been implemented, on one hand, through private, individualistic and unequal models driven by the market, and on the other hand by public, social and egalitarian models for correcting markets (Huber, 1996). Both these approaches are reflected in the food policy measures taken to reduce hunger and poverty.

The interpretation of poverty as subsistence refers to the fact that income is inadequate to cover basic minimum needs for maintaining physical efficiency. This argument was followed by the work of nutritionists to establish the so-called "poverty line". A family is considered poor if its income falls below this line. This approach has persisted since the postwar period and has been widely applied by international agencies, and it is still the criterion for measuring poverty in the United States (Townsend, 1993; FAO, 2006). Because expenditure on food is the most important component of subsistence incomes, policies designed from this approach sought mechanisms to provide food at low cost, either by purchasing it on the world market or by increasing agricultural productivity (Torres, 2003).

The first strategy resulted in welfare programs for the poor, such as food stamps, school lunches, and subsidies targeted at specific products. These measures may succeed in reducing hunger and poverty in the short term, but they tend to be temporary because making them permanent implies a high cost, or else the lack of funds makes them reversible (Kay, 2006). In fact, social spending in the region has been repeatedly cut, and in addition, bolstering the food supply with purchases from abroad can undermine financing capacity if there is instability in the prices of agricultural products (Hall, 1998). Another drawback is that it favors patronage and corruption (Huber, 1996).

The second strategy for enhancing agricultural productivity focused on sectors with productive potential, through the intensive use of inputs, which compromised sustainable development, and because it depended on returns from investment it did not guarantee attention to the needs of the poor. This output-maximizing focus is related to the notion that raising incomes is the way to resolve the problem of hunger and poverty, i.e., to focus on increasing the national wealth as the way to resolve the problem (Townsend, 1993).

One extension of the concept of subsistence is that of basic needs, which addresses the minimum requirements of private consumption, but also includes essential services provided by the community (drinking water, transportation, education, etc.). The problem with this approach lies in establishing the criteria for determining the elements that should be included. Through differences of constitution and location, people require different quantities of basic goods in order to satisfy the same needs, and so there is debate over the possibility of determining the basic human needs common to members of different cultures, and even to individuals within the same society.

The problem with this approach is that it does not make explicit the fundamental difference between needs and satisfiers. What changes, across time and across cultures, is the way or means by which the needs are satisfied (Max-Neef, 1993).

As noted in Chapter 1, the FAO, the World Bank, USDA, USAID and IFPRI have defined food security and formulated policies according to a basic food basket (Townsend 1993; Hall, 1998).

The social policy of food security relies on the notion of subsistence and/or basic needs. For Sen and Foster (1997), however, the key components of living standards and poverty are not goods, nor their characteristics, but rather the ability to do various things using those goods or their characteristics. Consequently, food security policy should start by considering the capacity of individuals and communities to function (Sen and Foster, 1997). For example, the supply of food does not reflect the individual's condition, i.e., his level of nutrition, or his level of utility, or the pleasure or the desire satisfied from consuming food. We must distinguish what the good does for the person from what the person does with the good (Cohen, 1993).

The relationship between income and capacities will be affected by people's age, by their gender, and by their social functions; by their location; by the epidemiological setting and other kinds of variations over which a person has limited or no control (Sen and Foster, 1997). In rural areas of LAC a high proportion of people are elderly or women and the men capable of working have left.

Policies focused on increasing productivity to raise incomes among the poor will not necessarily achieve the goal of food security, if they are not accompanied by pricing policy and adequate marketing channels for the output of family farms.

The concept of poverty as subsistence has been sharply criticized, because people are not only organisms that need to renew their energy sources, but social beings who must play various roles in society. Moreover, it is not easy to determine basic food needs, since food is socialized in all societies (Townsend, 1993). Consequently, policy in this area must consider the risk of opting for one food basket alone—which is that the impact on reducing hunger and poverty will be short-term or fleeting—in addition to the need to have the necessary resources to sustain programs of this kind.

5.2.2 Food sovereignty

To combat poverty we must enhance the capacities of individuals and not merely distribute goods (Sen and Foster, 1997). Beyond competition between people with different capacities there are many other factors that govern the circulation and appropriation of social wealth, such as power relationships and cultural traditions (Reygadas, 2002). The concept of food sovereignty points in this direction.

Food sovereignty combines a series of policies that go well beyond food production, as discussed in Chapter 1. Food sovereignty policy gives priority to local agricultural production for feeding the population, and access for farmers to natural resources, stressing autonomy for them in defining food and agriculture policy (Vía Campesina, 2003).

Policy measures take account not only of productive aspects but also those relating to the standard of living. There are experiences with poor indigenous and small-scale farming communities that have exploited market niches through certification schemes whereby they can offer specialized products and do not have to sell at spot prices (certification, internationally recognized specific protocols, etc.).[28]

As an instrument of food sovereignty policy, in the productive aspect, the creation of networks can correct market failures, as explained below. Oxfam, an NGO that fights hunger around the world, has drawn from its experience a list of measures for moving toward food sovereignty: (1) seek ways of enhancing agricultural productivity in a sustainable manner; (2) foster associations of NGOs and government; (3) promote capacity building; (4) include the participation of women; (5) have participatory extension systems; (6) have alternative sources of income; (7) respect rights to the land; (8) promote good nutrition practices; (9) understand regional food markets (Hall, 1998).

[28] For example in Mexico with organic coffee (Vanderhoff, 2005) or in Peru with organic bananas (Soldeville, 2005).

Among the proponents of food security, there are also groups that use the rule of "the right to food" (Glipo, 2003). To the extent that food sovereignty incorporates fundamental aspects of economic sovereignty, agrarian reform, women's rights and those of small farmers, it has become a broader platform for those seeking fundamental changes in the national and world order.

Following is a detailed discussion of policy measures that could lead toward the goal of food sovereignty.

5.2.3 Participation by women: The feminization of agriculture

According to official statistics, women produce 30% of the earnings from agriculture in South America, and account for 26% of the agricultural labor force, a proportion that is rising (Deere, 2005).

Consequently, efforts to alleviate rural poverty and improve food security will not have the hoped-for success unless they take into consideration the need to ensure women's access to productive resources. In this sense, as an alternative for local development, women must be given more flexible access to rural property, recognizing that most farms are still registered in the name of the man, regardless of the degree to which the woman participates in the management and work. The lack of land ownership limits women farmers' access to credit, since the land is generally taken as collateral.

Credit institutions should also be encouraged to change the ways they do business, by demonstrating to them that women can be fully creditworthy because they take seriously their obligation to repay, and because they are able to pursue productive undertakings with a mindset that is more open to change and to technological innovation adapted to the fluctuations in economic rules and markets. Another aspect to address in relation to this issue is the need to give women the chance to educate themselves, recognizing that an important sector of the adult rural female population remains functionally illiterate, meaning that they cannot incorporate themselves into the market. This is moreover a cultural factor, since males with little education achieve such incorporation. In this respect, guaranteeing equal education opportunities for males and females would help increase the productive potential of countries in LAC and would contribute positively to addressing the problem of poverty.

The inclusion of gender equity as a variable in development planning would be an important step toward giving women their proper place, and for overcoming what some experts have called the "feminization of poverty". Full and equitable participation for women and men in rural and agricultural development is an absolutely essential condition for eradicating food insecurity and rural poverty.

Improving household food security can only be achieved if female as well as male farmers have access to agricultural training and extension services (which have so far been geared primarily to men), and specifically to a good level of technological innovation in postharvest management, storage, quality, classification of products and standardization of packaging, optimization of processing and marketing. This would not only improve women's social status but would also allow them to enhance agricultural competitiveness, and facilitate access to food for all people, thereby reducing rural poverty (see Box 5-1).

5.2.4 Development and culture[29]

The LAC region is rich in ethnic and cultural diversity and in "agri-*cultures*" (see Chapter 1). Culture and development are closely related to agriculture (Sen, 2004). Yet development policies in the LAC region have tried to make small-scale peasant/indigenous farmers adapt their "agri-cultures" to models that are foreign to their reality and culture. Culture is indeed a central component that has been overlooked in the drive for development (Warren, 1992; PRATEC, 1993a,b; Warren et al., 1993; Hoage y Moran, 1998).

Informed by a Eurocentric[30] vision, development policies and the dominant AKST system have tended to favor conventional agriculture (Grillo, 1998). These policies, by promoting the mechanistic Western worldview, predominantly anthropocentric and unsustainable (see Table 1, Chapter 1), ignore the worldviews or cosmovisions (Gonzales, 1996, 1999; Valladolid, 1998, 2001; Toledo, 2001), knowledge, know-how and technologies of peasant and indigenous peoples[31] (more than 400 ethnic groups) and their respective agri-*cultures*. They thereby induce a process of marginalization, devaluation and erosion of peasant and indigenous knowledge and AKST systems and their respective resource management systems.[32]

The region's rural and agricultural development, and in particular its AKST system, has been closely associated from the outset with the financing and the models proposed by Western Europe and North America (Trigo et al., 1983a,b; Heissler, 1996), financed and supported by a transnational network of development agencies (USAID, CIDA, European cooperation), financial agencies (World Bank, IDB) multilateral organizations (FAO), international research systems and services (CGIAR) and regional cooperation (IICA). The system works with national and local research, education and agricultural extension systems (agricultural research in-

---

[29] For a definition of the concepts of development and cultures see Chapter 1. Development and culture as concepts and social practices are given particular definitions depending on the worldview (see Table 1, Chapter 1) and the theoretical paradigms of which they are components. In other words, there is no single definition of these concepts: indeed, there are as many definitions as there are cultures in the world and in LAC (more than 400 indigenous ethnic groups totaling more than 40 million people).

[30] See Chapter 1, footnote 7, p. 47.

[31] For further details, see World Forum on Agrarian Reform (2004).

[32] At the root of the conflict between conventional, output-oriented agriculture and indigenous-peasant agriculture we see that the cultures and societies that embrace them have two fundamentally different ways of knowing (epistemology), of being (ontology) and of relating to the world (cosmovision). The dominant liberal approach, which takes a mechanistic and positivistic view of the world, is to develop and modernize rural society through infrastructure (paved highways, improved roads), conventional agriculture, modern AKST, and the transfer of farming, forestry and fishing technology generated in first-world countries and adapted by local agricultural research institutes. This dominant process has not been balanced by any similar openness on the part of states in the region toward peasant-indigenous knowledge and AKST.

Box 5-1. Women in agriculture in the English-speaking Caribbean

During slavery women worked along the men in the cane fields and their contribution was seen as being equal to the man in the labor force. In more contemporary time, citing Barbados as an example, Barrow (1994) has noted that 61% of the total land holdings in Barbados is farmed by men and only 6% of land in size of 4 ha and more is farmed by women. With respect to holdings of 4 ha and more there is also a predominance of male ownership. It has been globally recognized that women have less access to land and are in the minority with respect to land ownership, they have fewer support services, earn less than men and are in more ways associated with subsistence agriculture. The English-speaking Caribbean is no exception. As a result of these constraints many women farm "family lands".

Family lands is a Caribbean phenomenon, in which there is co-ownership of the land and the rights to the land are transferred from generation to generation irrespective of place of residence of the descendants, birth order or gender. This form of land tenure is prevalent throughout the Caribbean especially in Jamaica, Saint Lucia, Dominica, St. Vincent and Grenada. This land tenure form is also more evident in Tobago than in Trinidad.

It has been observed in the Caribbean that unlike female farmers, male farmers have an additional form of gainful employment whereas the woman's other duties tend to be focused on housework and child rearing. In the case of the woman, the activity of farming and the performance of household duties are merged into one activity. It is the integration of the activity of farming and the conduct of household duties on the part of the woman that underlies the fact that the contribution of women in agriculture can be deemed invisible.

Historically the woman's place was seen to be in the home, and notwithstanding the fact that women engage in activities that generate income to supplement that of their mates, the research speaks to the family side of the Caribbean woman. This position begs the question as to how this phenomenon is accounted for in an agricultural census and by extension do agricultural policies speak specifically to the needs of Caribbean women in the sphere of agriculture. The perception still persists in the Caribbean that a woman cannot be a "true" farmer, with the capacity to contribute to a country's economy; that is the domain of the man. On the other hand, the new thinking is that women do farm and contribute to the nation's economy. In light of the fact that policies are supposed to be gender blind, they are now so formulated that the specific issues of the Caribbean woman in agriculture are subsumed in these gender-blind policies. On both fronts therefore, the contribution of the Caribbean woman to agriculture becomes disguised.

In Trinidad and Tobago in 1989, the Ministry of Agriculture, Land and Marine Resources formed the Organization of Professional Women in Agriculture. This organization is aimed, *inter alia*, at promoting the participation of women in agriculture in the decision making process at all levels of society, both locally and internationally and promoting the application of science and technology of agriculture for the welfare of society.

In spite of the attempts by organizations aimed at addressing issues specifically related to women in agriculture, the contribution of Caribbean women in agriculture is still not afforded the type of attention that it deserves mainly because the woman's place continues to be seen as being in the home and any agricultural pursuit is merely an extension of the family and the woman's attempt to supplement the daily meal. As such, issues of access to land, security of tenure and provision of support services remain unaddressed.

stitutes, national and regional universities) (Pimbert, 1994; Gonzales, 1996, 1999; Escobar, 1999).

Social, political and cultural marginalization. The dominant AKST system in the region is part and parcel of the dominant development and culture. Over the last 60 years, it has promoted the modern or conventional system of agriculture, while largely ignoring the other two existing systems in the region (peasant/indigenous and Agro ecological). It is only in recent years that the emerging processes of resistance, decolonization and cultural affirmation in the region have put forward the concepts of multiculturalism, intercultural knowledge, and "coloniality" of power (Quijano, 2000) and of knowledge (Lander, 2000) in an effort to explain other ways knowing, understanding and conducting agriculture, and the general features of local life as a whole (Warren, 1992; Leff and Carabias, 1993; Grillo, 1998; Agrawal, 1999; Delgado y Ponce, 1999; Huizer, 1999; Rist et al., 1999; Walsh, 2002, 2004; Ishizawa, 2006; Vía Campesina et al., 2006).

These concepts are renewing and deepening the epistemological, ontological and cosmological foundations of the AKST system adopted in the region over the last 60 years, and make it possible to revise them. But at the same time these concepts are still striving for incorporation into the AKST system. Policies for promoting AKST, by reproducing the political, social and cultural marginalization and devaluation of peasant and indigenous communities in their treatment by national society, have disregarded the languages, the cosmovisions, the knowledge and the technologies of peasants and indigenous people and local producers, as well as their systems for the on-site conservation of native crops and their wild relatives,[33] the management of natural

[33] Among other experiments in the local conservation and/or community management of natural resources is the "In-Situ Cultivation of Cultivars and Wild Relatives 2001-2005" project in Peru, financed by the Global Environment Facility (GEF), administered by the United Nations Development Program (UNDP), and implemented by the Instituto de

resources (hunting, gathering and fishing) and their forms of agriculture.

Intercultural and interethnic policies and policies of cultural affirmation directed toward an alternative AKST system would serve to capitalize on indigenous/peasant knowledge and AKST by incorporating them in their own terms—in other words, without attempting to validate them from some supposedly modern, scientific cognitive authority (Grillo, 1998; Agrawal, 1999), and as part of a process of food sovereignty and indigenous/peasant self-determination at the local, regional (e.g., a watershed), or national level. Such policies could in this way promote the revitalization and affirmation of indigenous/peasant culture that would contribute to IAASTD objectives.

To give effect to policies for strengthening indigenous/peasant systems of knowledge and AKST, it would be useful to assess the liberal[34] and/or neoliberal policies of governments, and the transnational network (based on financing and models from Western Europe and North America and sponsored by a network of agencies already mentioned) that supports and provides feedback to the region's AKST system (Escobar, 1995; Gonzales, 1996, 1999; Vía Campesina, 2006).

The rural development models and AKST systems adopted in the LAC region in the last 50 years continue to rely on a Eurocentric vision, transmitted via Europe and North America and their counterparts in the region[35]. Specific policies for institutional change and innovation have facilitated the adoption and adaptation of knowledge, institutions and technologies originating in Europe and North America. Given this situation, when it comes to nonconventional culture and agriculture, the direct and indirect impact of this dominant model has been of little benefit and has indeed eroded local and peasant/indigenous agri-*cultures* in LAC. The same holds for people's health, and the region's ecology and environment. In this context we must consider policies that reflect experience from the past and that encourage the integral participation of peasant/indigenous knowledge and AKST systems.

Agrarian reform (AR) and landholding are important issues for the region's agricultural development. Such is the heterogeneity of LAC, however, that these issues must be considered separately in each country. AR and landholding in the region are central factors associated with poverty, hunger, and the expulsion of small-scale peasant/indigenous farmers from the countryside to the city. Similarly, living conditions, identity, the environment, and sustainable development are being seriously affected within indigenous communities (Colchester, 2001). In general, in the context of the region's system of economic, political and social domination, landholding in the 20th century, during and after the oligarchic regimes, continues to show serious disparities and social divides (Van Dam, 1999; Baranyi et al., 2004). ARs and the associated policies for land redistribution and modernization of rural production relations, it must be noted, have tended temporarily to reduce social conflict and the demand for more land and justice on the part of peasants and indigenous people in the region.

It must also be noted that these ARs were designed on the basis of western premises and experience bound up with the liberal paradigm, and they had no cultural or environmental orientation appropriate to the great mass of peasants and indigenous people and for this reason in some countries there affects ran counter to the competitive development of farming.[36] This aspect could be reconsidered in future AR and landholding policies.

If agrarian reform and land distribution policies had been based on an appropriate cultural and environmental focus, in particular with respect to the peasant/indigenous sector, the results in terms of natural resource management could have been more sustainable and equitable.

Today there is tremendous pressure from the demand for land on the part of landless peasants and indigenous people, and for those who are trapped in the tilling of mini- and micro-plots the pressure is increasingly intense and is sparking social conflict in the countryside. This situation may require compiling reviewing and assessing the ARs that have been implemented, and proposing ARs that take account of the stakeholders, the specific features of the re-

Investigaciones de la Amazonia (IIAP), in collaboration with six Peruvian institutions (Ishizawa 2006, Valladolid 2005), http://www.insitu.org.pe/english.htm). In addition there are two indigenous research centers working on the management of biodiversity, El PEMANSKY, in southern Panama, and the Instituto Amazanga of the Organización de Pueblos Indígenas, in Puyo, Pastaza, in Ecuador (http://www.cdi.gob.mx/pnuma/c7_10.html).

[34] Liberal theory was developed in the 19th century in Western Europe and is associated with the "Age of Enlightenment". Since then, and particularly in the last 50 years, this theory has become the dominant paradigm in western or westernized countries. Although today this paradigm and the development theories based on it are in crisis, their hegemony is recognized worldwide (Harvey, 2007; Lander, 2000). The state and the development policies applied in the LAC region to date have had a liberal character, and more recently, generally since the late 1980s, have taken on a neoliberal character.

[35] The North American model of progress and of rural and agrarian development, as it developed through the 20th century, has shown many limitations and contradictions that have been highlighted in the literature (e.g., Berry, 1996; Gilbert et al., 2002). The question arises then, why do policies in the LAC region insist on trying to replicate the conventional agriculture model applied in North America under liberal or neoliberal models?

[36] The Western world there are three theories of development: the liberal theory, the Marxist theory, and the poststructuralist theory (Escobar, 2005). With the long-term dominance of Western European colonization, and later that of the United States, the paradigm of liberal theory expanded beyond the confines of those centers of political and economic power. During the 20th century, and especially in its second half, governments and policies in former colonial states, including those of the LAC region, took on a liberal character to various degrees. Since the 1980s, neoliberal thinking has been heavily adopted in government policies in the region.

source management systems, the crops involved sustainable development, and food sovereignty.

With respect to land rights, territory and indigenous peoples,[37] this issue is recognized to varying degrees by national constitutions[38] (Colchester, 2001) in the region as well as by international conventions[39] and international case law on human rights. For indigenous peoples, land and territory are closely tied to autonomy and self-determination (Via Campesina 2006, Van Dam, 1999), for which reason policy should be established to promote commercial or business undertakings that respect property rights over time.

In this respect, future agrarian reforms should have a clearer profile within the strategy of land distribution, especially when access to land must be created for vulnerable social groups such as indigenous peasants.

It must also be noted that, under current conditions, local indigenous communities must become more competitive in generating income as through the sustainable exploitation of natural resources. Yet there are many legal barriers that prevent communities from making better use of their communal lands, for example, and in any case their organizations can see no way to avoid the fragmentation of their property into "*minifundios.*" This situation calls for policies that will give communities the ability to enter the land market so as to grant concessions and to attract investment on the basis of rules and conditions established by the indigenous communities themselves. To this end, incentives could be established so that those who are no longer making use of the land can dispose of it and in this way help to reorder the size of properties.

The land and territory of indigenous peoples shows a tendency to shrink because of factors related to extractive economic activities, linked to the capitalist form of development supported by liberal and neoliberal policies of the LAC region (Deruyttere, 1997; Toledo et al., 2001). Under these conditions what is needed is to establish policies to defend these territories or in any case to foster negotiations to pay for the environmental services that indigenous communities provide in the course of conserving their ecosystems and cultures. The important point in this process of globalization and exploitation of natural resources is to find economic med denizens to recognize payment to those groups that act as the custodians of biodiversity in the various ecosystems.

Even so, depending on their level of organization and strength, autonomous processes (with or without external financial support) can be observed in various parts of the LAC region for restoring local peasant/indigenous agriculture as part of a process of decolonization and cultural affirmation (See Chapter 1: Figure 1-1).

Yet a great number of small farmers in the region are striving to restore their agricultures as part of their systems for managing local natural resources and as part of the process of decolonization and cultural affirmation (Grillo, 1998), as alternatives to the dominant society, culture, agriculture and AKST system. All of this suggests that a relevant policy issue from the development and culture viewpoint is to promote the strengthening of local cultures and know-how in rural communities, particularly those of indigenous origin, including the provision of financing to expand their contributions to strengthening regional and national AKST, from a democratic perspective, in the direction of promoting this component in the transition towards sustainable protective systems.

5.3 Policies for the Participation of Public and Private Stakeholders in the Development of AKST

Participation, understood as a democratic value that encourages citizens to collaborate in formulating and implementing public action, allows decisions to be taken on matters of collective interest (Kondo, 1996; Bañon, 2006), and the quality of that participation will depend on whether a suitable institutional, social and cultural conditions exist. A suitable legal framework, mechanisms for participation, and an appropriate degree of decentralization are the best institutional conditions for promoting participation. The history of the community, the types of leadership, its economic resources, political culture, local capacities and educational levels are the social and cultural conditions that will determine the degree of collective participation (Cartagena et al., 2005; Colomer, 2006).

Participation and decision making take place in different levels or spheres, in which stakeholders will have different perceptions, capacities, influence and roles. In the international sphere, the representatives of governments, multinational agencies and international NGOs work with macro information and take policy decisions that will affect the other levels; in the national sphere, governmental and private sector players and national leaders will take political decisions that affect the other two spheres, but especially the local level; and in the local sphere, communities and families work with information and decide on their resources, with direct influence on the application of policies taken at the other levels, especially at the national level.

One of the fundamental rules of national governments is to facilitate the relationship between the different levels of decision takers. While there may be a series of relationships between the public and private sectors, one challenge will be to move beyond the old schemes of vertical and hierarchical relations between government and society and to make networking a new form of relationship based on interchange and cooperation, thereby strengthening stakeholder participation in decision-making.

Bifarello (2002) explains that the concept of the "associative network" is useful for understanding the public-private societies in Latin America, and how stakeholders relate to each other through formal and informal societies. Associative networks are distinctive not only because they connect people around the taking of decisions, but also because of their multiplicity and their efficiency, and the fact

[37] This issue is part of the "ethnic question" (Stavenhagen 1990) or the "indigenous problem" (Quijano 2005). The situation calls into question the capacity of the state and of Latin American democracy to resolve satisfactorily the issue of land, territory, and self-determination of indigenous peoples

[38] For further detail see Colchester, 2001.

[39] Convention on Biological Diversity, United Nations Declaration on the Rights of Indigenous Peoples, ILO Convention 169 concerning Indigenous and Tribal Peoples in Independent Countries.

that the relationships they create tend to be based more on interdependence and collaboration than on competition.

Stakeholders at the different levels will insert themselves more efficiently into a cooperative network if the necessary conditions are in place to ensure equity in access to resources, information, training, etc. The following section describes the actions and roles of public and private stakeholders at the different levels of decision-making, for promoting their participation in the development of AKST.

5.3.1 International actions

The state and civil society have specific roles in the design and implementation of AKST agendas for countries of LAC. Policies are government principles for achieving goals for a specific population, and they go through the phases of statement and of practice (Perez-Ordoñez, 2005). The state must give effect to statements of intent by responding to demands through actions inscribed in a government agenda, while civil society must demand quality in government management and thereby contribute to democratic governance.

While policies also respond to the prevailing development models for the region, it is important to recall that political processes are the result of inter-linkages, exchange and dependency between interest groups and nation-states. This means that regional networking among governments could contribute to policy agendas for supporting the development of AKST in ways that will meet the specific needs of the region and its member countries. This networking will be based on the ties between the members of a social system structured by the existing connectivity among them. In other words, the greater the connectivity, the greater will be the interactions and the better the results (Wellman, 1987).

At this level, the decision-making spaces are forums, summits, conferences and international meetings, among others, in which the governments of the region participate along with multilateral organizations and international NGOs. Participation in decision-making should be directed at prioritizing AKST agendas that include the particular features and sensitivities of the region. A shift in the structures and social relations between nations that have contributed historically to the region's underdevelopment, based on inequitable access to science and technology for the poorest countries, could be addressed through networking as a regional bloc. This work could be based on pre-established regional agreements and on a clear understanding and vision of the problems, potentials and priorities at the national and regional levels with respect to generating AKST.

The particular ecological features associated with the region's cultural characteristics required a regional form of networking to address the shortage of science, technology and innovation and the recognition and use of local knowledge, especially in regions that have been the object of little comprehensive study such as the Amazon and the Caribbean, where studies have for the most part focused on conservation rather than on the people, the environment or development.

5.3.2 National actions

Under the government management model that has prevailed over the last two decades, the role of the state has been seen as institutionalizing governance through legal mechanisms such as creating new institution that will extend citizen rights. But it has also led to a reduction of government action for generating AKST, which has affected the most poverty-prone social sectors such as small-scale farmers.

Institutional reforms are key instruments for initiating changes in the relationship between government and society, but a new approach to government management based on working through cooperative networks will require humanizing those reforms, given the physical, natural and cultural complexity of the region. Some reforms that would help bring greater participation in the development of AKST are described below.

5.3.2.1 Suitable legal frameworks

Legal frameworks are instruments that protect civil society and can foster effective participation by the private sector in formulating policies and in other negotiations with the state. It is very important that the authorities be actively involved in applying the legal frameworks in support of AKST, but they must also have active support and participation from the grassroots, the academic world, and the private sector. Legal frameworks can only be effectively implemented if public and private stakeholders are included in their design, implementation and evaluation, and this can only be achieved through political will on the part of the authorities to decentralize certain roles and functions that will promote social co-responsibility. Following are some aspects to take into account for ensuring that legal frameworks will produce greater participation in the development of AKST.

- Guarantee the representativeness and legitimacy of the social base, so as to promote genuine participation and governance. Given the region's cultural heterogeneity, inclusion and respect for local forms of organization, representation and prioritizing demands will produce greater participation and commitment on the part of local stakeholders, based on grassroots social oversight.
- Orient the role of the academic sector (universities and research centers) toward the design and implementation of an AKST agenda appropriate to national needs, which will respond effectively to resolving concrete problems and will seek comparative advantages based on domestic agricultural potentials.
- Facilitate participation by the private sector and enhance its capacity to invest in innovation. The legal frameworks must guarantee a suitable institutional context and respond to the promotion of innovation.
- Ensure sufficient economic and technical resources for implementing the legal frameworks, so as to help make AKST development sustainable.

Recognition of the potentials of each of the social stakeholders involved in applying the legal frameworks, and including them in the respective spaces, will help ensure that genuine representatives will understand the particular features of the national-regional problem, and can negotiate critical issues or insert their priorities on the agenda at international forums, since the majority of the guidelines for public policies come from those spaces.

5.3.2.2 Effective mechanisms of participation

Having such mechanisms in place will not only guarantee inclusion of the various sectors related to science and technology, but will also ensure the definition, control and validation of government actions. Councils, committees, advisory bodies and other mechanisms of participation in defining AKST support policies have the advantage of bringing together players with diverse capacities, experience and expectations, an aspect that can be reinforced by applying a collaborative working philosophy, one that avoids internal power struggles over particular objectives and generates collective benefits through selective incentives.

These forums will be able to function if they have the necessary rules, if they can meet the demands of their participants, and if they have financing:

- Participatory mechanisms must be institutionalized from the outset in national legal frameworks and must be legitimized by social stakeholders.
- The setting of rules that engage participants, where the merger of formal and informal rules established by internal agreement can help the functioning of forums for taking decisions on AKST. Provided the legal framework is broad and allows these forums to be properly designed or adapted to local and regional realities, this will help not only to upgrade social capital but also to secure the participation and commitment of grass-roots players in the social oversight function.
- The inclusion of differentiated demands for the various kinds of users of science and technology could contribute to the priority objectives of the governments of Latin America and the Caribbean. On one hand, the inclusion of prioritized demands from the private sector and industry could help lift revenues from agricultural exports, increasing their share of GDP; and on the other hand, the inclusion of prioritized demands from small farmers could to a large extent resolve the problem of insecurity and food sovereignty in the countryside, and both actions would contribute to reducing hunger and poverty in the region.
- Governments should ensure financing for these institutionalized spaces, although a combination of public and private funding could guarantee their sustainability and efficiency.

5.3.2.3 Decentralizing the AKST system

Decentralizing the AKST system by delegating greater decision-making power under a new government approach to collaboration and networking could become a key instrument for the efficient design, execution and evaluation of the AKST agenda. The economic, social and political advantages of decentralizing AKST system are closely interlinked and can be summarized as follows:

- Decentralization produces economic advantages in the form of efficient expenditure. A number of studies show that there is greater efficiency of expenditure in decentralized systems than in centralized ones. As well, local revenues will rise, provided they are locally managed. Greater centralization means greater fiscal problems, and there is less fiscal vulnerability in decentralized systems (Wolman and McCormick, 1994).
- The social advantages flow from the fact that decentralized systems for AKST support contribute not only to the accountability of decision-makers and the quality of services, because of pressure from users, but also to active participation by various sectors, because it provides the opportunity for citizens at the local level to define, debate and decide an AKST agenda. Nevertheless, a decentralized AKST system also requires sound local capacities (technical and political), i.e., to make use of and strengthen human capital in order to ensure that priorities are set equitably, an aspect that will be addressed below.

The policy of promoting a decentralized system of AKST management will enhance the values of democratic governance. The implementation of innovative public policies by subnational governments is a characteristic of decentralized decision-making systems. As well, civil society participation in a decentralized AKST system will help to create co-responsibility for actions taken within the AKST system.

In the process of decentralization, the private sector should not be involved directly in policy formulation. Its role should come into effect downstream, working with the government to finance rural programs and projects. However, some caution is needed not to expand the private sector's role too far, for experience has demonstrated a relationship between fiscal decentralization and corruption. The private sector can be the voice of the community, to the extent that it identifies local needs and proposes viable solutions. In some cases, if there is a private business in the community, local residents can be informed of the qualifications needed for employment in that business.

Following are some of the areas in which the private sector should be permitted to participate:

- Fostering and strengthening linkages between all interested parties, including small local producers.
- Working with local government to ensure a positive impact on local development.
- Helping in the planning and implementation of local programs and projects, and sharing know-how and skills in the use of resources, financing and employment matters.
- Mobilizing local participation for defining priorities and how they are to be achieved.
- Advocating pro-poor and community initiatives that will be of benefit to all citizens.
- Creating associations to deliver local services such as electricity and rural roads.

These companies can include financial advisory services and micro-finance for local development. It is important to note that if the private sector is to participate in a decentralization process there must be appropriate institutional and administrative structures in place for handling funds. One of the lessons learned about the inclusion of the private sector as a partner with government and civil society is that each side must work with and support the others in order to achieve the objectives of decentralization.

5.3.2.4 Mechanisms for disseminating information

Information dissemination mechanisms should be developed in parallel with decentralization policies, for it is the quality and quantity of the information provided to civil society and the private sector that alone can guarantee that they participate and are well represented. It is important to remember that the diversity of local stakeholders demands a variety of means for disseminating information. The basic lines of policy for AKST support developed at the local level and those implemented in each area should start from comparative evaluations and a mutual understanding of the contribution of each as the basis for developing appropriate technologies locally. Information generated at this level, if it includes information about traditional knowledge and know-how, could be useful to decision makers as well as to the regional or national technical and academic bodies.

The legal rules with respect to AKST tend to be applied more successfully if they include mechanisms for disseminating information. A clear policy on information, disclosure and distribution of new findings in agriculture, science and technology will guarantee their proper use.

There should be policies to promote consensus-building and coordination between civil society, the state and the private sector as to the kind of information to be shared, which new discoveries should be publicized and when, and what contents should be revealed in light of the potential of civil society, so as to ensure smooth operation at all levels of decision-making. Given the existing asymmetries between social groups and players, actions should be planned to strengthen negotiation mechanisms so as to strike a fairer balance in the relative clout of the different sectors in setting AKST priorities. If information is clear and readily understandable by civil society organizations and rural people, this will contribute to the operational objective. To this end, specific policies need to be designed to promote access to information for marginalized rural groups.

Participatory methods could be seen as AKST support policies with a view to integrating rural communities into the technical information system, and at the same time integrating the academic sector into local knowledge and know-how, and in this way jointly to generate new knowledge, science and technology.

The kind of information disclosed to rural communities must consider their perspective, the integration of technical information and local knowledge, and the use of visual supports that are easy for them to understand, and civil society should be included in guidelines for planning and implementing projects and programs.

Finally, it is crucial to ensure the dissemination of successful experiments in applying the legal frameworks and implementing the AKST support policy agendas or generating new technologies and innovations. To this end there are many tools such as field tours, exchange of experience, farmer-to-farmer training, and local agricultural research committees.

5.3.2.5 Effective mechanisms for evaluating and monitoring policies

Such mechanisms are a vital condition to support the process of democratizing AKST. In Latin America and the Caribbean there are policies, programs and projects in place, but there are no mechanisms to evaluate their performance, measure their impact, or reformulate them in light of local needs. This weakness is glaringly evident in civil society, which in past decades was under the thumb of government. With the trend towards new forms of government based on collaboration and networking, it is time to consider some basic criteria for evaluating policies:

- Policies that involve civil society in their design and implementation can respond better to local AKST problems.
- Pilot projects implemented at the local level to test new knowledge and technologies could provide guidance for policy decisions in support of AKST.
- An inter-sectoral approach to policy, i.e., the review of policies in different areas that serve the same national objective and the repeal of those that do not fit the government agenda.
- In implementing policies there must be constant information on the roles and responsibilities of the stakeholders involved.

5.3.3 Local actions

There has been much discussion of the importance and the roles of civil society in a new approach to governance based on cooperation and networking. While civil society can alter in its favor the balance of power between state and society, it can also exert pressure for better government management or articulate interests by acting as intermediary. It is important to recognize as well that it has certain capacities to address these new challenges.

Grass-roots players (peasants and indigenous people) have developed certain knowledge, skills, abilities and other individual attributes relating to economic activities, which are recognized as human capital. As well, local societies have developed a series of social relations and rules for more effectively achieving common objectives, known as social capital.

The human and social capital of countries is highly important for democratization, and it is clear that in many countries these capacities are enriched through decentralization and local capacity building. Following are some key actions for dealing with these processes at the local level:

5.3.3.1 Strengthening local grassroots institutions

Working to reduce rural illiteracy and functional illiteracy and to enhance human capital will be an important task for governments in integrating the local grassroots sectors. As well, building technical and political capacities as part of the rural school curriculum could promote a democratic culture and improve local stakeholders' capacity for participation and negotiation.

Culturally appropriate training programs will be better accepted at the local level, where agriculture extension agents, indigenous or not, can become key links and can even serve as negotiators between the local and the government level with respect to AKST policy needs. Those agents must have capacities and skills based on experimental knowledge and learning. Moreover, considering that local technical capacities are weak in the face of innovations and market requirements, it is important to involve the local authorities in capacity building and technical assistance, as the only way to ensure co-responsibility, to strengthen their role, and to promote sustainability of the program.

5.3.3.2 Local capacity building

This picture of incipient representation and participation calls for parallel processes to develop capacities at all levels of society, with particular emphasis on rural dwellers who, sooner or later, will be making use of the AKST results developed in research centers, universities and elsewhere, and can then become active receivers, adapters or improvers of knowledge, science, technology and innovation in agriculture.

An important issue to address through a new form of government management is the effort to recognize and capitalize on local knowledge and know-how, which highlights the urgent need for intercultural education approaches, working through agents external to the communities as well as with the indigenous and peasant communities themselves. The protection of creativity rights and copyright through the intellectual property system hardly exists: in fact, this is not an appropriate mechanism for protecting the traditional knowledge of aboriginal communities and peoples because of the community nature of that knowledge.

5.4 Policies for the Sustainable Management of Production Systems

5.4.1 Sustainable management of production systems

The concept of sustainability is useful for integral rural development, because it treats agriculture as an economic, social and ecological system, the management of which is based on diversifying production over space and time. This approach must embrace all components of the land so as to improve its biological efficiency, maintain its productive capacity, conserve biodiversity, and generate conditions for the system to be self-regulating (Altieri, 1996).

Moreover, in cases of market-induced specialization, such systems must be managed with respect for agroecological principles if they are to be sustainable, whatever the size of the farm and the type of output. This agroecological approach should be a goal not only for small farmers and subsistence agriculture but for all production systems, even commercial ones, that are trying to move to sustainability and competitiveness.

5.4.1.1 The stages of transition

We start from the definition of the three types of agriculture given in Chapter 1 (conventional, traditional/indigenous, agroecological), characterizing the degree of sustainability of each type: the conventional system is dependent on the intensive use of industrial inputs; the traditional or present system makes little or no use of external (indigenous/forest) inputs; while the agroecological system uses resources generated within the system, with perhaps some alternative inputs. All these systems are in constant flux, depending on their components, functions and management. These different production systems contribute in different degrees to conservation of agrobiodiversity and of biodiversity in general, and they also contribute in different degrees to the internal food market and to the export market. Industrial, commercial agriculture systems, which are closely geared to the market, are more homogeneous but they are the ones that contribute least to maintaining biodiversity. By contrast, small peasant agriculture, despite its great limitations in farm size, has made the greatest contribution over time to the conservation, use and exploitation of biodiversity (Tapia, 1999; Caporal y Costabeber, 2004).

A number of conditions must be met if production systems are to move towards sustainable management:

- Diversified, multi-crop production, crop rotation, or a combination of systems (agro-sylvo-pastoral) managed over space and time.
- Meeting with the family's food needs and supplying the domestic market.
- Use of agroecological practices for efficient exploitation of natural resources available on the land.
- Reduced energy consumption in running the system (avoiding excessive mechanization and transportation distances, optimizing photosynthesis, etc.).
- Making proper use of the biomass byproducts of farming: stubble for cattle, ground cover, green manure, composting.
- Development of capacities based on local knowledge and proven technological innovations (see Chapter 4).

These conditions for transition must not affect levels of productivity and competitiveness of the different production systems. This situation implies a gradual conversion that will allow the restoration of soil fertility and functional biodiversity in agroecosystems. There may be a noticeable decline in yields while the ecological balances of the production system are being restored. During this time incentives may be needed for some producers until their systems recover their productivity.

If production systems are to shift towards ecological or organic farming, farmers will also need to receive a price for their output consistent with its quality, and this may be higher than the international market price. These products are now sold to a limited group of consumers, mainly abroad, with the capacity to pay a premium for them. Yet a number of studies show that this organic market can be expanded, and that it is possible to produce sufficient food without using chemicals to meet the nutritional needs of the world population. Ecological or organic farming is also becoming an important source of rural employment, and is thus contributing significantly to improving living standards. Figure 5-2 shows the three stages of transition, according to the state in which each production system finds itself initially.

1. Conventional systems

Conventional production systems, with their high use of chemical inputs, can move toward stage 1, "reduced use of chemical inputs", through greater efficiency in managing the system. Indeed, there are already various options that are being successfully applied such as sustainable low-

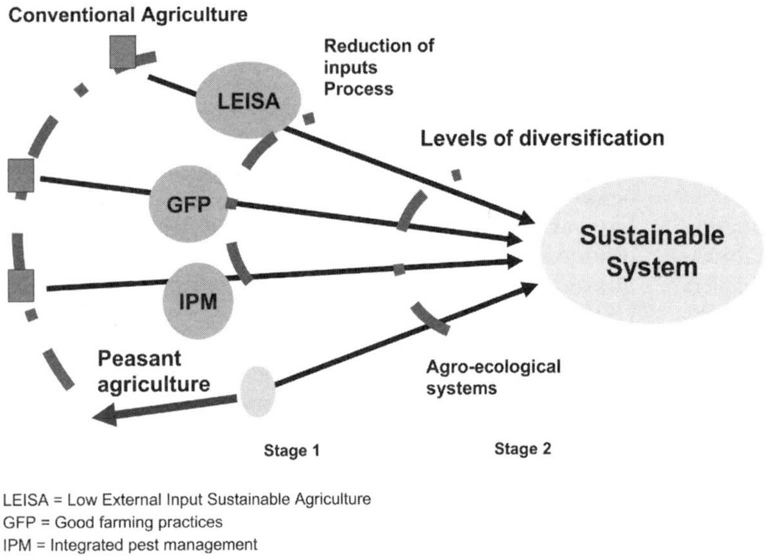

Figure 5-2. *Options for the transition to sustainability.* Source: Gomero and Velásquez, 2003.

external-input farming (Reijntjes et al., 1995),[40] integrated pest management (Cisneros, 1992), good agricultural practices (EUREPGAP, 2003), minimum tillage (PROCAS, 2001), and other practical models that enhance productive efficiency and reduce production costs. It is also possible that some of these production systems could move toward stage 2, "agroecological management", through a more profound change in system management and greater levels of product diversification in farming, livestock and forestry (Gomero, 2001; Willer and Yussef, 2004), as well as greater agrobiodiversity.

Shifting quality demands for food products in external markets and the certification mechanisms now in place may encourage these transitions. Incentive policies would be geared to progress in these stages of transition, assuming the conventional systems that do not follow these paths would then be made conditional upon other poverty reduction goals, such as employment creation, in order to receive subsidies.

It must also be recognized that some systems of cultivation, livestock rearing or plantations cannot be maintained without a package of agrochemical inputs, especially for combating insects and diseases in large-scale monocrop operations; in this case, they could be subject to the "polluter pays" principle. In turn, the revenues collected could be earmarked to promote further research in agroecology and in agrobiodiversity management.

[40] According to the Technical Advisory Committee of the Consultative Group for International Agricultural Research (TAC/CGIAR, 1998), "sustainable agriculture is the successful management of resources for agriculture to satisfy changing human needs while conserving natural resources".

2. Traditional systems

With respect to the peasant/indigenous systems that are already at stage 1 and that use little or nothing in the way of agrochemical inputs, they have two possibilities for evolution in their management:

1. Adopt the conventional production system by increasing farm size (economies of scale), standardizing plots, purchasing commercial seeds, and making greater use of external inputs (See Chapter 2).
2. Adopt sustainable agroecological systems, improving their integral management of available resources, their productive competitiveness and the quality of their output, which could be certified using a certificate of origin, or perhaps an "organic", "green" or "fair trade" certification (See Chapter 1).

In this second case, suitable incentives would be used to encourage the transition toward stage 2, rescuing local knowledge of agrobiodiversity management.

3. Sustainable system

The third and final stage of transition is the desired shift in production systems. These sustainable systems would rely on optimizing the natural processes of the productive system (such as photosynthesis, atmospheric nitrogen fixation, biological soil activity) and intensive labor input to reconcile environmental, economic and social objectives. This desired system also implies changes in the food system, together with other energy adaptations, particularly in the fuel used in the case of mechanization, reduced inequality of incomes and new social relations.

Figure 5-2 shows the different options for ecological transition, but it does not incorporate their economic impacts. For example, it does not show what incentives could be expected from the market so that conventional systems will begin the transition to stage 1: subsidizing the price of inputs, internalizing the costs of environmental degradation, and sustainable agriculture labeling are more realistic proposals than changing international commodity market prices to reflect the quality of products, when those prices have in fact been on a downward trend, with the exception of hydrocarbons.

Generally speaking, the proposed transitions move in the direction of reducing production costs, with no associated decline in productivity. They also assume more intensive use of labor, distributed over the year, which could have a positive impact on unmechanized farming and could help compensate for the seasonal nature of monocrop agriculture in the mechanized sector. This process of converting one production system to another usually requires a prior cost-benefit evaluation. That calculation must include the real costs of production, which will internalize the costs induced by environmental pollution (loss of biodiversity, damage to producers' health, contamination of waterways, etc.). Depending on the results, changes could be made in the systems to improve productivity and help reduce poverty, through concrete policies designed and applied by institutions in the agriculture sector (Gomero, 2001).

5.4.1.2 Policies to support sustainable management of production systems

The future challenges in moving towards more sustainable production models are enormous. Stakeholders will need to change their views about the value of agriculture in resolving problems of poverty, food security, and the conservation of agrobiodiversity. During these transitions, different levels of progress in different systems of production will coexist. It is clear that if policies are to support this process they will have to arbitrate in the development of highly differentiated technologies.

There is a global tendency to consume natural products, and governments, the private sector and civil society have an important task in promoting product niches. Consumers' concepts are changing, placing greater importance on food quality and safety. Special markets are emerging for products certified according to various concepts of differential quality. Similarly, rural employment can be boosted with a sound management structure for sustainable production systems. Yet such jobs are still precarious in many countries of the region, and the state must facilitate a process of formalization, so as to improve working conditions within production systems.

Policies are also needed to develop technologies that will contribute to the sustainable management of production systems. Those technologies will need to be differentiated and must respond to different geographic, ecological and social conditions. In addition, production systems must remain competitive. The use of latest-generation technologies such as genetic engineering can be adapted to the various demands of biosecurity and their use should be prohibited in some countries that are centers of biodiversity.

These changes should facilitate consumer access, in particular in the large cities, to high-quality products and in this way strengthen domestic markets. To this end, local products will need to be promoted, processed (into flour, cheese, sausages, dried or smoked foods, marmalades, etc.) and introduced into mass consumption by various means, such as school lunches. The proliferation of these processing firms, their size varying according to the market but oriented primarily to the domestic market, will have an impact on rural employment.

Other rural jobs should be promoted to increase the employment rate in the countryside and to give rural people a chance to find work at home and raise their incomes, and in this way allow them to stay on the land instead of migrating to the cities. Those jobs could be provided by family craft businesses, participation in the benefits of tourism (porters, guides, local accommodation, etc.), nonagricultural activities, or productive job-creating investments financed with remittances from abroad.

Various production systems have been developed throughout Latin America, and each has benefited from differentiated support policies: market-oriented conventional agriculture has received the greatest support in terms of subsidies and credit and technical assistance. This support has been used essentially to buy fertilizers, pesticides and hybrid seeds, and to a lesser extent farm machinery. This kind of government support has produced an economic and social divide between market-oriented industrial/commercial agriculture and small peasant farming, focused on the domestic market and food security. Policy initiatives to provoke sustainable management of production systems should consider the following aspects:

- Establish concrete policies for reducing fertilizer and pesticide use and promoting alternative technologies for the sustainable management of production systems.
- Encourage approaches such as "polluter pays" mechanisms to discourage the excessive use of chemical inputs, especially in intensive farming systems.
- Reform landholding and ownership, access to water, and the mass distribution of credit so poor farmers can (1) stabilize their production system and devote themselves exclusively to it, (2) find a more satisfactory ways of marketing their output, by organizing producers into groups, associations, and producers networks built around on productive chains.
- Develop markets and business opportunities for sustainably produced products, through certification mechanisms.
- Help producers develop the capacity to implement on a large scale production models such as ecological, organic, biological, biodynamic, or permaculture farming.
- Pursue policies to educate consumers on the importance of consuming sustainably produced food.
- Promote changes in urban consumer demand toward diversified food consumption and a change in quality standards, including food safety.
- Provide direct incentives with more effective support for the development of agroecological production systems, especially in the transition stage where there may be a risk that output and incomes will drop.
- Adopt financing policies to ensure that the objectives of sustainable management can be achieved.
- Encourage the development of technologies for sustainable management of the different production systems. These could rely on existing experience with sustainable agriculture in the region, with technical support based on:

- Maintenance or expansion of natural vegetation covers at the level of productive systems and territorial units.
- Proper soil management for conservation, maintenance of natural fertility, and erosion control.
- Protection of natural and second-growth forests, shrub nurseries or tree plantations within production systems.
- Crop diversification and rotation to avoid environmental and economic risks.
- Ecological and economic zoning of production areas and conservation to facilitate their efficient use.
- Establishment of protected areas in the form of biological corridors to make efficient use of beneficial wildlife (pest controllers), depending on local conditions.
- Integrated management of various aspects of production: control of pests and diseases, management of soil fertility, seed bank exchanges.

AKST can be devoted primarily or target more public money to small and medium-scale agriculture. This sector does not have the investment capacity of the big producers' associations that produce for industrial processing and that can co-finance research facilities and extension services. A lesser effort can be obtained from associations involved in some productive chains such as dairy, seed potatoes, wheat, fruit orchards, coffee, etc.

The process of developing technologies for managing productive systems has been exogenous: many sector-specific technologies have been introduced without any evaluation of their environmental impact. Many of them were developed under totally different ecological conditions, and when applied in other regions their performance has varied greatly. The assessment is that in some regions they have produced good results, while in others the impact was negative.

If technologies are to contribute to sustainability they must be ecologically appropriate, economically viable, and socially fair (Astier and Hollands, 2005). In this respect, AKST should consider the systemic management of production units in its future development and innovation. This will imply a paradigm shift at two levels: (1) taking account of farming-livestock interactions, agroforestry, integrated crop and livestock systems and the planting of trees on farms, and integrated management of soil fertility components; (2) taking account of agriculture's other roles.

AKST must also change to accompany these transitions at the university level, through a rapprochement between agronomy and ecology, and managing agricultural lands with the systemic focus.

To facilitate the evolution of knowledge in the management of productive systems (see Chapter 4) requires:

- Strengthening the human resource capacities of communities for developing appropriate technologies.
- Developing a common network of information and exchange of experience in managing productive systems, with scientific and technological support.
- Designing and implementing a national and regional platform for communication and technical information that will articulate agroecological data with sustainable management of production systems.

5.4.2 Biodiversity and intellectual property

The Millennium Ecosystems Assessment (MA) predicts that the continued degradation of ecosystems services will contribute substantially to the loss of biodiversity to the year 2050 with a consequent decline in the quality of environmental services, an aspect of particular concern for the objectives of reducing hunger and poverty (EEM, 2005). When it comes to formulating policies for managing ecosystems, there are two approaches: one of them is reactive, and most problems are addressed only after they have become obvious; in the other, ecosystem management is proactive and policies seek deliberately to maintain ecosystem services over the long term (EEM, 2005). In addition, environmental deterioration has reached the point where proactive measures must be taken to reduce the impact of climate change.

The available technology is focused on commercial crops, which require greater industrial inputs, and this perpetuates environmental deterioration. AKST policies have for the most part contributed to environmental degradation and the loss of biodiversity, and are threatening mankind's welfare through the reduction of phytogenetic resources,[41] which are the foundation of food sovereignty for many people. Policies to protect and conserve phytogenetic resources are a major consideration for achieving the IAASTD goals.

In this context, what is needed is a transformation of public awareness and international policy, and a determination to take measures to protect ecosystems throughout the planet, so as to defend basic services such as the secured supply of food and fresh water, and to protect against disasters.

The Earth Summit produced the Convention on Biological Diversity and the United Nations Framework Convention on Climate Change, but it is important to take into account sustainable farming practices in order to enhance food security for the world population and to help protect biologically diverse ecosystems. There must be better coordination between policies and actions; a study is needed of the economic benefits of biological diversity, the costs of its loss, and the costs that will flow from not taking protective measures, compared to the costs of effective conservation.

On the other hand, a framework of action is needed to reach commitment on reducing greenhouse gases (GHGs) under the Kyoto protocol, which expires in 2012. We have the responsibility of forging a global alliance to sustain life on earth, the principal objective of the Rio Summit, which established options for action in order to guarantee prevention, sustainable use, and equitable distribution of the benefits of biodiversity.

We may note that countries of the Third World are demanding that developed countries, which exploit their biological resources commercially, should provide them access to biotechnologies and the indispensable financing (Swaminathan, 2000). The trend of events demands that we change our rules governing intellectual property so that

[41] Phytogenetic resources refers to any genetic material of plant origin that is of real or potential value for food and agriculture; they are generally found in the seeds.

new varieties of seeds can be patented and protected as the basis of food and culture for local and indigenous community is an LAC. The existing system, based on individual and private property, is inadequate to protect the traditional rights of rural communities and of nations to their natural resources.

- Establishment of precautionary measures under the Cartagena Protocol (Article 10) prohibiting the transfer of genetically modified organisms (GMOs) among countries that are centers of origin or of genetic diversity.
- In countries of the region, limit production of GMO plants that have wild relatives and show botanical characteristics that could contaminate the gene pool (for example, the case of corn in Mesoamerica).
- Promote food safety research in cases of transgenic products that are consumed and produced in the region (for example, food safety studies are currently "rubber stamped", and there is no research geared to the particular conditions in the region).
- To protect human health and biodiversity from transgenic risks, governments should establish international standards for the clear, accurate and rigorous documentation and labeling of transgenic products in shipments of grain for human and animal consumption. Products that contain transgenics or derivatives, regardless of their final destination, should be recognizable as such by their labels, in order to respect the right of purchasers to choose freely. This label should identify the risks and enable the enforcement of biosecurity measures.
- From this perspective, there is also a need for policies to encourage those producers who contribute directly to genetic resource conservation as part of managing their productive systems.

The instruments for achieving such policies involve building capacities in biosecurity because modern biotechnology is still immature. All stakeholders need to know about its progress and provide continuous feedback. In addition, existing institutions devoted to biosecurity need to be strengthened and new ones created. When it comes to incorporating agrobiotechnology into the productive processes of small farmers, technical assistance is essential for assessing their risks and possibilities. The idea is not to go back to the old kind of extension services, where programs were designed in offices far from the people directly involved, but rather to strike a proper balance between the generation and validation of scientific and technical progress and the concrete demands of producers with less access to information and resources.

According to the Cartagena Protocol, states must establish a system of objective responsibility for the risks inherent in GMOs The sustainable management of biodiversity entails measures of economic compensation and reparations for damage to biodiversity (through oil spills, deforestation, pollution of water courses, release of GMOs into the environment, etc.), which is the basis of indigenous and peasant culture.

There is concern over the plundering of genetic resources located on the territory of various ethnic groups to make pharmaceuticals or other products that can be patented outside the country. This form of illegal appropriation of biological resources has been termed "bio-piracy" (Dutfield, 2004). Work is under way within the Convention on Biological Diversity to prepare an international system of Access and Benefit-Sharing (ABS). Yet communities still fear that under that scheme the benefits of such access will be shared only between governments and users (Einarsson, 2004). The distribution of benefits is thus a topic for debate. The best option would be to arrange channels of participation between the stakeholders involved so that collective rights to natural resources can be guaranteed.

Policy instruments would be designed to produce:

- Research for classifying plants of agrifood importance, so that those not yet classified and registered can be protected.
- Legal frameworks governing access to genetic resources, for example in the context of the Andean Community's Standard 391.
- Special regulatory frameworks to protect traditional knowledge about phytogenetic resources that will take account of the full scope of knowledge, as well as nontraditional records (oral history, for example) and systems for distributing the revenues generated by access to genetic resources.

While modern biotechnology developments constitute a competitive advantage for some countries in the region, as the growing of transgenic soybeans has done for Argentina, Paraguay and Brazil (albeit with sharp controversies and social tensions), recent advances in this leading-edge technology that allow use of food crops to produce pharmaceuticals, biofuels and plastics now pose a new threat to biodiversity. Not only could there be environmental impacts, but there is also a risk that products of this kind will pass into the food chain through uses that have nothing to do with human or animal consumption. For example, corn is the staple food of Mesoamerican cultures, and its use for producing pharmaceuticals and inedible industrial substances could affect directly the food security and safety of people in the region, without mentioning the effect on biodiversity in the center of origin (Galvez and Gonzalez, 2006).

The concern over producing biofuels from food crops is that it further threatens food security by increasing the price of foodstuffs, with the attendant impact on hunger and poverty.

When prices for biofuel crops rise, this does not necessarily benefit small-scale producers and peasants in developing countries, because they have no access to such markets, or market imperfections may deny them the benefits.

The idea here is not to discard biofuels production in the region, recognizing that, in some Caribbean countries for example where food must be imported, devoting farmland to biomass production for export could offer a way out of poverty. What is proposed, instead, is that the needed biomass should be derived from agricultural residues, from nonfood crops, and from animal wastes. The challenge is to ensure food security so that rural families can feed themselves and at the same time lift themselves out of poverty.

One possible alternative would be to adopt a policy that would prevent the use of food crops for other purposes, as has been done in the case of wheat.

5.4.3 Education and agricultural extension services

The promotion of education abroad, or indeed in the universities of LAC, is producing highly qualified personnel, but they are not necessarily equipped to address the problems of mega-diversity in agriculture, because their training may not pay much attention to the sustainable management of biodiversity or the care of genetic resources. Moreover, available infrastructure and human capital have been focused on boosting yields and production volumes, under an output- and export-oriented agricultural model.

The "demographic bonus"[42] is an advantage for countries of the region if they invest in human capital through education and scientific and technological development, in order to alleviate hunger and poverty. Yet because of heavy migration, the benefits of this "bonus" in coming decades could accrue to the countries that offer employment, which would lead to the loss of local talent and knowledge. One interesting proposal is to amend the study plans and programs in the agriculture professions to give priority to teaching agroecology. A government presence is justified in this field, when it is recognized that the knowledge involved must not be restricted to the kind sponsored by multinational companies that sell seeds, agrochemicals and farm machinery.

University training in agroecology needs to be strengthened through:

1. A holistic and interdisciplinary vision
2. Breaking down the walls between departments and faculties, so as to deal with such issues as:
   - Climate-soil-plant relationships;
   - Farming-livestock-forestry-fishing relationships;
   - Agroforestry, community woodlots;
   - Fertility management;
   - Systems analysis.
3. Allowing students to gain practical experience in the field.
4. Integrating scientific knowledge with peasant know-how in ethno-botany (knowledge of Amazon plants and ecosystems), household remedies, ways of organizing time and space, and the indigenous worldview. One way of restoring and capitalizing on peasant knowledge is to sift through it with the scientific knowledge at our disposal and subject it to reciprocal questioning. Participation by the rural poor in the design of projects will promote greater integration of traditional and scientific knowledge.

A necessary condition for achieving this is to integrate the university into its region and involve it in resolving producers' problems through coordination and cooperation with regional and local governments. In effect, AKST can be geared to small-scale producers and marginal rural sectors through reliance on a tripod of (1) publicly funded research, (2) public and private universities, and (3) networks of NGOs and other civil society players, including representatives of farmers, associations and unions.

The inequality of opportunities in education is a key element in perpetuating poverty, which impacts most heavily on children (Herrera, 2002). Considering the scant opportunities for primary and secondary education in rural areas, particularly for women, (1) greater emphasis should be placed on technical education that will meet a real labor need (securing value added for products, helping local governments to formulate development projects); these technical institutes are now supported by churches or by universities, (2) continuous education should be encouraged (from basic literacy through to specialized training). One measure that has proven to be effective in rescuing biodiversity and agrobiodiversity is to encourage self-training through the organization of agricultural fairs and competitions with prizes. If such initiatives could be generalized in networks this would help to collect and preserve the diversity of local seed populations (Raven, 2003). More generally, thought should be given to providing small farmers with opportunities to study new techniques, in accounting and marketing for example, in the course of their farming activity.

There are some new topics that should be included or pursued in specialized training curricula: (1) protection of genetic resources, bio-piracy, legal provisions and intellectual property; (2) food quality, standards, food labeling, guarantees for organic products, marks of origin for foods.

It is hard for producers, even if they have formed an association, to shoulder the costs of certification and traceability. Public support could be provided for this aspect in the form of loans.

AKST policies should develop a diversity of technological innovations, since the problems to be addressed are varied and are not all susceptible to the same response (FAO, 2004ab). More government spending on research and development and on agricultural extension services should be considered.

Innovation policies need to take account of cultural aspects. It has been documented that culture can influence or alter development policies that appear adequate without falling into the kind of cultural determinism that could lead to isolation and immobility (Sen, 2004).

Countries can be classed in three groups with respect to their AKST systems. The countries that are the biggest producers and exporters of food in the region, such as Argentina and Brazil, have maintained a public system of research and agricultural extension services, and Brazil in fact has a government research institute of international renown, EMBRAPA. Achievements in Mexico are more modest.

The Andean countries have abandoned their national research institutes under pressure from the IDB and the World Bank. They are left with very little possibility to pursue their own research on national genetic resources, in order to strengthen their food independence. In fact, much of the genetic research is being done outside the countries that are the centers of origin of germplasm. Among these two groups, other countries including Chile have privatized their research in important areas such as fruit, fish and lumber exports, with generally negative consequences for AKST objectives. In light of the results described in Chapter 2, policies should seek to foster creativity and to strengthen institutions, using public funds to encourage the emergence of technical assistance networks that involve local stakeholders, both public and private.

[42] Population of productive age

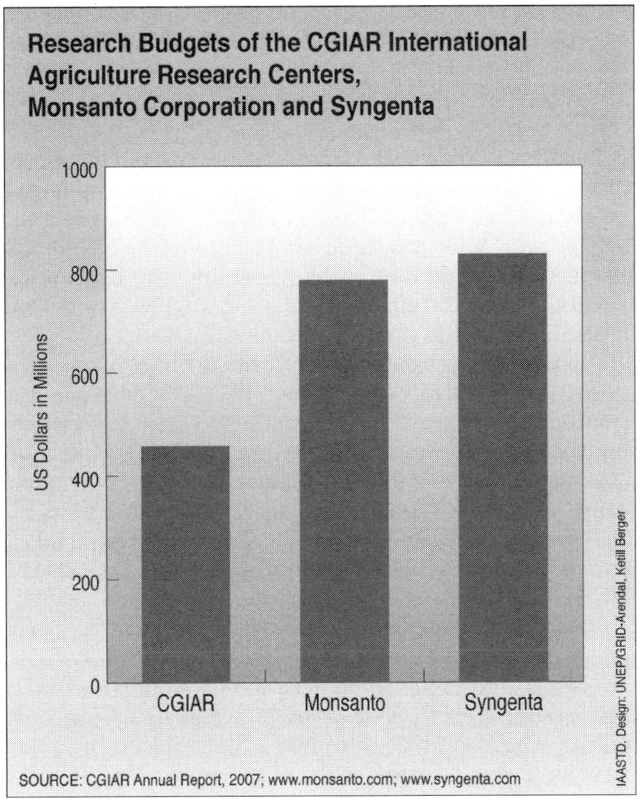

Figure 5-3. *Research Budgets of the CGIAR International Agricultural Research Centers, Monsanto Corporation, and Syngenta.*

Figure 5-3 illustrates the asymmetry in the agricultural research budgets of the Monsanto Corporation, the CGIAR international agricultural research centers, and national research programs in South America.

It is time to evaluate the private networks that have sought to replace the public sector in agricultural extension work. Agricultural extension services need to be adapted to changes in agriculture: the preponderant role of peasant women, part-time farmers who combine farming with other work, temporary migrations, and nonfarm rural jobs. Yet there is a certain contradiction between the holistic vision, which insists that agriculture extension must take account of all producers, in particular small-scale producers and all the activities of peasant families, and the fact that funding targeted at these groups is declining.

The solutions must be found in coordinating public efforts with private networks, under contracts that involve competition for public funding. The effectiveness of these private networks and their long-term impacts should be assessed, in light of their ambition to replace the public sector in agricultural extension work (See Chapter 2).

These problems explain why some local products are overlooked and do not receive sufficient support to penetrate national, regional or international markets. A portion of agricultural extension services is paid for by organizations of producers, when their crops serve as feedstocks for a processing industry: soybeans, sugarcane, cotton, coffee and to some extent milk. The problem arises in farming-livestock or multi-crop units. A better understanding of peasant organizations would help bring them into the networks that now exist or are being constituted. What the mono-crop associations are now doing through the organization of their productive chain should be extended to diversified producers including small farmers, but with government incentives.

5.4.4 Climate change

The global climate change that is affecting the planet is due to the release of greenhouse gases, which have increased significantly through massive use of fossil fuels. The root causes of this problem are the generation and consumption of energy in the form of coal or oil, automotive transport, and energy-intensive industrial processes. The burning of biomass in the forests is also harmful, not only because it releases carbon dioxide but also because it reduces the "carbon sink" that photosynthesis represents.

In 1990 the Intergovernmental Panel on Climate Change (IPPC) sounded the warning about rising atmospheric concentrations of carbon dioxide from human activities, leading to higher annual average temperatures accompanied by a changing climate. The greenhouse effect will be felt primarily in higher average world temperatures. This will affect all the processes that take place in the biosphere. The oceans will expand under the impact of warming, and the polar ice caps will melt, raising sea levels. Many low-lying coastal areas are at risk of disappearing under the sea. Preventing such occurrences will involve huge engineering costs (CONAM, 2006).

In general terms, the three broad issues that climate change management can and must address are the reduction of poverty and hunger, the improvement of competitiveness, and the achievement of sustainability. In order to do so effectively, climate change management must respond to the following challenges:

- How to reduce vulnerability to climate change (especially for the poorest population groups), the impact on production systems and infrastructure, and how to reap the potential benefits that climate change may offer;
- How to achieve energy and food security throughout the LAC region, through policies to mitigate the greenhouse effect and adapt to climate change;
- How to control emissions from deforestation, industry, and energy production;
- How to bring the LAC region into global policy, taking account of the benefits and impacts of climate change.

These challenges imply broadening the front against climate change to include all polluting countries (with mutual but differentiated responsibilities) and all the sectors involved (transportation in general, deforestation, etc.) (CONAM, 2006), through:

- Fostering innovation, which includes the application of existing technologies, and developing new technologies (in particular, active policies that take advantage of the normal replacement of equipment).
- Use and strengthening of market instruments (such as the trading in emission rights introduced by the EU).

- Efforts to adapt to climate change through preventive and curative measures geared to the regions and economic sectors most affected.

These elements could be given shape through the following actions:

- Ensure the immediate and effective enforcement of agreed policies, in order to achieve the Kyoto target of reducing greenhouse gas emissions by 8% below their 1990 level. Those measures are essentially the ones set forth in the Green Paper on energy security and the White Paper on European transport policy, as well as measures to promote climate-friendly technologies such as eco-technologies.
- Conduct community awareness campaigns to induce people to change their behavior.
- Intensify and target research to improve the understanding of climate change and its global and local fallout, and at the same time develop cost-effective strategies to mitigate climate change (especially in the areas of energy, transportation, agriculture and industry) as well as strategies for adapting to climate change.
- Strengthen scientific cooperation with countries beyond the region, and promote the transfer of climate-friendly technologies, and work with developing countries to prepare climate-friendly development policies and strengthen the adaptive capacities of the most vulnerable countries. The EU would in this way maintain its driving role in international negotiations in this area.

The European program on climate change entered a new phase in 2005, designed to determine the new measures that must be adopted in synergy with the Lisbon strategy, relating in particular to energy efficiency, renewable energy, transportation, and carbon capture and storage (CONAM, 2006).

Benefits and costs of the strategy. It is difficult to assess the costs of action. Those costs would reflect primarily the restructuring of transportation and production systems, as well as of energy use. On the other hand, those costs are bound to rise significantly if no action is taken by the other countries that are major producers of greenhouse gases. According to the commission, a less ambitious policy for combating climate change is not a sound alternative, for it would not achieve the objectives set and would imply additional costs due to climate change.

If dealing with climate change is to become a priority, it must be approached in the framework of the three broad issues of sustainable development, and there must be regular monitoring of its implementation, enforcement and reporting through suitable indicators. To this end, climate change management must focus on the following four lines of action:

- Pursue scientific and technological research to generate basic information in support of decisions and policies to mitigate the impacts of climate change.
- Establish mechanisms for outreach and active participation in the process of implementing the Climate Change Convention.
- Create mechanisms to facilitate the transfer of technology for mitigating the impacts of climate change.
- Strengthen inter-institutional, regional and international cooperation and forge strategic partnerships.

Policies for dealing with climate change demand the setting of national and regional priorities for reducing its impact in a concerted manner. These could be linked to programs and projects with targets for reducing GHG omissions in Latin America and the Caribbean. If this is to happen, the environment must be part of the political agenda of governments in the region (CONAM, 2006).

Governments could also reform their own organizational structures to promote the effective management of environmental issues. Those structures need to be made less bureaucratic and more participatory, they must have more concrete targets and the resources needed to meet commitments under the Climate Change Convention. Efforts are also needed to develop the technical and organizational capacities to address the problems created by GHG (CONAM, 2006).

Policies could also offer incentives to various social players and producers to attack the main causes of GHG omissions, and these would need to be accompanied by an active outreach and public awareness campaign. Clean technologies are also needed to address the problems of climate change, as an alternative to the main sources of GHG emissions. In this connection, research into alternative energy sources needs to be evaluated in light of social, environmental and economic variables.

5.5 Marketing and Market Access Policies

5.5.1 Access to international and regional markets

The access to the agriculture and agroindustrial markets of developed countries that protect domestic production should be based on strategies that recognize the competitive handicap of small farmers and peasants/indigenous producers in the region, as well as the specific impact that such agreements can have on the weakest sectors, and their differential impact on poverty. Implementing such policies will require absolute transparency in international and regional negotiations in LAC, from the initial stages of the negotiating process, and organizations of small-scale producers and peasants/indigenous farmers must acquire the capacities for monitoring those negotiations. To this end, they should have more opportunities for representation.

Another area that requires regulation is the growth of large-scale food distribution through supermarkets. Small-scale producers cannot compete with the oligopsonistic power of these companies to impose purchase prices, and government regulation is needed.

5.5.2 Active commercial policies for the domestic and international markets

These policies should be designed to generate market power through the creation of differentiated assets, for example by using different promotional instruments (designations of origin, internationally recognized protocols, eco-labeling, organic production, integrated production, etc.). The purpose here is to create specific assets that are differentiated from commodities, and in which small-scale and peasant/indigenous producers have special characteristics and advantages. This will also require appropriate institutional frameworks to promote these undertakings commercially (environmental standards, certification, etc.), and enhance their negotiating power vis-à-vis the "downstream" sectors in the marketing chain.

5.6 Financial Services for the Rural Economy

The availability of financial services is essential for supporting the AKST system's efforts to meet the IAASTD goals, and for the rural economy as a whole. Yet for more than a decade, and for various reasons, agricultural financing has been facing a dilemma in developing countries (FAO, 1996). Currently these issues are being examined by the Consultative Group to Assist the Poor (CGAP), an international consortium of 33 public and private development agencies working together to expand access to financial services for the urban and rural poor. For further information, see CGAP, 2003.

On one hand, there are the challenges of financing the investments needed so that the AKST system can enhance production in the rural sector, for which there is a strong incentive in light of the growing worldwide demand for food, spurred by demographic growth, particularly in low-income and densely populated countries. Here, the countries of LAC could expand their supply significantly. Within the region there are countries with large agricultural surpluses that could help meet this demand, but if their supply is to be sustained they will have to maintain the pace of investment. By contrast, there are other countries in the region with significant shortfalls in their food supply, and this is a point of national vulnerability. All of this suggests the need for financing policies for the AKST system that will recognize these contrasting situations in the region.

Other financial services are also important for the rural economy. A serious problem facing many rural towns, especially the smaller and more remote ones, is the lack of efficient, prompt and cost-competitive payment systems. Information and communication technologies and cellular telephony now offer broad possibilities in this area, but government policies are needed to create the technical conditions to finance the infrastructure that will make those resources usable.

On the other hand there are financial services that can reduce risks, both the general risks that arise from the uncertainties of day-to-day life, and those relating to crop losses through natural phenomena. The first requires an efficient and suitable savings system, while the second calls for the development of farm insurance systems with competitive costs and conditions appropriate to the activity.

Finally, if producers are to switch to new farming methods that will improve their lives, they will need to change the current organization of their production to one with higher capacities. This will require, among other things, financing in adequate amounts and at appropriate rates and terms. Overall, the rural sector in the region is in need of financial services to support its activities and in particular its investments in pursuit of development and sustainability at a time when the number of donor-supported farm credit programs has fallen. There are few indications that governments or commercial lenders are taking steps to compensate for the decline in funding for agricultural production, processing and marketing. As well, there are the new conditions governing international financial relations and the new macro-financial configurations prevailing in the region as a result of liberalization and deregulation during the 1990s.

On the first point, the current international financial environment is characterized by extraordinarily high liquidity and low interest rates, and at the same time there is still systemic weakness and instability in the international financial system that could threaten the progress achieved in national economies. Together with this, multilateral agencies are pushing market solutions to meet financing needs, and are restricting the scope of subsidies and transfers of public funds.

On the second point, most countries in the region currently enjoy a climate of relative price and exchange rate stability, fiscal balance or low government deficits, but at the same time financial services in many countries are provided predominantly by private entities that charge high real interest rates on loans to low-income sectors and offer them only limited services of other kinds, and at high cost.

The solution to the AKST financing problems in the rural sector is complex, not only because of the international and domestic context described above, but also because of the particular conditions of the sector in the various countries of Latin America. As noted in CGAP 2003, some of the key problems are:

- The thin demand for financial services;
- High information and transaction costs;
- Inadequate institutional capacity of rural lenders;
- The fact that much of farming activity is seasonal in nature, and that many crops take a long time to maturity;
- Risks relating specifically to cultivation of the land;
- Absence or insufficiency of usable collateral because of lack of clarity in ownership rights and institutional factors.

All of this must be viewed in the context of great heterogeneity in the conditions of the rural poor, and the productive possibilities of agriculture in different countries, and of regions within the same country, as well as their relations with the various national and international markets. Finally, within this heterogeneity there are great differences in local capacities for AKST in the different countries of the region.

In contrast to this complex and problematic situation, recent decades have seen a notable expansion in the possibilities for financial institutions geared to meeting the financial needs of poor and low-income groups, as much in terms of institutional organization as in sources of funding, the operating conditions of financial institutions, and the accessibility that the new ITCs offer rural people. There has been great progress in the capacity to offer low-cost financial products and risk cover for highly diverse situations.

If such institutions and systems are to be launched or

consolidated they must be supported and promoted with government strategies and policies for institutional development and seed capital. Care must be taken however to ensure that the risks inherent in this type of financing are not used as an excuse to charge excessively for these services. Moreover, there are certain kinds of risks that cannot be handled by the markets alone, and that will require public systems of guarantees or non-reimbursable funding.

In light of the foregoing, financing policies should address at least three priority aspects of support for AKST systems in the region: strengthening the capacities of those systems, and those of rural people and vulnerable groups, and providing funds to permit the transition of communities towards sustainable productive systems. For these various purposes there is a range of financing policies that can be considered, depending on the institutional context and development strategy adopted in a given country or region. These are considered here in relation to the three goals proposed.

5.6.1 Financing capacity building for AKST

In LAC as a whole and in the individual countries of the region, investment in AKST systems has been low, and this trend needs to be reversed through greater investment in various components of that system, in order to sustain its dynamics and reduce dependency on technological innovation from outside the region. Investment must be increased not only at the national level but also at the subregional and regional levels in order to capitalize on experience and minimize duplication of R&D effort. Since indigenous and agroecology systems have received virtually no financial support, and recognizing that agroecology systems in particular have made great progress over the last decade (e.g., in Cuba), investment in these systems could produce great rewards for the IAASTD goals in terms of supporting AKST, including specific technologies consistent with conditions in the different subregions of LAC, so they can be adapted to local needs. In particular, greater investment should be encouraged in:

- Strengthening agroecology programs in national and local universities and other educational institutions that will foster cultural diversity in LAC;
- Personnel training;
- Upgrading and maintaining research and outreach facilities;
- Maintenance of education centers for urban agriculture;
- Establishing education programs that will promote LAC values and culture.

To meet the objectives of strengthening capacities in the AKST system, the traditional approach of financing policies has been to work through national science and technology councils. Funds will be earmarked for agriculture, but the drive to develop AKST will be left for the most part in the hands of big transnational enterprises with robust R&D programs. Mexico is a typical case. From this perspective, the use of these financing policies for development and application of AKST will have an impact over the medium and long term, because it is subject to the reallocation of capital and labor that occurs through the play of supply and demand under market conditions.

On the other hand, with policies that stress sovereignty in a context of competition for hegemony in the international sphere, the government will maintain private financial markets for allocating funds, but may apply financing policies to sectors deemed strategic in order to maintain the supply of certain goods without depending on imports, for reasons of food security, for example. These funds can be mobilized by public or private banks or by trust funds. In this case, policies for financing AKST through national science and technology councils could involve the use of public or mixed funds to promote development in specific sectors for reasons of sovereignty. Brazil may be a typical case. The time needed for these policies on rural development and the agricultural application of AKST to have an impact on rural living conditions will depend on the intensity with which the government applies resources and efforts, in light of its strategies with respect to sovereignty objectives.

Within the new approach to public management, the government may assume that it has limited capacity to manage the use of funds devoted to strengthening AKST capacities in the country, and so it will encourage the emergence of nongovernmental public or mixed entities that will apply those funds to developing specific sectors. The impact of these policies will depend on the state's capacity to make those entities efficient, through various mechanisms of monitoring and accountability.

These policies have an impact on the sustainability of institutions and instruments, since accountability produces an incentive for these new public management entities to make more efficient use of public funds. It could also produce a "virtuous circle" in the application of funds, with progressive involvement of rural people in financial services and technological dissemination, if their development follows the precepts of decentralized private management, but with broad participation and local social oversight.

In these policies, the financing of AKST schemes is decentralized and in many cases involves both public and private funds, but there will be heavy influence exercised by medium- and large-scale producers in defining the institutional work agendas. Consequently, steps should also be taken to include small-scale producers and indigenous communities in managing and monitoring these entities to make sure their needs are addressed. All of this could translate into strengthening the capacities of the AKST system through further creation of decentralized technology centers run according to highly efficient private criteria and with an emphasis on environmental and biotechnology services and the promotion of human capital.

In a more systemic approach, the government could implement these financing policies by forming networks of research centers and institutions to articulate and socialize knowledge, while promoting activities at a scale adequate to assure specialization. As well, selective policies could be applied for financing AKST, through support for competitive networks, local environmental networks, networks promoting innovation, based on training and use of local resources, etc.

5.6.2 Financing to strengthen capacities of rural people and vulnerable groups

When it comes to financing policies for strengthening the capacities of rural people and vulnerable groups, these should promote employment in agricultural firms that foster sustainable production and the integration of small-scale producers into productive chains that operate in accordance with principles of sustainability and equity, and finally they should consolidate the efforts of indigenous communities by promoting their productive and organizational capacities within the context of their practices and cultures. The goal should be to enhance their productive capacities and thereby reduce poverty, exclusion and vulnerability. In conventional approaches, policies will be proposed for financing segments of the population living in poverty, particularly in the countryside, and this will be done with the help of multilateral agencies like the World Bank and IDB, through such programs as "*Oportunidades*" in Mexico. There will also be efforts to mobilize funds for these agencies to promote small businesses under market rules, for example through the IDB's MIF programs. The government will encourage financial innovation, so that rural producers will have the instruments to cover risks in the main agricultural products, farm insurance, etc.

Given the changes in financial systems, modern financial regulation is emerging with less emphasis on traditional schemes of official bank lending. Nevertheless, in this conventional perspective there is also a tendency to promote financing for rural development and agricultural production through policies to diversify financial systems by addressing the particular needs of rural people—specifically, legislative and regulatory reforms to strengthen different types of financial institutions in sectors such as microfinance, cooperatives, etc. From this conventional perspective, the impact of these financing policies will depend on articulation with other policies for promoting rural development, while the time needed for these policies to have an impact on rural development will depend on the involvement of the various intermediaries that will play roles in the new financial fabric.

In contrast, from the viewpoint of public management described above, financial policies will seek to strengthen these population groups with the central objective of establishing and strengthening rural financial markets, going beyond the former approach to government financial intervention that focused on development banks. This issue point out the necessity of developing new, nongovernmental public or mixed entities that mobilizes micro financing to the poor farmers. These institutions work with government funds or funds from multilateral development institutions (World Bank, IDB). They need to be articulated within the national institutional framework for sustaining macro-financial balance. This encourages the development of privately operated financing systems that can take the form of efficient local cooperatives. An example might be the *Fundaciones Produce* in Mexico. These financing approaches seek to promote efficiency in financing, and decentralized private management can sharply reduce transaction costs.

Viewed from the context of competition for hegemony, but operating under market rules, the government could apply financing policies to groups seeking to reinforce the presence of the national economy in the global context. Such policies could encourage the consolidation of producers' networks that would bring economies of scale and efficiency to the output of rural SMEs for the domestic market. The selection of sectors would be a mixed outcome, between favoring the most efficient ones and safeguarding the national economy, although under a "pick the winners" approach. The government promotes policies for financing the sectors, encouraging private financial intermediaries to channel funds to them, by offering government guarantees, etc. As well it will invest in infrastructure that will create positive externalities in these sectors. It promotes the policy of financing AKST through the establishment of internal market support networks and creating decentralized entities associated with those networks.

Approaching the financial issue more systematically, policies should try to encourage rural people and vulnerable groups to develop "popular" or grassroots financing institutions offering a full range of financial services (deposits and payment systems, savings, credit and insurance), operating with market efficiency and sustainability. These institutions should be developed in conjunction with local producers' networks (see FAO, 2004ab).

The financing of institutionalized forums for taking decisions and implementing the policy agenda for supporting AKST is an aspect that will contribute to their success. If financing for these activities can be made more independent of external cooperation, those agendas can be designed, implemented and evaluated more successfully, and they will contribute more to reducing hunger and poverty in the region.

A viable solution must recognize existing differences by creating comprehensive financial services for the indigent and for the creditworthy poor. The first group is unable to borrow, and they require specific solutions along the lines of the Grameen Bank in Bangladesh. The second group, on the other hand, can access financial services under certain assumptions, primarily the resolution of property rights, education, management capacities.

These policies for promoting institutions offering a full range of financial services will help generate decentralized financing networks of varying kinds, reflecting the varied conditions of different kinds of producers' networks, which will be supported with regulatory reforms and training policies for the efficient development of financial networks. These policies will promote local savings and financing capacities, and may trigger virtuous circles that will be differentiated according to the specific evolution of the various networks.

The greater efficiency of "popular" financial institutions of this kind is based on lower moral hazard, derived from specific knowledge of the borrowers, and lower transaction costs through local operation. However, it could require the government to provide offsetting policies and support for the weaker networks to help them stand on their own. Depending on these conditions, rural people and vulnerable groups will have better access to financing, and their communities will stand a better chance of survival in a context of progressive development.

In recent years, the potential for fostering grassroots financial institutions has been greatly enhanced by the emergence of information and communication technologies, which have made it possible to develop efficient and competitive rural financial networks that can achieve economies of scale and reduce the costs of producing and distributing financial products under market conditions. At the same time new, non-banking financial instruments have been developed: trust funds, investment funds, asset securitization, factoring, etc.

Assuming that the necessary IT and regulatory infrastructure is in place, these policies could have a swift impact on agricultural productivity and living standards by expanding access to low-cost financial services. This is all the more likely because the use of ICTs can help resolve the problem of access to financial services for rural people: the Internet and cellular phones make it possible to overcome the drawbacks of geographic isolation and can bring rural people into the banking system. Such policies would also provide tools for training rural people and producers through distance education.

The time that these policies will need to have an impact on development will be relatively short, if they are accompanied by other policies for training people in the use of ICTs, etc.

Nevertheless, all of this will have to be accompanied by a policy of investing in ICT access for rural sectors, if these policies for access to financial services are to work.

Finally, and no less importantly, it is clear that financial policies for improving conditions and capacities for rural people face the great challenge of promoting instruments and institutions for channeling remittances to support the development of regional and local financial services, in communities with heavy emigration rates. This should help to retain people in the countryside and boost employment through the development of family enterprises or small businesses.

5.6.3 Financial support programs for helping communities make the transition to a sustainable production system

An important aspect to consider in financial policies for supporting AKST systems has to do with the fact that in many parts of LAC the process must be launched under very backward conditions with pressing subsistence needs and no significant local resources. Consequently, these rural communities find it almost impossible to lift themselves out of their current condition and establish a productive system that is sustainable in both economic and environmental terms. Hence there is a need to offer financial support to farmers in order to make these transitions in an orderly and progressive way. That means formulating policies to provide structural funds through joint efforts by national, regional and local governments, so that communities can make the transition to a new configuration based on an agroecological system. Competitive funds should be established, to which multilateral agencies as well as national governments and regional institutions can contribute, in association with local development bodies.

References

Agrawal, A. 1999. Cultural and spiritual values of biodiversity. p. 177-180. *En* D. Posey (ed) UNEP's global biodiversity assessment volume. Cambridge Univ. Press, UK.

Altieri, M. 1996. Enfoques agroecológicos para el desarrollo de sistemas de producción sostenibles en los Andes. CIED, Perú.

Astier, M., y J. Hollands. 2005. Sustentabilidad y campesinado: Seis experiencias agroecológicas en Latinoamérica. ILEIA, GIRA, ICCO, México.

Bañón, R. 2006. Democracia y nueva gestión pública. *En* La participación en las administraciones públicas ¿cooperación o enfrentamiento? Univ. Politécnica de Valencia, España.

Baranyi, S., C.D. Deere, y M. Morales. 2004. Tierra y Desarrollo en América Latina. Perspectivas para la investigación sobre políticas. Available at www.idrc.ca/es/ev-67380-201-1-DO_TOPIC.html. North South Institute/IDRC, Ottawa.

Barrow, C. 1994. Advances in tropical agriculture in the 20th century and prospects for the 21st. TA 2000, Trinidad, 4-9 Sep1994.

Battiste, M., y J. Youngblood Henderson. 2000. Introduction. *In* Protecting indigenous knowledge and heritage. A global challenge. Purich's Aboriginal Issues Series. Purich Publ., Saskatchewan.

Berry, W. 1996. The unsettling of America: Culture and agriculture. Sierra Club Books, NY.

Bifarello, M. 2002. From delegation to participation the third sector and the state in associative networks Fifth Conf. Int. Society for Third-Sector Research, Cape Town, 7-11 July 2002. Univ. Nacional de Rosario, Argentina.

Caporal, R., y J. Costabeber. 2004. Agroecología e extensao rural. Contribuicoes para a promocao do desenvolvimento rural sustentable. MDA/SAF/DATER-IICA, Brasilia.

Cartagena, R., M. Parra, A. Burguete, y A. López. 2005. Participación social y toma de decisiones en los consejos municipales de los Altos de Chiapas. Revista Gestión y Política Pública. Vol. XIV, No. 2. Centro de Investigación y Desarrollo Económico de México CIDE, México.

CGAP. 2003. Servicios financieros para sectores rurales. Reseña para agencias de cooperación No 15 Octubre. Available at www.cgap.org/direct. CGAP, Washington DC.

Cisneros, F.H. 1992. El manejo integrado de plagas. Guía de Investigación CIP 7. CIP, Lima.

Cohen, G. 1993. Equality of What? On welfare, goods and capabilities. *In* A. Sen y M. Nussbaum. The quality of life. Clarendon Press, Oxford.

Colchester, M. 2001. A survey of indigenous land tenure. Land Tenure Service. FAO, Rome.

Colomer, A. 2006. Las condiciones de ciudadano y participación en las administraciones públicas. *En* La participación en las administraciones públicas ¿cooperación o enfrentamiento? Universidad Politécnica de Valencia, España.

CONAM. 2006. Perfil nacional sobre cambio climático en el Perú. Proyecto de Fortalecimiento de Capacidades para la Implementación de los Acuerdos Ambientales Globales, PNUD.

Deere, C. 2005. The feminization of agriculture? Economic restructuring in rural Latin America. UNRISD Occasional Paper 1. February. Available at www.unrisd.org/publications/opgp1.

Delgado, F.A., y B.D. Ponce. 1999. Cosmovisión Andina para un Desarrollo Rural Sustentable. Investigación, Interacción Social y Educación Superior en Bolivia. Available at www.agruco.org/pub/artic/1999/68.html.

Deruyttere, A. 1997. Indigenous peoples and

sustainable development. The role of the Interamerican Development Bank. IDB Forum of the Americas, Washington DC, 8 April 1997. No IND97-101. www.iadb.org/sds/IND/publication/publication_133_107_e.htm.

Dutfield, G. 2004. ¿Qué es la biopiratería? Memorias Taller de expertos sobre acceso a recursos genéticos y distribución de beneficios. Cuernavaca, octubre 24-27. México.

EEM. 2005. Evaluación de los Ecosistemas del Milenio 2005 Informe de Síntesis. Available at http://www.millenniumassessment.org//en/Products.Synthesis.aspx.

Einarsson, P. 2004. El régimen internacional de ABS. Memorias Taller de expertos sobre acceso a recursos genéticos y distribución de beneficios. Cuernavaca, octubre 24-27. México.

Escobar, A. 1995. Encountering development: the making and unmaking of the Third World. Princeton Univ. Press, NJ.

Escobar, A. 1999. Biodiversity. A perspective from within, Seedling, June 1999. Available at www.grain.org/seedling/?id=3#.

EUREPGAP. 2003. Control points and compliance criteria. English Version 2.0. Sept. FOOD PLUS, GmbH Cologne Germany.

FAO. 1996. Informe de la Cumbre Mundial sobre la Alimentación, 13-17 Noviembre. FAO, Roma.

FAO. 2004a. Las buenas prácticas agrícolas. Oficina Regional de la FAO para América Latina y el Caribe, Rome.

FAO. 2004b. Política de desarrollo agrícola. Available at http://www.fao.org/docrep/007/y5673s/y5673s00.htm#Contents. FAO, Rome.

FAO. 2006. Programa especial para la seguridad alimentaria. Available at http://www.fao.org/spfs/index_es.asp. FAO, Rome.

Foro Mundial sobre la Reforma Agraria. 2004. Available at http://www.fmra.org/.

Gálvez, A., y R.L. González. 2006. Cultivos biofarmacéuticos. Segundo Estudio de País. Conabio, México.

Gilbert, J., S. Wood, y G. Sharp. 2002. Who owns the land? Agricultural landownership by race/ethnicity. Rural America 17:55-62.

Glipo, A. 2003. Acuerdo sobre agricultura y soberanía alimentaria: Perspectivas de Mesoamérica y Asia. Global Issue Pap. No. 3. Available at http://www.boell.de/downloads/global/GIP%203%20Agricultura_span.pdf. Fundación Heinrich Böll, Berlin.

Gomero, L. 2001. Participación ciudadana para la institucionalidad de la agricultura ecológica. Red de Acción en Agricultura Alternativa (RAAA), Lima.

Gomero, L. y H. Velasquez. 2003. La agricultura ecologica: Una propuesta para el desarrollo de la agricultura en America Latina. V Congreso Nacional de la RAAA, Participacion Ciudadana para la Institucionalidad de la Agricultura Ecologica. RAAA, Lima.

Gonzales, T. 1996. Political ecology of peasantry, the seed, and non-governmental organizations in Latin America: A study of Mexico and Peru, 1940-1995. PhD thesis. Univ. Wisconsin, Madison.

Gonzales, T. 1999. The cultures of the seed in the Peruvian Andes. p. 193-205. *In* S.B. Brush (ed) Genes in the field: On-farm conservation of crop diversity. IPGRI, Rome.

González, E., y A. Ávila. 2005. Política Social y Pobreza hacia el Siglo XXI. *In* R. Cordera, C.Cabrera (eds) Superación de la pobreza y universalización de la política social. Facultad de Economía, UNAM.

Grillo, E. 1998. Development or cultural affirmation in the Andes? p. 124-145. *In* F. Apffel-Marglin with PRATEC (ed) The spirit of regeneration. Andean culture confronting western notions of development. Zed Books. New York.

Hall, D.O. 1998. Food security: what have sciences to offer? Available at http://www.icsu.org/2_resourcecentre/RESOURCE_list_base.php4?rub=7.

Harvey, D. 2007. Neoliberalism as creative destruction. Ann. Am. Acad. Polit. Social Sci. 610:22-44.

Heissler, M. 1996. R&D Cooperation between the EU and developing countries. Biotech. Dev. Monitor 26:12-14.

Herrera, J. 2002. Pobreza y desigualdad en el área andina. Introducción. Bulletin de l'Institut Francais d Etudes Andines. 31: 413-428.

Hoage, R.J., y K. Moran. 1998. Culture, the missing element in conservation and development. Kendall/Hunt, Dubuque.

Huber, E. 1996. Opciones de política social para América Latina: modelos neoliberales y socialdemócratas. p. 141-191. *En* Welfare status in transition, national adaptations in global economics. Sage Publ., London.

Huizer, G. 1999. People's spirit of resistance in Latin America. p.165-176. *En* B. Haverkort y W. Hiemstra (ed) Food for thought. Ancient visions and new experiments of rural people. COMPAS, Netherlands.

Ishizawa, J. 2006. What Next? Draft thematic paper. From Andean Cultural Affirmation to Andean Affirmation of Cultural Diversity-Learning with the Communities in the Central Andes. Available at www.dhf.uu.se/whatnext/papers_public/Ishizawa-Draft-01Sep2006.pdf.

Kay, C. 2006. Una reflexión sobre los estudios de pobreza rural y estrategias de desarrollo en la América Latina Rural. Revista ALASRU (Asociación Latinoamericana de Sociología Rural) No. 4, 29-76.

Kondo, Y. 1996. Leadership and participation. Human motivation: a key factor for management. Edited by Yoshio Kondo. Tokyo, Japan

Lander, E. 2000. La colonialidad del saber: eurocentrismo y ciencias sociales. Perspectivas latinoamericanas. Nepantla: Views from South 1.3 519-532.

Leff, E., y J. Carabias. 1993. Cultura y Manejo Sustentable de los Recursos Naturales. Vols I y II, Enrique Leff y Julia CArabias (Coordinadores) en Cultura y Manejo Sustentable de los Recursos Naturales. México: Universidad Nacional Autónoma de México, UNAM.

Lizarraga, T. 2002. Microempresa productora y comercializadora de plaguicidas y fertilizantes naturales en Cañete, Proyecto piloto demostrativo ambiental, Programa APGEP-SENREM, Convenio USAID-CONAM. Ejecutado por RAAA, Lima-Perú.

Max-Neef, M. 1993. Desarrollo a escala humana. Ed. Nordan-comunidad/Icaria editorial.

National Research Council. 1989. Alternative Agriculture. Washington, D.C.: National Academy Press.

Pérez-Ordóñez, D. 2005, Políticas públicas, poder local y participación ciudadana de los consejos de desarrollo urbano y rural, FLACSO, Guatemala.

Pimbert, M. 1994 The Need for Another Research Paradigm. Seedling 11:20-32.

PRATEC. 1993a. ¿Desarrollo o Descolonización en los Andes? Lima: Proyecto Andino de Tecnologías Campesinas, PRATEC. Lima, Perú.

PRATEC. 1993b. Afirmación Cultural Andina. Lima: Proyecto Andino de Tecnologías Campesinas, PRATEC. Lima, Perú.

PROCAS. 2001. Proyecto de Conservación de Suelos y Aguas en la Zona Andina de Colombia Campo para el futuro: cultivar sin arar sistemas sostenibles de producción. CAR, KFW, GTZ y Gobernación de Cundinamarca, Bogotá, Colombia.

Quijano, A. 2000. Coloniality of Power, Eurocentrism and Latin America. Nepantla: Views from the South 1(3):533-580.

Quijano, A. 2005. El "movimiento indígena" y las cuestiones pendientes en América Latina. Available at www.google.ca/search?sourceid=navclient&ie=UTF-8&rls=GGLF,GGLF:1969-53,GGLF:en&q=Quijano+La+cuestion+indigena.

Raven, P. 2003. The diversity of maize and teosinte: a global asset. Proceedings Workshop Gene Flow: What does it mean for Biodiversity and Centers of Origin, Pew Inititaive on Food and Biotechnology. México.

Reygadas, L. 2004. Las redes de la desigualdad: un enfoque multidimensional. Revista Política y Cultura, No. 22, pp. 7-25 UAM-X, México.

Rist, S., J. San Martin, y N. Tapia. 1999. Andean Cosmovision and Self-Sustained Development. Pp. 177-190. *In* B. Haverkort and W. Hoestra (eds) Food for Thought.

Ancient visions and new experiments of rural people. Bertus COMPAS.

Sen, A. 2004. ¿Cómo importa la cultura al desarrollo? Letras Libres, No. 71, noviembre. Available at http://www.letraslibres.com/index.php?art=9972.

Sen, A., y J. Foster. 1997. Space, Capability and Inequality. In On Economic Inequality. Clarendon Press, Oxford.

Soldevilla, S. 2005. Impacto económico, social y ambiental de la producción de banano orgánico en el Valle de El Chira, Perú. Tesis doctoral. Colegio de Postgraduados, México.

Stavenhagen, R. 1990. The Ethnic Question. United Nations University Press.

Swaminathan, M.S. 2000. Nueva ética para la biodiversidad. Available at: http://www.servindi.org/archivo/2007/2098

Tapia, M. 1999. Agrobiodiversidad en Los Andes. Fundación Frederich Ebert, Lima, Perú.

Toledo, V. 2001. Indigenous Peoples and Biodiversity. Encyclopedia of Biodiversity, Vol 3, pp. 451-463.

Toledo, V., P. Alarcón-Chaires, y P. Moguel. 2001. El Atlas Etnoecológico de México y Centro América: Fundamentos, Métodos y Resultados. *Etnoecológica*, Vol VI, No 8-9.

Torres, F. 2003. La visión teórica de la seguridad alimentaria como componente de la seguridad nacional. p. 15-21. *En* F. Torres (ed.) Seguridad alimentaria: seguridad nacional. UNAM/IIEc, Plaza y Valdés, México.

Towsend, P. 1993. Conceptualising Poverty, The Internacional Análisis of Poverty, Harvester Wheatsheaf, Londres.

Trejos, R., C. Pomareda, y J. Villasuso. 2004. Políticas e instituciones para la agricultura de cara al siglo XXI. IICA, Costa Rica.

Trigo, E, M. Piñeiro, y J. Sábato. 1983a. La Cuestión Tecnológica y la Organización de la Investigación Agropecuaria en América Latina. Desarrollo Económico 23:89(Abril-Junio):99-119.

Trigo, E, M. Piñeiro, y J. Sábato.1983b. Technology as a Social Issue: Agricultural Research Organization in Latin América. M. Pineiro y E. Trigo (eds), Technical Change and Social Conflict in Agriculture. Westview Press, Boulder, Colorado.

Valladolid, J. 1998. Andean Peasant Agriculture: Nurturing a Diversity of Life in the *Chacra*. p. 51-88. *En* The Spirit of Regeneration. Andean Culture Confronting Western Notions of Development. Frederique Apffel-Marglin with PRATEC (eds). Zed Books Ltd. New York.

Valladolid, J. 2001. Andean Cosmovision and the Nurturing of Biodiversity in the Peasant Chacra, p. 639-670. *En* J. Grim (ed) Indigenous traditions and ecology. The inter-being of Cosmology and Community. Harvard University Press.

Valladolid, J. 2005. Kawsay Mama. Madre Semilla. Proyecto In Situ. Serie Kawsay Mama 9. PRATEC. Lima, Perú.

Van Dam, C. 1999. La Tenencia de la Tierra en América Latina. El Estado del Arte de la Discusión en la Región. Available at www.grupochorlavi.org/php/doc/documentos/tenencia.html.

Vanderhoff, F. 2005. Excluidos hoy, protagonistas mañana. UCIRI, México.

Vía Campesina. 2003. ¿Qué significa soberanía alimentaria? Available at http://viacampesina.org/main_sp/index.php?option=com_content&task=view&id=78&Itemid=27.

Vía Campesina, S. Monsalve, FIAN International, y P. Rosset. 2006. Available at www.google.ca/search?hl=en&rls=GGLF%2CGGLF%3A196953%2CGGLF%3Aen&q=Agrarian+REform+in+the+context+of+food+sovereignty&meta.

Walsh, C. 2002. Interculturalidad, reformas constitucionales y pluralismo juridico. http://icci.nativeweb.org/boletin/36/walsh.html.

Walsh, C. 2004. Geopolíticas del Conocimiento, Interculturalidad y Descolonización. Available at http://icci.nativeweb.org/boletin/60/walsh.html.

Warren, D.M. 1992. Indigenous knowledge, biodiversity conservation and development. Keynote address at the International Conference on Conservation of Biodiversity in Africa: Local Initiatives and Institutional Roles, 30 August-3 September 1992, Nairobi, Kenya. Available at www.ciesin.columbia.edu/docs/004-173/004-173.html.

Warren, D.M., D. Brokensha, y L.J. Slikerveer. eds. 1993. Indigenous Knowledge Systems: The Cultural Dimension of Development. Kegan Paul International, London.

Wellman, B. 1987. El análisis estructural: del método y la metáfora a la teoría y la sustancia. Universidad de Toronto, Canadá.

Willer, H., y M. Yussefi. 2004. The World of Organic Agriculture: Statistics and Emerging Trends. International Federation of Organic Agriculture Movements (IFOAM), Bonn.

Wolman, H., y S. McCormick. 1994. The effect of decentralization on local governments. In R. Bennett (ed) Local governments and market decentralization: experiences in industrialized developing and farmer eastern bloc countries. United Nations University Press. Tokyo, Japan.

World Resources Institute. 1992. Global biodiversity strategy: guidelines for action to save, study, and use Earth's biotic wealth sustainably and equitably. WRI, Washington, DC.

Annex A

LAC Authors and Review Editors

Argentina

Javier Souza Casadinho • CETAAR-RAPAL
Hugo Cetrángolo • Universidad de Buenos Aires
Cecilia Gelabert • Universidad de Buenos Aires
Héctor Ginzo • Ministerio de Relaciones Exteriores, Comercio Internacional y Culto
Marcelo Regunaga • Universidad de Buenos Aires

Bolivia

Jorge Blajos • Fundación PROINPA
Edson Gandarillas • Fundación PROINPA
Ruth Pamela Cartagena • CIPCA Pando

Brasil

Flavio Dias Ávila • EMBRAPA
Dalva María da Mota • EMBRAPA
Antônio Gomes de Castro • EMBRAPA
Sergio Salles Filho • Universidad Estadual de Campinas (Unicamp)
Susana Valle Lima • EMBRAPA

Canadá

Tirso Gonzales • The University of British Columbia, Okanagan

Chile

Mario Ahumada • Comité Internacional de Planificación Regional para la Soberanía Alimentaria

Colombia

Inge Armbrecht • Universidad del Valle
Hernando Bernal • Universidad de la Amazonia de Colombia
Juan Cárdenas • Universidad de los Andes
Elsa Nivia • RAPALMIRA
Edelmira Pérez • Pontificia Universidad Javeriana de Bogotá

Costa Rica

Mario Samper • Instituto Interamericano de Cooperación para la Agricultura (IICA)

Estados Unidos

Jahi Michael Chappell • University of Michigan
Luis Fernando Chávez • Emory University
Celia Harvey • Conservation International
Eric Holt Jiménez • Food First/Institute for Food and Development Policy
Karen Luz • World Wildlife Fund
Ivette Perfecto • University of Michigan
David E. Williams • United States Department of Agriculture

Francia

Dominique Hervé • L'Institut de Recherche pour le Développement (IRD)
Patrick Lavelle • L'Institut de Recherche pour le Développement (IRD)

Ghana

Claudio Bragantini • EMBRAPA

México

Rosa Luz González Aguirre Universidad Autónoma Metropolitana Unidad Azcapotzalco
Michelle Chauvet • Universidad Autónoma Metropolitana Unidad Azcapotzalco
Amanda Gálvez • Universidad Nacional Autónoma de México
Jesús Moncada • Independent
Celso Garrido Noguera • Universidad Autónoma Metropolitana
Roberto Saldaña • SAGARPA

Nicaragua

Falguni Guharay • Servicio de Información Mesoamericana sobre Agricultura Sostenible
Carlos J. Pérez • Earth Institute
Ana Cristina Rostrán • UNAN-León
Jorge Irán Vásquez • Unión Nacional de Agricultores y Ganaderos (UNAG)

Panamá

Julio Santamaría • INIAP

Perú

Clara G. Cruzalegui • Ministerio de Agricultura y Ganadería
Luis A. Gomero • Red de Acción en Alternativas al Uso de Agroquímicos (RAAA)

República Dominicana

Rufino Pérez-Brennan • ALIMENTEC S.A.

Trinidad y Tobago

Salisha Bellamy • Ministry of Agriculture, Land & Marine Resources
Ericka Prentice-Pierre • Agriculture Sector Reform Programme (ASRP), IBD

Uruguay

Luis Carlos Paolino • Laboratorio Tecnológico del Uruguay (LATU)
Lucía Pitalluga • Instituto de Economía, Universidad de la República

Annex B

Peer Reviewers

Argentina
Marcelo Huarte • Proyecto Nacional de Papa y ALAP
Víctor H. Trucco • Asociación Argentina de Consorcios Regionales de Experimentación Agrícola

Benin
Peter Neuenschwander • International Institute for Tropical Agriculture

Bolivia
Peter Cronkleton • Center for International Forestry Research

Brasil
Government of Brasil
Alexandre Cardoso • Embrapa
Serguei Franco de Camargo • Universidade do Estado do Amazonas
Álvaro Macedo da Silva • Embrapa
Francisco Reifschneider • Embrapa
Angela Weber • TRANSFORMAR
Luís Fernando Stone • Embrapa Arroz e Feijão

Canada
Joann Jaffe • University of Regina

Chile
Claudio Barriga • Agronegocios Latinoamericanos Chile Ltd

Colombia
Daniel Castillo • Javeriana University, Bogota
María Hersilia Bonilla • Ministerio de Agricultura y Desarrollo Rural

Costa Rica
Carlos Manuel Araya Fernández • Escuela de Ciencias Agrarias, Universidad Nacional
Bert Kohlmann • Universidad EARTH
Dora Lorena Ocrospoma Ramírez • IICA

Cuba
Avelino G. Suarez Rodriguez • Institute of Ecology and Systematic, Cuban Environmental Agency

Kenya
María Eugenia Arreola • United Nations Environment Programme
Christian Borgemeister • International Centre of Insect Physiology and Ecology (ICIPE)

Mexico
Francisca Acevedo • CONABIO
Edit Antal • UNAM
Alejandro Blanco-Labra • Centro de Investigación y de Estudios
Marco Antonio Galindo • National Agricultural Council
Agustín López Herrera • Universidad Autónoma Chapingo
Armando Paredes • Consejo Nacional Agropecuario
Marcelo Signorini • Comisión Federal para la Protección contra Riesgos Sanitarios

Peru
César Bravo • INIA

Poland
Ursula Soltysiak • AgroBio Test

Sweden
Ulf Herrström • Ulf Herrström Konsult AB

United Kingdom
Philip Bubb • UNEP World Conservation Monitoring Centre
Daniela Rocha • UNEP World Conservation Monitoring Centre

United States
Miguel Altieri • Sociedad Cientifica LatinoAmericana de Agroecologia (SOCLA)
Kerry Byrnes • US Agency for International Development
Luis Fernando Chaves • University of Michigan
Indira Janaki Ekanayake • World Bank
Doug Gurian-Sherman • Food and Environment Program, Union of Concerned Scientists
Michael Hansen • Consumer Policy Institute, Consumers Union
Yurie Hoberg • The World Bank
Richard Levins • Harvard School of Public Health
Margaret Reeves • Pesticide Action Network, North America
Matt Rooney • US Department of State
Sara Scherr • Ecoagriculture Partners
Doreen Stabinsky • College of the Atlantic
John Vandermeer • University of Michigan

Uruguay
Claudia Karez • UNESCO Montevideo
Diego Martino • CLAES

Annex C
Glossary

Agriculture A linked, dynamic social-ecological system based on the extraction of biological products and services from an ecosystem, innovated and managed by people. It thus includes cropping, animal husbandry, fishing, forestry, biofuel and bioproducts industries, and the production of pharmaceuticals or tissue for transplant in crops and livestock through genetic engineering. It encompasses all stages of production, processing, distribution, marketing, retail, consumption and waste disposal.

Agricultural biodiversity Encompasses the variety and variability of animals, plants and microorganisms necessary to sustain key functions of the agroecosystem, its structure and processes for, and in support of, food production and food security.

Agricultural extension Agricultural extension deals with the creation, transmission and application of knowledge and skills designed to bring desirable behavioral changes among people so that they improve their agricultural vocations and enterprises and, therefore, realize higher incomes and better standards of living.

Agricultural innovation Agricultural innovation is a socially constructed process. Innovation is the result of the interaction of a multitude of actors, agents and stakeholders within particular institutional contexts. If agricultural research and extension are important to agricultural innovation, so are markets, systems of government, relations along entire value chains, social norms, and, in general, a host of factors that create the incentives for a farmer to decide to change the way in which he or she works, and that reward or frustrate his or her decision.

Agricultural population The agricultural population is defined as all persons depending for their livelihood on agriculture, hunting, fishing or forestry. This estimate comprises all persons actively engaged in agriculture and their non-working dependants.

Agricultural subsidies Agricultural subsidies can take many forms, but a common feature is an economic transfer, often in direct cash form, from government to farmers. These transfers may aim to reduce the costs of production in the form of an input subsidy, e.g., for inorganic fertilizers or pesticides, or to make up the difference between the actual market price for farm output and a higher guaranteed price. Subsidies shield sectors or products from international competition.

Agricultural waste Farming wastes, including runoff and leaching of pesticides and fertilizers, erosion and dust from plowing, improper disposal of animal manure and carcasses, crop residues and debris.

Agroecological Zone A geographically delimited area with similar climatic and ecological characteristics suitable for specific agricultural uses.

Agroecology The science of applying ecological concepts and principles to the design and management of sustainable agroecosystems. It includes the study of the ecological processes in farming systems and processes such as: nutrient cycling, carbon cycling/sequestration, water cycling, food chains within and between trophic groups (microbes to top predators), lifecycles, herbivore/predator/prey/host interactions, pollination etc. Agroecological functions are generally maximized when there is high species diversity/perennial forest-like habitats.

Agroecosystem A biological and biophysical natural resource system managed by humans for the primary purpose of producing food as well as other socially valuable nonfood goods and environmental services. Agroecosystem function can be enhanced by increasing the planned biodiversity (mixed species and mosaics), which creates niches for unplanned biodiversity.

Agroforestry A dynamic, ecologically based, natural resources management system that through the integration of trees in farms and in the landscape diversifies and sustains production for increased social, economic and environmental benefits for land users at all levels. Agroforestry focuses on the wide range of work with trees grown on farms and in rural landscapes. Among these are fertilizer trees for land regeneration, soil health and food security; fruit trees for nutrition; fodder trees that improve smallholder livestock production; timber and fuelwood trees for shelter and energy; medicinal trees to combat disease; and trees that produce gums, resins or latex products. Many of these trees are multipurpose, providing a range of social, economic and environmental benefits.

AKST Agricultural knowledge, science and technology (AKST) is a term encompassing the ways and means used to practice the different types of agricultural activities, and including both formal and informal knowledge and technology.

Alien Species A species occurring in an area outside of its historically known natural range as a result of intentional or accidental dispersal by human activities. Also referred to as introduced species or exotic species.

Aquaculture The farming of aquatic organisms in inland and coastal areas, involving intervention in the rearing process to enhance production and the individual or corporate ownership of the stock being cultivated. Aquaculture practiced in a marine environment is called mariculture.

Average Rate of Return Average rate of return takes the whole expenditure as given and calculates the rate of return to the global set of expenditures. It indicates whether or not the entire investment package was successful, but it does not indicate whether the allocation of resources between investment components was optimal.

Biodiversity The variability among living organisms from all sources including, inter alia, terrestrial, marine and other aquatic ecosystems and the ecological complexes of which they are part; including diversity within species and gene diversity among species, between species and of ecosystems.

Bioelectricity Electricity derived from the combustion of biomass, either directly or co-fired with fossil fuels such as coal and natural gas. Higher levels of conversion efficiency can be attained when biomass is gasified before combustion.

Bioenergy (biomass energy) Bioenergy is comprised of bioelectricity, bioheat and biofuels. Such energy carriers can be produced from energy crops (e.g., sugar cane, maize, oil palm), natural vegetation (e.g., woods, grasses) and organic wastes and residues (e.g., from forestry and agriculture). Bioenergy refers also to the direct combustion of biomass, mostly for heating and cooking purposes.

Biofuel Liquid fuels derived from biomass and predominantly used in transportation. The dominant biofuels are ethanol and biodiesel. Ethanol is produced by fermenting starch contained in plants such as sugar cane, sugar beet, maize, cassava, sweet sorghum or beetroot. Biodiesel is typically produced through a chemical process called trans-esterification, whereby oily biomass such as rapeseed, soybeans, palm oil, jatropha seeds, waste cooking oils or vegetable oils is combined with methanol to form methyl esters (sometimes called "fatty acid methyl ester" or FAME).

Bioheat Heat produced from the combustion of biomass, mostly as industrial process heat and heating for buildings.

Biological Control The use of living organisms as control agents for pests, (arthropods, nematodes mammals, weeds and pathogens) in agriiculture. There are three types of biological control:

Conservation biocontrol: The protection and encouragement of local natural enemy populations by crop and habitat management measures that enhance their survival, efficiency and growth.

Augmentative biocontrol: The release of natural enemies into crops to suppress specific populations of pests over one or a few generations, often involving the mass production and regular release of natural enemies.

Classical biocontrol: The local introduction of new species of natural enemies with the intention that they establish and build populations that suppress particular pests, often introduced alien pests to which they are specific.

Biological Resources Include genetic resources, organisms or parts thereof, populations, or any other biotic component of ecosystems with actual or potential use or value for humanity.

Biotechnology The IAASTD definition of biotechnology is based on that in the Convention on Biological Diversity and the Cartagena Protocol on Biosafety. It is a broad term embracing the manipulation of living organisms and spans the large range of activities from conventional techniques for fermentation and plant and animal breeding to recent innovations in tissue culture, irradiation, genomics and marker-assisted breeding (MAB) or marker assisted selection (MAS) to augment natural breeding. Some of the latest biotechnologies, called "modern biotechnology", include the use of *in vitro* modified DNA or RNA and the fusion of cells from different taxonomic families, techniques that overcome natural physiological reproductive or recombination barriers.

Biosafety Referring to the avoidance of risk to human health and safety, and to the conservation of the environment, as a result of the use for research and commerce of infectious or genetically modified organisms.

Blue Water The water in rivers, lakes, reservoirs, ponds and aquifers. Dryland production only uses green water, while irrigated production uses blue water in addition to green water.

BLCAs Brookered Long-term Contractual Arrangements (BLCAs) are institutional arrangements often involving a farmer cooperative, or a private commercial, parastatal or a state trading enterprise and a package (inputs, services, credit, knowledge) that allows small-scale farmers to engage in the production of a marketable commodity, such as cocoa or other product that farmers cannot easily sell elsewhere.

Catchment An area that collects and drains rainwater.

Capacity Development Any action or process which assists individuals, groups, organizations and communities in strengthening or developing their resources.

Capture Fisheries The sum (or range) of all activities to harvest a given fish resource from the "wild". It may refer to the location (e.g., Morocco, Gearges Bank), the target resource (e.g., hake), the technology used (e.g., trawl or beach seine), the social characteristics (e.g., artisanal, industrial), the purpose (e.g., (commercial, subsistence, or recreational) as well as the season (e.g., winter).

Carbon Sequestration The process that removes carbon dioxide from the atmosphere.

Cellulosic Ethanol Next generation biofuel that allows converting not only glucose but also cellulose and hemi-cellulose—the main building blocks of most biomass—into ethanol, usually using acid-based catalysis or enzyme-based reactions to break down plant fibers into sugar, which is then fermented into ethanol.

Climate Change Refers to a statistically significant variation in either the mean state of the climate or in its variability, persisting for an extended period (typically decades or longer). Climate change may be due to natural internal processes or external forcing, or to persistent anthropogenic changes in the composition of the atmosphere or in land use.

Clone A group of genetically identical cells or individuals that are all derived from one selected individual by vegetative propagation or by asexual reproduction, breeding of completely inbred organisms, or forming genetically identical organisms by nuclear transplantation.

Commercialization The process of increasing the share of income that is earned in cash (e.g., wage income, surplus production for marketing) and reducing the share that is earned in kind (e.g., growing food for consumption by the same household).

Cultivar A cultivated variety, a population of plants within a species of plant. Each cultivar or variety is genetically different.

Deforestation The action or process of changing forest land to non-forested land uses.

Degradation The result of processes that alter the ecological characteristics of terrestrial or aquatic (agro)ecosystems so that the net services that they provide are reduced. Continued degradation leads to zero or negative economic agricultural productivity.

For loss of *land* in quantitative or qualitative ways, the term *degradation* is used. For water resources rendered unavailable for agricultural and non-agricultural uses, we employ the terms *depletion* and *pollution*. *Soil* degradation refers to the processes that reduce the capacity of the soil to support agriculture.

Desertification Land degradation in drylands resulting from various factors, including climatic variations and human activities.

Domesticated or Cultivated Species Species in which the evolutionary process has been influenced by humans to meet their needs.

Domestication The process to accustom animals to live with people as well as to selectively cultivate plants or raise animals in order to increase their suitability and compatibility to human requirements.

Driver Any natural or human-induced factor that directly or indirectly causes a change in a system.

Driver, direct A driver that unequivocally influences ecosystem processes and can therefore be identified and measured to different degrees of accuracy.

Driver, endogenous A driver whose magnitude can be influenced by the decision-maker. The endogenous or exogenous characteristic of a driver depends on the organizational scale. Some drivers (e.g., prices) are exogenous to a decision-maker at one level (a farmer) but endogenous at other levels (the nation-state).

Driver, exogenous A driver that cannot be altered by the decision-maker.

Driver, indirect A driver that operates by altering the level or rate of change of one or more direct drivers.

Ecoagriculture A management approach that provides fair balance between production of food, feed, fuel, fiber, and biodiversity conservation or protection of the ecosystem.

Ecological Pest Management (EPM) A strategy to manage pests that focuses on strengthening the health and resilience of the entire agro-ecosystem. EPM relies on scientific advances in the ecological and entomological fields of population dynamics, community and landscape ecology, multi-trophic interactions, and plant and habitat diversity.

Economic Rate of Return The net benefits to all members of society as a percentage of cost, taking into account externalities and other market imperfections.

Ecosystem A dynamic complex of plant, animal, and microorganism communities and their nonliving environment interacting as a functional unit.

Ecosystem Approach A strategy for the integrated management of land, water, and living resources that promotes conservation and sustainable use in an equitable way.

An ecosystem approach is based on the application of appropriate scientific methodologies focused on levels of biological organization, which encompass the essential structure, processes, functions, and interactions among organisms and their environment. It recognizes that humans, with their cultural diversity, are an integral component and managers of many ecosystems.

Ecosystem Function An intrinsic ecosystem characteristic related to the set of conditions and processes whereby an ecosystem maintains its integrity (such as primary productivity, food chain biogeochemical cycles). Ecosystem functions include such processes as decomposition, production, pollination, predation, parasitism, nutrient cycling, and fluxes of nutrients and energy.

Ecosystem Management An approach to maintaining or restoring the composition, structure, function, and delivery of services of natural and modified ecosystems for the goal of achieving sustainability. It is based on an adaptive, collaboratively developed vision of desired future conditions that integrates ecological, socioeconomic, and institutional perspectives, applied within a geographic framework, and defined primarily by natural ecological boundaries.

Ecosystem Properties The size, biodiversity, stability, degree of organization, internal exchanges of material and energy among different pools, and other properties that characterize an ecosystem.

Ecosystem Services The benefits people obtain from ecosystems. These include provisioning services such as food and water; regulating services such as flood and disease control; cultural services such as spiritual, recreational, and cultural benefits; and supporting services such as nutrient cycling that maintain the conditions for life on Earth. The concept "ecosystem goods and services" is synonymous with ecosystem services.

Ecosystem Stability A description of the dynamic properties of an ecosystem. An ecosystem is considered stable if it returns to its original state shortly after a perturbation (resilience), exhibits low temporal variability (constancy), or does not change dramatically in the face of a perturbation (resistance).

Eutrophication Excessive enrichment of waters with nutrients, and the associated adverse biological effects.

Ex-ante The analysis of the effects of a policy or a project based only on information available before the policy or project is undertaken.

Ex-post The analysis of the effects of a policy or project based on information available after the policy or project has been implemented and its performance is observed.

Ex-situ Conservation The conservation of components of biological diversity outside their natural habitats.

Externalities Effects of a person's or firm's activities on others which are not compensated. Externalities can either hurt or benefit others—they can be negative or positive. One negative externality arises when a company pollutes the local environment to produce its goods and does not compensate the negatively affected local residents. Positive externalities can be produced through primary education—which benefits not only primary school students but also society at large. Governments can reduce negative externalities by regulating and taxing goods with negative externalities. Governments can increase positive externalities by subsidizing goods with positive externalities or by directly providing those goods.

Fallow Cropland left idle from harvest to planting or during the growing season.

Farmer-led Participatory Plant Breeding Researchers and/or development workers interact with farmer-controlled, managed and executed PPB activities, and build on farmers' own varietal development and seed systems.

Feminization The increase in the share of women in an activity, sector or process.

Fishery Generally, a fishery is an activity leading to harvesting of fish. It may involve capture of wild fish or the raising of fish through aquaculture.

Food Security Food security exists when all people of a given spatial unit, at all times, have physical and economic access to sufficient, safe and nutritious food to meet their dietary needs and food preferences for an active and healthy life, and that is obtained in a socially acceptable and ecologically sustainable manner.

Food Sovereignty The right of peoples and sovereign states to democratically determine their own agricultural and food policies.

Food System A food system encompasses the whole range of food production and consumption activities. The food system includes farm input supply, farm production, food processing, wholesale and retail distribution, marketing, and consumption.

Forestry The human utilization of a piece of forest for a certain purpose, such as timber or recreation.

Forest Systems Forest systems are lands dominated by trees; they are often used for timber, fuelwood, and non-wood forest products.

Gender Refers to the socially constructed roles and behaviors of, and relations between, men and women, as opposed to sex, which refers to biological differences. Societies assign specific entitlements, responsibilities and values to men and women of different social strata and subgroups.

Worldwide, systems of relation between men and women tend to disadvantage women, within the family as well as in public life. Like the hierarchical framework of a society, gender roles and relations vary according to context and are constantly subject to changes.

Genetic Engineering Modifying genotype, and hence phenotype, by transgenesis.

Genetic Material Any material of plant, animal, microbial or other origin containing functional units of heredity.

Genomics The research strategy that uses molecular characterization and cloning of whole genomes to understand the structure, function and evolution of genes and to answer fundamental biological questions.

Globalization Increasing interlinking of political, economic, institutional, social, cultural, technical, and ecological issues at the global level.

GMO (Genetically Modified Organism) An organism in which the genetic material has been altered anthropogenically by means of gene or cell technologies.

Governance The framework of social and economic systems and legal and political structures through which humanity manages itself. In general, governance comprises the traditions, institutions and processes that determine how power is exercised, how citizens are given a voice, and how decisions are made on issues of public concern.

Global Environmental Governance The global biosphere behaves as a single system, where the environmental impacts of each nation ultimately affect the whole. That makes a coordinated response from the community of nations a necessity for reversing today's environmental decline.

Global Warming Refers to an increase in the globally averaged surface temperature in response to the increase of well-mixed greenhouse gases, particularly CO_2.

Global Warming Potential An index, describing the radiative characteristics of well-mixed greenhouse gases, that represents the combined effect of the differing times these gases remain in the atmosphere and their relative effectiveness in absorbing outgoing infrared radiation. This index approximates the time-integrated warming effect of a unit mass of a given greenhouse gas in today's atmosphere, relative to that of carbon dioxide.

Green Revolution An aggressive effort since 1950 in which agricultural researchers applied scientific principles of genetics and breeding to improve crops grown primarily in less-developed countries. The effort typically was accompanied by collateral investments to develop or strengthen the delivery of extension services, production inputs and markets and develop physical infrastructures such as roads and irrigation.

Green Water Green water refers to the water that comes from precipitation and is stored in unsaturated soil. Green water is typically taken up by plants as evapotranspiration.

Ground Water Water stored underground in rock crevices and in the pores of geologic materials that make up the Earth's crust. The upper surface of the saturate zone is called the water table.

Growth Rate The change (increase, decrease, or no change) in an indicator over a period of time, expressed as a percentage of the indicator at the start of the period. Growth rates contain several sets of information. The first is whether there is any change at all; the second is what direction the change is going in (increasing or decreasing); and the third is how rapidly that change is occurring.

Habitat Area occupied by and supporting living organisms. It is also used to mean the environmental attributes required by a particular species or its ecological niche.

Hazard A potentially damaging physical event, phenomenon and/or human activity, which my cause injury, property damage, social and economic disruption or environmental degradation.

Hazards can include latent conditions that may represent future threats and can have different origins.

Household All the persons, kin and non-kin, who live in the same or in a series of related dwellings and who share income, expenses and daily subsistence tasks. A basic unit for socio-cultural and economic analysis, a household may consist of persons (sometimes one but generally two or more) living together and jointly making provision for food or other essential elements of the livelihood.

Industrial Agriculture Form of agriculture that is capital-intensive, substituting machinery and purchased inputs for human and animal labor.

Infrastructure The facilities, structures, and associated equipment and services that facilitate the flows of goods and services between individuals, firms, and governments. It includes public utilities (electric power, telecommunications, water supply, sanitation and sewerage, and waste disposal); public works (irrigation systems, schools, housing, and hospitals); transport services (roads, railways, ports, waterways, and airports); and R&D facilities.

Innovation The use of a new idea, social process or institutional arrangement, material, or technology to change an activity, development, good, or service or the way goods and services are produced, distributed, or disposed of.

Innovation system Institutions, enterprises, and individuals that together demand and supply information and technology, and the rules and mechanisms by which these different agents interact.

In recent development discourse agricultural innovation is conceptualized as part and parcel of social and ecological organization, drawing on disciplinary evidence and understanding of how knowledge is generated and innovations occur.

In-situ Conservation The conservation of ecosystems and natural habitats and the maintenance and recovery of viable populations of species in their natural habitats and surroundings and, in the case of domesticated or cultivated species, in the surroundings where they have developed their distinctive properties and were managed by local groups of farmers, fishers or foresters.

Institutions The rules, norms and procedures that guide how people within societies live, work, and interact with each other. Formal institutions are written or codified rules, norms and procedures. Examples of formal institutions are the Constitution, the judiciary laws, the organized market, and property rights. Informal institutions are rules governed by social and behavioral norms of the society, family, or community. Cf. Organization.

Integrated Approaches Approaches that search for the best use of the functional relations among living organisms in relation to the environment without excluding the use of external inputs. Integrated approaches aim at the achievement of multiple goals (productivity increase, environmental sustainability and social welfare) using a variety of methods.

Integrated Assessment A method of analysis that combines results and models from the physical, biological, economic, and social sciences, and the interactions between these components in a consistent framework to evaluate the status and the consequences of environmental change and the policy responses to it.

Integrated Natural Resources Management (INRM) An approach that integrates research of different types of natural resources into stakeholder-driven processes of adaptive management and innovation to improve livelihoods, agroecosystem resilience, agricultural productivity and environmental services at community, eco-regional and global scales of intervention and impact. INRM thus aims to help to solve complex real-world problems affecting natural resources in agroecosystems.

Integrated Pest Management The procedure of integrating and applying practical management methods to manage insect populations so as to keep pest species from reaching damaging levels while avoiding or minimizing the potentially harmful effects of pest management measures on humans, non-target species, and the environment. IPM tends to incorporate assessment methods to guide management decisions.

Intellectual Property Rights (IPRs) Legal rights granted by governmental authorities to control and reward certain products of human intellectual effort and ingenuity.

Internal Rate of Return The discount rate that sets the net present value of the stream of the net benefits equal to zero. The internal rate of return may have multiple values when the stream of net benefits alternates from negative to positive more than once.

International Dollars Agricultural R&D investments in local currency units have been converted into international dollars by deflating the local currency amounts with each country's inflation ration (GDP deflator) of base year 2000. Next, they were converted to US dollars with a 2000 purchasing power parity (PPP) index. PPPs are synthetic exchange rates used to reflect the purchasing power of currencies.

Knowledge The way people understand the world, the way in which they interpret and apply meaning to their experiences. Knowledge is not about the discovery of some finale objective "truth" but about the grasping of subjective culturally conditioned products emerging from complex and ongoing processes involving selection, rejection, creation, development and transformation of information. These processes, and hence knowledge, are inextricably linked to the social, environmental and institutional context within which they are found.

Scientific knowledge: Knowledge that has been legitimized and validated by a formalized process of data gathering, analysis and documentation.

Explicit knowledge: Information about knowledge that has been or can be articulated, codified, and stored and exchanged. The most common forms of explicit knowledge are manuals, documents, procedures, cultural artifacts and stories. The information about explicit knowledge also can be audio-visual. Works of art and product design can be seen as other forms of explicit knowledge where human skills, motives and knowledge are externalized.

Empirical knowledge: Knowledge derived from and constituted in interaction with a person's environment. Modern communication and information technologies, and scientific instrumentation, can extend the 'empirical environment' in which empirical knowledge is generated.

Local knowledge: The knowledge that is constituted in a given culture or society.

Traditional (ecological) knowledge: The cumulative body of knowledge, practices, and beliefs evolved by adaptive processes and handed down through generations. It may not be indigenous or local, but it is distinguished by the way in which it is acquired and used, through the social process of learning and sharing knowledge.

Knowledge Management A systematic discipline of policies, processes, and activities for the management of all processes of knowledge generation, codification, application and sharing of information about knowledge.

Knowledge Society A society in which the production and dissemination of scientific information and knowledge function well, and in which the transmission and use of valuable experiential knowledge is optimized; a society in which the information of those with experiential knowledge is used together with that of scientific and technical experts to inform decision-making.

Land Cover The physical coverage of land, usually expressed in terms of vegetation cover or lack of it. Influenced by but not synonymous with land use.

Land Degradation The reduction in the capability of the land to produce benefits from a particular land use under a specific form of land management.

Landscape An area of land that contains a mosaic of ecosystems, including human-dominated ecosystems. The term cultural landscape is often used when referring to landscapes containing significant human populations.

Land Tenure The relationship, whether legally or customarily defined, among people, as individuals or groups, with respect to land and associated natural resources (water, trees, minerals, wildlife, and so on).

Rules of tenure define how property rights in land are to be allocated within societies. Land tenure systems determine who can use what resources for how long, and under what conditions.

Land Use The human utilization of a piece of land for a certain purpose (such as irrigated agriculture or recreation). Land use is influenced by, but not synonymous with, land cover.

Leguminous Cultivated or spontaneous plants which fix atmospheric nitrogen.

Malnutrition Failure to achieve nutrient requirements, which can impair physical and/or mental health. It may result from consuming too little food or a shortage or imbalance of key nutrients (e.g., micronutrient deficiencies or excess consumption of refined sugar and fat).

Marginal Rates of Return Calculates the returns to the last dollar invested on a certain activity. It is usually estimated through econometric estimation.

Marker Assisted Selection (MAS) The use of DNA markers to improve response to selection in a population. The markers will be closely linked to one or more target loci, which may often be quantitative trait loci.

Minimum Tillage The least amount possible of cultivation or soil disturbance done to prepare a suitable seedbed. The main purposes of minimum tillage are to reduce tillage energy consumption, to conserve moisture, and to retain plant cover to minimize erosion.

Model A simplified representation of reality used to simulate a process, understand a situation, predict an outcome or analyze a problem. A model can be viewed as a selective approximation, which by elimination of incidental detail, allows hypothesized or quantified aspects of the real world to appear manipulated or tested.

Multifunctionality In IAASTD, multifunctionality is used solely to express the inescapable interconnectedness of agriculture's different roles and functions. The concept of multifunctionality recognizes agriculture as a multi-output activity producing not only commodities (food, feed, fibers, agrofuels, medicinal products and ornamentals), but also non-commodity outputs such as environmental services, landscape amenities and cultural heritages (See Global SDM Text Box)

Natural Resources Management Includes all functions and services of nature that are directly or indirectly significant to humankind, i.e., economic functions, as well as other cultural and ecological functions or social services that are not taken into account in economic models or not entirely known.

Nanotechnology The engineering of functional systems at the atomic or molecular scale.

Net Present Value (NPV) Net present value is used to analyze the profitability of an investment or project, representing the difference between the discounted present value of benefits and the discounted present value of costs. If NPV of a prospective project is positive, then the project should be accepted. The analysis of NPV is sensitive to the reliability of future cash inflows that an investment or project will yield.

No-Till Planting without tillage. In most systems, planter-mounted coulters till a narrow seedbed assisting in the placement of fertilizer and seed. The tillage effect on weed control is replaced by herbicide use.

Obesity A chronic physical condition characterized by too much body fat, which results in higher risk for health problems such as high blood pressure, high blood cholesterol, diabetes, heart disease and stroke. Commonly it is defined as a Body Mass Index (BMI) equal to or more than 30, while overweight is equal to or more than 25. The BMI is an idex of weight-for-height and is defined as the weight in kilograms divided by the square of the height in meters (kg/m^2).

Organic Agriculture An ecological production management system that promotes and enhances biological cycles and soil biological activity. It is based on minimal use of off-farm inputs and on management practices that restore, maintain and enhance ecological harmony.

Organization Organizations can be formal or informal. Examples of organizations are government agencies (e.g., police force, ministries, etc.), administrative bodies (e.g., local government), nongovernmental organizations, associations (e.g., farmers' associations) and private companies (firms). Cf. with Institutions.

Orphan Crops Crops such as tef, finger millet, yam, roots and tubers that tend to be regionally or locally important for income and nutrition, but which are not traded globally and receive minimal attention by research networks.

Participatory Development A process that involves people (population groups, organizations, associations, political parties) actively and significantly in all decisions affecting their lives.

Participatory Domestication The process of domestication that involves agriculturalists and other community members actively and significantly in making decisions, taking action and sharing benefits.

Participatory Plant Breeding (PPB) Involvement of a range of actors, including scientists, farmers, consumers, extension agents, vendors, processors and other industry stakeholders—as well as farmer and community-based organizations and non-government organization (NGOs) in plant breeding research and development.

Participatory Varietal Selection (PVS) A process by which farmers and other stakeholders along the food chain are involved with researchers in the selection of varieties from formal and farmer-based collections and trials, to determine which are best suited to their own agroecosystems' needs, uses and preferences, and which should go ahead for finishing, wider release and dissemination. The information gathered may in turn be fed back into formal-led breeding programs.

Pesticide A toxic chemical or biological product that kills organisms (e.g., insecticides, fungicides, weedicides, rodenticides).

Poverty There are many definitions of poverty.

Absolute Poverty: According to a UN declaration that resulted from the World Summit on Social Development in 1995, absolute poverty is a condition characterized by severe deprivation of basic human needs, including food, safe drinking water, sanitation facilities, health, shelter, education and information. It depends not only on income but also on access to services.

Dimensions of Poverty: The individual and social characteristics of poverty such as lack of access to health and education, powerlessness or lack of dignity. Such aspects of deprivation experienced by the individual or group are not captured by measures of income or expenditure.

Extreme Poverty: Persons who fall below the defined poverty line of US$1 income per day. The measure is converted into local currencies using purchasing power parity (PPP) exchange rates. Other definitions of this concept have identified minimum subsistence requirements, the denial of basic human rights or the experience of exclusion.

Poverty Line: A minimum requirement of welfare, usually defined in relation to income or expenditure, used to identify the poor. Individuals or households with incomes or expenditure below the poverty line are poor. Those with incomes or expenditure equal to or above the line are not poor. It is common practice to draw more than one poverty line to distinguish different categories of poor, for example, the extreme poor.

Private Rate of Return The gain in net revenue to the private firm/business divided by the cost of an investment expressed in percentage.

Processes A series of actions, motions, occurrences, a method, mode, or operation, whereby a result or effect is produced.

Production Technology All methods that farmers, market agents and consumers use to cultivate, harvest, store, process, handle, transport and prepare food crops, cash crops, livestock, etc. for consumption.

Protected Area A geographically defined area which is designated or regulated and managed to achieve specific conservation objectives as defined by society.

Public Goods A good or service in which the benefit received by any one party does not diminish the availability of the benefits to others, and/or where access to the good cannot be restricted. Public goods have the properties of non-rivalry in consumption and non-excludability.

Public R&D Investment Includes R&D investments done by government agencies, nonprofit institutions, and higher-education agencies. It excludes the private for-profit enterprises.

Research and Development (R&D) Organizational strategies and methods used by research and extension program to conduct their work including scientific procedures, organizational modes, institutional strategies, interdisciplinary team research, etc.

Scenario A plausible and often simplified description of how the future may develop based on explicit and coherent and internally consistent set of assumptions about key driving forces (e.g., rate of technology change, prices) and relationships. Scenarios are neither predictions nor projections and sometimes may be based on a "narrative storyline". Scenarios may be derived from projections but are often based on additional information from other sources.

Science, Technology and Innovation Includes all forms of useful knowledge (codified and tacit) derived from diverse branches of learning and practice, ranging from basic scientific research to engineering to local knowledge. It also includes the policies used to promote scientific advance, technology development, and the commercialization of products, as well as the associated institutional innovations. *Science* refers to both basic and applied sciences. *Technology* refers to the application of science, engineering, and other fields, such as medicine. *Innovation* includes all of the processes, including business activities that bring a technology to market.

Shifting Cultivation Found mainly in the tropics, especially in humid and subhumid regions. There are different kinds; for example, in some cases a settlement is permanent, but certain fields are fallowed and cropped alternately ("rotational agriculture"). In other cases, new land is cleared when the old is no longer productive.

Slash and Burn Agriculture A pattern of agriculture in which existing vegetation is cleared and burned to provide space and nutrients for cropping.

Social Rate of Return The gain to society of a project or investment in net revenue divided by cost of the investment, expressed by percentage.

Soil and Water Conservation (SWC) A combination of appropriate technology and successful approach. Technologies promote the sustainable use of agricultural soils by minimizing soil erosion, maintaining and/or enhancing soil properties, managing water, and controlling temperature. Approaches explain the ways and means which are used to realize SWC in a given ecological and socioeconomic environment.

Soil Erosion The detachment and movement of soil from the land surface by wind and water in conditions influenced by human activities.

Soil Function Any service, role, or task that a soil performs, especially: (a) sustaining biological activity, diversity, and productivity; (b) regulating and partitioning water and solute flow; (c) filtering, buffering, degrading, and detoxifying potential pollutants; (d) storing and cycling nutrients; (e) providing support for buildings and other structures and to protect archaeological treasures.

Staple Food (Crops) Food that is eaten as daily diet.

Soil Quality The capacity of a specific kind of soil to function, within natural or managed ecosystem boundaries, to sustain plant and animal productivity, maintain or enhance water and air quality, and support human health and habitation. In short, the capacity of the soil to function.

Subsidy Transfer of resources to an entity, which either reduces the operating costs or increases the revenues of such entity for the purpose of achieving some objective.

Subsistence Agriculture Agriculture carried out for the use of the individual person or their family with few or no outputs available for sale.

Sustainable Development Development that meets the needs of the present without compromising the ability of future generations to meet their own needs.

Sustainable Land Management (SLM) A system of technologies and/or planning that aims to integrate ecological with socio-economic and political principles in the management of land for agricultural and other purposes to achieve intra- and intergenerational equity.

Sustainable Use of Natural Resources Natural resource use is sustainable if specific types of use in a particular ecosystem are considered reasonable in the light of both the internal and the external perspective on natural resources. "Reasonable" in this context means that all actors agree that resource use fulfils productive, physical, and cultural functions in ways that will meet the long-term needs of the affected population.

Technology Transfer The broad set of deliberate and spontaneous processes that give rise to the exchange and dissemination of information and technologies among different stakeholders. As a generic concept, the term is used to encompass both diffusion of technologies and technological cooperation across and within countries.

Terms of Trade The *international terms* of trade measures a relationship between the prices of exports and the prices of imports, this being known strictly as the barter terms of trade. In this sense, deterioration in the terms of trade could have resulted if unit prices of exports had risen less than unit prices for imports. The *inter-sectoral terms of trade* refers to the terms of trade between sectors of the economy, e.g., rural & urban, agriculture and industry.

Total Factor Productivity A measure of the increase in total output which is not accounted for by increases in total inputs. The total factor productivity index is computed as the ratio of an index of aggregate output to an index of aggregate inputs.

Tradeoff Management choices that intentionally or otherwise change the type, magnitude, and relative mix of services provided by ecosystems.

Transgene An isolated gene sequence used to transform an organism. Often, but not always, the transgene has been derived from a different species than that of the recipient.

Transgenic An organism that has incorporated a functional foreign gene through recombinant DNA technology. The novel gene exists in all of its cells and is passed through to progeny.

Undernourishment Food intake that is continuously inadequate to meet dietary energy requirement.

Undernutrition The result of food intake that is insufficient to meet dietary energy requirements continuously, poor absorption, and/or poor biological use of nutrients consumed.

Urban and Peri-Urban Agriculture Agriculture occurring within and surrounding the boundaries of cities throughout the world and includes crop and livestock production, fisheries and forestry, as well as the ecological services they provide. Often multiple farming and gardening systems exist in and near a single city.

Value Chain A set of value-adding activities through which a product passes from the initial production or design stage to final delivery to the consumer.

Virtual Water The volume of water used to produce a commodity. The adjective "virtual" refers to the fact that most of the water used to produce a product is not contained in the product. In accounting virtual water flows we keep track of which parts of these flows refer to green, blue and grey water, respectively.

The real-water content of products is generally negligible if compared to the virtual-water content.

Waste Water "Grey" water that has been used in homes, agriculture, industries and businesses that is not for reuse unless it is treated.

Watershed The area which supplies water by surface and subsurface flow from precipitation to a given point in the drainage system.

Watershed Management Use, regulation and treatment of water and land resources of a watershed to accomplish stated objectives.

Water Productivity An efficiency term quantified as a ration of product output (goods and services) over water input.

Expressions of water productivity. Three major expressions of water productivity can be identified: (1) the amount of carbon gain per unit of water transpired by the leaf or by the canopy (photosynthetic water productivity); (2) the amount of water transpired by the crop (biomass water productivity); or (3) the yield obtained per unit amount of water transpired by the crop (yield water productivity).

Agricultural water productivity relates net benefits gained through the use of water in crop, forestry, fishery, livestock and mixed agricultural systems. In its broadest sense, it reflects the objectives of producing more food, income, livelihood and ecological benefits at less social and environmental cost per unit of water in agriculture.

Physical water productivity relates agricultural production to water use—more crop per drop. Water use is expressed either in terms of delivery to a use, or depletion by a use through evapotranspiration, pollution, or directing water to a sink where it cannot be reused. Improving physical water productivity is important to reduce future water needs in agriculture.

Economic water productivity relates the value of agricultural production to agricultural water use. A holistic assessment should account for the benefits and costs of water, including less tangible livelihood benefits, but this is rarely done. Improving economic water productivity is important for economic growth and poverty reduction.

Annex D
Acronyms, Abbreviations and Units

ABS Access and Benefit Sharing
AIDS Acquired immune deficiency syndrome
AKST Agricultural knowledge, science and technology
AR agrarian reform
billion one thousand million
Bt soil bacterium *Bacillus thuringiensis* (usually refers to plants made insecticidal using a variant of various *cry* toxin genes sourced from plasmids of these bacteria)
C carbon
CAN Andean Community of Nations
CARDI Caribbean Agricultural Research and Development Institute
CARICOM Caribbean Commercial Community (13 members)
CATIE Tropical Agriculture Research and Higher Education Center
CBD Convention on Biological Diversity
CEPAL Economic Commissions for Latin America and the Caribbean
Cepredenac Center for the Coordination for the Prevention of Natural Disasters in Central America
CGAP Consultative Group to Assist the Poor
CGIAR Consultative Group on International Agricultural Research
CIAT International Center for Tropical Agriculture
CIDA Canadian International Development Agency
CIFAA Committee for Inland Fisheries and Aquaculture of Africa
CIFOR Center for International Forestry Research
CIMMYT International Maize and Wheat Improvement Center
CIP International Potato Center
CLADEHL Comisión Latinoamericana por los Derechos y Libertades de los Trabajadores y Pueblos
CLADES Latin American Consortium for Agroecology and Development
COFUPRO Coordinadora Nacional de las fundaciones PRODUCE, A.C.
CONAM National Environmental Commission
CORPOICA Colombian Agricultural Research Institute
DANIDA Danish International Development Agency
DIA Dirección de Investigación Agrícola, Uruguay
DNA deoxyribonucleic acid
EARTH Escuela Agrícola de la Región del Trópico Húmedo
ECLAC Economic Commission for Latin America
EMBRAPA Brazilian Agricultural Research Institute
EPA US Environmental Protection Agency
EU European Union
EurepGAP Euro-Retailer Produce Working Group (EUREP)
FAO Food and Agriculture Organization of the United Nations
FAPRI Food and Agricultural Policy Research Institute
FIOCRUZ Fundación Oswaldo Cruz
FONTAGRO Regional Fund for Agricultural Technology
FORAGRO Regional Forum for Agricultural Research and Technological Development
FUMIAF A.C. Mexican Foundation for Agricultural and Forestry Research
GATT General Agreement on Trade and Tariffs
GDP Gross domestic product
GEF Global Environment Facility
GEO Global Environment Outlook
GFAR Global Forum on Agricultural Research
GHG greenhouse gas
GIS geographic information system
GM genetically modified/genetic modification
GMO genetically modified organism
GNP Gross National Product
GTZ German Agency for Technical Cooperation
ha hectare (10^4 m^2)
HACCP Hazard Analysis and Critical Control Points
IAASTD International Assessment of Agricultural Knowledge, Science and Technology for Development
IABA Inter-American Board of Agriculture
IAvH Institute of Alexander von Humboldt

ICCARD	International Commission on Central American Reconstruction and Development
ICP	integrated crop management
ICRAF	World Agroforestry Centre
IDB	Inter-American Development Bank
IDEAM	Instituto de Hidrología, Meteorología y Estudios Ambientales de Colombia
IDRC	International Development Research Center (Canada)
IEA	International Energy Agency
IFAD	International Fund for Agricultural Development
IFOAM	International Foundation for Organic Agriculture
IFPRI	International Food Policy Research Institute
IIAP	Institute for Amazon Investigations
IICA	Inter-American Institute for Cooperation on Agriculture
ILRI	International Livestock Research Institute
INIFAP	National Forestry, Agricultural and Livestock Research Institute (Mexico)
INTA	National Agricultural Technology Institute (Argentina)
IPCC	Intergovernmental Panel on Climate Change
IPGRI	Bioversity International
IPM	integrated pest management
ISNAR	International Service for National Agricultural Research
IUCN	World Conservation Union
LAC	Latin America and the Caribbean
MA	Millennium Ecosystem Assessment
MAELA	Latin America Agroecological Movement
MERCOSUR	Common Market of the South
MIF	The Fund for Multilateral Investments of the Inter-American Development Bank
MRL	maximum residue limit
NAFTA	North American Free Trade Agreement
NARI	National Agricultural Research Institute
NGO	nongovernmental organization
OECD	Organization of Economic Cooperation and Development
ONU	Organización Mundial de las Naciones Unidas
PAHO	Pan American Health Organization
PAN	Pesticide Action Network International
PCCMCA	Central American Cooperative Program for the Improvement of Crops and Animals
PID	Participatory innovation and development
POP	Persistent organic pollutant
PRATEC	Projecto Andino de Technologias Campesinas
PROCI	Cooperative agricultural research program
PROCIANDINO	Cooperative Research and Technology Transfer Program for the Andean Subregion
PROCICARIBE	Caribbean Agricultural Science and Technology Networking System for the CARDI countries plus Suriname
PROCINORTE	Cooperative Research and Technology Transfer Program for the Northern Region
PROCISUR	Cooperative Program for the Development of Agricultural Technology in the Southern Cone
PROCITROPICOS	Cooperative Research and Technology Transfer Program for the South American Tropics
PROMECAFE	Cooperative Program for the Technological Development and Modernization of Coffee Cultivation in Central America and the Dominican Republic
QPM	Quality Protein Maize
R&D	research and development
RALLT	Red por una América Latina Libre de Transgénicos
RAP-AL	Pesticide Action Network Latin America
RNA	ribonucleic acid
RR	Roundup Ready
SICTA	Central American Agricultural Technology Integration System
SPFS	Special Program for Food Security
TAC	Technical Advisory Council to CGIAR
TNC	The Nature Conservancy
tonne	10^3 kg (metric ton)
UN ECLAC	United Nations Economic Commission for Latin America
UNCED	UN Conference on Environment and Development
UNDP	United Nations Development Program
UNEP	United Nations Environment Programme
UNESCO	United Nations Educational, Scientific and Cultural Organization
UPOV	International Union for the Protection of New Varieties of Plants
USA-GAP	US Good Agricultural Practices
USAID	US Agency for International Development
USDA	US Department of Agriculture
WB	World Bank
WHO	World Health Organization
WTO	World Trade Organization
WWF	World Wildlife Fund

Annex E
Steering Committee for Consultative Process and Advisory Bureau for Assessment

Steering Committee

The Steering Committee was established to oversee the consultative process and recommend whether an international assessment was needed, and if so, what was the goal, the scope, the expected outputs and outcomes, governance and management structure, location of the secretariat and funding strategy.

Co-chairs

Louise Fresco, Assistant Director General for Agriculture, FAO
Seyfu Ketema, Executive Secretary, Association for Strengthening Agricultural Research in East and Central Africa (ASARECA)
Claudia Martinez Zuleta, Former Deputy Minister of the Environment, Colombia
Rita Sharma, Principal Secretary and Rural Infrastructure Commissioner, Government of Uttar Pradesh, India
Robert T. Watson, Chief Scientist, The World Bank

Nongovernmental Organizations

Benny Haerlin, Advisor, Greenpeace International
Marcia Ishii-Eiteman, Senior Scientist, Pesticide Action Network North America Regional Center (PANNA)
Monica Kapiriri, Regional Program Officer for NGO Enhancement and Rural Development, Aga Khan
Raymond C. Offenheiser, President, Oxfam America
Daniel Rodriguez, International Technology Development Group (ITDG), Latin America Regional Office, Peru

UN Bodies

Ivar Baste, Chief, Environment Assessment Branch, UN Environment Programme
Wim van Eck, Senior Advisor, Sustainable Development and Healthy Environments, World Health Organization
Joke Waller-Hunter, Executive Secretary, UN Framework Convention on Climate Change
Hamdallah Zedan, Executive Secretary, UN Convention on Biological Diversity

At-large Scientists

Adrienne Clarke, Laureate Professor, School of Botany, University of Melbourne, Australia
Denis Lucey, Professor of Food Economics, Dept. of Food Business & Development, University College Cork, Ireland, and Vice-President NATURA
Vo-tong Xuan, Rector, Angiang University, Vietnam

Private Sector

Momtaz Faruki Chowdhury, Director, Agribusiness Center for Competitiveness and Enterprise Development, Bangladesh
Sam Dryden, Managing Director, Emergent Genetics
David Evans, Former Head of Research and Technology, Syngenta International
Steve Parry, Sustainable Agriculture Research and Development Program Leader, Unilever
Mumeka M. Wright, Director, Bimzi Ltd., Zambia

Consumer Groups

Michael Hansen, Consumers International
Greg Jaffe, Director, Biotechnology Project, Center for Science in the Public Interest
Samuel Ochieng, Chief Executive, Consumer Information Network

Producer Groups

Mercy Karanja, Chief Executive Officer, Kenya National Farmers' Union
Prabha Mahale, World Board, International Federation Organic Agriculture Movements (IFOAM)
Tsakani Ngomane, Director Agricultural Extension Services, Department of Agriculture, Limpopo Province, Republic of South Africa
Armando Paredes, Presidente, Consejo Nacional Agropecuario (CNA)

Scientific Organizations

Jorge Ardila Vásquez, Director Area of Technology and Innovation, Inter-American Institute for Cooperation on Agriculture (IICA)
Samuel Bruce-Oliver, NARS Senior Fellow, Global Forum for Agricultural Research Secretariat
Adel El-Beltagy, Chair, Center Directors Committee, Consultative Group on International Agricultural Research (CGIAR)
Carl Greenidge, Director, Center for Rural and Technical Cooperation, Netherlands
Mohamed Hassan, Executive Director, Third World Academy of Sciences (TWAS)
Mark Holderness, Head Crop and Pest Management, CAB International
Charlotte Johnson-Welch, Public Health and Gender Specialist and Nata Duvvury, Director Social Conflict and Transformation Team, International Center for Research on Women (ICRW)
Thomas Rosswall, Executive Director, International Council for Science (ICSU)
Judi Wakhungu, Executive Director, African Center for Technology Studies

Governments

Australia: Peter Core, Director, Australian Centre for International Agricultural Research

China: Keming Qian, Director General Inst. Agricultural Economics, Dept. of International Cooperation, Chinese Academy of Agricultural Science

Finland: Tiina Huvio, Senior Advisor, Agriculture and Rural Development, Ministry of Foreign Affairs

France: Alain Derevier, Senior Advisor, Research for Sustainable Development, Ministry of Foreign Affairs

Germany: Hans-Jochen de Haas, Head, Agricultural and Rural Development, Federal Ministry of Economic Cooperation and Development (BMZ)

Hungary: Zoltan Bedo, Director, Agricultural Research Institute, Hungarian Academy of Sciences

Ireland: Aidan O'Driscoll, Assistant Secretary General, Department of Agriculture and Food

Morocco: Hamid Narjisse, Director General, INRA

Russia: Eugenia Serova, Head, Agrarian Policy Division, Institute for Economy in Transition

Uganda: Grace Akello, Minister of State for Northern Uganda Rehabilitation

United Kingdom Paul Spray, Head of Research, DFID

United States: Rodney Brown, Deputy Under Secretary of Agriculture and Hans Klemm, Director of the Office of Agriculture, Biotechnology and Textile Trade Affairs, Department of State

Foundations and Unions

Susan Sechler, Senior Advisor on Biotechnology Policy, Rockefeller Foundation

Achim Steiner, Director General, The World Conservation Union (IUCN)

Eugene Terry, Director, African Agricultural Technology Foundation

Advisory Bureau

Non-government Representatives

Consumer Groups

Jaime Delgado • Asociación Peruana de Consumidores y Usuarios
Greg Jaffe • Center for Science in the Public Interest
Catherine Rutivi • Consumers International
Indrani Thuraisingham • Southeast Asia Council for Food Security and Trade
Jose Vargas Niello • Consumers International Chile

International Organizations

Nata Duvvury • International Center for Research on Women
Emile Frison • CGIAR
Mohamed Hassan • Third World Academy of Sciences
Mark Holderness • GFAR
Jeffrey McNeely • World Conservation Union (IUCN)
Dennis Rangi • CAB International
John Stewart • International Council of Science (ICSU)

NGOs

Kevin Akoyi • Vredeseilanden
Hedia Baccar • Association pour la Protection de l'Environment de Kairouan
Benedikt Haerlin • Greenpeace International
Juan Lopez • Friends of the Earth International
Khadouja Mellouli • Women for Sustainable Development
Patrick Mulvaney • Practical Action
Romeo Quihano • Pesticide Action Network
Maryam Rahmaniam • CENESTA
Daniel Rodriguez • International Technology Development Group

Private Sector

Momtaz Chowdhury • Agrobased Technology and Industry Development
Giselle L. D'Almeida • Interface
Eva Maria Erisgen • BASF
Armando Paredes • Consejo Nacional Agropecuario
Steve Parry • Unilever
Harry Swaine • Syngenta (resigned)

Producer Groups

Shoaib Aziz • Sustainable Agriculture Action Group of Pakistan
Philip Kiriro • East African Farmers Federation
Kristie Knoll • Knoll Farms
Prabha Mahale • International Federation of Organic Agriculture Movements
Anita Morales • Apit Tako
Nizam Selim • Pioneer Hatchery

Government Representatives

Central and West Asia and North Africa

Egypt • Ahlam Al Naggar
Iran • Hossein Askari
Kyrgyz Republic • Djamin Akimaliev
Saudi Arabia • Abdu Al Assiri, Taqi Elldeen Adar, Khalid Al Ghamedi
Turkey • Yalcin Kaya, Mesut Keser

East and South Asia and the Pacific

Australia • Simon Hearn
China • Puyun Yang
India • PK Joshi
Japan • Ryuko Inoue
Philippines • William Medrano

Latin America and Caribbean

Brazil • Sebastiao Barbosa, Alexandre Cardoso, Paulo Roberto Galerani, Rubens Nodari
Dominican Republic • Rafael Perez Duvergé
Honduras • Arturo Galo, Roberto Villeda Toledo
Uruguay • Mario Allegri

North America and Europe

Austria • Hedwig Woegerbauer
Canada • Iain MacGillivray
Finland • Marja-Liisa Tapio-Bistrom
France • Michel Dodet
Ireland • Aidan O'Driscoll, Tony Smith
Russia • Eugenia Serova, Sergey Alexanian
United Kingdom • Jim Harvey, David Howlett, John Barret
United States • Christian Foster

Sub-Saharan Africa

Benin • Jean Claude Codjia
Gambia • Sulayman Trawally
Kenya • Evans Mwangi
Mozambique • Alsácia Atanásio, Júlio Mchola
Namibia • Gillian Maggs-Kölling
Senegal • Ibrahim Diouck

Annex F
Secretariat and Cosponsor Focal Points

Secretariat
World Bank
Marianne Cabraal, Leonila Castillo, Jodi Horton, Betsi Isay, Pekka Jamsen, Pedro Marques, Beverly McIntyre, Wubi Mekonnen, June Remy

UNEP
Marcus Lee, Nalini Sharma, Anna Stabrawa

UNESCO
Guillen Calvo

With special thanks to the Publications team: Audrey Ringler (logo design), Pedro Marques (proofing and graphics), Ketill Berger and Eric Fuller (graphic design)

Regional Institutes
Sub-Saharan Africa – African Centre for Technology Studies (ACTS)
Ronald Ajengo, Elvin Nyukuri, Judi Wakhungu

Central and West Asia and North Africa – International Center for Agricultural Research in the Dry Areas (ICARDA)
Mustapha Guellouz, Lamis Makhoul, Caroline Msrieh-Seropian, Ahmed Sidahmed, Cathy Farnworth

Latin America and the Caribbean – Inter-American Institute for Cooperation on Agriculture (IICA)
Enrique Alarcon, Jorge Ardila Vásquez, Viviana Chacon, Johana Rodríguez, Gustavo Sain

East and South Asia and the Pacific – WorldFish Center
Karen Khoo, Siew Hua Koh, Li Ping Ng, Jamie Oliver, Prem Chandran Venugopalan

Cosponsor Focal Points

GEF	Mark Zimsky
UNDP	Philip Dobie
UNEP	Ivar Baste
UNESCO	Salvatore Arico, Walter Erdelen
WHO	Jorgen Schlundt
World Bank	Mark Cackler, Kevin Cleaver, Eija Pehu, Juergen Voegele

Index

Page references followed by *f*, *t*, and *b* indicate figures, tables, and boxes, respectively.

About Island Press

Since 1984, the nonprofit Island Press has been stimulating, shaping, and communicating the ideas that are essential for solving environmental problems worldwide. With more than 800 titles in print and some 40 new releases each year, we are the nation's leading publisher on environmental issues. We identify innovative thinkers and emerging trends in the environmental field. We work with world-renowned experts and authors to develop cross-disciplinary solutions to environmental challenges.

Island Press designs and implements coordinated book publication campaigns in order to communicate our critical messages in print, in person, and online using the latest technologies, programs, and the media. Our goal: to reach targeted audiences—scientists, policymakers, environmental advocates, the media, and concerned citizens—who can and will take action to protect the plants and animals that enrich our world, the ecosystems we need to survive, the water we drink, and the air we breathe.

Island Press gratefully acknowledges the support of its work by the Agua Fund, Inc., Annenberg Foundation, The Christensen Fund, The Nathan Cummings Foundation, The Geraldine R. Dodge Foundation, Doris Duke Charitable Foundation, The Educational Foundation of America, Betsy and Jesse Fink Foundation, The William and Flora Hewlett Foundation, The Kendeda Fund, The Andrew W. Mellon Foundation, The Curtis and Edith Munson Foundation, Oak Foundation, The Overbrook Foundation, the David and Lucile Packard Foundation, The Summit Fund of Washington, Trust for Architectural Easements, Wallace Global Fund, The Winslow Foundation, and other generous donors.